ADVANCED POLYMER STRUCTURES

Chemistry for Engineering Applications

ADVANCED POLYMER STRUCTURES

Chemistry for Engineering Applications

Edited by
Omar Mukbaniani, DSc
Tamara Tatrishvili, PhD
Marc Jean M. Abadie, DSc

APPLE ACADEMIC PRESS

First edition published 2024

Apple Academic Press Inc.
1265 Goldenrod Circle, NE,
Palm Bay, FL 32905 USA

760 Laurentian Drive, Unit 19,
Burlington, ON L7N 0A4, CANADA

CRC Press
2385 NW Executive Center Drive,
Suite 320, Boca Raton FL 33431

4 Park Square, Milton Park,
Abingdon, Oxon, OX14 4RN UK

Library and Archives Canada Cataloguing in Publication

Title: Advanced polymer structures : chemistry for engineering applications / edited by Omar Mukbaniani, DSc, Tamara Tatrishvili, PhD, Marc Jean M. Abadie, DSc.
Names: Mukbaniani, O. V. (Omar V.), editor. | Tatrishvili, Tamara, editor. | Abadie, Marc J. M., editor.
Description: First edition. | Includes bibliographical references and index.
Identifiers: Canadiana (print) 20230225624 | Canadiana (ebook) 20230225640 | ISBN 9781774913017 (hardcover) | ISBN 9781774913024 (softcover) | ISBN 9781003352181 (ebook)
Subjects: LCSH: Polymers. | LCSH: Polymerization. | LCSH: Composite materials. | LCSH: Nanostructures. | LCSH: Green chemistry. | LCSH: Polymers in medicine.
Classification: LCC QD381 .A38 2023 | DDC 547/.7—dc23

Library of Congress Cataloging-in-Publication Data

CIP data on file with US Library of Congress

ISBN: 978-1-77491-301-7 (hbk)
ISBN: 978-1-77491-302-4 (pbk)
ISBN: 978-1-00335-218-1 (ebk)

About the Editors

Omar Mukbaniani, DSc
Professor, Ivane Javakhishvili Tbilisi State University, Faculty of Exact and Natural Sciences, Department of Chemistry; Chair of Macromolecular Chemistry; Director of the Institute of Macromolecular Chemistry and Polymeric Materials at TSU, Tbilisi, Georgia

Omar Mukbaniani, DSc, was a Professor at Ivane Javakhishvili Tbilisi State University (TSU), Faculty of Exact and Natural Sciences, Department of Chemistry; Chair of Macromolecular Chemistry, Tbilisi, Georgia. He was also the Director of the Institute of Macromolecular Chemistry and Polymeric Materials at TSU, and a member of the Academy of Natural Sciences of Georgia. For several years, he was a member of the advisory board of the journal *Proceedings of Ivane Javakhishvili Tbilisi State University* (Chemical Series) and contributing editor of the journals *Polymer News, Polymers Research Journal,* and *Chemistry and Chemical Technology.* His research interests included polymer chemistry, polymeric materials, and the chemistry of organosilicon compounds. He is the author of more than 480 publications, 25 books and monographs, and 10 inventions. He created the International Symposium on Polymers and Advanced Materials ICSP&AM in 2007, which takes place every two years in Georgia. In 2018, he was chair of the 26th World Annual Forum on Advanced Materials PolyChar 26.

Tamara Tatrishvili, PhD
Assistant Professor, Department of Chemistry, Tbilisi State University; Director, TSU Faculty of Exact and Natural Sciences, Institute of Macromolecular Chemistry and Polymeric Materials; Main Specialist, TSU Faculty of Exact and Natural Sciences, The Office of Academic Process Management

Tamara Tatrishvili, PhD, is an Assistant Professor at Ivane Javakhishvili Tbilisi State University, Department of Chemistry; the Main Specialist in the Office of Academic Process Management (Faculty of Exact and Natural Sciences) at Ivane Javakhishvili Tbilisi State University; as well as Director of the Institute of Macromolecular Chemistry and Polymeric Materials at TSU, Tbilisi, Georgia; DAAD alumni; and a member of Georgian Chemical Society. Her research interests include polymer chemistry, polymeric materials, and chemistry of silicon-organic compounds. She is the executive editor of *Journal of the Georgian Chemical Society*. Dr. Tatrishvili has authored more than 190 scientific publications, 12 books, and monographs.

Marc J. M. Abadie, DSc
Professor Emeritus, ICGM, University of Montpellier, CNRS, ENSCM, Montpellier, France

Marc J. M. Abadie, DSc, is an Emeritus Professor at the University Montpellier, France. He was head of the Laboratory of Polymer Science and Advanced Organic Materials – LEMP/MAO. He is currently the "Michael Fam" Visiting Professor at the School of Materials Sciences and Engineering, Nanyang Technological University (NTU), Singapore. His present activity concerns high-performance polymers for PEMFCs, composites, nanocomposites, UV/EB coatings, and biomaterials. He has published 11 books and holds 11 patents. He has advised approximately 95 MS and 52 PhD students with whom he has published over 402 papers. He has more than 40 years of experience in polymer science with 10 years in industry (IBM, USA MOD, UK, and SNPA/Total, France). He created the International Symposium on Polyimides and High-Temperature Polymers, a.k.a. Step I, in the 1980s which takes place every three years in Montpellier, France.

Contents

In Memoriam

Prof. Omar V. Mukbaniani, DSc
Ivane Javakhishvili Tbilisi State University, Georgia

Professor Omar Mukbaniani (1948–2022), Doctor of Chemical Sciences, was a Professor at Ivane Javakhishvili Tbilisi State University, where he was also Head Chairperson of Macromolecular Chemistry and Head of the Department of Chemistry. He also served as the Director of the Institute of Macromolecular Chemistry and Polymeric Materials at Ivane Javakhishvili Tbilisi State University.

Professor Omar Mukbaniani was a member of the Academy of Natural Sciences of Georgia. For several years, he served as a member of the advisory board and the editorial board of the *Journal Proceedings of Ivane Javakhishvili Tbilisi State University* (Chemical Series) and contributing editor of the journal *Polymer News*, the *Polymers Research Journal*, and *Chemistry and Chemical Technology*.

The research interests of Professor Omar Mukbaniani included polymer chemistry, polymeric materials, and the chemistry of organosilicon compounds. He developed methods of precision synthesis to build block, graft, and comb-type structures, and also studied the mechanisms of reactions leading to these polymers. Some of his effort was devoted to the synthesis of various types of functionalized silicon polymers, copolymers, block copolymers, and composites.

Prof. Omar Mukbaniani, with his colleagues Prof. Marc Jean M. Abadie and Dr. Vazha Tskhovrebashvili, was the founder of the "Caucasian Symposium on Polymers and Advanced Materials" (ICSP&AM). The first conference started in 2007, and ICSP&AM 7 was held in July 2021.

Prof. Omar Mukbaniani was the main author who came up with the idea to publish the full texts of the conference abstracts in Apple Academic Press. This book is the fifth edition.

Unfortunately, Professor Omar Mukbaniani is not with us now. His absence is a great loss to his colleagues and for the entire chemical community in general. We will pursue Omar's work in both the organization of ICMP&AM and the publication of the symposium book, where Omar will continue to be the editor.

Dr. Tamara Tatrishvili, PhD
Assistant Professor, Director of the Institute of Macromolecular Chemistry and Polymeric Materials, Ivane Javakhishvili Tbilisi State University, Tbilisi, Georgia

Prof. Marc Jean M. Abadie, DSc
ICGM, Université de Montpellier, CNRS, ENSCM, Montpellier, France

Contributors

Marc J. M. Abadie
ICGM, University of Montpellier, CNRS, ENSCM, Montpellier, France

G. N. Abaszade
Institute of Polymer Materials, Azerbaijan National Academy of Sciences, Sumgait, Azerbaijan

R. A. Akhmedova
Institute of Polymer Materials, Azerbaijan National Academy of Sciences, Sumgait, Azerbaijan

T. I. Alkhazov
Institute of Polymer Materials, Azerbaijan National Academy of Sciences, Sumgait, Azerbaijan

Zurab Amiridze
Petre Melikishvili Institute of Physical and Organic Chemistry, Tbilisi State University, Tbilisi, Georgia

Koba Amirkhanashvili
Petre Melikishvili Institute of Physical and Organic Chemistry, Tbilisi State University, Tbilisi, Georgia

Natela Ananiashvili
Ivane Javakhishvili Tbilisi State University, R. Agladze Institute of Inorganic Chemistry and Electrochemistry. Tbilisi, Georgia

L. Arabuli
University of Georgia, Tbilisi, Georgia

R. V. Asadov
Institute of Polymer Materials, Azerbaijan National Academy of Sciences, Sumgait, Azerbaijan

O. B. Askerov
Institute of Polymer Materials, Azerbaijan National Academy of Sciences, Sumgait, Azerbaijan

E. T. Aslanova
Institute of Polymer Materials, Azerbaijan National Academy of Sciences, Sumgait, Azerbaijan

Marina Avaliani
Ivane Javakhishvili Tbilisi State University, R. Agladze Institute of Inorganic Chemistry and Electrochemistry, Tbilisi, Georgia

Nana Barnovi
R. Agladze Institute of Inorganic Chemistry and Electrochemistry, Ivane Javakhishvili Tbilisi State University, Tbilisi, Georgia

M. Berulava
Sukhumi State University, Tbilisi, Georgia

George Bezarashvili
Faculty of Exact and Natural Sciences, Ivane Javakhishvili Tbilisi State University, Tbilisi, Georgia

Bendeliani Bezhan
Laboratory of Cryogenic Technique and Technologies, Ilia Vekua Sukhumi Institute of Physics and Technology, Tbilisi, Georgia

Manana Burdjanadze
Petre Melikishvili Institute of Physical and Organic Chemistry, Ivane Javakhishvili Tbilisi State, Tbilisi, Georgia

Efkan Çatıker
Ordu University, Department of Chemistry, Ordu, Turkey

Vaja Chagelishvili
R. Agladze Institute of Inorganic Chemistry and Electrochemistry, Ivane Javakhishvili Tbilisi State University, Tbilisi, Georgia

A. Z. Chalabiyeva
Institute of Polymer Materials, Azerbaijan National Academy of Sciences, Sumgait, Azerbaijan

Ketevan Chikovani
R. Agladze Institute of Inorganic Chemistry and Electrochemistry, Ivane Javakhishvili Tbilisi State University, Tbilisi, Georgia

M. Chikovani
Kutaisi Akaki Tsereteli State University, Department of Chemistry, Kutaisi, Georgia

Nanuli Dolaberidze
Petre Melikishvili Institute of Physical and Organic Chemistry, Tbilisi State University, Tbilisi, Georgia

M. Donadze
Faculty of Chemistry and Metallurgy, Georgian Technical University, Tbilisi, Georgia

Vusala Dostuyeva
Institute of Polymer Materials, Azerbaijan National Academy of Sciences, Sumgait, Azerbaijan

D. Dzanashvili
R. Agladze Institute of Inorganic Chemistry and Electrochemistry, Ivane Javakhishvili Tbilisi State University, Tbilisi, Georgia

V. A. Dzhafarov
Institute of Polymer Materials, Azerbaijan National Academy of Sciences, Sumgait, Azerbaijan

D. Dzidziguri
Ivane Javakhishvili Tbilisi State University, Faculty of Exact and Natural Sciences, Tbilisi, Georgia

K. Dzuliashvili
Agricultural University of Georgia, Tbilisi, Georgia

Chkhaidze Ekaterina
Department of Chemical and Biological Technologies, Georgian Technical University, Tbilisi, Georgia

Luba Eprikashvili
Petre Melikishvili Institute of Physical and Organic Chemistry, Ivane Javakhishvili Tbilisi State, Tbilisi, Georgia

Nana Esakia
Faculty of Exact and Natural Sciences, Ivane Javakhishvili Tbilisi State University, Tbilisi, Georgia

Manana Gachechiladze
Ivane Javakhishvili Tbilisi State University, R. Agladze Institute of Inorganic Chemistry and Electrochemistry. Tbilisi, Georgia

Ioseb Gejadze
Ivane Javakhishvili Tbilisi State University, Al. Tvalchrelidze Caucasian Institute of Mineral Resources, Tbilisi, Georgia

Mehdi Ghamami
Department of Mechanical Engineering, Isfahan University of Technology, Isfahan, Iran

Shahriar Ghammamy
Department of Chemistry, Faculty of Science, Imam Khomeini International University, Ghazvin, Iran

R. Gigauri
TSU – R. Agladze Institute of Inorganic Chemistry and Electrochemistry, Tbilisi, Georgia

Kr. Giorgadze
Ivane Javakhishvili Tbilisi State University, Department of Chemistry, Tbilisi, Georgia

Antia Giorgi
Department of Medical Chemistry, Tbilisi State Medical University, Tbilisi, Georgia

Giorganashvili Giorgi
Laboratory of Cryogenic Technique and Technologies, Ilia Vekua Sukhumi Institute of Physics and Technology, Tbilisi, Georgia

Przemysław Gnatowski
Department of Polymer Technology, Faculty of Chemistry, Gdańsk University of Technology, Gdańsk, Poland

Nazi Goliadze
Ivane Javakhishvili Tbilisi State University, Tbilisi, Georgia

Dgebuadze Guram
Laboratory of Cryogenic Technique and Technologies, Ilia Vekua Sukhumi Institute of Physics and Technology, Tbilisi, Georgia

M. Sh. Gurbanov
Institute of Polymer Materials, Azerbaijan National Academy of Sciences, Sumgait, Azerbaijan

Farzad Haghighi
Department of Chemistry, Faculty of Science, Imam Khomeini International University, Ghazvin, Iran

Józef T. Haponiuk
Department of Polymer Technology, Faculty of Chemistry, Gdańsk University of Technology, Gdańsk, Poland

Seyed Hamed Hosseini
Department of Chemistry, Faculty of Science, Imam Khomeini International University, Ghazvin, Iran

Lim Kah Hui
Materials Science and Engineering, Nanyang Technological University, Singapore

Metskhvarishvili Ioseb
Laboratory of Cryogenic Technique and Technologies, Ilia Vekua Sukhumi Institute of Physics and Technology, Tbilisi, Georgia

Berdzenishvili Irine
Department of Chemical and Biological Technologies, Georgian Technical University, Tbilisi, Georgia

Ch. O. Ismailova
Institute of Polymer Materials, Azerbaijan National Academy of Sciences, Sumgait, Azerbaijan

Helena Janik
Department of Chemical Faculty, Polymer Technological Department, Gdańsk University of Technology, Gdańsk, Poland

Jobin Joy
Department of Polymer Technology, Faculty of Chemistry, Gdańsk University of Technology, Poland; School of Energy Materials, Mahatma Gandhi University, Kerala, India; Department of Mechanical Engineering, Muthoot Institute of Science and Technology, Varikoli, Kerala, India

Haponiuk Jozef
Gdansk University of Technology, Gdansk, Poland

A. Ya. Kagramanova
Institute of Polymer Materials, Azerbaijan National Academy of Sciences, Sumgait, Azerbaijan

Manana Kekenadze
Faculty of Exact and Natural Sciences, Ivane Javakhishvili Tbilisi State University, Tbilisi, Georgia

Ümit Keleş
Ordu University, Department of Chemistry, Ordu, Turkey

Khimersen Khuangul
JSC Institute of Chemical Sciences (named after A.B. Bekturov), Almaty, Kazakhstan; Abai Kazakh National Pedagogical University, Almaty, Kazakhstan

Bela Khutsishvili
Petre Melikishvili Institute of Physical and Organic Chemistry, Tbilisi State University, Tbilisi, Georgia

L. Khvichia
GTU – Institute of Hydrogeology and Engineering Geology, Tbilisi, Georgia

Teimuraz Kordzakhia
Petre Melikishvili Institute of Physical and Organic Chemistry, Ivane Javakhishvili Tbilisi State, Tbilisi, Georgia

Giorgadze Kristine
Department of Chemistry, Faculty of Exact and Natural Sciences, Ivane Javakhishvili Tbilisi State University, Tbilisi, Georgia

Justyna Kucińska-Lipka
Department of Polymer Technology, Faculty of Chemistry, Gdańsk University of Technology, Gdańsk, Poland

Rusudan Lazashvili
Faculty of Exact and Natural Sciences, Ivane Javakhishvili Tbilisi State University, Tbilisi, Georgia

Vitali Lipik
Materials Science and Engineering, Nanyang Technological University, Singapore

Nino Lominadze
Faculty of Exact and Natural Sciences, Ivane Javakhishvili Tbilisi State University, Tbilisi, Georgia

Levan Londaridze
Ivane Javakhishvili Tbilisi State University, Faculty of Exact and Natural Sciences, Departments of Chemistry, Tbilisi, Georgia; Institute of Macromolecular Chemistry and Polymeric Materials, Ivane Javakhishvili Tbilisi State University, Tbilisi, Georgia

Metskhvarishvili Magda
Department of Microprocessor and Measurement Systems, Faculty of Informatics and Control Systems, Georgian Technical University, Tbilisi, Georgia

Vera Maisuradze
Ivane Javakhishvili Tbilisi State University, Al. Tvalchrelidze Caucasian Institute of Mineral Resources, Tbilisi, Georgia

N. Makhaldiani
Faculty of Chemistry and Metallurgy, Georgian Technical University, Tbilisi, Georgia

Giorgi Makharadze
Ivane Javakhishvili Tbilisi State University, Tbilisi, Georgia

Tamar Makharadze
Ivane Javakhishvili Tbilisi State University, R. Agladze Institute of Inorganic Chemistry and Electrochemistry, Tbilisi, Georgia

F. M. Mamedaliyeva
Institute of Polymer Materials, Azerbaijan National Academy of Sciences, Sumgait, Azerbaijan

B. A. Mamedov
Institute of Polymer Materials, Azerbaijan National Academy of Sciences, Sumgait, Azerbaijan

Bakhtiyar Mammedov
Institute of Polymer Materials, Azerbaijan National Academy of Sciences, Sumgait, Azerbaijan

Aynura Mammedova
Institute of Polymer Materials, Azerbaijan National Academy of Sciences, Sumgait, Azerbaijan

Siradze Manana
Department of Chemical and Biological Technologies, Georgian Technical University, Tbilisi, Georgia

Hanna J. Maria
School of Energy Materials, Mahatma Gandhi University, Priyadarsini Hills, Kottayam, Kerala, India

Tamaz Marsagishvili
Ivane Javakhishvili Tbilisi State University, R. Agladze Institute of Inorganic Chemistry and Electrochemistry, Tbilisi, Georgia

S. S. Mashayeva
Institute of Polymer Materials, Azerbaijan National Academy of Sciences, Sumgait, Azerbaijan

Marina Matchavariani
Ivane Javakhishvili Tbilisi State University, R. Agladze Institute of Inorganic Chemistry and Electrochemistry. Tbilisi, Georgia

Nato Mirdzveli
Petre Melikishvili Institute of Physical and Organic Chemistry, Tbilisi State University, Tbilisi, Georgia

Lela Mirtskhulava
Department of Computer Science, Ivane Javakhishvili Tbilisi State University, Tbilisi, Georgia

Golnoosh Mivehchi
Department of Chemistry, Faculty of Science, Imam Khomeini International University, Ghazvin, Iran

Omar Mukbaniani
Department of Macromolecular Chemistry, Ivane Javakhishvili University, Tbilisi, Georgia; Institute of Macromolecular Chemistry and Polymeric Materials, Ivane Javakhishvili University, Tbilisi, Georgia

Lia Nadareishvil
Ivane Javakhishvili Tbilisi State University, Tbilisi, Georgia

Marine Nadirashvili
Ivane Javakhishvili Tbilisi State University, Al. Tvalchrelidze Caucasian Institute of Mineral Resources, Tbilisi, Georgia

Yagub Nagiyev
Azerbaijan National Academy of Sciences, Institute of Catalysis and Inorganic Chemistry (named after academician M. Nagiyev), Baku, Azerbaijan

G. Nakhutsrishvili
Ivane Javakhishvili Tbilisi State University, Faculty of Exact and Natural Sciences, Tbilisi, Georgia

M. R. Nasuhbeyoğlu
Ordu University, Department of Chemistry, Ordu, Turkey

Manana Nijaradze
Petre Melikishvili Institute of Physical and Organic Chemistry, Tbilisi State University, Tbilisi, Georgia

N. N. Nikoleishvili
Department of Exact and Natural Sciences, Ivane Javakhishvili Tbilisi State University. Tbilisi, Georgia

D. R. Nurullayeva
Institute of Polymer Materials, Azerbaijan National Academy of Sciences, S. Vurgun Str. 124, Az5004, Sumgait, Azerbaijan

N. Ochkhikidze
Agricultural University of Georgia, Tbilisi, Georgia

Miriam Pacler
Department of Polymer Technology, Faculty of Chemistry, Gdańsk University of Technology, Gdańsk, Poland

Tamar Petriashvili
Ivane Javakhishvili Tbilisi State University, Al. Tvalchrelidze Caucasian Institute of Mineral Resources, Tbilisi, Georgia

K. M. Praveen
Department of Mechanical Engineering, Muthoot Institute of Science and Technology, Varikoli, Ernakulam, Kerala, India

Marina Rukhadze
Faculty of Exact and Natural Sciences, Ivane Javakhishvili Tbilisi State University, Tbilisi, Georgia

M. Rusia
Ivane Javakhishvili Tbilisi State University, Department of Chemistry, Tbilisi, Georgia

S. A. Rzayeva
Institute of Polymer Materials, Azerbaijan National Academy of Sciences, Sumgait, Azerbaijan

A. A. Salimova
Institute of Polymer Materials, Azerbaijan National Academy of Sciences, Sumgait, Azerbaijan

Zurab Samkharadze
Ivane Javakhishvili Tbilisi State University, R. Agladze Institute of Inorganic Chemistry and Electrochemistry, Tbilisi, Georgia

Sh. A. Samsoniya
Department of Exact and Natural Sciences, Ivane Javakhishvili Tbilisi State University. Tbilisi, Georgia

L. Shamanauri
R. Dvali Institute of Machine Mechanics, Tbilisi, Georgia

Elena Shapakidze
Ivane Javakhishvili Tbilisi State University, Al. Tvalchrelidze Caucasian Institute of Mineral Resources, Tbilisi, Georgia

Tinatin Sharashenidze
Petre Melikishvili Institute of Physical and Organic Chemistry, Ivane Javakhishvili Tbilisi State, Tbilisi, Georgia

Alexandre N. Sobolev
Center for Microscopy, Characterization, and Analysis, University of Western Australia, Perth, Australia

Poya Soroshian
Department of Chemistry, Faculty of Science, Imam Khomeini International University, Ghazvin, Iran

Seyedeh Saba Tabatabaei
Department of Chemistry, Faculty of Science, Imam Khomeini International University, Ghazvin, Iran

Nino Takaishvili
Faculty of Exact and Natural Sciences, Ivane Javakhishvili Tbilisi State University, Tbilisi, Georgia

Jumadilov Talkybek
JSC Institute of Chemical Sciences (named after A.B. Bekturov), Almaty, Kazakhstan

Tamara Tatrishvili
Department of Macromolecular Chemistry, Ivane Javakhishvili University, Tbilisi, Georgia; Institute of Macromolecular Chemistry and Polymeric Materials, Ivane Javakhishvili University, Tbilisi, Georgia

E. Tavdishvili
Ivane Javakhishvili Tbilisi State University, Faculty of Exact and Natural Sciences, Tbilisi, Georgia

Lobzhanidze Tea
Department of Chemistry, Faculty of Exact and Natural Sciences, Ivane Javakhishvili Tbilisi State University, Tbilisi, Georgia

Sabu Thomas
School of Energy Materials, Mahatma Gandhi University, Kottayam, Kerala, India

Gulnara Todradze
Al. Tvalchrelidze Caucasian Institute of Mineral Resources, Tbilisi, Georgia

M. V. Trapaidze
Department of Exact and Natural Sciences, Ivane Javakhishvili Tbilisi State University. Tbilisi, Georgia

Giorgi Tsintskaladze
Petre Melikishvili Institute of Physical and Organic Chemistry, Ivane Javakhishvili Tbilisi State, Tbilisi, Georgia

Vladimer Tsitsishvili
Petre Melikishvili Institute of Physical and Organic Chemistry, Tbilisi State University, Tbilisi, Georgia

Gabunia Vakhtang
Laboratory of Cryogenic Technique and Technologies, Ilia Vekua Sukhumi Institute of Physics and Technology, Tbilisi, Georgia

Mariam Vibliani
R. Agladze Institute of Inorganic Chemistry and Electrochemistry, Ivane Javakhishvili Tbilisi State University, Tbilisi, Georgia

Marine Zautashvili
Petre Melikishvili Institute of Physical and Organic Chemistry, Ivane Javakhishvili Tbilisi State, Tbilisi, Georgia

Eldar Zeynalov
Azerbaijan National Academy of Sciences, Institute of Catalysis and Inorganic Chemistry (named after academician M. Nagiyev), Baku, Azerbaijan

Nani Zhorzholiani
Petre Melikishvili Institute of Physical and Organic Chemistry, Tbilisi State University, Tbilisi, Georgia

Abbreviations

AgZ	silver-containing zeolite
AI	artificial intelligence
AIBN	2,2-azobis(2-methylpropionitrile)
ALARA	as low as reasonably achievable
ANN	artificial neural networks
AOP	advanced oxidation processes
AROP	anionic ring-opening polymerization
B_4C	boron carbide
BET	Brunauer-Emmett-teller
BP	benzophenone
BR	butyl rubber
CdO	cadmium oxide
CEC	composite electrochemical cofting
CFA	confirmatory factor analysis
CO	cotton oil
COD	chemical oxygen demand
CRS	chronic sinusitis
CuZ	copper-containing zeolite
DE	dye exhaustion
DFT	density functional theory
DHB	2,5-dihydroxybenzoic acid
DL	deep learning
DLS	dynamic light scattering
DSC	differential scanning calorimetry
DTBP	di-tret-butyl peroxide
d-TFA	deutero trifluoroacetic acid
DTG	differential thermogravimetry
ECO	epoxidized cotton oil
EDS	electron beam specimens
EFA	exploratory factor analysis
ESS	endoscopic sinus surgery
EVA	ethylene-vinyl acetate
FA	fulvic acids
FDA	Food and Drug Administration

FDM	fused deposition modeling
FESEM	field emission scanning electron microscopy
FTIR	Fourier transform infrared
FTIR-ATR	FTIR-attenuated total reflectance
GCNN	graph convolutional neural networks
GNN	graph neural network
hBN	hexagonal boron nitride
HDPE	high-density polyethylene
HEMA	2-hydroxyethyl methacrylate
HTP	hydrogen-transfer polymerization
ICME	integrated computational materials engineering
IFS	iterated function system
IUPAC	The International Union of Pure and Applied Chemistry
KNO_3	potassium nitrate
LDPE	low-density polyethylene
LG	liquid glass
MF	mometasone furoate
ML	machine learning
MLPs	multilayer perceptron
MW	molecular weight
MWD	molecular weight distribution
NPs	nanoparticles
NR	natural rubber
NRL	natural rubber latex
ONA	oligonaphthylamine
PA	propane acid
PAA	polyacrylic acid
PA-g-mBA	poly(acrylic acid-g-α-methyl-β-alanine)
PCL	polycaprolactone
PETP	polymer-polyethylene terephthalate
PFTNA	pulsed fast thermal neutron activation
pHEMA	poly hydroxyethyl methacrylate
PLA	polylactide
PLGA	poly(lactide-co-glycolide)
PmBA	poly(α-methyl β-alanine)
PPA	peroxypropionic acid
r-PVC	recycled polyvinyl chloride
SEM	scanning electron microscope
SG	sol-gel

SLS	selective laser sintering
SSR	solid-state reaction
TEMPO	2,2,6,6-tetramethyl-1-piperidinyloxy
TFA	trifluoroacetic acid
TFA-$CDCl_3$	trifluoroacetic acid/deutero chloroform
TG/DTA	thermogravimetric and differential thermal analyzer
TGA	thermo-gravimetric analysis
TGI	thermogravimetric index
TPNR	thermoplastic natural rubber
TPU	thermoplastic polyurethane
UHMWPE	ultra-high molecular weight polyethylene
VMQ	vinyl silicone rubber
VOCs	volatile organic compounds
WC	Wilsmeyer complex
WPC	wood-polymer composites
XRD	X-ray powder diffraction
XRED	X-ray energy dispersive
ZnZ	zinc-containing zeolite

Preface

Polymers are studied in fields as diverse as polymer science (polymer chemistry and polymer physics), biophysics, biochemistry, and more generally, in materials science and engineering.

Polymer matrix composites (PMCs) or nanocomposites (PMNCs) are widely used in high-tech material structures such as in the automotive, marine, and aerospace industries. Their impact on physical and mechanical performance is mainly due to the reinforcing agent, fiber (glass, carbon, aramid), or nanofiber (MMT, CNTs, graphene, etc.), and also to a perfect mastery of the matrix/reinforcement interface.

For several decades, polymer scientists have invested in research in the field of biomedicine with the aim of mimicking natural systems.

Selected papers presented at the 7th International Caucasian Symposium on Polymers and Advanced Materials, a.k.a. ICSP&AM 7, are collected in this book. This conference took place in Tbilisi, Georgia, on July 27–30, 2021, at the Ivane Javakhishvili Tbilisi State University.

The goal and originality of the book are to collect interdisciplinary papers on the state of knowledge of each topic under consideration through a combination of overviews and original, unpublished, recent research.

The general purpose of this book is to consider polymer science, such as advanced polymers, composites, and nanocomposites, and the role of polymers in the progress of green chemistry and medicine.

The book is divided into five parts: Part I: Polymer Synthesis and Application; Part II: Materials and Properties; Part III: Composites and Nanostructures; Part IV: Sustainable and Green Chemistry; and Part V: Constitutional Systems for Medicine.

Each part of the book contains really interesting papers that will be helpful and informative for all who work in the field of macromolecular chemistry and polymeric materials.

This book is addressed to academics, scientists, engineers, and medical technologists of institutes, research centers, and universities.

PART I

Polymer Synthesis and Application

CHAPTER 1

Biodegradable Hydrogels by UV Curing

LIM KAH HUI,[1] VITALI LIPIK,[1] and MARC J. M. ABADIE[2]

[1]*Materials Science and Engineering, Nanyang Technological University, Singapore*

[2]*Institute Charles Gerhardt Montpellier (ICGM), Department of Chemistry of Materials, Nanostructures Materials, Nanostructures, Materials for Energy, Pole Chemistry Balard Research, Campus CNRS, Montpellier, France*

ABSTRACT

UV curing technology has been developed for over 50 years. It has a number of advantages over thermal reactions that have made it the method of choice in 3D network structures. Hydrogels have been a branch of polymers which has been extensively researched due to favorable properties such structural integrity with high water uptake, making hydrogels good drug carriers, biodegradable hydrogels have also been used in branches such as tissue engineering and vascular grafts.

The chapter presents a crosslinking pathway of synthesis and properties of biodegradable hydrogel by UV curing of diacrylate polyethylene glycol (PEGDMA) mixed to 2-hydroxyethyl methacrylate (HEMA). The reaction kinetics and optimization of the chemical reaction were examined, and a swelling test was conducted to evaluate the ability of the synthesized hydrogels to hold and store water.

In Memoriam
Prof Vazha Tskovrebashvili (2021)
Prof Omar Mukbaniani (2022)

Advanced Polymer Structures: Chemistry for Engineering Applications. Omar Mukbaniani, DSc, Tamara Tatrishvili, PhD & Marc Jean M. Abadie, DSc (Eds.)

1.1 INTRODUCTION

A specific branch of biomaterials, in the form of biodegradable polymeric materials have been of interest in the biomaterial industries in the past few decades [1]. In particular, the class of biodegradable hydrogels have been extensively researched upon. Hydrogels are three-dimensionally cross-linked polymeric networks that are capable of absorbing and retaining huge amounts of water without leaking it and dissolution instantaneously [2]. Much interest lies in the study of biodegradable hydrogels due to superior properties such as their high-water update, structural integrity, which make them ideal drug carriers. Biodegradable hydrogels are commonly studied for their uses, such as scaffolds in tissue engineering, vascular grafts, or drug control and delivery systems [3].

Biodegradable hydrogels have been used widely in different applications of medicine and esthetics [4]. In more recent times, biomaterials include not only used in medical devices such as implants including mechanical heart valves, joint implants, heart stents, and many other applications [5], but also are used in a wide range of applications such as tissue engineering to create synthetic replacements for biological tissues and sutures to help speed up wound healing processes or drug delivery [6, 7].

The high water content of hydrogels makes them a close resemblance to natural living tissue more than other classes of synthetic biomaterials. Contact lenses are an example of hydrogels used in everyday lives [8]. In the medical community, biodegradable hydrogels have been used in the delivery of macro-molecules for drugs, artificial skin, and even linings for artificial hearts [9–11].

Studies have explored several methods of biodegradable hydrogel synthesis [12] since the first biomedical application of poly hydroxyethyl methacrylate (pHEMA) by Wichterle and Lim in 1961 [13]. Many monomers or natural polymers [14] have also been discovered for use in the synthesis of biodegradable hydrogels since then [15–18]. Finally, UV irradiations were used for the manufacture of water-soluble films [19].

1.2 MATERIALS AND EXPERIMENTAL METHODS

1.2.1 MATERIALS

- **Photosensitive Formulation:**
  - **Photoinitiator: Irgacure® 651**

- The radical photoinitiator: by intramolecular photocleavage the photolyze of Irgacure® 651 (2,2-dimethoxy-2-phenyl acetophenone DMPA) produce radicals (Figure 1.1) [20].

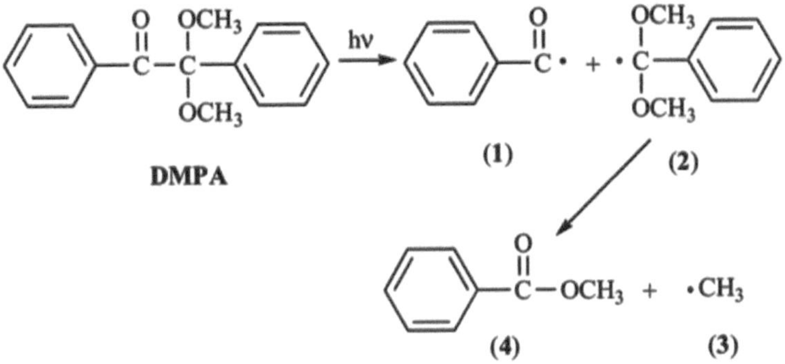

FIGURE 1.1 Photolyze of Irgacure® 651.
Note: Only radicals (1) and (3) are reactive, radical (2) is stabilized by isomery to give inactive methyl benzoate (4).

The most important intrinsic drawbacks of the radically induced radiation curing technology is that the radicals formed are inhibited by the presence of oxygen which limits the propagation of the chains by degradation and formation of hydroperoxidation in chains and the yellowing of the product.

- **Monomer/Oligomer:**
  - The chemical formulation is constituted by a mixture of two acrylates:
    - one monofunctional monomer *vz.* hydroxy-ethyl methacrylate (HEMA) (Figure 1.2).

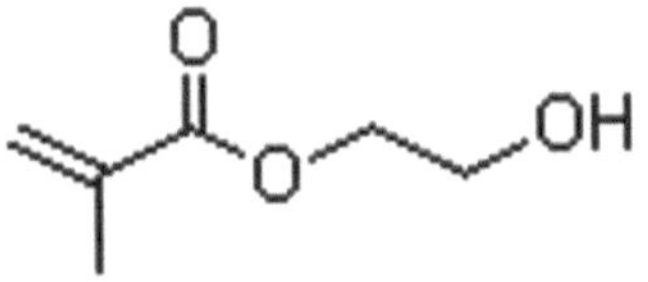

FIGURE 1.2 HEMA having an average functionality $f = 2$ (giving rise to chains propagation in 2 directions) and a thermoplastic.

HEMA is biodegradable, making it a viable choice as one of the chemicals used for hydrogel. HEMA-based hydrogels have also been extensively researched upon due to superior qualities such as their non-toxicity, and biocompatibility. HEMA is extensively used in the optical industry, HEMA-based hydrogels are used as contact lenses, and they are also used more recently in ocular drug delivery systems. The widespread use of HEMA in forming hydrogels makes HEMA a safe and appropriate choice in the synthesis of a hydrogel. However, as HEMA is a monofunctional acrylate, synthesizing poly-HEMA will only result in a linear polymer structure which will not be able to hold water and swell.

- one oligomer *vz.* tetra ethyleneoxide di-methacrylates tetra-EG (TEGDMA) also known as dimethacrylated polyethylene glycol PEG (PEGDMA) (Figure 1.3).

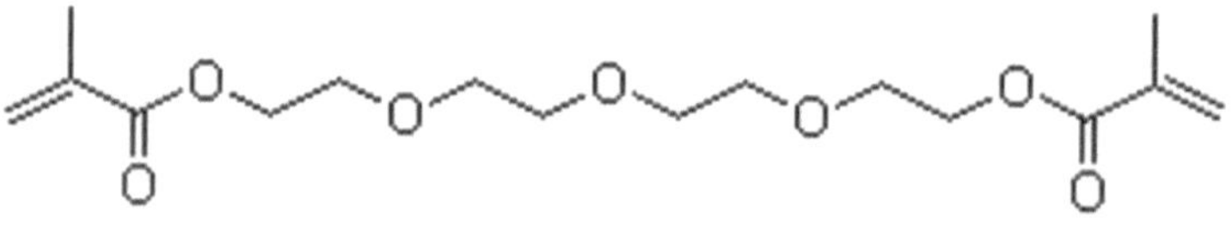

FIGURE 1.3 Tetra-EG having an average functionality $f = 4$ (giving rise to a chain propagation in four directions) and 3D structure thermoset.

PEG is soluble in water and generally considered to be inert and possess low toxicity to humans. PEG is also known to lengthen the biological half-life and reduce the immunogenicity of high molecular weight (MW) substance. The many advantages of PEG make it an ideal monomer in the formation of a biodegradable hydrogel [21–23].

The mixture Tetra-EG and HEMA giving rise to a 3D network [24] where the *Mc* (molar mass between two crosslinks) depend of the ratio Tetra-EG/HEMA. If the concentration of HEMA increases, Tetra-EG being constant, the cell size will increase too and therefore the swelling increases.

1.2.2 EXPERIMENTAL METHODS

1.2.2.1 MECHANISMS OF PHOTOPOLYMERIZATION

In the presence of monomer/oligomer and co-solvent, photoinitiation takes place to form either a linear chain or a three-dimensional crosslinked network [25], depending on the average functionality of the monomers/oligomers/co-solvent being used.

$$Photoinitiator \overset{hv}{\Rightarrow} Active\ species \xrightarrow{monomers} Polymer\ chains$$

The overall mechanism for free radical chain photopolymerization can be generally described as:

1. **Initiation:** Generation of two active species $R°$ by homolytique dissociation of the photoinitiator *PI*

$$PI \xrightarrow{kd} 2R^{\bullet}$$

where; *kd* being the rate coefficient for the dissociation of the photo-initiator followed by the reaction between $R°$ previously formed and monomer M_1 producing new active species $M_1°$.

$$R^{\bullet} + M_1 \xrightarrow{k1} M_1^{\bullet}$$

where; k_1 being the rate coefficient for the initiation step.

2. **Propagation:** Chain extension by successive addition of monomer molecules (*M*) to the monomer active species ($M_1°$) formed in the initiation step.

$$M_1^{\bullet} + (n-1)\, M \xrightarrow{kp} M_n^{\bullet}$$

where; k_p being the rate coefficient for the propagation step.

3. **Termination:** In the last step radical combine or disproportionate to terminate the chain growth and form polymer molecules.
   - by combination between two active species:

$$M_n^{\bullet} + M_m^{\bullet} \xrightarrow{ktc} M_{n+m}$$

   where; k_{tc} being the rate coefficient for the combination step.
   - by disproportionation where hydrogen atom is transferred from one chain to another:

$$M_n^{\bullet} + M_m^{\bullet} \xrightarrow{ktd} M_n + M_m$$

where; k_{td} being the rate coefficient for the combination step.

➤ ***Remark:*** In the crosslinking reactions, the recombination reaction is first favored, but as the viscosity of the middle increases during the cross-linking, the dismutation reaction is rather privileged.

1.2.2.2 KINETICS OF PHOTOPOLYMERIZATION

Kinetics were established by using Differential Photo-calorimetry DPC [26–28]. Calculations of the kinetics of photopolymerization of the acrylates are based on the Sestak and Berggren SB Eqn. (1) [28].

$$R_T = \left(\frac{da}{dt}\right)_{t,T} = k(T)a^m(1-a)^n(-\ln[1-a])^p \tag{1}$$

where; a is the degree of monomer conversion; k is the rate coefficient; m is the order of the initiation reaction; n is the order of propagation reaction; and p is the order of termination reaction.

In order to simplify the SB equation, we consider only the outset of the polymerization process. In so doing, the value of p in the SB equation can be taken as 0 as we are far away from the termination reaction. A simplified autocatalytic kinetic equation can thus be obtained which gives us the following rate Eqn. (2):

$$\frac{d\alpha}{dt} = k\alpha^m(1-\alpha)^n \tag{2}$$

Finally, if the reaction follows nth order kinetics, the general equation of rate will be as follows:

$$\frac{d\alpha}{dt} = k(1-\alpha)^n$$

Just like the kinetics analysis for polymerization of acrylates, the values of k and n can be determined from a ln curve–ln plot of $\frac{d\alpha}{dt}$ *vs.* $[\alpha^{m/n}(1-\alpha)]$, Eqn. (3):

$$\ln\left(\frac{d\alpha}{dt}\right) = \ln k + n\ln \alpha^{m/n}(1-\alpha) \tag{3}$$

Both the autocatalytic and nth order models were applied, and it was found that the reaction appeared to be better modeled by the autocatalytic representation. While applying the autocatalytic equation, the value of n was kept fixed at 1.5 and the values of $m + n$ at 2. Maintaining fixed values of n and $m + n$ gave more consistent results [29].

The monomer conversion of the polymerization reaction is given as:

$$a = \frac{\Delta H_{exp}}{\Delta H_T}$$

where; ΔH_{exp} is the experimental enthalpy of reaction (Jg^{-1}); and ΔH_T is the theoretical enthalpy (Jg^{-1}).

The theoretical enthalpy was calculated through the following relationship:

$$\Delta H_T = \left(\frac{\Delta H_{Tf}}{M}\right)[f]$$

where; [f] is the number of functional groups of each of the monomers; ΔH_{Tf} is the theoretical enthalpy of the acrylates used in the reactions, 19.2 kcal/mol ≈ 80 kJ/mol [33]; M is the total molar mass of the sample. The theoretical enthalpy, molar mass and number of functional groups of the chemicals used are listed in Table 1.1.

TABLE 1.1 Theoretical Enthalpy, Molar Mass, and Number of Functional Groups

Chemical	Δ H (Theoretical) (kJ/mol)	Molar Mass (g/mol)	Number of Functional Groups [f]
PEGDMA	80	330.37	2
HEMA	80	130.14	1

Since the proportion of chemicals are in 50–50 weight percent, substituting the values into Eqn. (2), will give us a value of the calculated theoretical enthalpy of the system:

$$\Delta HT = 0.5 \times \frac{(80 \times 2)kJ}{330} + 0.5 \times \frac{80kJ}{130}$$

$$\Delta HT = 550 J / g$$

1.2.2.3 ACTIVATION ENERGY

The rate coefficient k, can be expressed by Arrhenius Eqn. (4):

$$k_{(T)} = Ze^{-\left(\frac{E_a}{RT}\right)} \tag{4}$$

where; Z is the frequency (collisions) factor; E_a is the activation energy ($Jmol^{-1}$); R is the ideal gas constant (8.314 $Jmol^{-1}K^{-1}$); and T is the temperature (K).

1.2.2.4 DENSITY OF CROSSLINKING X_C

A 3D network structure is formed when the monomers crosslink. The network consists of "infinite" molar mass molecules, which are insoluble, therefore the characterization of crosslinked polymers cannot be determined by molar mass determination, but by the degree of crosslinking, $X_{c,}$ and molar mass between the crosslinks, M_c that can be calculated with the following Eqn. (5):

$$M_c = \frac{M_0}{(f_0 - 2)} \tag{5}$$

where; M_0 is the average molar mass of the different constituents; f_0 is the functionality of the molecule (number of links formed, not to be confused with the number of functional groups [f]).

For example, if the monomer is monofunctional, [f] = 1 and $f_0 = 2$, Mc is infinite which indicates that there is no crosslink and therefore the polymer is linear (thermoplastic).

If several components are used, the molar mass M_0 can be calculated according to Eqn. (6):

$$M_0 = \frac{n_1 M_1 + n_2 M_2 + \ldots + n_i M_i}{n_1 + n_2 + \ldots + n_i} \tag{6}$$

and the functionality of the molecule f_0 with multiple components by Eqn. (7):

$$f_0 = \frac{n_1 f_1 + n_2 f_2 + \ldots + n_i f_i}{n_1 + n_2 + \ldots + n_i} \tag{7}$$

As there are only 2 components used in the ratio 50/50 in weight in the crosslinking reaction, the equations can be simplified into the following expressions:

$$M_0 = \frac{n_{TetraEGDA} M_{TetraEGDA} + n_{HEMA} M_{HEMA}}{n_{TetraEGDA} + n_{HEMA}}$$

$$f_0 = \frac{n_{TetraEGDA} f_{TetraEGDA} + n_{HEMA} f_{HEMA}}{n_{TetraEGDA} + n_{HEMA}}$$

The degree of crosslinking, X_c can be calculated by taking the reciprocal of the M_c value.

1.3 RESULTS AND DISCUSSION

1.3.1 KINETIC STUDIES OF UV CURABLE BIODEGRADABLE HYDROGELS

A typical exotherm plot of PEGDMA and HEMA photopolymerized by Irgacure 651 is obtained by Differential Photocalorimetry DPC (Figure 1.4).

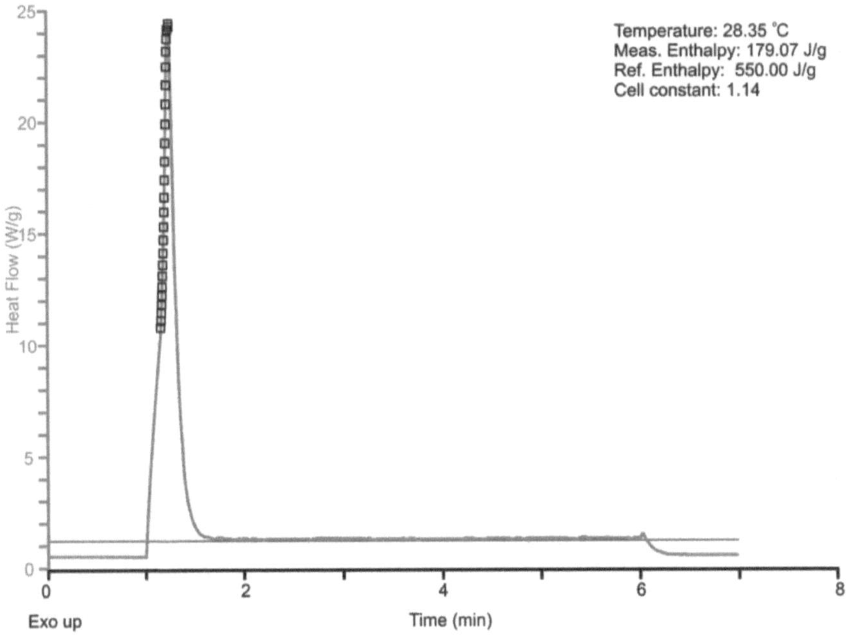

FIGURE 1.4 A typical plot of the exothermic crosslinking reaction by DPC.

The total area under the exotherm gives the enthalpy of the reaction. At the 1-minute mark is when the UV lamp is switched on; a sharp increase in the heat flow is observed. This implies that the crosslinking reaction is taking place due to the fact that polymerization reactions are highly exothermic. The straight line obtained after approximately at 1.5 minutes shows that the reaction has stopped, and the system has reached equilibrium. The higher equilibrium line obtained implies that an exothermic reaction has taken place. One can obtain the induction time, time taken to reach the peak maximum, percentage conversion from the exotherm plot with the analysis software.

DPC was conducted on PEGDMA and HEMA in 4 weight percent photoinitiator individually to determine the reaction rate coefficient before mixing of the 2 chemicals. PEGDMA yielded a rate coefficient value of 2.56 and HEMA yielded a value of 1.43. Subsequently, 4 weight percent photoinitiator dissolved in mixtures of PEGDMA and HEMA of 50/50 weight percentage. The experiments were carried out in triplicate and the average results were taken for more accurate values. The results of the graphs at different temperatures are shown in Tables 1.2 and 1.3.

TABLE 1.2 Kinetic Parameters of the Second-Order Autocatalytic Model for the Photopolymerization of PEGDMA-HEMA 50/50 with 4 wt.% Irgacure® 651 Obtained by TA Specialty Library Analysis

Temperature (°C)	Enthalpy (J/g)	Peak Maximum (s)	Reacted at Peak (%)	Rate Coefficient (k)	Conversion (%)
30	126.7	1.24	34.1	4.13 ± 1.68	23
40	196.74	1.21	33.53	4.36 ± 0.58	36
50	176.65	1.19	27.43	5.96 ± 2.98	30
60	166.06	1.17	20.87	4.84 ± 0.91	32
70	190.94	1.15	17.87	6.12 ± 1.40	35

From the table of results obtained, one can see that an increase in the temperature also resulted in a shorter time to reach peak maximum. Conversion rate was calculated from Eqn. (1). Average conversion rates range from 23–36% within 5 minutes of UV exposure. The reaction rates obtained for the mixture showed a visible increase as compared to the individual systems of PEGDMA and HEMA individually. This shows a synergetic reaction of the two chemicals to produce an increased reaction polymerization which is desirable.

According to the Arrhenius equation [Eqn. (4)], an increase in the temperature will lead to a corresponding increase in the reaction rate coefficient k as seen in the results obtained.

TABLE 1.3 Table of ln k and 1,000/T

1,000/T	ln k
3.30	1.42
3.19	1.47
3.00	1.58
2.92	1.81

The plot of the ln k and the reciprocal temperature (K) in Figure 1.5 displays a linear relationship between ln k and the reciprocal temperature of the experiment, suggesting that the reaction rate coefficient *k*, is in accordance with the Arrhenius equation as assumed in Eqn. (4).

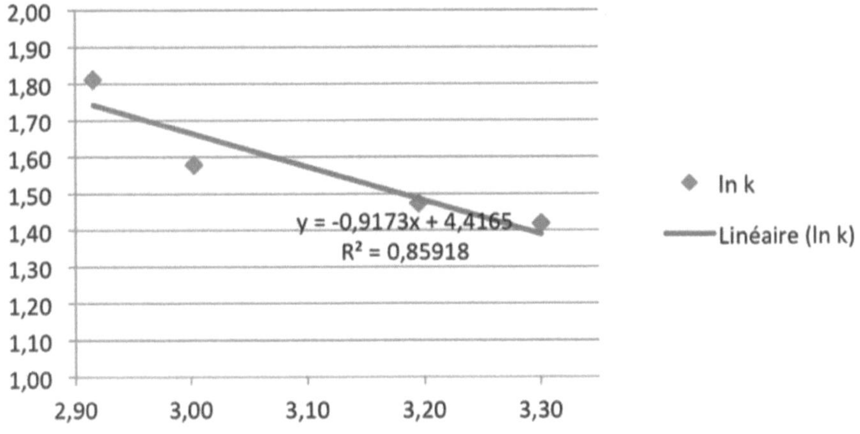

FIGURE 1.5 Graph of ln k vs. 1,000/T.

The 50°C mark was conspicuously omitted from the plot due to its misfit from the trend obtained. The original plot (including 50°C) had an R-squared value of 0.62, suggesting a poor fit to the points, the revised plot in Figure 1.6 showed an improved R-squared value of 0.86, which was a significantly better fit to all the points. The misfit of points at the 50°C temperature can be attributed to a reasonably high standard deviation of rate coefficient values obtained. This deviation could be further attributed to the specific amounts of chemicals placed for each experiment. The high viscosity of the mixed chemicals meant that it was extremely difficult to weigh the exact amount of chemicals onto the experiment pan but rather an approximated value of 2.0

± 0.5 mg as mentioned earlier in the experimental procedure which is likely to have affected the fit of the plot.

An estimated value of activation energy (Ea), 7.62 kJ/mol was obtained. The low activation energy obtained suggests a high reaction rate due to a lower energy barrier to overcome for reaction to take place. This is also in accordance with the sharp peaks obtained in a typical DPC exotherm of the system suggesting that there is a higher reaction rate.

1.3.2 SWELLING RATIO

The swelling ratio of a PEGDMA and HEMA system of 50 weight percent each is 15% which is a low swelling ratio. It was hypothesized that the low swelling ratio is due to the high crosslinking density that limited water penetration. Therefore, by decreasing the weight percentage of difunctional PEGDMA, a more open structure with fewer crosslinks will be obtained, allowing for water to penetrate the network better and improve the swelling ratio [10, 30].

The experimental conditions are held the same as that of the earlier PEGDMA-HEMA (50/50), except that the weight percentages are different. The results of the modified weight percentages can be seen in Table 1.4.

TABLE 1.4 Swelling Ratio

PEGDMA Weight Percent (%)	HEMA Weight Percent (%)	Average Swelling Ratio (%)
50.0	50.0	15.0
10.0	90.0	15.3
5.0	95.0	29.6
2.5	97.5	39.7

The swelling ratio of the system can be calculated according to the Eqn. (8):

$$Swelling\ Ratio(\%) = \frac{W_{swollen} - W_{dry}}{W_{dry}} \times 100 \quad (8)$$

where $W_{swollen}$ is the weight of the hydrogel when it is swollen, and W_{dry} is the initial weight of the hydrogel before swelling.

From Figure 1.6, it can be seen that with a decrease in weight percentage of PEGDMA, there is a significant increase in the average swelling ratio of the hydrogels calculated from Eqn. (8). This can be justified from the calculation of the degree of crosslinking of the different systems. The small increase in swelling ratio of 50 weight percent of PEGDMA and 10 weight percent of PEGDMA suggested that even at 10 weight percent, the cross-linked network is still too dense for water permeation into the 3D network of the hydrogel for swelling.

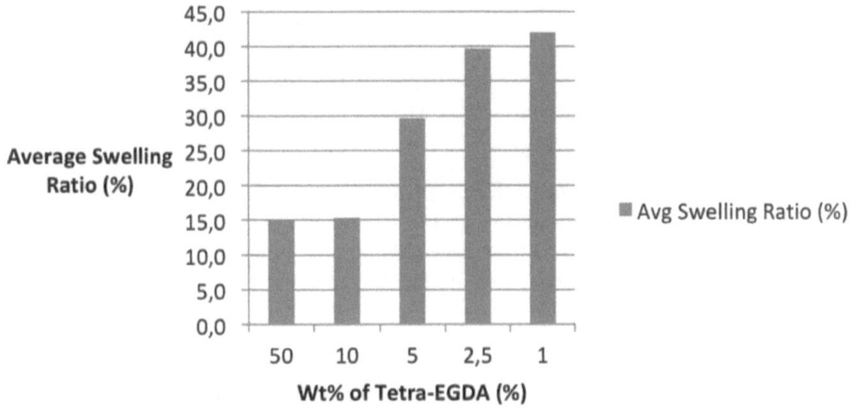

FIGURE 1.6 Average swelling ratio of different weight percent concentrations of PEGDMA.

It also appears that at low weight percent value of PEGDMA in the system, between 2.5 wt.%, and 1.0 wt.%, there is a negligible increase in the swelling ratios. This suggests that the system has reached its maximum swelling point and is very unlikely to increase in the swelling with further decrease in the PEGDMA concentration. The PEGDMA/HEMA hydrogel system showed a maximum swelling ratio of 45–50%.

1.3.3 DEGREE OF CROSSLINKING XC

Table 1.5 shows the tabulation of results compile from the calculation of the degree of crosslinking. Parameters f_1 and f_2 are the functionalities of PEGDMA and HEMA, respectively. The functionality of a system is defined as the number of bonds each double bond forms upon reaction. Each double bond opens up to form 2 bonds when reacted. This is not to be confused with the number of functional groups present in each chemical used in the DPC calculations earlier.

TABLE 1.5 Calculation of f_1 and f_2 Are the Functionalities of PEGDMA and HEMA, Respectively*

Quantity of PEGDMA (10^{-3}.mol)	Quantity of HEMA (10^{-3}.mol)	f_1	f_2	f_0	M_0 (g/mol)	M_c (g/mol)	X_c (mol/g)
7.57	0	4	x	4	330.37	165	6.05×10^{-3}
7.57	19.21	4	2	2.56	186.74	333	3.00×10^{-3}
1.51	34.58	4	2	2.15	138.52	923	1.08×10^{-3}
0.76	36.50	4	2	2.04	134.22	3 355	0.30×10^{-3}
0.38	37.46	4	2	2.02	132.15	6 607	0.15×10^{-3}
0.15	38.04	4	2	2.01	130.93	13 093	0.10×10^{-3}
0	38.42	x	2	2	130.14	∞	0

* f_0 the functionality of the mixture calculated from the Eqn. (7), M_0 the average molar mass of the mixture calculated from Eqn. (6), M_c the average molar mass between crosslinking's of PGDMA and HEMA calculated from Eqn. (5) and X is the degree of crosslinking of the system calculated by taking the reciprocal of the Mc value.

It can be seen from the degree of crosslinking X_c in Table 1.5 that a decrease in the weight percentage of PEGDMA, results in a decrease of degree of crosslinking. Lowering the amount of PEGDMA system decreases the number of difunctional groups in the system, since PEGDMA acts as the only potential crosslinker in the reaction, less crosslinking occurs. Lower degree of crosslinking, X_c implies an increase in the 3D mesh size of the hydrogel, allowing for water to penetrate the crosslinked network.

For pure PGDMA (alone without HEMA) it crosslinks and X_c = 6.05×10^{-3}. Increasing the quantity of HEMA in the mixture leads to a decrease of the degree of crosslinking, going to high density (6.05×10^{-3}) to low density (0.15×10^{-3}) of the 3D network structure. But if we consider HEMA alone (difunctional monomer, one unit of double bond), Mc is infinite, therefore the system does not crosslink and chains tend to be linear (thermoplastic).

1.4 CONCLUSION

The mixture of PEGDMA and HEMA, crosslinked with a UV photo-initiator yields an hydrogel. It can be concluded that the chemical crosslinking reaction of the PEGDMA/HEMA system occurred from the consistent exothermic curve obtained from the DPC. The crosslinking reaction is assumed to follow an Arrhenius equation as proven in the reaction kinetics results obtained showing a linear relationship between the ln k *vs.* $1/T$ curve. We have been able to determine the actual activation energy of the system from the exotherms obtained and calculation of the enthalpies.

We have also proven that it is possible to alter the swelling ratio of the PEGDMA/HEMA hydrogel by changing the crosslinking monomers ratio. An increase in difunctional PEGDMA results in high crosslinking density and lower water uptake, while lowering the difunctional groups and increasing the monofunctional HEMA in the system will increase the 3D mesh size of the polymer network and increase the swelling ratios. The maximum swelling ratio of the system appears to be approximately 45–50% swelling, and this implies that it has superior stability in the swollen state characterized by a high solid content, instead of a high swelling capacity hydrogel typified by a gel with low solid content.

KEYWORDS

- **3-dimensional networks**
- **activation energy**
- **differential photo-calorimetry DPC**
- **hydrogels**
- **kinetics**
- **multifunctional monomers/oligomers**
- **photosensible formulation**
- **relation structure/properties**
- **swelling**

REFERENCES

1. Ratner, B. D., & Bryant, S. J., (2004). Biomaterials: Where we have been and where we are going. *Annu. Rev. Biomed. Eng., 6*, 41–75.
2. Bahram, M., Mohseni, N., & Moghtader, M., (2016). An introduction to hydrogels and some recent applications, Chapter 2. In: *Emerging Concept in Analysis and Applications of Hydrogels*. Intech. https://dx.doi.org/10.5772/64301.

3. Wen, Z., Xing, J., Yang, C., Yuying, L., & Fu Jun, F., (2013). Degradable natural polymer hydrogels for articular cartilage tissue engineering. *J. Chem. Technol. Biotechnol., 88*, 327–339.
4. Langer, R., & Tirrell, D. A., (2004). Designing materials for biology and medicine. *Nature, 428*, 487–492.
5. Mathur, A. M., Moorjani, S. K., & Scranton, A. B., (1996). Methods for synthesis of hydrogel networks: A review. *J. Macromol. Sci. Part C Polym. Rev., 36*, 405–430.
6. Gulsen, D., & Chauhan, A., (2005). Dispersion of microemulsion drops in HEMA hydrogel: A potential ophthalmic drug delivery vehicle. *International Journal of Pharmaceutics, 292*, 95–117.
7. Raj Singh, T. R., Laverty, G., & Donnelly, R., (2018). *Design, Synthesis and Application in Drug Delivery and Regenerative Medicine.* CRC Press.
8. Michalek, J., Hobzova, R., Pradny, M., & Duskova-Smrckova, M., (2010). Hydrogen contact lenses. In: *Book: Biomedical Applications of Hydrogen Handbook.* doi: 10.1007/978-1-4419-5919-5_16.
9. Ahmed, E. M., (2015). Hydrogel: Preparation, characterization, and applications: A review. *J. Adv. Res., 6*, 105–121. https://doi.org/10.1016/j.jare.2013.07.006.
10. Marí-Buyé, N., O'Shaughnessy, S., Colominas, C., Semino, C. E., Gleason, K. K., & Borros, S., (2009). Functionalized, swellable hydrogel layers as a platform for cell studies. *Advanced Functional Materials, 19*, 1276–1286.
11. Moreau, D., (2016). *Design and Characterization of Hydrogel Films and Hydrogel-Ceramic Composites for Biomedical Applications* (p. 21). PhD Thesis. Mines Paris-tech.
12. Casadio, Y. S., Brown, D. H., Chirila, T. V., Kraatz, H. B., & Baker, M. V., (2010). Biodegradation of poly(2-hydroxyethyl methacrylate) (PHEMA) and poly{(2-hydroxyethyl methacrylate)-co-[poly(ethylene glycol) methyl ether methacrylate]} hydrogels containing peptide-based cross-linking agents. *Biomacromolecules, 11*, 2949–2959.
13. Wichterle, O., & Lim, D., (1960). Hydrophilic gels for biological use. *Nature, 185*, 117, 118. doi: 10.1038/185117a.
14. Shan, S., Sun, X. F., Xie, Y., Li, W., & Ji, T., (2021). High-performance hydrogel adsorbent based on cellulose, hemicellulose, and lignin for copper(ii) ion removal. *Polymers, 13*, 3063.
15. Laftah, W. A., Hashim, S., & Ibrahim, A. N., (2011). Polymer hydrogels: A review. *Polym. Technol. Eng., 50*, 1475–1486.
16. Danek, C., (2022). Recent advances and future challenges in the additive manufacturing of hydrogels. *Polymers, 14*, 494. https://doi.org/10.3390/polym14030494 https://www.mdpi.com/journal/polymers.
17. Huang, L. J., Lee, W. J., & Chen, Y. C., (2022). Bio-based hydrogel and aerogel composites prepared by combining cellulose solutions and waterborne polyurethane. *Polymers, 14*, 204–215. https://www.mdpi.com/journal/polymers (accessed on 02 January 2022).
18. Pepelanova, I., Kruppa, K., Scheper, T., & Lavrentieva, A., (2018). Gelatin-methacryloyl (GelMA) hydrogels with a defined degree of functionalization as a versatile tool kit for 3D cell culture and extrusion bioprinting. *Bioengineering (Basel), 5*, 55–60.
19. Shirahama, H., & Lee, B. H., (2018). Preparation of photocurable hydrogels (Chapter 15). In: Raj, S. T. R., Laverty, G., & Donnelly, R., (eds.), *Design, Synthesis and Application in Drug Delivery and Regenerative Medicine* (pp. 265–283). CRC Press.

20. Abadie, M. J. M., & Voytekunas, V. Y., (2004). New trends in UV curing. *Eurasian Chem. Tech. Journal, 6,* 67–77.
21. Moussa, K., & Decker, C., (1993). Semi-interpenetrating polymer networks synthesis by photocrosslinking of acrylic monomers in a polymer matrix. *Journal of Polymer Science, Part A: Polymer Chemistry, 31*, 2633–2642.
22. Shia, J., Yaub, L., Dingab, J., PEG-based thermosensitive and biodegradable hydrogels. *Acta Biomaterialia, 128*, 42–59.
23. Asmussen, S. V., Gomez, M. L., & Vallo, C. I., (2018). Novel hydrogels based on a high-molar-mass water-soluble di-methacrylate monomer. *Polymer International, 67*, 606–614.
24. Chan, K., & Gleason, K. K., (2005). Initiated chemical vapor deposition of linear and cross-linked poly(2-hydroxyethyl methacrylate) for use as thin-film hydrogels. *Langmuir, 21*, 8930–8939.
25. Pamedytytė, V., Abadie, M. J. M., & Makuska, R., (2002). Photopolymerization of N,N-dimethyl-amino-ethyl-methacrylate studied by photocalorimetry. *Journal of Applied Polymer Science, 86*, 579–588.
26. Appelt, B. K., & Abadie, M. J. M., (1988). Calorimetric characterization of photosensitive materials. *Polymer Eng. Sci., 28*, 367–371.
27. Seghier, Z., Diby, A., Voytekunas, V. Y., Cheang, P., & Abadie, M. J. M., (2008). Effect of filler type, content and size on the UV photocuring dental materials. *Chemistry & Chemical Technology, 2*, 15–18.
28. Sestak, J., & Berggren, G., (*1971*). Study of the kinetics of the mechanism of solid-state reactions at increasing temperatures. *Therm Acta, 3*, 1–11.
29. Abadie, M. J. M., De Almeida, Y. M. B., & Carrera, L. C. M., (1996). Photopolymerization of multifunctional acrylates (MAM) using trans-10, 11-dibromodibenzosuberone as a radical photoinitiator. *European Polymer Journal, 32*, 1355–1360.
30. Omidian, H., Park, K., Kandalam, U., & Rocca, J. G., (2010). Swelling and mechanical properties of modified HEMA-based super porous hydrogels; *Journal of Bioactive and Compatible Polymers, 25*, 483–497.

CHAPTER 2

Free Radical Addition of Polyhaloidolefins to α-Pyrrolidone and N-methylpyrrolidone

A. Z. CHALABIYEVA, D. R. NURULLAYEVA, and B. A. MAMEDOV

Institute of Polymer Materials, Azerbaijan National Academy of Sciences, Sumgait, Azerbaijan

ABSTRACT

The radical addition reactions of α-pyrrolidone and N-CH_3-pyrrolidone to tetrachloroethylene and trichloroethylene in the presence of ditretbutyl peroxide (DTBP) have been carried out. It has been revealed that polyhaloidolefins with α-pyrrolidone form 5-(polychlorovinyl)-pyrrolidones, and N-CH_3-pyrrolidone – the mixture of 3- and 5-(polychlorovinyl-N-CH_3-pyrrolidones. It has been shown that the absence or presence of a hydrogen bond between the reacting substrates is responsible for the regioselectivity of the addition; the one-stage method of the preparation of 3- and 5-(polychlorovinyl)-pyrrolidones has been developed and the optimal conditions of the synthesis of adducts 1:1 (T – 150–155°C, duration – 5–6 h, molar ratio of addend:unsaturated compound:DTBP = 10÷30:1:0.1÷0.25) have been found.

2.1 INTRODUCTION

A large number of publications has been devoted to the chemical conversions of α-pyrrolidone and its reactivity in various reactions, and a considerable part of the papers includes the development of synthesis methods of the known

Advanced Polymer Structures: Chemistry for Engineering Applications. Omar Mukbaniani, DSc, Tamara Tatrishvili, PhD & Marc Jean M. Abadie, DSc (Eds.)

natural substances and synthetic drugs on the basis α-pyrrolidone and its derivatives [1–3]. The analysis of the literature data on the chemical conversions of α-pyrrolidone and its derivatives indicates their high reactivity in various chemical reactions and the possibility of reaction behavior on several reaction centers. The structural peculiarities of molecules containing unique heterocyclic structure, combining C^3H- and NH acid centers and carbonyl group in molecule, open up large prospects for active participation of such compounds in the chemical conversions [4, 5].

The direct N-alkylation of 2-pyrrolidones with halogen alkyls allows to obtain a series of various N-alkylsubstituted 2-pyrrolidones with the aim of finding new pharmacological active substances and revealing the bond "structure of the substance-biological activity."

We have previously investigated some peculiarities and addition reaction products of α- and N-CH_3-pyrrolidone to acrylonitrile, allyl alcohol and vinyl acetate initiated with di-tret-butyl peroxide (DTBP) [6, 7]. The analysis of forming adducts 1:1 showed that they are the mixture of two isomers. The addition reaction of acrylonitrile to α-pyrrolidone leads to the formation of only C^5-adduct with yield 82%, and in a case of allyl alcohol the mixture of isomers is formed at ratio 3:1.

Continuing investigations in the field of synthesis, study of the structure and properties of nitrogen and halide-containing lactams, the free radical addition of α-pyrrolidone and N-CH_3-pyrrolidone to tetrachloroethylene and trichloroethylene in the presence of DTBP has been carried out.

2.2 EXPERIMENTAL METHODS AND MATERIALS

2.2.1 SYNTHESIS OF 3,-5-(DICHLOROVINYL)-N-CH_3-PYRROLIDONE (IV)

All experiments were carried out in glass equipment at atmospheric pressure. A solution of mixture of 10.95 g (0.075 mol) of DTBP and 39.47 g (0.3 mol) of trichloroethylene was uniformly added to 297.2 g (3 mol) of N-CH_3-pyrrolidone during the reaction. As a result, a large excess of N-CH_3-pyrrolidone in relation to the unsaturated compound was achieved, which favored the formation of adducts 1:1 and hindered the telomerization reaction behavior. The reaction was carried out in an excess of the initial lactamese at temperature 155°C for 6 h at molar ratio 30÷20:1:0.1÷0.25% (DTBP).

Similarly, the adducts (I–III) have been obtained. The purification of the synthesized compounds was carried out by vacuum distillation of an excess of the initial pyrrolidone. The obtained products have been characterized by data of elemental analysis, IR and PMR spectroscopy (Table 2.1).

2.3 RESULTS AND DISCUSSION

Lactams, being cyclic amides, are perspective intermediates in the synthesis of some biologically active substances, and are also of interest as compounds with a number of valuable properties allowing them to be used in pharmacology. With the aim of search of a convenient and effective method of the synthesis of polyhalides-containing lactams, the free radical addition reaction of α-pyrrolidone and N-CH_3-pyrrolidone to tetra- and trichloroethylene in the presence of DTBP has been carried out. The reactions were carried out in an excess of the initial lactams on the scheme:

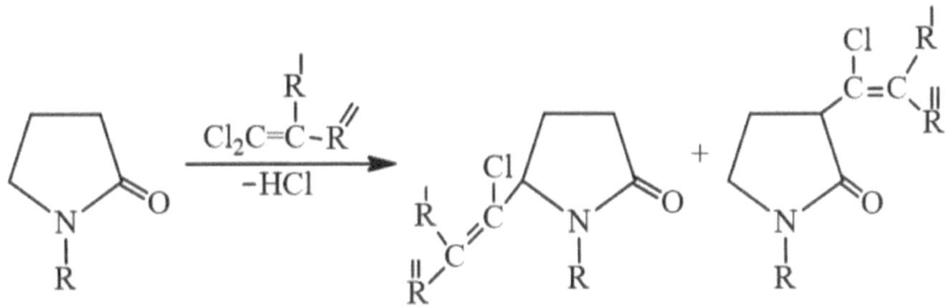

R=H, CH_3; $\overset{I}{R}$=H,Cl; $\overset{II}{R}$=Cl

On completion of the reaction, the end products were purified by vacuum distillation. The yields of lactams (I-IV) were 65–75%.

The obtained derivatives (I-IV) have been characterized by data of elemental analysis, IR and PMR spectroscopy.

There have been revealed the absorption bands in the IR spectra of compounds (CCl_3), γ, cm^{-1}: 1,605 (for fragment of $CCl{=}CCl_2$), 1,640 (C=C); 2,960 and 2,970 (CH_3), 2,926 and 2,850 (CH_2), 3,250 (N–H in lactame); NMR ^1H spectrum ($CDCl_3$)δ, ppm: 1.1–4.5 m (CH_2), 1.4–2 m (–CH–); 4.0–5.5 m (=CH); 3.4 s (C^5H_2), 2.3–2.5 s (C^3H_3), 2.0c (C^4H_2); 2.8c (N-CH_3). The chemical shifts of protons connected with the nitrogen atom are less

TABLE 2.1 Physical Constants of the Synthesized Adducts I–IV

№ of compounds	Compounds	B.p. (mm merc.c.)	d^{20}_{4}	n^{20}_{D}	Calculated (%) / Found (%)			
					C	H	N	Cl
I	5-(trichlorovinyl)-	145–147/0.5	1.3350	1.4925	33.0	2.8	6.53	49.6
	α-pyrrolidone				35.8	2.75	6.70	48.5
II	5-(dirichlorovinyl)-	140–142/0.5	1.2930	1.4450	40.22	3.35	7.80	39.66
	α-pyrrolidone				40.31	3.27	7.85	38.20
III	3,-5-(trichlorovinyl)-	103–105/1	1.3056	1.5600	43.52	4.14	7.25	36.78
	N-CH_3-pyrrolidone				43.10	4.35	7.51	37.50
IV	3,-5-(dichlorovinyl)-	101–102/1	1.2760	1.5230	43.3	3.9	7.8	36.6
	N-CH_3-pyrrolidone				45.6	3.81	8.10	36.5

characteristic. This is mainly explained by their tendency to formation of intra- and intermolecular hydrogen bonds.

It has been revealed that tetrachloroethylene with α-pyrrolidone in the presence of DTBP forms 5-(trichlorovinyl)-pyrrolidone. At the same time, trichloroethylene and tetrachloroethylene with N-CH_3-pyrrolidone form the mixture of 3- and 5-(dichlorovinyl)- and 3- and 5-(trichlorovinyl)-N-CH_3-pyrrolidones at ratio 1:3. It has been detected by PMR spectroscopy that the absence or presence of a hydrogen bond between the reacting substrates (intramolecular hydrogen bond δ 4.5–9 ppm) is responsible for the regioselectivity of the addition [9].

Since the main task of these investigations was the development of effective methods of the synthesis of adducts 1:1, the quantitative yields of these compounds were determined in the selection of the initiator and in the condition of the reaction carrying out.

The searches of optimal process conditions with the aim of preparation of the formation of adducts 1:1 (with maximum yields), we have revealed the influence of the molar ratio of addend: unsaturated compound and a quantity of used DTBP on the yield of purposeful products. Based on the obtained data, the optimal reaction conditions: T – 150–155°C, duration – 5–6 h, molar ratio addend: unsaturated comp.: DTBP = 10 ÷ 30:1:0.1 ÷ 0.25 have been found. The fact that the use of DTBP leads to the formation of only 5-trichlorvinyl-2-pyrrolidones with high yields, it can be concluded that the process proceeds regioselectively and noticeably with high rate.

The exact quantitative ratio of isomers was determined by a method of GLC using the example of adduct 1:1 of N-CH_3-pyrrolidone with tetrachloroethylene. In a total yield of (III) 72%, 5-(trichlorovinyl)-N-CH_3 pyrrolidone is formed in a quantity of 52%, and 3-(trichlorvinyl)-N-CH_3-pyrrolidone – in a quantity of 20%. It can be concluded from the obtained data that the stabilizing effect of the nitrogen atom is greater than the carbonyl group.

For estimation of the influence of a quantity of peroxide on the yield of adducts 1:1 of α-pyrrolidone with tetrachloroethylene, the quantity of DTBP was varied from 5 to 35 mol.% in relation to tetrachloroethyleneone. An increase of a quantity of peroxide, at a constant molar ratio of α-pyrrol idone:tetrachloroethylene, equal to 20:1, leads to an increase of the adduct yield 1:1 (from 65% to 77.5) with simultaneous increase of a quantity of "residue." In Figure 2.1, the dependence of the yield of 5-(trichlorovinyl)-α-pyrrolidone on the quantity of DTBP per tetrachloroethylene has been shown.

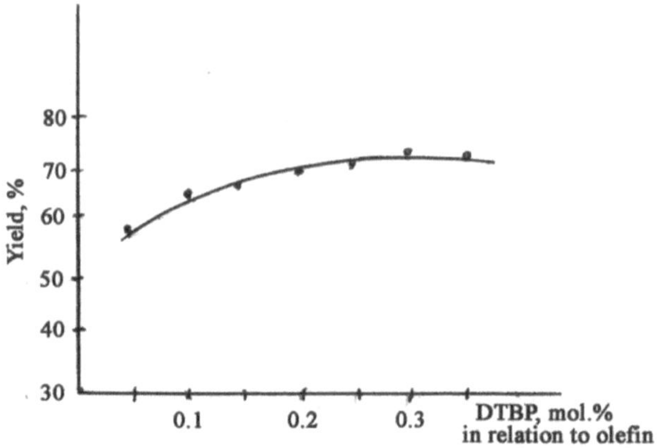

FIGURE 2.1 Dependence of yield of 5-(trichlorovinyl)-α-pyrrolidone per peroxide.

2.4 CONCLUSION

1. The free radical addition reactions of α-pyrrolidone and N-CH_3-pyrrolidone to tetrachloroethylene and trichloroethylene in the presence of DTBP have been carried out. It has been established that the products of these reactions are the mixtures of 3- and 5-(chlorovinyl)-pyrrolidones, and the latter ones are formed in the predominant quantity.
2. The effective one-stage method of the preparation of 3- and 5-(chlorovinyl)-pyrrolidones has been developed and the optimal conditions of the synthesis of adducts 1:1 (t – 150–155°C, duration – 5–6 h, molar ratio addend: unsaturated compound: DTBP 10 ÷30:1÷ 0.1–0.25) have been found.

KEYWORDS

- **ditretbutyl peroxide**
- **lactam cycle**
- **N-methylpyrrolidone**
- **regioselectivity**
- **tetrachloroethylene**
- **trichloroethylene**
- **α-pyrrolidone**

REFERENCES

1. Berestovitskaya, V. M., Vasilyeva, O. S., & Zobachova, M. M., (2002). İzv. RGPU N.A. A.İ. Gertsen. Ser. *Natural and Exact Sciences, 2*, 133–144.
2. Shmaryan, M. İ., Klimova, N. V., Morozov, İ. S., & Lavrova, L. N., (1996). Pat. Russia 270311204. RZhKh., 9032П.
3. Gaffar, A., Alflito, & Williams, M. (1962). Pat. USA 103005. *Antiplaque, Anti-Gingivitis, Anti-Carus Oral Composition* / РЖХ, 19962, 90193 П.
4. Sybramanyam, R., & Gu, B. (1962). Pat. USA 568701. *Composition of Alkylpyrrolidone for Solubilizing Anti-Bacterials.* РЖХ, г. 60391 П.
5. Shmaryan, M. I., Klimova, N. V., Morozov, I. S., & Lavrova, L. N., (1996). Pat. Russia 2703112/04.5-*(1-Adatantil) 2-Pyrrolidone with Psychotropic Activity and Method of its Preparation* (p. 9032). RZhKh.
6. Chalabiyeva, A. Z., (2021). Free radical addition of alkenes to α-pyrrolidone and N-CH_3-pyrrollidone. *Modern Problems of Chemistry. Mat. Republican Scientific Conference* (p. 145–147). Sumgait.
7. Ishenko, N. Ya., & Chalabiyeva, A. Z., (2015). Free radical telomerization of allylglycidyl ether with N-CH_3-pyrrollidone. *Mat. Eurasian Union of Scientists of the International Scientific-Practical Conference* (pp. 2, 76–78). "Modern concepts of scientific investigations," Moscow.
8. Saygili, N., Altunbaş, A., & Yeşilada, A., (2006). *Turk. Chem., 30*, 125–130.
9. Gordon, A., & Ford, R., (1974). *Chemist's Satellite* (p. 541). M.; Mir., p. 541.

CHAPTER 3

Synthesis and Investigation of Maleinized Oligopropylene

VUSALA DOSTUYEVA, BAKHTIYAR MAMMEDOV, and
AYNURA MAMMEDOVA

Institute of Polymer Materials, Azerbaijan National Academy of Sciences, Sumgait, Azerbaijan

ABSTRACT

Maleinized oligopropylene was synthesized by reacting oligopropylene macromonomer with maleic anhydride. The influence of the process conditions on the course of the reaction was studied, and the optimal reaction conditions were found. The radical copolymerization reaction of maleic anhydride with oligopropylene macromonomer was carried out, and the regularities of the copolymerization reaction and the composition and structures of the obtained products were determined. It was found that copolymerization of a pair of oligopropylene macromonomer + maleic anhydride monomers in the presence of benzoyl peroxide forms copolymer macromolecules with an alternative structure.

3.1 INTRODUCTION

Currently, new scientific developments in the field of high MW compounds play an important role in solving many problems in medicine, agriculture, and the food industry. Naturally, many technologies have been developed to produce new antibacterial polymer materials that are non-toxic, highly active, and durable [1, 2].

Advanced Polymer Structures: Chemistry for Engineering Applications. Omar Mukbaniani, DSc, Tamara Tatrishvili, PhD & Marc Jean M. Abadie, DSc (Eds.)

Due to the purposeful functionalization of polymers, it is possible to achieve a change in their physicochemical and mechanical properties, as well as the formation of antibacterial properties [2]. One of the effective methods for obtaining antibacterial polymeric materials is their physical modification. Thus, biologically active additives are introduced in the process of extrusion of polymeric materials. Many organic and inorganic compounds, oligomers, and polymers are used as biologically active additives. Some conditions are put forward for these additives, one of which is their harmlessness to living beings. From this point of view, salicylic acid and its derivatives, as well as oligomers containing maleic anhydride and salicylic groups, are of great interest [3, 4].

Some works [4–8] provide information about the preparation of high-impact and antibacterial polymer materials by modifying polymeric materials such as polyethylene, polypropylene, acrylonitrile-butadiene-styrene, and Noryl (a mixture of polyphenylene oxide and polystyrene).

The copolymerization reactions of acetylsalicylic acids and their methacryloyl, allyl, and vinyl esters with such monomers as maleic anhydride, methyl methacrylate, and styrene have been studied [9–12]. As a result of the reactions of functionalization of polyethylene and polypropylene macromonomers with acetylsalicylic acids, oligoolefin esters of acetylsalicylic acids were synthesized, and their properties were studied [6, 13]. The radical copolymerization reactions of polyethylene and polypropylene macromonomers with acetylsalicylic acids and their methacryloyl and allyl ethers were carried out, and antibacterial block copolymers were obtained.

The presented chapter is devoted to the study of the regularities and products of the reactions of the maleinization of the oligopropylene macromonomer (OPMM) and the copolymerization of the macromonomer with maleic anhydride.

3.2 EXPERIMENTAL METHODS AND MATERIALS

3.2.1 MALEINIZATION OF OLIGOPROPYLENE MACROMONOMERS

The ampoules are filled with 0.1 g/mol (50 g) of oligopropylene macromonomer, 0.1 g/mol (9.80 g) of maleic anhydride, 0.07 g of the initiator – benzoyl peroxide and 100 ml of dimethylformamide. The ampoule is frozen in a mixture of "acetone-ice," sucked off in a vacuum, and blown through with nitrogen. The last operation is repeated several times. After welding, the neck of the ampoule is placed in a thermostat and heated at 70°C for 6 hours.

The ampoule is removed from the thermostat, cooled, and precipitated in methanol or ethanol. The resulting white powdery product is washed several times in a precipitator and dried in a vacuum dryer at 40°C until constant weight (72% yield).

3.2.2 *ANALYSIS METHODS OF MALEINIZED OLIGOPROPYLENE*

IR spectra of monomers and copolymers were recorded on an Agilent Cary 630 FTIR spectrometer from Agilent Technologies in the range of 600–4,000 cm^{-1}. From the samples that did not form a high-quality thin layer, transparent tablets were obtained by pressing them in a vacuum under a fine powder mixture with ZnSe, and the IR spectra were recorded. The intrinsic viscosity of the copolymers was determined using an Ubbeloda viscometer in a decane solution. Copolymerization constants were calculated using the Mayo-Lewis equation.

3.3 RESULTS AND DISCUSSION

A method has been developed for the preparation of oligomers containing a fragment of maleic anhydride in the terminal group by maleation reactions of oligopropylene macromonomers. It was found that, in contrast to the reactions of maleation of polyolefins by known methods, in the maleization of oligopropylene macromonomers, maleic anhydride binds only to the double bond in the terminal group of alpha-oligopropylene.

Alpha-maleated oligoolefins were obtained by the addition of oligomers with an average MW of 400–600, in other words, macromonomers of an oligoolefin into maleic anhydride under special conditions in the presence of benzoyl peroxide initiators. Maleinized oligopropylene containing 8–15 wt.% malein group has the following properties: intrinsic viscosity – 0.22 g/100 ml, melting point 169°C, dissolves in all proportions in dimethylformamide. In contrast to the maleinization reactions of polyolefins, in the maleinization reaction of the oligopropylene macromonomer, depending on the conditions, the binding of maleic anhydride is observed only with the terminal group of alpha-oligopropylene. This fact was confirmed by analysis of the IR spectra of the samples (Figure 3.1). In the IR spectra of maleized oligomers, absorption bands were recorded that can be attributed to the C=O (1,725 and 1,780 cm^{-1}), C-O-C (1,055 and 1,150 cm^{-1}) bonds, and absorption bands characteristic methyl group of oligopropylene units (2,869 and 2,916 cm^{-1}) were identified.

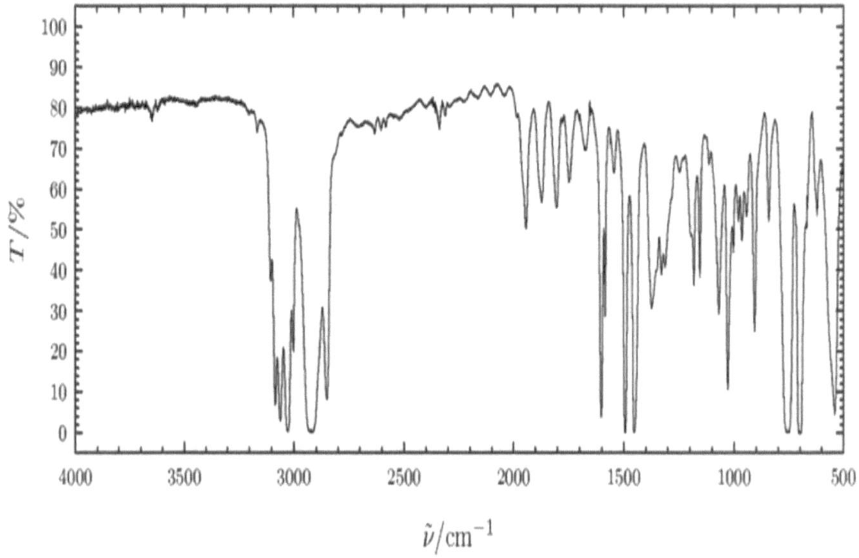

FIGURE 3.1 IR spectrum of maleinized oligopropylene.

The influence of such factors as temperature, the ratio of the components involved in the reaction, the reaction time on the yield of the maleinization reaction of oligopropylene macromonomer was studied, and the optimal reaction conditions were determined. Studies have shown that the optimal conditions for maleinization of OPMM are as follows: T = 65–75°C; OPMM:MA=1:1; t = 6 hours.

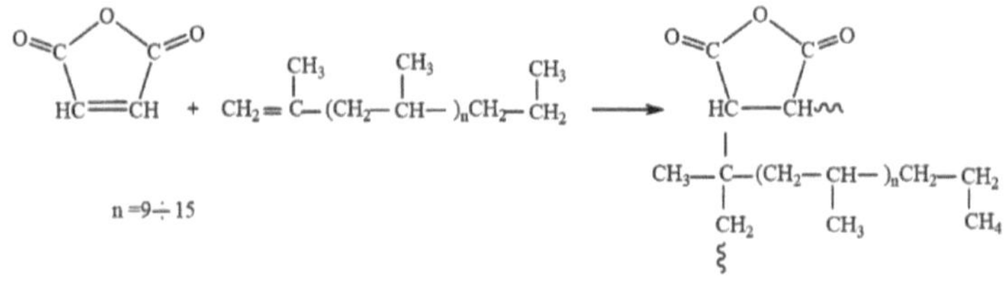

For a pair of monomers OPMM + MA r_1 = 002 and r_2 = 0.03, the closeness of the values of the relative activity of monomers in copolymerization reactions to zero proves that the copolymers have an alternative (sequential) structure (Table 3.1).

TABLE 3.1 Initial Ratios of Monomers and the Composition of the Unit of the Obtained Copolymers During the Radical Copolymerization Reaction of Oligopropylene Macromonomer (M_1) with Maleic Anhydride (M_2)

M_1	M_2	Conversion Level (%)	m_1	m_2
10	90	5.0	50.4	49.6
25	75	8.0	50.6	49.4
50	50	8.5	49.7	50.3
75	25	9.5	51.0	49.0
90	10	9.8	51.2	48.8

3.4 CONCLUSION

A new method for producing maleinized oligopropylene has been developed. In contrast to the previous methods, as a result of the interaction of the oligopropylene macromonomer with maleic anhydride, a new oligopropylene macromonomer was obtained, the regularities of the process were studied, and the structure was determined. Radical copolymerization of oligopropylene macromonomer with maleic anhydride was carried out, and an antibacterial copolymer with an alternative structure was synthesized.

KEYWORDS

- **alpha-maleated oligoolefins**
- **alpha-oligopropylene**
- **antibacterial polymer additives**
- **oligopropylene macromonomer**
- **polyethylene**
- **polypropylene**
- **salicylic acid**

REFERENCES

1. Dontsova, E. P., Zharnenkova, O. A., Snezhko, A. G., & Uzdensky, V. B., (2014). Polymer materials with antimicrobial properties. *Plastic, 1, 2*, 30–35.
2. Shtilman, M. I., (1998). Polymers in biologically active systems. *Soros Educational Journal, 5*, 48–53.

3. Lisina, S. V., & Brel, A. K., (2006). Synthesis and pharmacological activity of new derivatives of salicylic acid and aspirin as potential drugs. *Advances in Modern Natural Science, 11*, 95–96.
4. Oromiehie, A., Ebadi-Dehaghani, H., & Mirbagheri, S., (2014). Chemical modification of polypropylene by maleic anhydride: Melt grafting, characterization and mechanism. *International Journal of Chemical Engineering and Applications, 5*(2), 117–122.
5. Rasulzadeh, N. Sh., & Safarova, Q. M., (2016). The obtaining of potential antibacterial polyethylene composites and research of their properties. *Proceedings of Young Scientists, 14*, 43–46.
6. Ibadov, E. A., & Rasulzadeh, N. Sh., (2016). The obtaining of polyethylene macromonomers and methacryloyl salicylates copolymers and research of their properties. *Proceedings of Young Scientists, 14*, 61–64.
7. Dostuyeva, V. M., & Rasulzade, N. Sh., (2018). Investigation of preparation of esters of reactions of polypropylene macromonomers with salicylic acid. *Young Scientist, 2*, 40–43.
8. Dostuyeva, V. M., & Rasulzade, N. Sh., (2020). Synthesis and research of N-oligoalkyl morpholine. *International Scientific and Technical Conference of Young Scientists Innovative Materials and Technologies* (pp. 272, 273). Minsk.
9. Rasulzade, N. Sh., Dostuyeva, V. M., Bakhshaliyeva, K. F., & Muradov, P. Z., (2019). Preparation and investigation of antibacterial materials on the basis of oligopropylene ester of salicylic acid and polypropylene. *Processes of Petrochemistry and oil Refining, 20*(3), 284–290.
10. Rasulzade, N. Sh., Azizov, A. H., Safarova, G. M., & Rasulov, N. Sh., (2017). Study of copolymerization reactions of allyl- and vinyl acetylsalicylates with maleic anhydride. *Azerbaijan Chemical Journal, 2*, 34–37.
11. Rasulzade, N. Sh., Azizov, A. H., & Safarova, G. M., (2017). *Investigation of Copolymerization Reactions of Acetylsalicylic Acid with Methyl Methacrylate of Allyl and Vinyl Esters* (Vol. 2, pp. 28–33). News of Sumgayit State University.
12. Rasulzadeh, N. Sh., Azizov, A. H., İbadov, E. A., Zeynalova, S. G., & Rasulov, N. Sh., (2017). The research of antibacterial properties of methyl methacrylate and methacryloyl salicylate copolymers. *Azerbaijan Chemical Journal, 3*, 17–20.
13. Rasulzadeh, N. Sh., & Ibadov, E. A., (2017). The synthesis and properties of acrylic and methacrylic ether of salicylic acid. *International Journal of Research Studies in Science, Engineering and Technology, 4*(3), 1–3.
14. Rasulzade, N. Sh., & Safarova, G. M., (2017). *Synthesis and Study of Potential Biologically Active Oligoalkyl Esters of Acetylsalicylic Acid* (Vol. 5, No. 38, pp. 76–79. Eurasian Union of Scientists.

CHAPTER 4

Synthesis and Characterization of Poly(Acrylic Acid-g-α-Methyl-β-Alanine)

EFKAN ÇATIKER and M. R. NASUHBEYOĞLU

Ordu University, Department of Chemistry, Ordu, Turkey

ABSTRACT

Oligomeric poly(α-methyl-β-alanine) was obtained from methacrylamide by base-catalyzed hydrogen-transfer polymerization (HTP). The oligomeric moiety was determined to have olefinic end-groups to be used as functional groups for further polymerization, namely the oligomer was used as macromonomer in a polymerization used "graft through" approach. For this purpose, free radical (co)polymerization of acrylic acid and the macromonomer was performed to obtain poly(acrylic acid) copolymers grafted by α-methyl-β-alanine fragments. The structure of the copolymer was clarified by FTIR and ^{1}H-NMR techniques. To evaluate the effects of the aliphatic amide branches on the thermal properties of poly(acrylic acid) thermal analyzes of the copolymer were performed by using DSC and TGA methods. As expected, the glass transition of poly(acrylic acid) shifted in the lower temperatures by about 20C with introducing the α-methyl-β-alanine fragments.

4.1 INTRODUCTION

PAA is widely used as a superabsorbent and dispersant in daily life. In literature, PAA has been reported as materials for soft contact lenses [1] and matrices [2] for oral and transdermal systems. Although PAA has features such as biodegradable nature, non-toxic behavior, biocompatibility, etc.,

Advanced Polymer Structures: Chemistry for Engineering Applications. Omar Mukbaniani, DSc, Tamara Tatrishvili, PhD & Marc Jean M. Abadie, DSc (Eds.)

for use as a biomaterial [3], some mechanical properties like impact strength, hardness, flexural modulus require modification for special applications. The optimum mechanical strength of PAA can be achieved by its copolymers obtained by partial hydrolysis of poly(tert-butyl acrylate) [4] and controlled radical polymerization [5]. Controlled radical polymerization enables to be synthesized block copolymers of PAA including polystyrene, poly(2-vinyl pyridine), and poly(methyl methacrylate) [6]. Polyacrylic acid (PAA) is a common reactive synthetic polymer used in grafting onto various natural polymers such as cellulose [7], carrageenan [8], and cacia gum [9]. Some branched copolymers of PAA [10–14] are also prepared by different polymerization methods to modify its properties like pH response [10, 12], phase behavior [10], micelle morphology [15, 16].

Macromonomers have importance in macromolecular architecture to design novel and special macromolecules. Hydrogen-transfer polymerization (HTP) is a useful route that enables the preparation of an oligomeric heterochain backbone with functional end-groups [17]. The oligomeric moieties can be used as macromonomers to obtain some graft copolymers by the "graft through" approach [18]. Moreover, these oligomeric structures can be functionalized (activated) chemically to achieve novel block copolymers [19].

In this study, HTP and FRP were performed consecutively to obtain poly(acrylic acid-g-α-methyl--alanine) (PA-g-mBA). Poly(α-methyl β-alanine) [PmBA] oligomer with olefinic end-group as macromonomer was synthesized by base-catalyzed HTP. PmBA macromonomer and acrylic acid were used to obtain PA-g-mBA using the "graft through" strategy. Structural and thermal characterizations of the novel graft copolymer were performed in detail.

4.2 EXPERIMENTAL METHODS AND MATERIALS

4.2.1 MATERIALS

Methacrylamide (99% Fluka), formic acid (99.9%, Carlo Erba), methanol (99.9%, Sigma-Aldrich), and [2,2-azobis(2-methylpropionitrile) (AIBN), Sigma-Aldrich] were commercially obtained and used without purification. Acrylic acid (99%, Sigma) was passed through a column filled with neutral alumina to remove the inhibitor before use.

4.2.2 INSTRUMENTATION

^{1}H-NMR spectra were recorded using Bruker Ultra-Shield Plus, ultra-long hold time 400 NMR spectrometers. FTIR-attenuated total reflectance (FTIR-ATR) spectra were detected with Shimadzu IRaffinity-1 spectrometer in the range of 600–4,000 cm^{-1}. DSC measurements were conducted at a rate of 10°C/min from – 50°C to 250°C under nitrogen atmosphere using TA Instruments (DSCQ2000 model). TGA measurements were conducted using TA Instruments (SDTQ600 model). The samples were heated at a rate of 10°C/min from 25°C to 600°C under nitrogen atmosphere.

4.2.3 SYNTHESIS OF OLIGOMERIC POLY(A-METHYL B-ALANINE) (PMBA)

About 0.10 moles of methacrylamide and 0.04 moles of *t*BuONa were added into a round bottom balloon under argon flux. The balloon was inserted into an oil bath equipped with a magnetic stirrer. The temperature of the bath was adjusted to 110°C. Approximately 2 hours after the methacrylamide starts to melt at 105°C, the reaction mixture is completely solidified. Excess methanol was added into the reaction medium to extract the by-products and residues of the monomer and initiator. The methanol insoluble fraction was filtered, washed with methanol, and dried for characterization. The yield was determined gravimetrically as 89.2% (wt.).

4.2.4 SYNTHESIS OF POLY(ACRYLIC ACID-G-A-METHYL B-ALANINE) (PA-G-MBA)

About 0.01 moles of acrylic acid, 0.35 g of PmBA, and 3 mg of AIBN were placed in a 10-mL round bottom flask under argon flux. Then, 3 mL of formic acid as a solvent was injected into the flask. The flask was heated to 70°C in an oil bath equipped with a magnetic stirrer. The reaction mixture became so viscous in an hour that the magnetic bar stopped to stir. The reaction mixture was cooled to room temperature and added excess methanol to precipitate possible unreacted PmBA.

4.3 RESULTS AND DISCUSSION

4.3.1 SYNTHESIS OF OLIGOMERIC POLY(A-METHYL B-ALANINE) (PMBA)

As well known, the initiator (tBuONa) behaves as a strong base rather than as a nucleophile in reactions. Thus, the initiation step involves a proton abstraction from the amide group of the monomer [20]. Hence, it was expected that HTP of methacrylamide yielded with an oligomeric PmBA with olefinic end-group as given in Scheme 4.1. The oligomeric PmBA was prepared previously [17] and regarded as a macromonomer for further polymerization [21].

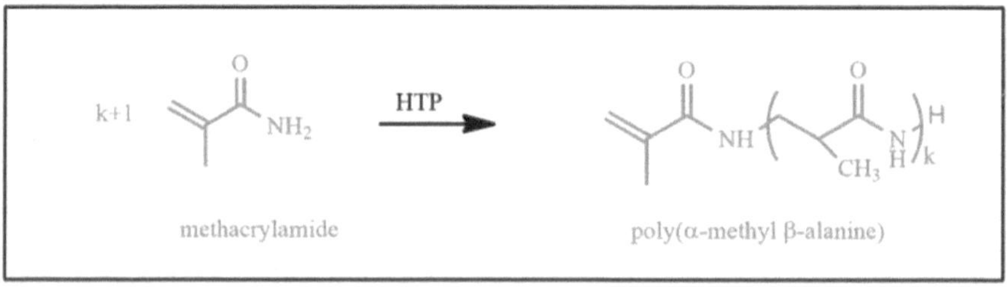

SCHEME 4.1 Synthesis outline of oligomeric poly(α-methyl β-alanine) (PmBA).

Structural characterization of the oligomeric PmBA was performed by using FTIR spectroscopy and MALDI mass spectrometry. FTIR spectrum of PmBA in Figure 4.1 shows characteristic secondary amide bands such as amide I band at 1643.35 cm^{-1} associated with the C=O stretching vibration and amide II band at 1543.05 cm^{-1} result from the N-H bending vibration. The band at 3286.70 cm^{-1} is also an indicator of secondary amide N-H stretching vibration.

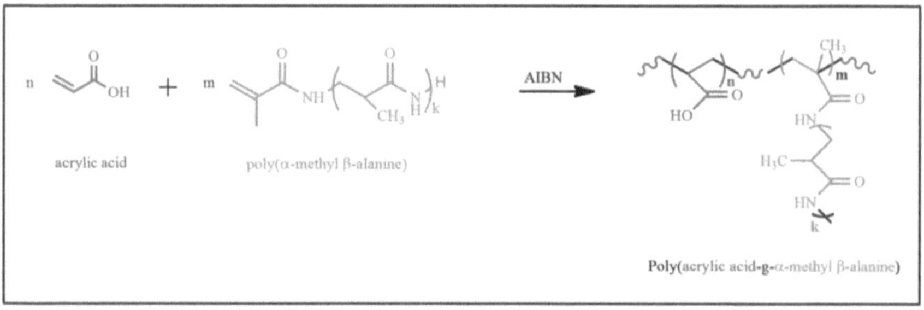

FIGURE 4.1 FTIR spectrum of poly-α-methyl-β-alanine macromonomer.

An expanded view of the MALDI mass spectrum of oligomeric PmBA was illustrated in Figure 4.2. The mass differences between the consecutive signals were measured as about 85 Da, which is equal to the mass of α-methyl β-alanine. The signal with m/z 1112.4 Da belongs to the chain $[M_{13}\ Li]^{+}$. With a simple calculation, the mass of the molecule is equal to the $[(13 \times 85) + 7]$, 1112 Da.

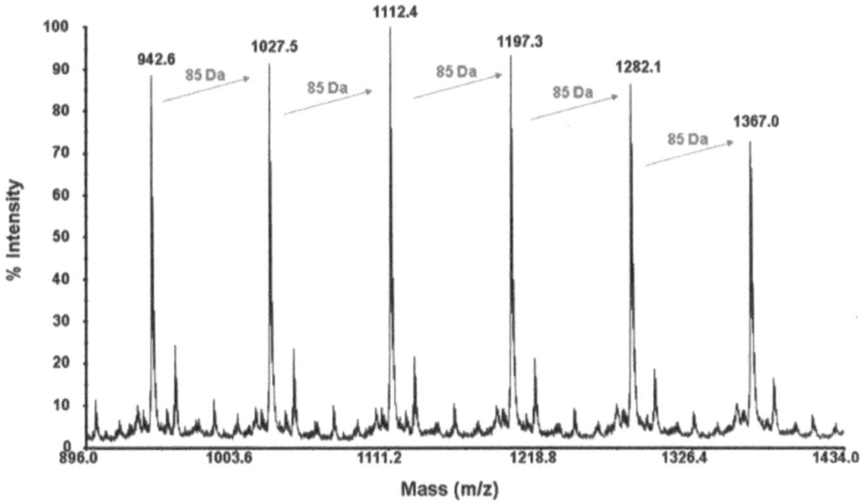

FIGURE 4.2 An expanded view of m/z the 890–1,434 region of MALDI-MS spectrum of the PmBA oligomer.

FTIR spectrum of the PA-g-mBA in Figure 4.3 was recorded and compared to that of PmBA. In addition to the existence of PmBA bands in the spectrum of PA-g-mBA is expected result, the shifts in the wavelength values of mBA bands (1643.35 to 1627.92 cm^{-1} and 3286.70 to 3348.42 cm^{-1}) proves that these structures (acrylic acid and mBA) are miscible in molecular level [22]. This result may be regarded as a simple indicator of graft copolymer formation.

4.3.1.1 SYNTHESIS OF POLY(ACRYLIC ACID-G-A-METHYL B-ALANINE) (PA-G-MBA)

The product obtained from the free radical polymerization of acrylic acid and mBA shown in Scheme 4.2 was called PA-g-mBA. It was determined as completely soluble in methanol. Since the PmBA oligomer is not soluble in methanol the phenomenon was attributed to the formation of the graft copolymer. The characterization of PA-g-mBA samples was performed by using FTIR, ^{1}H-NMR, DSC, and TGA methods.

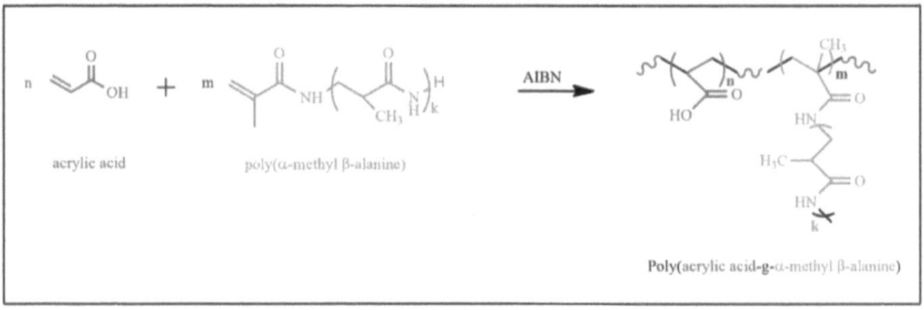

SCHEME 4.2 The structure of poly(acrylic acid-g-α-methyl β-alanine) (PA-g-mBA).

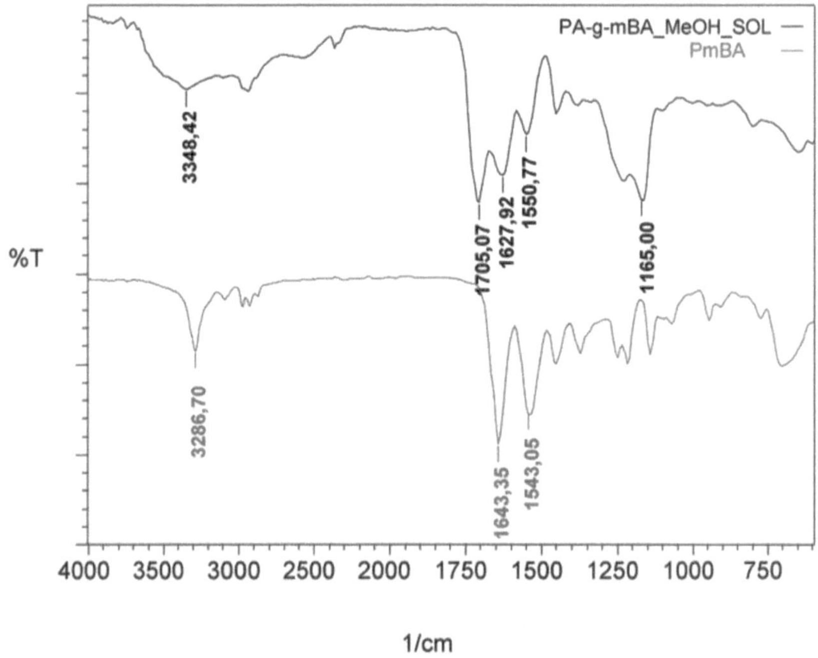

FIGURE 4.3 FTIR spectra of poly-α-methyl-β-alanine macromonomer and PA-g-mBA_MeOH_SOL.

The ^{1}H-NMR spectrum of the PA-g-mBA sample in deuterated methanol was recorded and given in Figure 4.4. The signals at 1.12, 2.59, 3.28, and 8.09 ppm belong to the protons in the mBA units [17]. Their chemical shift values and relative intensities are consistent with their chemical structure. Similarly, the signals at 1.73, 2.45, and 4.33 ppm were assigned to the protons in acrylic acid units. Relative intensities of the acrylic acid and mBA protons are also found to be consistent with the amounts (0.01 mol acrylic acid and 0.35 g mBA) used in the polymerization.

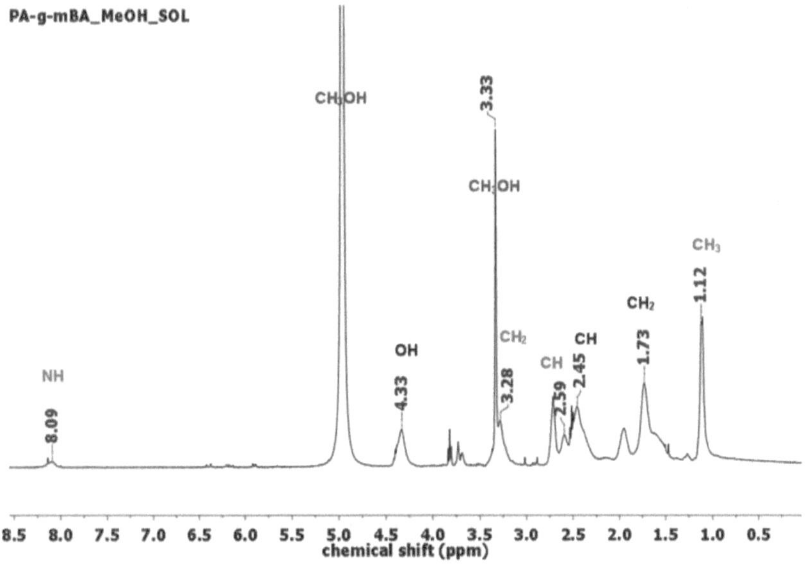

FIGURE 4.4 ^{1}H-NMR spectrum of (PA-g-mBA).

The glass transition temperature (T_g) of neat poly(acrylic acid) was reported many times [23–25] between 75°C and 126°C. However, Tg = 101°C is generally preferred. DSC thermogram of the PA-g-mBA sample in Figure 4.5 yielded a T_g at 83.32°C. The shift in the Tg value may result from the hindrance of acrylic acid interactions by the grafts and hence increase in free volume.

The TGA curve of the graft copolymer was recorded. Then, DTG curves in Figure 4.6 were created by taking the derivative of the curves. Three stages of degradation were detected. The first (around 200°C) and third (around 400°C) stages of these degradations correspond to the degradation of acrylic acid units [25, 26], while the second (around 300°C) stage corresponds to the degradation of α-methyl-β-alanine units [27]. It is too common to observe multi-step decomposition behavior for graft or block copolymers as if the components of the copolymers are alone [28, 29].

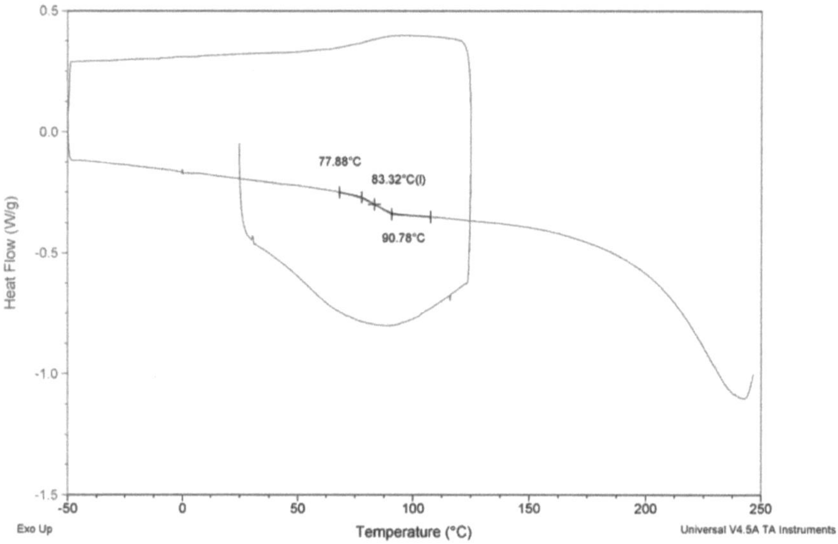

FIGURE 4.5 DSC thermogram of PA-g-mBA.

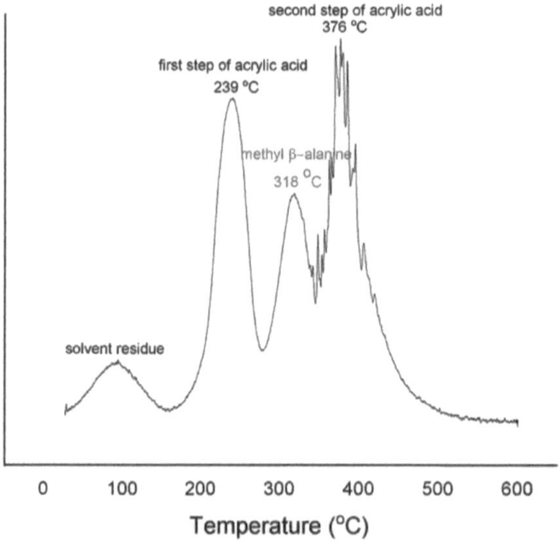

FIGURE 4.6 DTG curve of the PA-g-mBA sample.

4.4 CONCLUSIONS

In the study, poly(acrylic acid-g-α-methyl-β-alanine) was able to synthesize via the "graft through" strategy for the first time. In other words, α-methyl-β-alanine moieties were introduced to the acrylic acid segments successfully. The structure of the novel copolymer was confirmed by FTIR and ^{1}H-NMR spectroscopy. Glass transition temperature of the poly(acrylic acid-g-α-methyl-β-alanine) was found to be lower than that of poly(acrylic acid), which was assigned to increase in the free volume by the introduction of α-methyl-β-alanine moieties. In the future, the strategy of "grafting through" may be applied for many monomers to obtain graft copolymer with modified thermal/mechanical properties.

ACKNOWLEDGMENT

The study has been supported by the Ordu University Scientific Research Project Coordination (ODUBAP) with project number B-2112.

KEYWORDS

- **graft copolymer**
- **graft through**
- **hydrogen-transfer polymerization**
- **poly (2-vinyl pyridine)**
- **poly(acrylic acid-g-α-methyl β-alanine)**
- **poly(methyl methacrylate)**

REFERENCES

1. Kopecek, J., (2009). Hydrogels: From soft contact lenses and implants to self-assembled nanomaterials. *Journal of Polymer Science Part A: Polymer Chemistry, 47*(22), 5929–5946. https://doi.org/https://doi.org/10.1002/pola.23607.
2. Pandey, S. P., Tripti, S., Vinod, K. D., Dinesh, K. M., Rahul, M., & Rakesh, K. T., (2019). Chapter 4 – use of polymers in controlled release of active agents. In: Rakesh, K. B. T., (ed.), *Advances in Pharmaceutical Product Development and Research: Basic Fundamentals of Drug Delivery* (pp. 113–172). Tekade. Academic Press. https://doi.org/https://doi.org/10.1016/B978-0-12-817909-3.00004-2.

3. Qi, Y., Zhou, Y., Alex, F., Brian, N. H., Montserrat, E., Maria-Pau, G., & Conrado, A., (2018). Effects of molecular weight and concentration of poly(acrylic acid) on biomimetic mineralization of collagen. *ACS Biomaterials Science & Engineering, 4*(8), 2758–2766. https://doi.org/10.1021/acsbiomaterials.8b00512.
4. Sütekin, S. D., & Olgun, G., (2020). Preparation of poly(tert-butyl acrylate)-poly(acrylic acid) amphiphilic copolymers via radiation-induced reversible addition–fragmentation chain transfer mediated polymerization of tert-butyl acrylate. *Polymer International, 69*(8), 693–701. https://doi.org/https://doi.org/10.1002/pi.6004.
5. Matyjaszewski, K., & James, S., (2005). Controlled/living radical polymerization. *Materials Today, 8*(3), 26–33. https://doi.org/10.1016/S1369-7021(05)00745-5.
6. Kahveci, M. U., Yagci, Y., Avgeropoulos, A., & Tsitsilianis, C., (2012). 6.13-well-defined block copolymers. In: Krzysztof, M., & Martin, B. T., (eds.), *Polymer Science: A Comprehensive Reference Möller* (pp. 455–509). Amsterdam: Elsevier. https://doi.org/https://doi.org/10.1016/B978-0-444-53349-4.00171-0.
7. Gürdağ, G., Gamze, G., & Saadet, Ö., (2001). Graft copolymerization of acrylic acid onto cellulose: Effects of pretreatments and crosslinking agent. *Journal of Applied Polymer Science, 80*(12), 2267–2272. https://doi.org/https://doi.org/10.1002/app.1331.
8. Srivastava, A., & Kumar, R., (2013). Synthesis and Characterization of Acrylic Acid-g-κ-Carrageenan) Copolymer and Study of Its Application. *International Journal of Carbohydrate Chemistry,* 2013, Article ID 892615, 1–8. https://doi.org/10.1155/2013/892615.
9. Abdel-Bary, E. M., & Elbedwehy, A. M., (2018). Graft copolymerization of polyacrylic acid onto acacia gum using erythrosine–thiourea as a visible light photoinitiator: Application for dye removal. *Polymer Bulletin, 75*(8), 3325–3340. https://doi.org/10.1007/s00289-017-2205-x.
10. Liu, S., Xia, L., Fang, L., Yu, F., Yijuan, W., & Juan, Y., (2008a). Phase behavior of temperature- and PH-sensitive poly(acrylic acid-g-N-isopropylacrylamide) in dilute aqueous solution. *Journal of Applied Polymer Science, 109*(6), 4036–4042. https://doi.org/https://doi.org/10.1002/app.28602.
11. Nam, J., Eunsoo, K., Rajeev, K. K., Yeonho, K., & Tae-Hyun, K., (2020). A conductive self-healing polymeric binder using hydrogen bonding for Si anodes in lithium ion batteries. *Scientific Reports, 10*(1), 14966. https://doi.org/10.1038/s41598-020-71625-3.
12. Shibanuma, T., Takashi, A., Kohei, S., Naoya, O., Akihiko, K., Yasuhisa, S., & Teruo, O., (2000). Thermosensitive phase-separation behavior of poly(acrylic acid)-graft-poly(N,N-dimethylacrylamide) aqueous solution. *Macromolecules, 33*(2), 444–450. https://doi.org/10.1021/ma9915374.
13. Fu, Q., Zachary, R. G., Art Van, D. E., & Robert, H. P., (2016). Phase behavior of aqueous poly(acrylic acid-g-TEMPO). *Macromolecules, 49*(13), 4935–4939. https://doi.org/10.1021/acs.macromol.6b00977.
14. La Fuente, J. L. De, Wilhelm, M., Spiess, H. W., Madruga, E. L., Fernández-Garcia, M., & Cerrada, M. L., (2005). Thermal, morphological and rheological characterization of poly(acrylic acid-g-styrene) amphiphilic graft copolymers. *Polymer, 46*(13), 4544–4553. https://doi.org/10.1016/j.polymer.2005.03.076.
15. Li, Y., Yaqin, Z., Dong, Y., Yongjun, L., Jianhua, H., Chun, F., Sujuan, Z., Guolin, L., & Xiaoyu, H., (2010). PAA-g-PPO amphiphilic graft copolymer: Synthesis and diverse micellar morphologies. *Macromolecules, 43*(1), 262–270. https://doi.org/10.1021/ma901526j.

16. Li, Y., Yaqin, Z., Dong, Y., Chun, F., Sujuan, Z., Jianhua, H., Guolin, L., & Xiaoyu, H., (2009). Well-defined amphiphilic graft copolymer consisting of hydrophilic poly(acrylic acid) backbone and hydrophobic poly(vinyl acetate) side chains. *Journal of Polymer Science Part A: Polymer Chemistry, 47*(22), 6032–6043. https://doi.org/https://doi.org/10.1002/pola.23646.
17. Çatıker, E., Olgun, G., & Bekir, S., (2018). Novel hydrophobic macromonomers for potential amphiphilic block copolymers. *Polymer Bulletin, 75*(1), 47–60. https://doi.org/10.1007/s00289-017-2014-2.
18. Çolakoğlu, G. N., Efkan, Ç., Temel, Ö., & Ergül, M., (2021). Synthesis and characterization of brush-type polyβ-alanine-grafted polymethyl methacrylate using 'grafting through' method. *Chemical Papers*. https://doi.org/10.1007/s11696-021-01908-0.
19. Savaş, B., Çatıker, E., Öztürk, T. & Meyvacı E., (2021). Synthesis and characterization of poly(α-methyl β-alanine)-poly(ε-caprolactone) tri arm star polymer by hydrogen transfer polymerization, ring-opening polymerization and "click" chemistry. *Journal of Polymer Research, 28*(2), 30. https://doi.org/10.1007/s10965-020-02367-z.
20. Trossarelli, L., Guaita, M., & Camino, G., (1969). Research on the mechanism of base-catalyzed hydrogen-transfer polymerization. *Journal of Polymer Science Part C: Polymer Symposia, 22*(2), 721–727. https://doi.org/https://doi.org/10.1002/polc.5070220214.
21. Savaş, B., Çatıker, E., Öztürk, T., & Meyvacı E., (2021). Synthesis and characterization of poly(methyl methacrylate-g-α-methyl-β-alanine) copolymer using "Grafting Through" method. *Journal of Polymer Research, 28*, 194. https://doi.org/10.1007/s10965-021-02551-9.
22. Alex, S., Le Thanh, H., & Vocelle, D., (1992). Studies of the effect of hydrogen bonding on the absorption and fluorescence spectra of all-trans-retinal at room temperature. *Canadian Journal of Chemistry, 70*(3), 880–887. https://doi.org/10.1139/v92-117.
23. Tan, T. T. M., & Bernd, M. R., (1996). Molecular modeling of polymers, 3. Prediction of glass transition temperatures of poly(acrylic acid), poly(methacrylic acid) and polyacrylamide derivatives. *Macromolecular Theory and Simulations, 5*(3), 467–475. https://doi.org/https://doi.org/10.1002/mats.1996.040050306.
24. Ruiz-Rubio, L., José, M. L., Leyre, P., Nerea, R., & Elena, B., (2014). Polymer–polymer complexes of poly(N-isopropylacrylamide) and poly(N,N-diethylacrylamide) with poly(carboxylic acids): A comparative study. *Colloid and Polymer Science, 292*(2), 423–430. https://doi.org/10.1007/s00396-013-3086-7.
25. Maurer, J. J., Eustace, D. J., & Ratcliffe, C. T., (1987). Thermal characterization of poly(acrylic acid). *Macromolecules, 20*(1), 196–202. https://doi.org/10.1021/ma00167a035.
26. Eisenberg, A, Yokoyama, T., & Emma, S., (1969). Dehydration kinetics and glass transition of poly(acrylic acid). *Journal of Polymer Science Part A-1: Polymer Chemistry, 7*(7), 1717–1728. https://doi.org/https://doi.org/10.1002/pol.1969.150070714.
27. Çatıker, E., & Erol, S., (2014). Blends of poly(3-hydroxybutyrate) with poly(β-alanine) and its derivatives. *Journal of Applied Polymer Science, 131*(13). https://doi.org/https://doi.org/10.1002/app.40484.
28. Çatıker, E., Temel, Ö., Mehmet, A., & Bekir, S., (2019). Synthesis and characterization of novel ABA type poly(ester-ether) triblock copolymers. *Journal of Polymer Research, 26*(5), 123–132. https://doi.org/10.1007/s10965-019-1778-5.

29. Çatıker, E., Temel, Ö., Mehmet, A., & Bekir, S., (2020). Synthesis and characterization of the ABA-type poly(ester-ether-ester) block copolymers. *Journal of Macromolecular Science, Part A: Pure and Applied Chemistry, 57*(8), 600–609. https://doi.org/10.1080/10601325.2020.1745080.

CHAPTER 5

Synthesis and Characterization of Poly(β-Alanine-co-3-Hydroxybutyrate) Through HTP and AROP

EFKAN ÇATIKER and ÜMIT KELEŞ

Ordu University, Department of Chemistry, Ordu, Turkey

ABSTRACT

In the study, acrylamide as the highest polymerizable monomer through hydrogen-transfer polymerization (HTP), and β-butyrolactone as a comonomer (mole % of BBL; 10, 25, 50, and 75) were used to synthesize a novel poly(ester-amide). Compositions, average molar masses, and thermal properties of the copolymers were elicited by using elemental and spectroscopic analyzes (FTIR and NMR), mass spectrometry (MALDI), and thermal analyzes (DSC and TGA), respectively. The copolymers were found to have compositions different from the feed ratio applied but close when the data obtained from elemental analysis were evaluated in detail. The results obtained from different methods to determine the copolymer compositions were found to be consistent with each other. The highest average mass of 6,000 g/mol was reached for the copolymers prepared. Glass transition temperature (T_g) shifts between 0°C and 10C in the DSC thermograms of the copolymers proves the existence of ester blocks in the main chains. DTG thermograms exhibit two-step thermal decomposition shifts centered at about 240°C and 340°C that also support the existence of two chemically distinct blocks in the copolymer samples.

Advanced Polymer Structures: Chemistry for Engineering Applications. Omar Mukbaniani, DSc, Tamara Tatrishvili, PhD & Marc Jean M. Abadie, DSc (Eds.)

5.1 INTRODUCTION

Synthetic and natural polymeric materials [1–3] are used in biotechnology, medicine, and pharmacy as implants, biosensors, bioadhesives, orthopedic/dental materials, tissue/cell culture scaffold materials, diagnostic test kits, components in drug delivery systems, disposable medical materials, and wound healing materials. This wide range of use of polymeric materials is possible thanks to the variety of chemical structure and composition, the change of topological and mechanical properties, and the ability to convert them into different shapes (fiber, membrane, film, gel, particle, microsphere, and even spongy).

Recently, there is a need for functional biocompatible polymers for use in tissue engineering, medicine, gene therapy, and drug delivery systems. Thus, the modification of existing polymeric biomaterials and the design of novel synthetic polymeric materials have become of interest. The limitation in the diversity of natural polymeric biomaterials and the difficulty of their chemical modification are beginning to force researchers to design new synthetic polymeric materials. The novel aliphatic polyesters [4, 5], polyphosphoesters [6, 7], poly(ester-amides) [8–10], polyanhydrides [11], poly(ester urethanes) [12, 13] are the most emphasized synthetic polymer groups.

Base-catalyzed hydrogen transfer polymerization (HTP) is a particular type of anionic addition polymerization that was discovered in the production of nylon-3 (poly-β-alanine) from acrylamide about half a century ago [14]. HTP includes a monomer containing a vinyl group and an acidic proton, a nucleophilic initiator (also a catalyst), and an aprotic solvent. An inhibitor to prevent free radical vinyl polymerization might be required if the polymerization is carried out at elevated temperatures. Some monomers such as (meth)acrylamide, acrylic acid, and some of their derivatives were polymerized via HTP to obtain polyamides [15, 16], polyesters [17], and poly(ester ethers) [18]. Acrylamide has been reported to yield the corresponding HTP product with highest molar mass [14, 19], thus it may be regarded as the most polymerizable monomer with HTP.

In a study [20], two potential monomers of HTP were used at the same time. Otsu et al. [20] used acrylamide and acrylamide derivatives such as methacrylamide, crotonamide, and tiglinamide as the comonomer, and their compositions in the copolymer were determined. Another study [21] has reported the anionic copolymerization of acrylamide and ε-caprolactone. In this study, poly[acrylamide-co-(ε-caprolactone)] products were synthesized with different ratios of acrylamide and ε-caprolactone, and then hydrolytic and enzymatic degradability of the copolymers were investigated. Apart from these two studies [20, 21], there are no other studies involving copolymer synthesis by hydrogen transfer (co) polymerization.

Within the scope of this work, it was planned to synthesize a novel poly(ester amide) by using acrylamide known as the highest polymerizable monomer with HTP and β-butyrolactone as comonomer. The β-butyrolactone is the monomer of poly(3-hydroxybutyrate), a well-known polyester. That is, the copolymer aimed to be synthesized is intended to contain both peptide units, such as β-alanine and a biocompatible/biodegradable ester unit, such as 3-hydroxybutyrate (3-HB). Briefly, the β-alanine/3-HB copolymers in different compositions were aimed to synthesize by hydrogen transfer (co) polymerization.

5.2 EXPERIMENTAL METHODS AND MATERIALS

5.2.1 MATERIALS

Acrylamide (Acros, 99% +) was recrystallized in acetone. Then, it was dried in a vacuum oven at room temperature. β-butyrolactone (Sigma-Aldrich, 98%) was distilled in vacuo on calcium hydride (~2 mmHg). Calcium hydride (Sigma-Aldrich, 99.99%), TEMPO (Acros Organic, 98%), sodium tert-butoxide (Sigma-Aldrich, 97%), diethyl ether (Sigma-Aldrich, 99.7%), 15-crown-5-ether (Sigma-Aldrich, 98%), and acetone (Macron Fine Chemicals, 99.9%) were commercially available.

5.2.2 SYNTHESIS OF POLY(B-ALANINE-CO-3-HB)

Reaction parameters applied in the copolymer synthesis are summarized in Table 5.1. Sodium tert-butoxide (*t*BuONa) and 2,2,6,6-tetramethyl-1-piperidinyloxy (TEMPO) were used as an initiator and an inhibitor, respectively. 15-crown-5 ether which is effective for sodium ions is used as a co-catalyst. The numbers of moles of *t*BuONa and 15-crown-5 ether were adjusted as 1% of the total number of moles of the monomers. The amount of TEMPO was applied as 0.1% of the total number of moles of the monomers. The given amounts of reactants listed in Table 5.1 were added into Schlenk tubes with a magnetic bar using Schlenk techniques (air-free technique) [22]. The bulk(co)polymerization of acrylamide and β-butyrolactone were carried out in an argon atmosphere at 90°C.

TABLE 5.1 Reaction Parameters*

Sample Code	X_{M1}	X_{M2}	Yield (%)
75% BBL	0.25	0.75	73.4
50% BBL	0.50	0.50	68.2
25% BBL	0.75	0.25	68.7
10% BBL	0.90	0.10	88.9

*X_{M1}: mole fraction of acrylamide; X_{M2}: mole fraction of β-butyrolactone; temperature: 90°C; and time: 24 h.

The polymerization was terminated 24 hours after by adding diethyl ether (slightly acidified with acetic acid). The resulting mixtures were stirred on a magnetic stirrer overnight and then filtered through a suitable porous Gooch crucible using a vacuum pump. The filtrates were found to have consisted of unreacted monomers, oligomers, and initiator residues. The residue in the crucible was dried in a vacuum oven at room temperature overnight, weighed, and the yield was calculated.

5.2.2 CHARACTERIZATION

5.2.3.1 CHEMICAL STRUCTURE

Structural characterization of the β-alanine/3-HB copolymers in different compositions was achieved by using FTIR, ^{1}H-NMR, MALDI-MS, and elemental analysis. The FTIR spectrum was recorded with 32 scans in the range 800–4,000 cm^{-1} by ATR technique using Shimadzu IRaffinity-1. For NMR analysis, solutions of the copolymers were prepared using trifluoroacetic acid (TFA), TFA/$CDCl_3$, and deutero trifluoroacetic acid (d-TFA) as solvents. ^{1}H-NMR spectra were recorded using Bruker AVANCE III 400 MHz NMR Spectrometer, given in the range 0–10 ppm. NMR data were processed using MestReNova 6.0.2–5,475 software. The Vario Micro-Cube model elemental analyzer was used for the determination of carbon, hydrogen, and nitrogen percentages. The molar mass distributions of the copolymers were acquired by the MALDI-TOF Mass spectrometry. MALDI mass spectra were recorded with MALDI-TOF (Applied Biosystems, USA) with Voyager-DETM PRO coupled to a nitrogen UV-laser operating at 337 nm. Spectra were obtained by linear and positive ion mode with an average of 500 pulses. 2,5-dihydroxybenzoic acid (DHB) was used as the MALDI matrix. The matrix was prepared in a mixture of THF:ACN (1:1, v/v) at a concentration of 20 mg/mL. The cationizing agent (LiTFA, 10 mg/mL THF) was added to the matrix solution at 1.0% (v/v) of the total solution. About 1.0 μL of the sample was dropped to the MALDI target. The air-dried sample was analyzed.

5.2.3.2 THERMAL ANALYZES

Thermal analyzes were carried out at 10°C/min under a nitrogen atmosphere using TA Instruments Q2000 and Q600. The second scan was initiated from – 80°C to the polymer decomposition temperature.

5.3 RESULTS AND DISCUSSION

5.3.1 SYNTHESIS OF POLY(B-ALANINE-CO-3-HB)

Melt polymerizations of the acrylamide/β-butyrolactone mixtures (75% to 10% BBL as moles) were carried out at four different feed rates. The polymerization reactions were terminated after the mixture became completely solid or viscous liquid. The yields of each copolymer were determined gravimetrically and given in Table 5.2. As seen, the products are obtained with yields between 68.2% and 88.9%.

The initiation [15, 19] and propagation step of the polymerization in the presence of strong bases was given in Scheme 5.1.

Similarly, in the presence of nucleophilic compounds, β-butyrolactone yields the corresponding poly(3-HB) via anionic ring-opening polymerization (AROP) mechanism. Two different mechanisms have been proposed depending on the alkalinity and the nucleophilic strength of the initiator [23]. Since *t*BuONa was used as an initiator being a strong base in this study, it was predicted that the mechanism of initiation underwent through the removal of the α-proton and then ring-opening occurred through the olefinic bond formation. Possible copolymerization pathway was given in Scheme 5.2. Here, living end-groups of the poly-β-alanine chains behave as anionic initiator for ring-opening polymerization of β-butyrolactone. Alternative propagation pathways are also possible and have been discussed elsewhere [21].

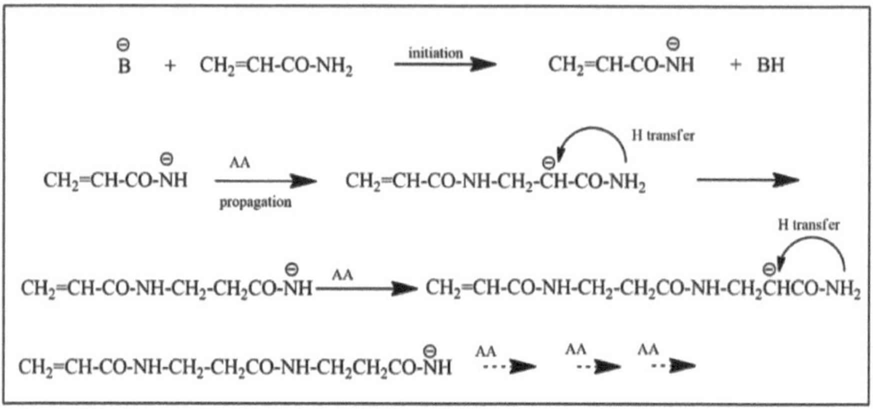

SCHEME 5.1 Initiation and propagation mechanism of acrylamide via HTP.

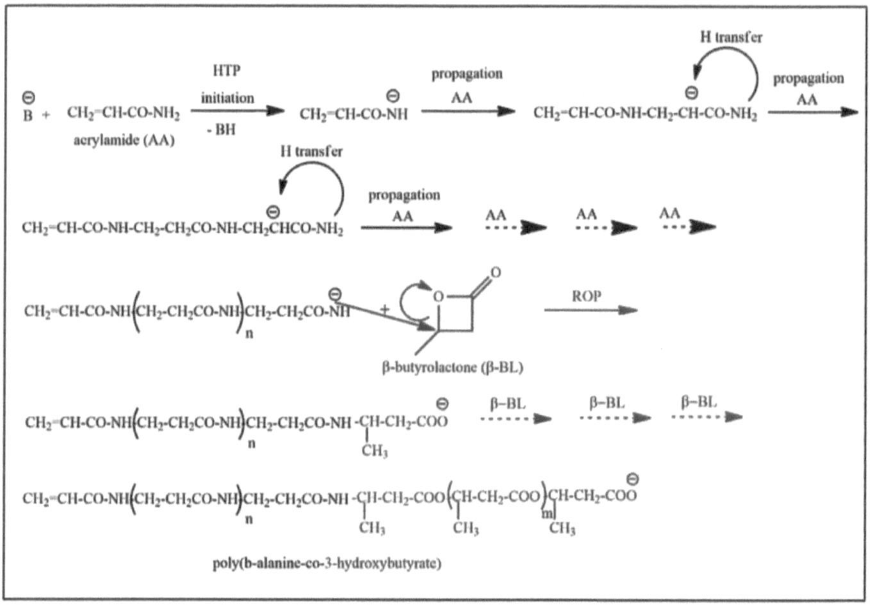

SCHEME 5.2 Possible copolymerization pathway for formation of poly(β-alanine-co-3-hydroxybutyrate).

5.3.2 CHARACTERIZATION OF POLY(Β-ALANINE-CO-3-HB)

Characterizations of the products were carried out by FTIR, ^{1}H-NMR, elemental analysis, MALDI-MS, DSC, and TGA analysis, respectively.

5.3.2.1 FTIR SPECTROSCOPY

FTIR spectroscopy is the most used method in qualitative analysis of organic compounds because it is the cheapest and easiest method. Figure 5.1 shows the FTIR spectra of the copolymers synthesized with various feeding rates (75, 50, 25, and 10% BBL). The characteristic bands belonging to the β-alanine and 3-HB units were observed in each spectrum. For example, the sharp bands which belong to the ester-carbonyl stretching [24] at about 1,728 cm^{-1} are remarkable. Besides, the bands at about 1,628 cm^{-1} belong to the secondary amide-carbonyl stretching vibrations [25–27]. As easily can be seen, the intensities of the bands discussed above are proportional to the feed rates applied.

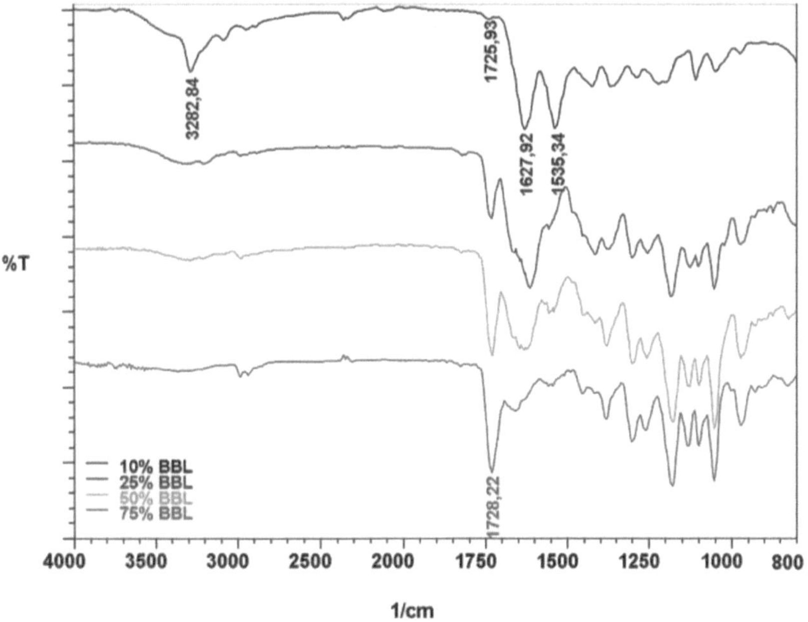

FIGURE 5.1 FTIR spectra of poly(β-alanine-co-3-HB) samples obtained different feed ratio (mole % of BBL: 10% BBL, 25% BBL, 50% BBL, and 75% BBL).

5.3.2.2 NMR SPECTROSCOPY

NMR spectroscopy is the most frequently used method that provides the most information for detailed structural analysis of organic compounds.

^{1}H-NMR spectra of the copolymers in various solvent systems such as d-TFA, TFA, and trifluoroacetic acid/deutero chloroform (TFA-$CDCl_3$) were recorded and given in Figure 5.2. The only proton possessed by TFA is not shown as it has a very high chemical shift (about 12 ppm). The $CDCl_3$ signal is observed at about 7.28 ppm and does not coincide with the proton signals present in the copolymer. In the spectrum in Figure 5.2(A), the signal at chemicals shifts of 1.37, 2.82, and 5.42 ppm were assigned to methyl, methylene, and methine protons in 3-HB units [28], respectively. However, the signal at 2.82 ppm overlapped with the signal of the $COCH_2$ [25, 29] group in the β-alanine unit, and therefore its intensity is higher than expected. The signal observed at 3.71 ppm belongs to the CH_2 [29] group adjacent to the NH in the β-alanine unit. The above-mentioned signals of 3-HB and β-alanine units were observed in the NMR spectra of the other three copolymers in Figure 5.2(B–D). However, what is different is the relative proportions of the protons from 3-HB to β-alanine units. That is, the only difference is the ratio of the ester and the amide units in the copolymer.

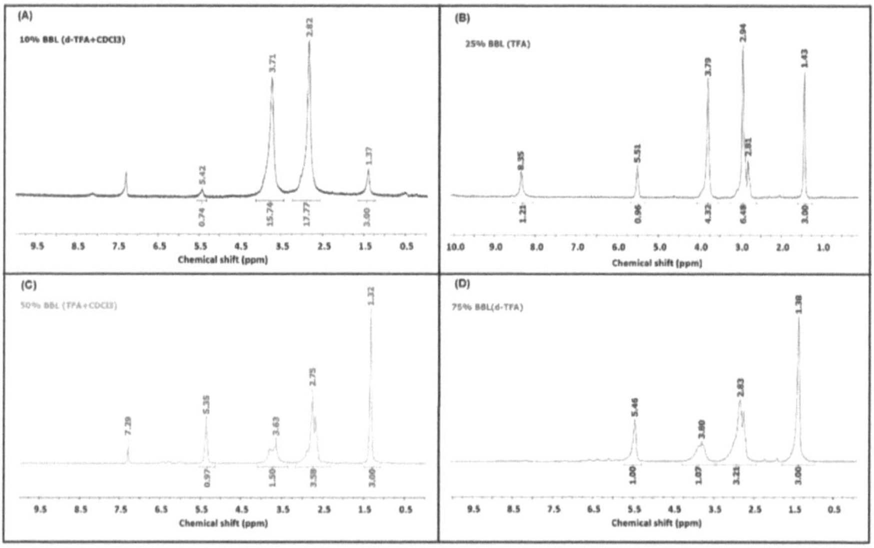

FIGURE 5.2 ^{1}H-NMR spectra of poly(β-alanine-co-3-HB) samples obtained different feed ratio as mole percentages (A) 10% BBL; (B) 25% BBL; (C) 50% BBL; and (D) 75% BBL.

5.3.2.3 ELEMENTAL ANALYSIS ON THE COMPOSITION

Elemental analysis is the most sensitive method to determine the elemental composition of an organic compound. Since the nitrogen atoms originate only from the β-alanine units in the copolymers, the determination of nitrogen percentages provides reliable information about the compositions of the copolymers. The compositions of the copolymers were therefore calculated using the elemental analysis data and the equations given below. The results were summarized in Table 5.2. The mass of β-alanine (m_{BA}) in 100 g of the sample can be found by Eqn. (1). Here, N% is the percentage of nitrogen found experimentally in the sample, and 19.71 is the percentage of nitrogen in pure poly-β-alanine.

$$m_{BA} = \frac{N\%}{19.71} \times 100 \quad (1)$$

Assuming 100 g of the sample, the number of moles of β-alanine (n_{BA}) found in the sample can therefore be calculated by Eqn. (2). Here, 71.08 is the mass of the β-alanine unit.

$$\left(n_{BA}\right) = \frac{\left(\frac{N\%}{19.71}\right)}{71.08} \times 100 \quad (2)$$

The percentages of β-alanine as the mole in the samples can be found by Eqn. (3). Here, 86.06 is the molar mass of 3-HB. The 3-HB% (mole) given in Table 5.2 is calculated by subtracting the β-alanine percentage from 100.

$$\beta - alanine\ \%\left(mole\right) = \frac{n_{BA}}{n_{BA} + \frac{\left(100 - m_{BA}\right)}{86.06}} \times 100 \quad (3)$$

TABLE 5.2 Elemental Analysis Data and the Calculated 3-HB Contents of the Copolymers

BBL Feed (mole,%)	C (mass, %)	H (mass, %)	N (mass, %)	3-HB (mole, %)
75	54.11	7.17	6.03	65.2
50	53.18	7.21	9.61	46.5
25	52.08	7.26	13.91	25.6
10	50.97	7.29	18.22	6.3

The values in the first and last column (feed rates of BBL and present amount in the copolymers) do not fully coincide, although parallelism is observed. The deflections are due to the unreacted monomer and oligomeric structures formed in the isolation step which passes into the ether phase.

5.3.2.4 MALDI MASS SPECTROMETRY

The samples were analyzed using the MALDI-TOF-MS instrument to monitor the presence of 3-HB-β-alanine copolymer species. The signals of singly charged copolymer ions are obtained in the MALDI-MS spectrum ranging from m/z 2,000 to m/z 10,200 as shown in the corner of Figure 5.3, which is acquired in positive ion mode.

An expanded view of m/z the 4,740–4,980 region of the MALDI mass spectrum of the copolymer used 50% (mole) BBL as a comonomer is illustrated in Figure 5.3 that includes signals for the sodiated ions of the copolymer species. The differences between consecutive signals are about m/z 86.1 and m/z 71.1 corresponding to the masses of 3-HB ($C_4H_6O_2$) and β-alanine (C_3H_5NO) repeating units of the copolymer, respectively. Signals of copolymer ions having various intensities are shown in the expanded region of the MALDI mass spectrum in various stoichiometry ranging from $C_{31}N_{29}$ to $C_{36}N_{26}$ while C and N representing 3-HB and β-alanine units, respectively. The data obtained from MALDI-TOF-MS analyzes of the copolymer sample show that the β-alanine/3-HB copolymer has been synthesized successfully by confirming the targeted chemical structure with high mass accuracy [30–32]. Figure 5.3 also has the full range of the MALDI mass spectrum to show the molar mass distribution. When the m/z values and distributions given in the spectra are taken into consideration, the average molar mass of 5,000–6,000 $gmol^{-1}$ can be predicted.

5.3.2.5 DIFFERENTIAL SCANNING CALORIMETRY (DSC)

Differential scanning calorimetry (DSC) thermograms of the copolymer samples in Figure 5.4 were recorded to determine the thermal transition temperatures and their dependencies on the copolymer compositions. The most remarkable result in the thermograms is the clear glass transitions assigned to the 3-HB units [33]. For the 75%, 50%, and 25% BBL samples, the T_g values were determined as 6.04, 7.80, and 7.26°C, respectively. T_g transition was not observed for the 10% BBL sample possibly due to the dominancy of β-alanine units in the copolymer. Although the T_g reported in the literature [25, 34] for homo poly-β-alanine is around 73–124°C, it is hard to determine due to the high crystalline structure of poly-β-alanine [35]. The Tg values assigned to the 3-HB units, are so close to that of the homo poly(3-HB), indicate that 3-HB blocks exist in the copolymer chains. A similar evaluation had been reported for poly(β-alanine-co-caprolactone) [21].

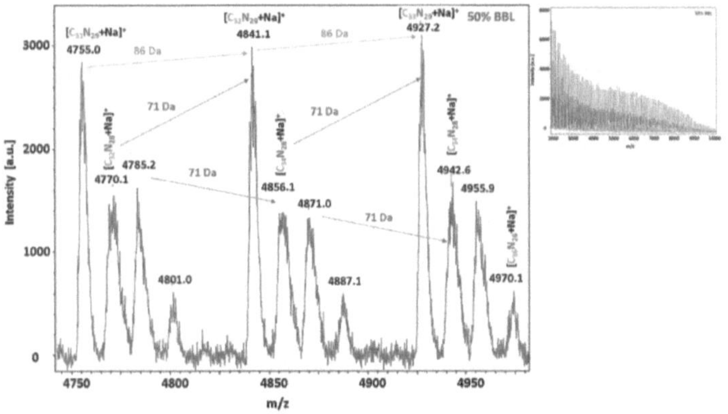

FIGURE 5.3 Expanded m/z 4740–4980 region of MALDI mass spectrum of the 50% BBL sample. The notation C_xN_y is used for poly(β-alanine-co-3-HB) chains, while C and N representing 3-HB and β-alanine repeating units, respectively ('x' and 'y' subscripts indicate the number of repeat units).

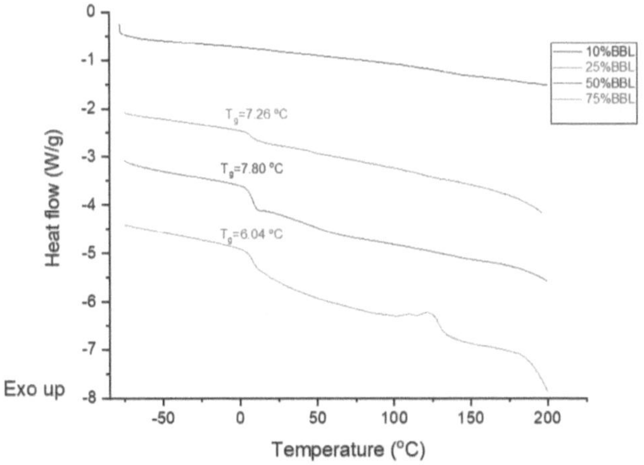

FIGURE 5.4 DSC thermogram of poly(β-alanine-co-3-HB) samples obtained with different feed ratios of BBL (10% BBL, 25% BBL, 50% BBL, and 75% BBL).

5.3.2.6 THERMOGRAVIMETRIC ANALYSIS (TGA)

Thermogravimetric analysis is a widely used method to demonstrate the thermal stability of polymeric materials. Furthermore, the ratio of additives, copolymer composition, decomposition kinetics, and decomposition mechanism can be determined by TGA [36–38]. In this study, thermogravimetric analysis was conducted to obtain information about both the thermal stability and the copolymer composition. Differential thermogravimetry (DTG) curves of the poly(ester amides) with different compositions were represented in Figure 5.5.

As seen, all poly(ester amide) are thermally stable up to 180°C. After that, DTG curves reveal similar two-stage decomposition for all studied copolymers. The first decomposition steps in the curves, which range from ~180°C to ~275°C, were assigned to the ester blocks [39, 40] in the copolymers. Then the second decomposition step, attributed to the decomposition of the β-alanine blocks [25, 26], ended up at ~420°C. Since the areas under the bands are proportional to the mass of the degraded species, these results seem to be consistent with the results of NMR and elemental analysis. Another noteworthy result is that as the ester ratio in the copolymer increases, the thermal decomposition of the ester units (except 10% BBL) increases 226.02 to 259.74°C and the thermal decomposition temperature of the amide units decreases 362.32 to 334.69°C.

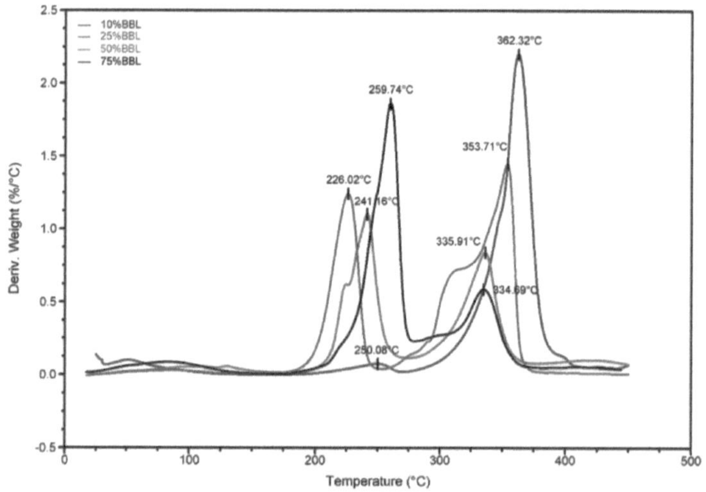

FIGURE 5.5 DTG thermograms of poly(β-alanine-co-3-HB) sample obtained with different feed ratios of BBL (mole % of BBL; 10% BBL, 25% BBL, 50% BBL, and 75% BBL).

5.4 CONCLUSION

Poly(β-alanine-co-3-HB) composed of opposite character segments (soft and hard) was assumed to be biocompatible and biodegradable to be used as biomaterials. The β-alanine (–CH_2-CH_2-CO-NH–) unit as a hard segment [41], and a relatively soft and readily biodegradable [42, 43] 3-HB [–CH(CH_3)-CH_2-CO-O–] were preferred for this purpose. Considering this information, acrylamide, and β-butyrolactone monomers in different proportions (mole % of BBL; 75, 50, 25, and 10) were used to evaluate the effect of composition on the thermal properties of the copolymer. The yields were calculated as ~70 to ~90% depending on the feed ratios.

Spectroscopic analyzes (FTIR and ^{1}H-NMR) revealed that β-alanine and 3-HB units occurred. MALDI mass analysis was used to prove that the units were in the same chains, namely, a novel copolymer was obtained. Although not fully calculated with the aid of MALDI mass spectra, the average number of molar masses of copolymers obtained with 75, 50, 25, and 10 BBL feeds were in the range of 2,000–6,000. By using elemental analysis data, the relationship between the BBL feed rates and the 3-HB contents in the copolymers was determined to exhibit parallelism. The DSC analyzes of the 75, 50, 25% samples revealed that a glass transition occurred at around that of homo poly(3-HB), which is the indication of the 3-HB blocks in the copolymers. This approach was also supported by the DTG curves with two-stage decompositions belonging to the 3-HB and β-alanine blocks.

Considering all the results above, it was concluded that the poly(β-alanine-co-3-HB)'s having different contents of β-alanine and 3-HB blocks were obtained. Although the low molar mass polymers have also an application area, the molar mass of the copolymers must be elevated to extend the potential use area. To achieve a higher molar mass, parameters such as initiator type (such as organometallic compounds), polymerization temperature, polymerization technique (solution), and type of solvent can be considered. Furthermore, since the sensitivity of anionic polymerization to moisture and oxygen is very high, the reactants used as monomers, initiators, co-catalysts, and inhibitors must be extremely pure and the reaction system must be completely hermetic.

ACKNOWLEDGMENT

The study has been supported by the Scientific and Technological Research Council of Turkey (TUBITAK) with project number 116Z542.

KEYWORDS

- **anionic ring-opening polymerization**
- **deutero trifluoroacetic acid**
- **hydrogen-transfer polymerization**
- **mass spectrometry**
- **poly(ester amide)**
- **ring-opening polymerization**

REFERENCES

1. Razak, S. I. A., Noor, F. A. S., & Wan, A. A. R., (2012). Biodegradable polymers and their bone applications: A review. *International Journal of Engineering & Sciences, 12*(1), 31–49. http://www.ijens.org/IJBAS Vol 12 Issue 01.html (accessed on 02 January 2022).
2. Brovold, M., Joana, I. A., Pla-Palacín, I., Sainz-Arnal, P., Sánchez-Romero, N., Jesus, J. R., Helen, A., et al., (2018). Naturally-derived biomaterials for tissue engineering applications. *Advances in Experimental Medicine and Biology, 1077*, 421–449. https://doi.org/10.1007/978-981-13-0947-2_23.
3. Aidun, A., (2020). A brief review of application of natural biomaterials. *Journal of Bioengineering Research, 2*(2). https://doi.org/10.22034/jbr.2020.109409.
4. Arvanitoyannis, I., Atsuyoshi, N., Eleni, P., Norioki, K., & Noboru, Y., (1996). Synthesis and degradability of a novel aliphatic polyester based on l-lactide and sorbitol: 3. *Polymer, 37*(4), 651–660. https://doi.org/10.1016/0032-3861(96)83152-1.
5. Ghosh, S., Khastgir, D., Anil, K. B., & Mukunda, P. G., (2000). Thermal degradation and ageing of segmented polyamides. *Polymer Degradation and Stability, 67*(3), 427–436. https://doi.org/10.1016/S0141-3910(99)00141-X.
6. Kraicheva, I., Vodenicharova, E., Shenkov, S., Tashev, E., Tosheva, T., Tsacheva, I., Kril, A., et al., (2014). Synthesis, characterization, antitumor activity and safety testing of novel polyphosphoesters bearing anthracene-derived aminophosphonate units. *Bioorganic and Medicinal Chemistry, 22*(2), 874–882. https://doi.org/10.1016/j.bmc.2013.12.001.
7. Yilmaz, Z. E., & Christine, J., (2016). Polyphosphoesters: New trends in synthesis and drug delivery applications. *Macromolecular Bioscience, 16*(12), 1745–1761. https://doi.org/10.1002/mabi.201600269.
8. Fonseca, A. C., Maria, H. G., & Pedro, N. S., (2014). Biodegradable poly(ester amide)s – A remarkable opportunity for the biomedical area: Review on the synthesis, characterization and applications. *Progress in Polymer Science, 39*(7), 1291–1311. https://doi.org/10.1016/j.progpolymsci.2013.11.007.
9. Zeng Fu, R., Jing, X., Lin, H. S., Jimei, M., Hong, J., & Zi Long, L., (2020). Copolymers of ϵ-caprolactone and ϵ-caprolactam: Via polyesterification: Towards sequence-controlled poly(ester amides). *Polymer Chemistry, 11*(6), 1211–1219. https://doi.org/10.1039/c9py01388a.

10. Rodriguez-Galan, A., Lourdes, F., & Jordi, P., (2011). Degradable poly(ester amide)s for biomedical applications. *Polymers, 3*(1), 65–99. https://doi.org/10.3390/polym3010065.
11. Gunatillake, P. A., Raju, A., & Gadegaard, N., (2003). Biodegradable synthetic polymers for tissue engineering. *European Cells and Materials, 5*, 1–16. https://doi.org/10.22203/eCM.v005a01.
12. Zeng, J. B., Yi Dong, L., Qun, Y. Z., Ke Ke, Y., Xiu Li, W., & Yu Zhong, W., (2009). A novel biodegradable multiblock poly(ester urethane) containing poly(L-lactic acid) and poly(butylene succinate) blocks. *Polymer, 50*(5), 1178–1186. https://doi.org/10.1016/j.polymer.2009.01.001.
13. Liu, X., Yiran, X., Lulu, L., Dongmei, Z., & Zhaosheng, H., (2018). Synthesis of a novel biomedical poly(ester urethane) based on aliphatic uniform-size diisocyanate and the blood compatibility of PEG-grafted surfaces. *Journal of Biomaterials Applications, 32*(10), 1329–1342. https://doi.org/10.1177/0885328218763912.
14. Breslow, S. D., George, E. H., & Albert, S. M., (2002). Synthesis of poly-β-alanine from acrylamide. A novel synthesis of β-alanine. *Journal of the American Chemical Society, 79*(14), 3760–3763. https://doi.org/10.1021/ja01571a039.
15. Rozenberg, B. A., (2004). Proton-transfer anionic polymerization of vinyl monomers. *Designed Monomers and Polymers, 7*(1, 2), 135–150. https://doi.org/10.1163/156855504322890106.
16. Kumar, K., Pubali, A., Kranthikumar, T., Venkanna, A., & Krishnamoorthi, S., (2015). Synthesis, characterization, and *in vitro* drug release study of 3-Arm poly-β-alanine. *Journal of Applied Polymer Science, 132*(25), 1–6. https://doi.org/10.1002/app.42124.
17. Saegusa, T., Shiro, K., & Yoshiharu, K., (1974). Hydrogen-transfer polymerization of acrylic acid to poly(β-propiolactone). *Macromolecules, 7*(2), 256–258. https://doi.org/10.1021/ma60038a021.
18. Yang, H. J., Chen, Q. C., Yong, K. Z., Jin, F. H., Yi, Y. S., Li, J., Wen, Y. H., et al., (2020). Hybrid copolymerization via the combination of proton transfer and ring-opening polymerization. *Chinese Journal of Polymer Science (English Edition), 38*(3), 231–239. https://doi.org/10.1007/s10118-020-2341-x.
19. Gur'eva, L. L., Tkachuk, A. I., Dzhavadyan, E. A., Estrina, G. A., Surkov, N. F., Sulimenkov, I. V., & Rozenberg, B. A., (2007). Kinetics and mechanism of the anionic polymerization of acrylamide monomers. *Polymer Science – Series A, 49*(9), 987–999. https://doi.org/10.1134/S0965545X07090052.
20. Otsu, T., (1972). Hydrogen transfer polymerization with anionic catalysts and the problem of anionic lsomerization polymerization. *Journal of Macromolecular Science, Part C, 6*(2), 237–283. https://doi.org/10.1080/15321797208073415.
21. Kawasaki, N., Atsuyoshi, N., Takashi, H., Yasukatsu, M., Noboru, Y., & Sei-Ichi, A., (2001). "Studies on poly[acrylamide-co-(e-caprolactone)]: Synthesis, characterization and biodegradability. *Macromol. Chem. Phys., 202.*
22. Hadjichristidis, N., & Akira, H., (2015). *Anionic Polymerization: Principles, Practice, Strength, Consequences and Applications*, 1–1082. https://doi.org/10.1007/978-4-431-54186-8.
23. Kurcok, P., Marek, K., Krystian, H., & Zbigniew, J., (2002). Anionic polymerization of beta-lactones initiated with alkali-metal alkoxides: Reinvestigation of the polymerization mechanism. *Macromolecules, 25*(7), 2017–2020. https://doi.org/10.1021/ma00033a027.
24. Uzunlar, A., & Efkan, Ç., (2018). Synthesis, characterization and biodegradation of poly(?-methyl ?-propiolactone). *Hacettepe Journal of Biology and Chemistry, 3*(46), 435–443. https://doi.org/10.15671/hjbc.2018.251.

25. Güven, O., Bekir, S., & Efkan, Ç., (2014). Synthesis and characterization of nylon 3. *Hacettepe Journal of Biology and Chemistry, 1*(42), 41–47. https://doi.org/10.15671/hjbc.201442210818.
26. Gudeangadi, P. G., Kei, U., Ayaka, T., Kayo, T., Hiroyasu, M., Kousuke, T., Hitoshi, M., & Keiji, N., (2020). Poly(alanine-nylon-alanine) as a bioplastic: Chemoenzymatic synthesis, thermal properties and biological degradation effects. *Polymer Chemistry, 11*(30), 4920–4927. https://doi.org/10.1039/d0py00137f.
27. Cui, X., Weihua, L., & Deyue, Y., (2004). Investigation on odd-odd nylons based on undecanedioic acid: 1. Synthesis and characterization. *Polymer International, 53*(11), 1729–1734. https://doi.org/https://doi.org/10.1002/pi.1546.
28. Das, R., Arundhati, P., Paul, A. K., & Paul, A. K., (2017). *Poly(3-Hydroxybutyrate) Production by Seed Endophytic Bacillus Spp. of Oleaginous Plants, 3*(6), 56–66. https://doi.org/10.9790/264X-03065666.
29. Yang, H., Junpeng, Z., Manqing, Y., Stergios, P., & Guangzhao, Z., (2011). Nylon 3 synthesized by ring opening polymerization with a metal-free catalyst. *Polymer Chemistry, 2*(12), 2888–2892. https://doi.org/10.1039/c1py00334h.
30. Hong, J., Donghyun, C., Taihyun, C., Woo, S. S., & Doo, S. L., (2003). Characterization of poly(ethylene oxide)-b-poly(L-lactide) block copolymer by matrix-assisted laser desorption/ionization time-of-flight mass spectrometry. *Polymer Society of Korea 341 Macromolecular Research*, p. 11.
31. Savaş, B., Efkan, Ç., Temel, Ö., & Ergül, M., (2021). Synthesis and characterization of poly(α-methyl β-alanine)-poly(ε-caprolactone) tri arm star polymer by hydrogen transfer polymerization, ring-opening polymerization and 'click' chemistry. *Journal of Polymer Research, 28*(2), 1–10. https://doi.org/10.1007/s10965-020-02367-z.
32. Çatıker, E., Temel, Ö., Mehmet, A., & Bekir, S., (2020). Synthesis and characterization of the ABA-type poly(ester-ether-ester) block copolymers. *Journal of Macromolecular Science, Part A: Pure and Applied Chemistry, 57*(8), 600–609. https://doi.org/10.1080/10601325.2020.1745080.
33. Czerniecka-Kubicka, A., Wiesław, F., Marek, J., Wojciech, B., Pilch-Pitera, B., Marek, P., & Iwona, Z., (2017). Thermal properties of poly(3-hydroxybutyrate) modified by nanoclay. *Journal of Thermal Analysis and Calorimetry, 128*(3), 1513–1526. https://doi.org/10.1007/s10973-016-6039-9.
34. Morgenstern, U., & Berger, W., (1994). Studies on the synthesis of block copolymers containing polyamide-3 as one constituent. *Polymer, 35*(11), 2438–2442. https://doi.org/10.1016/0032-3861(94)90785-4.
35. Masamoto, J., Kiichiro, S., & Chihiro, O., (1970). *Of Nylon 3*, 8*, 1703–1711.
36. Liu, N. A., & Fan, W. C., (1999). Critical consideration on the freeman and Carroll method for evaluating global mass loss kinetics of polymer thermal degradation. *Thermochimica Acta 338*(1, 2), 85–94. https://doi.org/10.1016/s0040-6031(99)00197-5.
37. Carrasco, F., Dionisi, D., Martinelli, A., & Majone, M., (2006). Thermal stability of polyhydroxyalkanoates. *Journal of Applied Polymer Science, 100*(3), 2111–2121. https://doi.org/10.1002/app.23586.
38. McNeill, I. C., & Sadeghi, S. M. T., (1990). Thermal stability and degradation mechanisms of poly(acrylic acid) and its salts: Part 1-poly(acrylic acid). *Polymer Degradation and Stability, 29*(2), 233–246. https://doi.org/10.1016/0141-3910(90)90034-5.
39. Aoyagi, Y., Koichi, Y., & Yoshiharu, D., (2002). *Thermal Degradation of Poly [(R)-3-Hydroxybutyrate], Poly [e-Caprolactone], and Poly [(S)-Lactide], 76*, 53–59.

40. Wang, J., Man, K. C., & Yongli, M., (2002). Miscibility and morphology in crystalline/amorphous blends of poly (caprolactone)/poly(4-vinylphenol) as studied by DSC, FTIR, and 13C solid state NMR. *Polymer, 43*(4), 1357–1364. https://doi.org/10.1016/S0032-3861(01)00673-5.
41. Scavuzzo, J., Shogo, T., Shiwang, C., Hao, L., Min, G., Joseph, P. K., Shinichi, S., Stephen, Z. D. C., & Li, J., (2015). Supramolecular elastomers: Self-assembling star-blocks of soft polyisobutylene and hard oligo(β-alanine) segments. *Macromolecules, 48*(4), 1077–1086. https://doi.org/10.1021/ma502322n.
42. Manna, A., & Paul, A. K., (2000). Degradation of microbial polyester poly(3-hydroxybutyrate) in environmental samples and in culture. *Biodegradation, 11*(5), 323–329. https://doi.org/10.1023/A:1011162624704.
43. Alejandra, R. C., Calafell, M. M., & Marqués, C. M. S., (2012). Enzymatic degradation of poly(3-hydroxybutyrate) by a commercial lipase. *Polymer Degradation and Stability, 97*(11), 2473–2476. https://doi.org/10.1016/j.polymdegradstab.2012.07.042.

PART II
Materials and Properties

CHAPTER 6

Recapitulation of Earlier and Recent Studies Carried Out in Open Multi-Component Systems Focused on the Synthesis of Condensed Phosphates

MARINA AVALIANI

Ivane Javakhishvili Tbilisi State University, R. Agladze Institute of Inorganic Chemistry and Electrochemistry, Tbilisi, Georgia

ABSTRACT

The presented chapter provides a compilation of our works carried out during the last few years. This is a sort of summary, analysis, and generalization of the experiences that we have been implemented over a period of time. Earlier and recent research conducted in open systems, such as three-component systems $M^{III}_2O_3$-P_2O_5-H_2O and multi-component systems M^I_2O-$M^{III}_2O_3$-P_2O_5-H_2O revealed the possibilities of synthesis of the many various condensed normal and/or double phosphates, so-called inorganic polymers. Studies of their properties, structure, and comparison with achievements in the domain of inorganic polymer's chemistry were carried out. It was discovered: condensed phosphates of polyvalent metals hold a number of quite interesting, useful, and appreciable properties, which explains the various prospects of their applications. It is necessary pointed that nearly always for each trivalent metal there is a series of isomorph and/or iso-structural compounds. Thus, high thermal stability, elevated content of phosphorus – these properties have caused various applications as raw components for the

Advanced Polymer Structures: Chemistry for Engineering Applications. Omar Mukbaniani, DSc, Tamara Tatrishvili, PhD & Marc Jean M. Abadie, DSc (Eds.)

manufacture of phosphates glasses, the use of crystalline and non-crystalline ultra-phosphates in quantum electronics, applications as nanomaterials, etc.

The presented work is focused on the investigation on phases' formation in the systems Ga_2O_3-P_2O_5-H_2O and In_2O_3-P_2O_5-H_2O at temperature range between 50°C and 600°C. The study of three-component systems was conducted to develop methods for the synthesis and growth of single crystals of condensed phosphates and to determine the solubility of compounds, taking into account the influence of various factors. Phase identification was given in accordance with standard data of the International Center for Diffraction Database of the American Society for Testing and Materials – ASTM. The elemental composition of condensed compounds was examined by gravimetric analysis. Phosphorus was determined by precipitation using a molybdate reagent and a quinoline solution, gallium, and indium were determined by the hydroxyquinoline precipitation method. Concisely, this work covers a selection of primary aspects of condensed phosphates, including history, physical properties, chemical theory, synthesis data and applications and reviews the evolution of the synthesis conditions' of a number of widely studied condensed materials with different matrices. It was also defined that synthesized acidic triphosphates of Ga and In – $M^{III}H_2P_3O_{10}$.(1–2)H_2O are finest ion exchange agents.

6.1 INTRODUCTION

The chemistry of inorganic compounds of phosphorous, concretely composites under the name condensed phosphates has advanced intensively in the last few years for the reason that, first: for the progress in this domain of inorganic chemistry and for the cause that phosphate compounds are most suitable for further development of the field of inorganic polymers, and, second: they are finding ever increasing practical application as nanomaterials, materials for quantum electronics, as fertilizers, detergents, and as materials used in engineering and construction [1–12]. Is also known about the practice of application of the mentioned phosphate's systems of hardening which are used in fine arts also by painter and scientist O. Pavlov. The procedure of thermo-phosphate painting is very interesting and in fact is a new trend in the art avowed as "Thermo-phosphate Pictorial Art"[13].

It is recognized that condensed rare earth and/or polyvalent metals' phosphates are predestined to play an even more important role in our "high-tech" society in the future than they have played in the past. These conclusions were based upon the trend of increasing applications resulting from the electronic structures of these materials that lead to their unusual optical, magnetic, electrical, and chemical properties impressively adaptable to the demands now being placed on materials [11–14].

A little bit of history: in fact many innovative and original researches in the field of phosphate's chemistry, notably in the field of chemistry of condensed phosphates of rare earths and/or tri- and polyvalent metals has begun in the 19th century; The German School of chemistry has been very forceful and dynamic in the domain of phosphates. Starting from the innovative works of Berzelius, Clarke, and Graham, German chemists elaborated and developed, in the course of this century, various technologies for the synthesis of large numbers of condensed phosphates. A great number of discoverer investigations of this epoch is really treasured and – by the opinion of well-known scientist, founder of systematization and classification of condensed phosphates, Professor A. Durif – are really very valuable and appreciated papers [12]. In the XX century, more serious attention of researchers was dedicated as well to the chemistry of phosphates. Later on, during XX century chemistry of phosphates, particularly chemistry of condensed phosphates developed considerable rapidly, by advantage of the development of methods of analysis and also of the prominence and significant application of phosphates materials in various domains.

The significant properties: elevated thermal stability, and important content of phosphorus – these preconditions have caused their application as raw components for manufacture of phosphates glasses; the use of crystalline and non-crystalline ultraphosphates in quantum electronics are predetermined by their specific properties. Between a large scales' of condensation schemes taking place in phosphoric anions, one of them leads to the arrangement/configuration of cyclic substances, or oligomeric or to polymeric compounds' formation. The analogous entities/salts, for a longtime nominated metaphosphates now are identified, recognized, and entitled such as cyclophosphates.

Various cyclophosphates with different formula were synthesized and described in chemical literature for the period of last 30 years, but a number of the defined entities have not been investigated from a structural point of view to date and the character of the anion remains to be definitively confirmed [12, 14–16].

In accordance with the nomenclature elaborated and proposed by Prof. Durif, prior groups are: cyclotriphosphates, cyclotetraphosphates, cyclopentaphosphates, cyclohexaphosphates, cyclooctaphosphates, cyclodecaphosphates, cyclododecaphosphates [12]. In their interesting publications [12,

14], authors review the current state of discovered condensed phosphates, examine recent developments in X-ray structural techniques, and report essential results obtained through their use. Reciting Professor M-T. Averbouch-Pouchot, "The suggested classification is not revolutionary, with respect to condensed phosphates for instance, but establishes clear boundaries between some categories of compounds such as adducts and hetero-poly-anion-compounds" [16].

In this introduction, we would like to say two words about some spheres of application of condensed phosphates. The domains are very variable, such as: raw materials for creation of phosphates glasses, thermo-resistant materials, effective applying fertilizers, detergents, cement substances, ion-exchange materials and also catalytic agents [13, 15, 17–19]. The composition and thermal properties, as well as the vibrational and luminescent properties of compounds determine their use in quantum electronics; The bio-materials appears on the base of hydroxi-apatite and polyphosphates; Important researches regarding double, triple, polymeric, and substituted phosphates, where oxygen's atoms are interchange by nitrogen, fluorine, and sulfur's atoms are accomplished [7, 9, 13, 14, 20]. The phosphate's binding agents, phosphate-binders, and laser materials are supplanted/replaced by biomaterials, on the base of polyphosphates and hydroxyl apatite. Many authors in their works, In particular, Acad. I. Tananaev, always underlined the vital role of hydroxyl apatite, such as main corposant of the bio organisms comparable by their great importance with the DNA. Rare earth polyphosphates are interesting materials also and bear potential applications. Aspects of condensed phosphates' synthesis conditions and their use is enough huge and cannot be exposed in one chapter. But we would like to underline the several important works [13, 20–23]. Our earlier and recent researches conducted in the open systems, containing main components: $M^{III}{}_2O_3$-P_2O_5-H_2O and $M^I{}_2O$-$M^{III}{}_2O_3$-P_2O_5-H_2O describes the possibilities of synthesis more of 85 discovered and newly synthesized normal and double phosphates – so-called inorganic polymers [15, 20, 21, 24–29].

6.2 EXPERIMENTAL METHODS AND MATERIALS

6.2.1 OBJECTS OF THE EXPERIMENTAL STUDY

The main goal of our work was the study of interaction, novel phases' formation and the solubility study in the systems M_2O_3-P_2O_5-H_2O at temperature range 50–600°C. A major task was also to examine the temperature range of the solubility of trivalent metal oxides in polyphosphoric acid and to detail the phase equilibrium in these three component systems.

Other important objectives were: (a) an analysis by different methods of newly obtained compounds; and (b) the determination and evaluation of several properties of the synthesized condensed phosphates.

6.2.2 MATERIALS

In the presented work, we review solely and exclusively the interaction, phase's formation, and the solubility in the three-component systems $M^{III}{}_2O_3$-P_2O_5-H_2O, where M^{III} are Ga and In at 50–600°C.

The initial components are Ga_2O_3, In_2O_3, and ortho-phosphoric acid (85%). Studied systems can be presented, such as $M^{III}{}_2O_3$–P_2O_5–H_2O, where $M^{III}{}_2O_3$ are Ga_2O_3 and In_2O_3. The molar ratio of initial components P/M^{III} almost always was constants $n=P/M^{III} = 20$ for each experimentation and consequently for each temperature range, but sometimes-depending of the tasks, it was varied in the range from 10 to 20. The temperature scale of synthesis was equal between 50°C and 600°C. Duration of the synthesis process up to 7 days to 80–90 days, in correlation with the temperature regime.

6.2.3 CHARACTERISTICS OF THE METHODS

6.2.3.1 SYNTHESIS METHOD

Depending on the previous experiments targets' the above-mentioned compounds, such as H_3PO_4 of percentage 85%, trivalent metal oxides Ga_2O_3 or In_2O_3 were mixed in a special carbon-glass or platinum crucible during some munities. A mixture of these reagents was used for phase's formation and heated at temperatures between 50°C and 600C. Depending of the envisaged and desired condensed compounds, in our experimental studies the duration of the synthesis process was very variable, up 5–6 days to 4–12 weeks. After mixing gallium oxide, or indium oxide with ortho-phosphoric acid and after placing this mixture in the thermostatically controlled shaft furnace at a predetermined temperature, the process of phases' formation is started. Dimensions of the pit stove: diameter = 10 cm, height = 50–55 cm. The melting solution must be mixed permanently, using oval glass-carbon stirrer connected to a motor with rotational speed 60–70 rpm. The furnace shaft was covered with a lid with holes for the stirrer and the thermocouple. After completion of crystallization (the examination of crystals by microscope and other methods were utilized), the crystalline phases, formed in the poly-phosphoric acid melts were extracted with half-litter of hot distilled water. Then the single crystals must be washed with ethanol and acetone, then dried in the thermostat or in the air. The elemental composition of the newly synthesized condensed compounds was examined by gravimetric analysis. Phosphorus was determined by precipitation using a molybdate mixture and a quinoline solution. Three valent metals were determined by the hydroxyquinoline precipitation method [16–18, 24].

6.2.3.2 PAPER CHROMATOGRAPHY

Sufficient stability of polymeric phosphates makes it able to categorize and classify them by the method of paper chromatography. Thus the condensation degree of the synthesized phosphates was checked by paper chromatography. For the chromatographic analysis, the obtained crystalline samples were decomposed with H-form cation type ion-exchanger. In these experiments, 0.1 g of the substance was contacted for ca. Around 1 h with an aqueous suspension of 2 g (KU-2, Dowex) type sulfonated cation exchanger at ~0C. After the decomposition of the main mass of the substance according to the following scheme: cationite-H (solid) + M-phosphate (solid) → cationite-M (solid) + H-phosphate (liquid) the solution, containing phosphoric acids was neutralized with sodium bicarbonate and chromatographed on Filtrak FN-11 paper (Germany) by acidic CCl_3-COOH-CH_3COOH-CH_3OH solvent. Chromatograms were sprayed with ammonium molybdate $(NH_4)_2MoO_4$ 5% solution and irradiated with ultraviolet light ($\lambda = 400$ nm) [15–18].

6.2.3.3 THERMO-GRAVIMETRIC ANALYSIS (TGA), IR SPECTROSCOPY, AND ROENTGEN-PHASE'S ANALYSIS

The technique of paper chromatography, accompanying with IR spectroscopy, thermo-gravimetric analysis (TGA) and X-ray diffraction measurements, allowed us to study the formation process and composition of many normal and/or acidic samples [13–19]. For TGA it was used a Derivatograph Q1500-D with a heating rate of 10 degrees/min., in air atmosphere and maximum temperature of 1,280°C (sometimes to 1,600°C). The powder diffraction intensity data collections were performed on a DRON-3M diffractometer, with anodic Cu–K_α radiation in the range of 2θ = 10–60°, detector's speed 2° min^{-1}, lattice spacing d_α/n in Å, and I/I_0 – is the relative intensity (used model/standard data – by ASTM: American Society for Testing and Materials) [20, 24].

6.2.2.4 DETERMINATION OF PHOSPHORUS, GALLIUM, AND INDIUM

A 1–2 g sample of the liquid phase was hydrolyzed by boiling for 1–2 hours with concentrated hydrochloric acid. Then the solution was diluted to 50 mL and analyzed for gallium and indium content.

The solid phase was washed from the melt with water to neutral reaction of the wash water and dried with acetone and ether. A 0.15 g sample was dissolved by boiling with KOH, then was further hydrolyzed by heating with HCl and diluted up to 50 mL. The mixture of liquid and solid phases was dissolved similarly and hydrolyzed during 1–2 hours. Then, the analyzed solutions were brought to a volume of 50 ml and chemically analyzed for the quantitative content of P, Ga, and/or in Refs. [25–29].

6.2.2.4.1 Determination of Phosphorus

P was determined by precipitation using a quinoline solution [14, 24]. To 5 mL of liquid phase solution, diluted by 200 mL of distilled water was added 25 mL molybdate reagent and precipitated by adding, drop by drop, a quinoline solution. The precipitate was filtered after one and a half to two hours by using a porous crucible G4 and dried at 130–140°C during 1 hour. Conversion factor for P is 0.0321.

6.2.2.4.2 Determination of Gallium

A number of variants of proposed gravimetric methods were compared. Due to our numerous experiments, the influence of different factors, such as: temperature, quantity of buffer solutions and reagent-precipitant, initial pH of the solution, mixing time of the initial components, etc., on the accuracy of the results were detailed and carefully studied. Experimental data regarding the sedimentation process: the influence of buffer and precipitator volumes, solution pH before deposition, sedimentation time, solution pH after deposition, and other factors were carefully analyzed. So, gallium and indium were determined by the ortho-oxyquinoline precipitation method. It can be summarized as follows: approximately 3–5 mL oxyquinoline's solution was added to a 15 mL aliquot of the test solution, was heated to 70–80°C and after the ammonia's 1:1 solution was added drop by drop until complete precipitation [30, 31]. The solution above the precipitate has a yellowish color. The precipitate was left in a water bath for about 2 hours, then set aside and after 12 hours filtered through a porous G3 or G4 crucible. Sludge's drying temperature is 120–130°C. Conversion factor for Ga is 0.1389.

6.2.2.4.3 Determination of Indium

About 2 g ice-cold acetic acid sodium acetate, 2 mL ice-cold acetic acid were added to the analyzed solution and diluted with 150–200 mL hot water. Precipitation was carried out with a small excess of oxyn-acetate injected dropwise at 70–80°C and continuous stirring. After settling for 2 hours, the sediments were filtered through G3 or G4 porous crucible. The sediment was washed with cold water until the yellow color of the wash water disappeared. Conversion factor for In is 0.2099.

6.3 RESULTS AND DISCUSSION

6.3.1 STUDY OF INTERACTIONS IN THE SYSTEMS M_2O_3-P_2O_5-H_2O AT TEMPERATURE RANGE 50–600°C

For investigation of interaction in the systems M_2O_3-P_2O_5-H_2O (as mentioned above where M^{III} are trivalent metals Ga and/or In) the thermostatically controlled shaft furnace was used. The studies were carried out in open vessels communicating with the atmosphere – there was air access to the analyzed phases. At constant pressure of water vapor in the gaseous phase, the equilibrium of liquid, solid, and gaseous phases in this system is monovariant and depends only on temperature changes. It was found that the actually observed variations in atmospheric humidity practically do not affect the results of the experiment.

The following components: trivalent metal oxides – Ga_2O_3 or In_2O_3 and ortho phosphoric acid (85%) were placed in a round-bottom glassy carbon crucible of approximately volume 70–100 mL and heated between temperatures range from 50 to 600°C. Molar ratios $n=P/M^{III}$ was permanent for each experiment, but depending on the tasks, varied in the following range: from 10 to 20. On the contrary, the duration of the synthesis process was very variable, up 6–7 days to 28–90 days, depending on the experiment's temperature and the main goal of the synthesis process. The crystals formed in the solution-melts were removed by extraction of 30–35 g samples, was washed with 700–750 mL of distilled water, and after were rinsed with ethanol and ether. For the analysis of the liquid phase and a mixture of phases, the samples were taken with a quartz capillary and placed in a covered glass container. In order to monitor the extent to which the equilibrium state was reached, samples of the liquid phase were taken periodically to analyze the content of phosphorus and gallium or indium, respectively. The equilibrium state was considered to be reached when the same results were obtained. In this case, the stirrer was turned off, and then samples were taken for analysis.

6.3.2 THE SOLUBILITY STUDY IN THE SYSTEM GA_2O_3-P_2O_5-H_2O

The main results of the study of solubility in tree-component system, containing gallium are shown in Table 6.1 and Figure 6.1. The solubility curve consists of two branches corresponding to the crystallization of acidic gallium triphosphate $GaH_2P_3O_{10}$ in the temperature range from 140 (145°C) to 220°C and polyphosphate $Ga(PO_3)_3$ – above 225°C. The eutonic is located between 220°C and 225°C at the phosphorus oxide P_20_5 concentration approximately of 81%. The solubility of mentioned compounds is around 1% and decreases as the temperature rises. No other phases were detected in the system Ga_2O_3-P_2O_5-H_2O at higher temperatures. In the region of temperatures below 145°C where crystallization of acidic gallium orthophosphate could be projected, we were unable to obtain a saturated melt due to the high solubility of the solid phase and the tendency to form metastable viscous supersaturated solutions. X-ray phase analysis shows that triphosphate is released in two forms: I and II modifications. The formation of I or II modification is apparently not temperature dependent, but is determined by the ratio of the components at the moment (particular time) of crystal nucleation.

As shown by our researches at a P/Ga ratio equal to 5, the form II crystallizes, at a higher phosphorus content it is the hydrated form $GaH_2P_3O_{10}H_2O$ which generally appears. By the Skreinmakers method, it was found that in the melt both forms are waterless (the rays cross exactly on the composition $Ga_2O_33P_2O_52H_2O$, which corresponds to the formula $2GaH_2P_3O_{10}$, but at the subsequent washing with water as well as at contact with moisture of air one of the forms, particularly – form I absorbs about one or two moles of water. Our studies have shown that triphosphate of gallium $GaH_2P_3O_{10}H_2O$ has very evident ion-exchange properties, but form II does not hold these features.

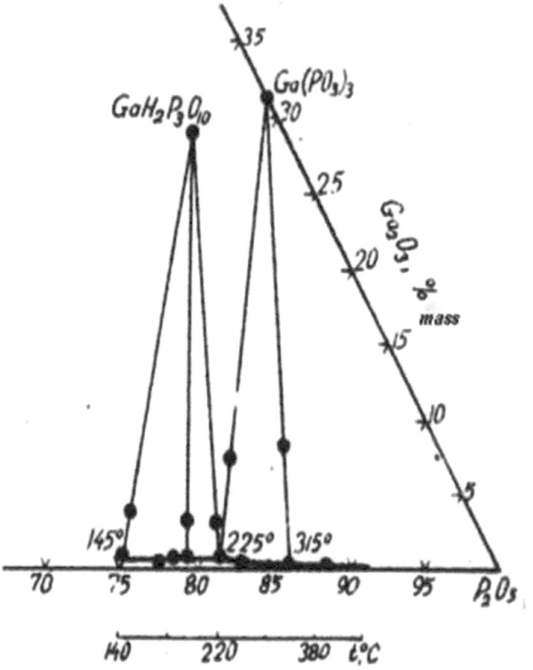

FIGURE 6.1 The solubility in the system Ga_2O_3-P_2O_5-H_2O at 50–600°C.

We have developed an appropriate method for obtaining the ion-exchange form of gallium acid triphosphate – a new ion-exchange material for subsequent use. The optimal synthesis technique was proposed: the oxide of gallium Ga_2O_3 should be mixed with phosphoric acid in the proportion P_2O_5/Ga_2O_3: from 10 to 20, mixed and heated in a platinum or glass-carbon crucible during one week at 180°C. Obtained crystals should be washed from the melt solution with water and dried in air.

One of the possibilities to prove that $GaH_2P_3O_{10}H_2O$ is the ion-exchanger: by addition of this phase to the solution of KCl, in this case pH of solution decreases to about ~1. We also performed ion-exchanger's properties of mentioned triphosphate by the potentiometric titration of $GaH_2P_3O_{10}H_2O$ – form 1 by 0.1 N solution of NaOH.

The flexure on the titration curve indicates the neutralization of two hydrogen ions and the formation of triphosphate of sodium $Na_2GaP_3O_{10}H_2O$. The amount of alkali solution used for the neutralization process is related to calculate theoretically. Further, when treating of $Na_2GaP_3O_{10}H_2O$ with a solution of HCl – appears again $GaH_2P_3O_{10}H_2O$.

After repeated titration of recently obtained $GaH_2P_3O_{10}H_2O$ with NaOH solution, the results were completely reproducible.

Table 6.1 assumed the results of solubility and phase's formation in the system Ga_2O_3-P_2O_5-H_2O at 50–600°C.

6.3.3 THE SOLUBILITY STUDY IN THE SYSTEM IN_2O_3-P_2O_5-H_2O

We reviewed the interaction and solubility in the system In_2O_3-P_2O_5-H_2O at 50–600°C and studied the formation of the solid phases. In this three-component system in the range of temperatures ~100–175°C non-crystallizing supersaturated melts were formed, solidifying at times into glass. By detailing: in fact, indium triphosphate (form I) has also been obtained under nearly similar conditions (at 180–185°C, at molar ratio P/In=5 (or 7.5), but the results are not always reproducible, and the crystallization range of this compound is much narrower, compared to the same gallium compound. Sometimes the melt in which this crystalline phase is found changes to a glassy phase. Beginning from temperature ~195°C and up to 550–590C the indium polyphosphate $In(PO_3)_3$ was crystallized. This long-chain polymeric compound is extremely insoluble in polyphosphoric acid melts. We have established that at ~195–200°C its solubility is equal to ~0.04% (in terms of In_2O_3) and further decreases as the temperature rises.

Polyphosphate $In(PO_3)_3$ crystallized as a C-form, which is isomorphic to the analogous compound of gallium and aluminum.

6.4 CONCLUSIONS

We reviewed the interaction and phase's formation in the three-component systems $M^{III}{}_2O_3$-P_2O_5-H_2O, where M^{III} are Ga and In at 50–600°C.

We have studied the solubility in the systems Ga_2O_3-P_2O_5-H_2O and In_2O_3-P_2O_5-H_2O in a fairly large temperature range. The formation of the acidic triphosphates $GaH_2P_3O_{10}H_2O$ (form I) and $GaH_2P_3O_{10}$ (form II) and their crystallization areas were determined. Polyphosphates of trivalent metals, such as: polyphosphate of gallium – $Ga(PO_3)_3$ form-C and polyphosphate of

TABLE 6.1 Solubility and Phase's Formation in the System Ga_2O_3-P_2O_5-H_2O at 50–600°C

t (C)	Composition of Liquid Phase (%)		Composition of Mixtures – Liquid and Solid Phases (%)		Composition of Solid Phase, Washed with Water (%)			Formula of Compound
	P_2O_5	Ga_2O_3	P_2O_5	Ga_2O_3	P_2O_5	Ga_2O_3	P_2O_5/Ga_2O_3	
50	–	–	–	–	–	–	–	Metastable viscous supersaturated solutions
100	–	–	–	–	–	–	–	Metastable phase
145	75.40	0.75	74.25	4.00	64.74	28.98	2.98	
160	77.20	0.30	–	–	65.60	28.8	3.02	
170	77.98	0.64	–	–	62.50	26.70	3.08	$GaH_2P_3O_{10}$
200	78.93	0.82	77.60	3.00	61.06	27.30	2.89	
218–220	80.90	0.44	79.80	2.80	–	–	–	
225	81.00	0.24	78.39	6.67	69.78	30.44	3.02	
230	82.17	0.19	–	–	70.20	29.00	3.08	
265	82.70	0.16	–	–	69.61	30.50	3.01	$Ga(PO_3)_3$
315	85.40	0.09	82.11	8.38	69.40	30.40	0.00	
380	88.00	Not found	–	–	69.50	30.50	3.01	
480	92.00	Not found	–	–	69.50	30.50	3.01	
580	92.00	Not found	–	–	69.50	69.50	3.01	

indium – $In(PO_3)_3$ form-C have been obtained. Mentioned compounds are isomorphic to the analogous polyphosphate of aluminum.

It has been established and confirmed by scientific experiments that acidic triphosphate $GaH_2P_3O_{10}H_2O$ is the best ion-exchanger and indeed is the new material that could be used in the future. We have developed an appropriate method for obtaining the ion-exchange form of gallium acid triphosphate – a new ion-exchange material for subsequent use.

Indium triphosphate $InH_2P_3O_{10}(1\text{-}2)H_2O$ was also obtained under almost similar conditions. But quite often the melt turns into a glassy phase. Synthesized condensed compounds were wholly studied by chemical analysis, also are observed by X-ray structural techniques and thermal analysis.

ACKNOWLEDGMENTS

I would like to express my gratitude to Dr. Galina M. Balagina for her great contribution to this work. I would especially like to thank Professor Omar Mukbanian for the encouragement and important advice in writing this chapter.

KEYWORDS

- **condensed phosphates**
- **crystallization**
- **inorganic polymer**
- **ion exchanger**
- **multi-component systems**
- **polyphosphate**
- **triphosphate**

REFERENCES

1. Lalena, J. N., Cleary, D. A., Carpenter, E. E., & Dean, N. F., (2008). Inorganic Materials Synthesis and Fabrication (pp. 5–314). John Wiley & Sons, Inc., New Jersey, Canada, Springer.
2. Strutynska, N. Y., Bondarenko, M. A., Ogorodnyk, I. V., Zatovsky, I. V., Slobodyanik, N. S., Baumer, V. N., & Puzan, A. N., (2015). Interaction in the molten system

Rb_2O-P_2O_5-TiO_2-NiO. Crystal structure of the langbeinite-related $Rb_2Ni_{0.5}Ti_{1.5}(PO_4)_3$. Cryst. Res. Technol., 50(7), 549–555.
3. Briche, S., Zambon, D., Chadeyron, G., Boyer, D., Dubois, M., & Mahiou, R., (2010). Comparison of yttrium polyphosphate $Y(PO_3)_3$ prepared by sol–gel process and solid state synthesis. J. Sol-Gel. Sci. Technol., 55, 41–51. doi: 10.1007/s10971-010-2211-z.
4. Palkina, K. K., Masloboev, V. A., & Sazhenkov, A. Y., (2007). Crystal structures of polyphosphates $MCs_5(PO_3)_8$ (M = Pr, Eu and Gd). Russ. J. of Inorg. Chem., 52(12), 1817–1821.
5. Aneli, J., Zaikov, G., & Mukbaniani, O., (2011). Physical principles of the conductivity of electrical conducting polymer composites: A review. Chem. & Chem. Technol., 5, 75–87.
6. Melnikov, P., Guirardi, A. L., Secco, M. A. C., & Nogueira De, A. E., (2008). Study of the trivalent elements polyphosphates by thermal analysis. J. Therm. Anal. & Calor., 94, 163–167.
7. Pfitzner, A., & Seidlmayer, S., (2009). Synthesis and structure determination of $AgScP_2Se_6$, $AgErP_2Se_6$ and $AgTmP_2Se_6$. Z. Anorg. Allg. Chem., 635, 704–707.
8. Oudahmane, A., Daoud, M., Tanouti, B., Avignant, D., & Zambon, D., (2009). The type IV polymorph of $KEu(PO_3)_4$. Acta Cryst., E65, 91, 92.
9. Möncke, D., (2018). Manoel Da, C. B. N. M. C., Bradtmüller, H., Buzatto De, S. G., Rodrgues, A. M., et al., (2018). $NaPO_3$–AlF_3 glasses: Fluorine evaporation during melting and the resulting variations in structure and properties. J. of Chem. Technol. & Metall., 53, 1047–1060.
10. Wiglusz, R. J., Pazik, R., Lukowiak, A., & Strek, W., (2011). Synthesis, structure, and optical properties of $LiEu(PO_3)_4$ nanoparticles. Inorg. Chem., 50, 1321–1330.
11. Gschneidner, K. A. Jr., & LeRoy, E., (1987). Handbook on the Physics and Chemistry of Rare Earths, 9(B5), 5–60. Elsevier Science Publisher.
12. Durif, A., (2005). The development of cyclophosphate crystal chemistry. Solid State Sci., 7, 760–766.
13. Tananaev, I. V., (1980). New data on the phosphates of polyvalent metals. Pure & Appl. Chem., 52, 1099–1115.
14. Durif, A., (2014). Crystal Chemistry of Condensed Phosphates (3th edn, pp. 5–498). Plenum Press.
15. Avaliani, M. A., (1982). PhD Thesis/dissertation. N. Kurnakof Instit. of Gener. & Inorg. Chem., 3–182.
16. Averbouch-Pouchot, M. Th., & Durif, A., (1996). Topics in Phosphate Chemistry (pp. 174–209). World Scientific.
17. Bandurkin, G. A., Chudinova, N. N., Palkina, K. K., Balagina, G. M., & Komova, M. G., (2008). Russ. J. Inorg. Chem., 53, 1077–1087.
18. Oudahmane, A., Avignant, D., & Zambon, D., (2010). Dipotassium dialuminium cyclo-octaphosphate. Acta Crystallographica, Structure Reports Online, 2, i49, i50.
19. Nalina, M., Ribeiro, S. J. L., Messaddeq, Y., Schneider, J., & Donoso, P., (2002). Scandium fluorophosphate glasses: A structural approach. Acad. Des Sciences/Éditions Scientifiques et Médicales, 915–920.
20. Tananaev, I. V., Grunze, I., & Chudinova, N. N., (1984). Prior directions and results in the domain on condensed phosphates' chemistry. J. Neorg. Mater., 20(6), 887–900.
21. Avaliani, M. A., (2011). About development in chemistry and application of condensed phosphates. Proceedings of Intern. Conf. GEOHET, TSU, Tbilisi, Georgia., 1, 162, 163.

22. Kulakovskaya, T. V., Vagabov, V. M., & Kulaev, I. S., (2012). Inorganic polyphosphates in industry, agriculture and medicine: Modern state and outlook process. Biochem., 47, 2–10.
23. Tananaev, I. V., (1987). Some aspects of the chemistry of phosphates and their practical application. Problems of Chem. and Chem. Technol. Nauka, 1, 4–65.
24. Avaliani, M., Chagelishvili, V., Shapakidze, E., Gvelesiani, M., Barnovi, N., Kveselava, V., & Esakia, N., (2019). Crystallization fields of condensed scandium-silver and gallium-silver phosphates. Eur. Chem. Bull., 5, 164–170.
25. Avaliani, M. A., (2021). Investigation and thermal behavior of double condensed phosphates of gallium, scandium and silver. Bull. of Intern. Nucl. Inform. Syst. INIS-IAEA., 52, 31, 32.
26. Avaliani, M. A., Todradze, G., Shapakidze, Ukleba, M., Chikovani, N., Vibliani, M., & Magradze, G., (2020). Optimization of the gravimetric method for determination of trivalent metals using oxiquinoline. Proceed. Intern. Conf. on Analyt. Chem. Modern Trends. Kiev, Ukraune. http://kcacmt.univ.kiev.ua/en (accessed on 02 January 2022).
27. Avaliani, M., Gvelesiani, M., Barnovi, N., Purtseladze, B., & Dzanashvili, D., (2016). New investigations of poly-component systems. J. Proc. Georgian Acad. Sci. Chem. Ser., 42, 308–311.
28. Avaliani, M., (2016). Main types of condensed phosphates synthesized in open systems from solution-melts of phosphoric acids. Nano Studies, 1, 135–138.
29. Avaliani, M., Shapakidze, E., Barnovi, N., Dznashvili, D., Todradze, G., Kveselava, V., & Gongadze, N., (2019). Regio-controlled synthesis of double condensed oligo-, poly- and cyclo-phosphates, their characterization and possibilities of applications, due to their solid-state properties. J. Nano Studies-Eur. Chem. Bull., 19, 273–284.
30. Avaliani, M., Chagelishvili, V., Barnovi, N., & Esakia, N., (2019). Condensed phosphates as inorganic polymers and various domain of their applications; Chapter 8. In: Materials Science, Composite Materials Engineering: Modeling and Technology (pp. 107–116). APPLE ACADEMIC PRESS, Canada, Toronto/USA, New Jersey. doi: 10.13140/RG.2.2.31613.82408.
31. Avaliani, M., Shapakidze, E., Chagelishvili, V., Barnovi, N., & Esakia, N., (2021). Condensed phosphates: New inorganic polymers with a variety of applications and improvement of their gravimetric determination methods: II; Chapter 18. In: Book Advanced Materials, Polymers, and Composites (pp. 255–276). APPLE ACADEMIC PRESS, Canada, Toronto/USA. doi: 10.13140/RG.2.2.20531.37924. https://www.routledge.com/Advanced-Materials-Polymers-and-Composites-New-Research-on-Properties/Mukbaniani-Abadie-Tatrishvili/p/book/9781771889513 (accessed on 02 January 2022).

CHAPTER 7

Influence of Initial States on the Electrochemical Behavior of Industrial Ionites in the Interpolymer System Lewatit CNPLF-AB-17-8

JUMADILOV TALKYBEK,[1] KHIMERSEN KHUANGUL,[1,2] and HAPONIUK JOZEF[3]

[1]*JSC Institute of Chemical Sciences (Named After A.B. Bekturov), Almaty, Kazakhstan*

[2]*Abai Kazakh National Pedagogical University, Almaty, Kazakhstan*

[3]*Gdansk University of Technology, Gdansk, Poland*

ABSTRACT

The electrochemical behavior of commercia ionites in the interpolymer system lewatit CNPLF-AB-17-8 ion exchangers under various initial states was analyzed by the methods of electrical conductivity and pH-meters. When considering ion exchangers in swollen states, with an increase in the ratio from cation exchanger to anion exchanger, significant changes are observed. The highest values of electrical conductivity are observed at 5:1 and 0:6, where at 5:1 the electrical conductivity of interpolymer systems is 2.5 times greater than at the minimum point at 5:1 (lewatit:anion exchanger), and about 2.0 times greater at 0:6, respectively. Due to the weak dissociation of H_2O, the equilibrium will shift to the right. According to the Le Chatelier's principle, additional H^+ ions appear in the aquatic environment, the concentration of which increases to a ratio of 5:1. In the case of Lewatit

Advanced Polymer Structures: Chemistry for Engineering Applications. Omar Mukbaniani, DSc, Tamara Tatrishvili, PhD & Marc Jean M. Abadie, DSc (Eds.)

< AB-17-8, another equilibrium sets in, in which OH^- groups are spent to neutralize H^+ ions. To restore the concentration of OH^- ions, the anion exchanger AB-17-8 additionally dissociates. The transition region at a ratio of 2:4 indicates that AB-17-8 is a stronger polyelectrolyte than lewatit. An analysis of the dependence of electrical conductivity on the initial ratio of acidic and basic components and on time shows significant changes in the studied parameter up to 2.5 times, which indicates a possible change in the conformational state and flexibility of internodal chains under the influence of electrostatic interactions. In the interpolymer system $CNPLF_{dry}$-AB-17-8 dry for the anion exchanger AB-17-8, a significant increase in the value of electrical conductivity is observed. Moreover, with an increase in the time of interaction, their value increases. The minimum values are reflected in the equimolar state 3:3 (cation exchanger:anion exchanger). The difference between the minimum and maximum values reaches significant values. For example, at 0.08 hours, the χ value increases from 2 to 7, which is a 3.5-fold increase in electrical conductivity. The maximum electrical conductivity for the system $CNPLF_{dry}$–AB-17-8_{sw} is observed at a ratio of cation exchanger:anion exchanger 5:1. The minimum points are at ratios of 4:2 and 2:4. A distinctive feature of this system is the large difference between the maximum and minimum values, the values of which are higher than in the case of $CNPLF_{sw}$–AB-17-8_{sw}.

In the $CNPLF_{sw}$–AB-17-8 dry system, the maximum areas were established at ratios of 3:3 and 2:4. The values of the maximum electrical conductivity are observed during the interaction of 30 hours. Minimal χ is observed at 0.08 hours.

Thus, the obtained experimental results on χ and pH measurements for interpolymer systems consisting of dry and swollen ion exchangers indicate that changes in the initial state of one of the components significantly change the electrochemical behavior of ion exchangers and interpolymer systems, which affects the applied properties of industrial ion exchangers.

7.1 INTRODUCTION

At present, sorption technologies use various macromolecular sorbents [1]. An important part of them is aimed at the selection and separation of rare earth metals [2]. In hydrometallurgy, sorption technologies are based on the use of ion exchange resins [3–7]. Ion exchangers are insoluble, solid polymers that swell to some extent in electrolyte solutions and organic solvents [8]. As objects of the study widely used industrial ion-exchange resins lewatit

CNPLF (cation exchanger) and AB-17-8 (anion exchanger) were chosen as macromolecular sorbents.

Lewatit CNP LF is a Food grade, weakly acidic, macroporous cation exchange resin based on crosslinked polyacrylate. Along with activated carbon, it is used in drinking water cartridges, decarbonization, and softening of drinking water due to its high exchangeability, excellent mechanical and chemical stability. This resin is economically viable for decarbonization of drinking water and liquids in contact with food or used in food production, as it does not require large amounts of regenerating agents [9]. Anion exchanger AB-17-8 is a strong base ion exchange resin with a gel structure. It is used in softening technology and water demineralization. It is characterized by good osmotic stability, high chemical resistance to alkalis, acids, oxidants, insoluble in water and organic solvents [10–13].

7.2 EXPERIMENTAL METHODS AND MATERIALS

7.2.1 EQUIPMENT

Conductometer MARK 603 (Russia) and pH-meter Metrohm 827 pH-Lab (Switzerland) were used to measure the pH and specific electrical conductivity of aqueous solutions.

7.2.2 MATERIALS

The study was performed in an aqueous solution. Acid and base CNPLF and AB-17-8 industrial ion exchangers with functional groups were used in work. CNPLF –AB-17-8 interpolymer pairs were formed on the basis of these ion exchangers.

7.2.3 EXPERIMENTS

Experiments were carried out at room temperature. Studies of the interpolymer system were carried out as follows: each dry ion exchanger was placed in separate polypropylene meshes, which were then placed together in an aqueous medium in one beaker. Swelling time of ion exchangers is 24 hours, interaction time is 30 hours. The specific electrical conductivity and pH of solutions were monitored in the presence of ion exchangers in the solution.

7.3 RESULTS AND DISCUSSION

7.3.1 *ELECTROCHEMICAL PROPERTIES OF THE INTERPOLYMER SYSTEM CNP LF–AB-17-8*

Previous studies have shown that the conformational state of interstitial bonds in the structure of polymers plays an important role in the sorption capacity [14]. Using the methods of electrical conductivity and pH-meter, it is possible to estimate the electrochemical action of ion exchangers during remote interaction. Electrochemical properties, in turn, affect the conformational state of ion exchangers. In this regard, for the analysis of the electrochemical state of ion-exchange macromolecules studied lewatit CNPLF and AB-17-8 in the swollen state. Figure 7.1(a) shows significant changes in the specific molecular ratios of the cation exchanger and the anion exchanger. First of all, it should be noted that there is no additivity between the electrical conductivity and the ratio of CNPLF and AB-17-8. The highest values are observed in the ratio 5:1 and 0:6, where the electrical conductivity of interpolymer systems in a ratio of 5:1 is 2.5 times higher than the minimum point of 5:1 (lewatit: anionite), in a ratio of 0:6–2.0 times higher. In the case of lewatit > AB-17-8 the following process takes place:

$$–COOH + N – OH \rightarrow –COO^{-} + N^{+} + H_2O$$

As H_2O is weakly dissociated, the equilibrium shifts to the right. According to the principle of Le Chatelier, additional H^+ ions are formed in the aqueous medium, the concentration of which increases to a ratio of 5:1. For AB-17-8 anion exchange resin dissociation is lost.

Lewatit <AB-17-8 has a different equilibrium, where OH^- groups are used to neutralize H^+ ions. The 2:4 ratio indicates that AB-17-8 is a much stronger polyelectrolyte than lewatit.

It is known that the electrical conductivity reflects a change in the total concentration of ions (regardless of charge). Comparison of electrical conductivity and pH allows to determine the cause of changes in electrochemical parameters, because the change in pH reflects the dynamics of changes in the concentration of H^+ ions, and χ is the sum of H^+ and OH^- ions.

Figure 7.1(b) shows the time change in the electrical conductivity of the primary cation exchanger and the anion exchanger in the ratio of 5: 1, 4:2, 3:3, 2:4, 1:5 mol lewatit CNPLF and AB-17-8. And we see an increase in the value of χ in the ratio 4:2 and 1:5. Thus, the analysis of the dependence of the specific conductivity on the initial ratio of acid and base components and time showed that the studied parameter changes up to 2.5 times, which indicates a possible change in the conformational state and elasticity of interstitial chains under the influence of electrostatic interactions.

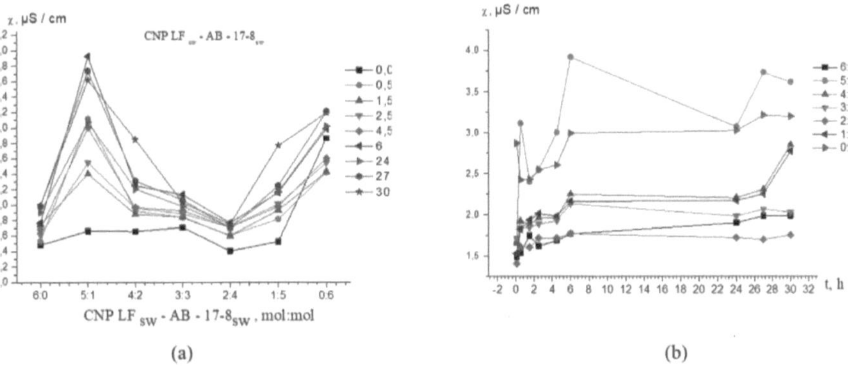

FIGURE 7.1 Time dependence of the specific electrical conductivity of aqueous solutions in the presence of interpolymer systems $CNPLF_{sw}$–AB-17-8_{sw}.

The results of measuring the concentration of H^+ ions are shown in Figure 7.2. As can be seen in the figure, the pH value in the 5: 1 ratio (cation exchanger:anion exchanger) decreases over time from 7.8 in 0.08 hours to 5.9 in 30 hours. These changes in pH indicate an increase in the concentration of H^+ ions in the time zone under consideration. The difference between pH 7.8 and 5.9 is equal to the logarithmic scale ΔpH = 1.9, so large changes from numbers are poorly visible. These results indicate that the specific electrical conductivity of the system at this time falls, and the concentration of the proton grows.

Absolute values of specific electrical conductivity in the ratio 5: 1, where the content of cation exceeds 5 times the anion, significantly higher than in the field of relation 2:4–0:6.

Figure 7.2(b) shows the time dependence of pH values in different ratios of cation exchange resin and anion exchange resin. The obtained results show a sharp increase in the pH of the aqueous medium in the presence of the interpolymer system in 2.5 hours (except 1.5 hours) in the initial time range. Then the pH drops sharply and rises again. The initial increase in pH, which indicates a decrease in the concentration of hydrogen ions, may be primarily due to the disruption of intergenerational associations caused by the loss of protons and the conformational transformation of the interstitial joints of ion exchangers. When the associations are removed, the H^+ ions again form associations with the H^+ ions of the ion exchangers. There is no significant change in the interaction time up to 24 hours, only a sharp decrease from 24 hours to 30 hours. In this picture, large changes occur in the ratio of (cation exchanger:anion exchanger)5:1. The minimum change in pH was recorded at 6:0.

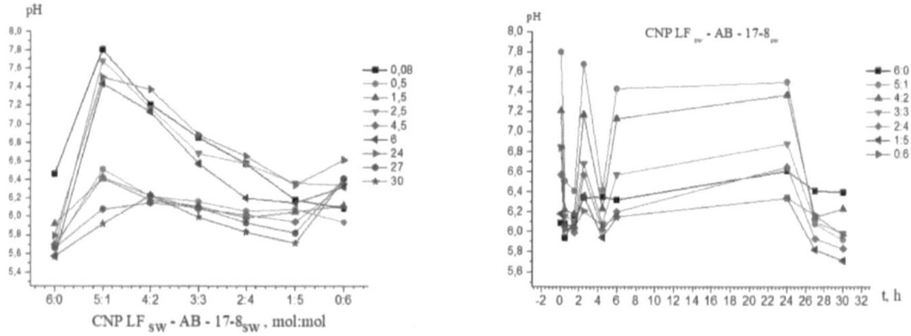

(a) (b)

FIGURE 7.2 Time dependence of pH values of aqueous solutions in the presence of interpolymer systems $CNPLF_{sw}$–AB-17-8_{sw}.

The next stage of the study is the study of pH and χ dependence of different ratios of interpolymer systems in the condensed state. Figure 7.3 shows the dependence of electrical conductivity on the ratio CNPLF–AB-17-8 at different interactions. Changes in the electrical conductivity for swollen lewatit CNPLF and swollen AB-17-8 in the previous figure are also found here. At a ratio of 5: 1, we notice a slight increase in electrical conductivity. There is an increase in the electrical conductivity of the AV-17-8 anion exchange resin. As the interaction time increases, their value increases. The minimum values are expressed in the equimol state in the ratio 3: 3 (cation exchanger:anion exchanger). The difference between the minimum and maximum values is significant. For example, at 0.08 hours, the value of χ increases from 2 to 7, which means that the electrical conductivity increases by 3.5 times. A similar increase in electrical conductivity is observed at other times of interaction.

The time dependence of the electrical conductivity of ion exchanger resins in different ratios (Figure 7.3(b)) shows that the behavior of anion exchanger differs significantly from other systems. The specific electrical conductivity of the AB-17-8_{dry} anion exchanger is in the range of 4–7 μS/cm, and for CNPLF$_{dry}$ cation exchangers and interpolymer systems in the range of 2 to 5 μS/cm.

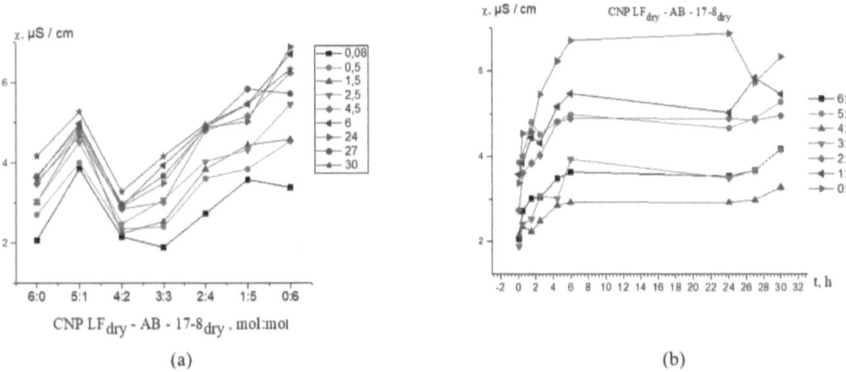

FIGURE 7.3 Time dependence of the specific electrical conductivity of aqueous solutions in the presence of dry interpolymer systems CNPLF and AB-17-8.

Analysis of pH curves depending on the ratio of cation:anion exchange resin for the considered CNPLF$_{dry}$-AB-17-8$_{dry}$ system shows that the pH values differ significantly from the results of electrical conductivity (Figure 7.4). The pH of the cation exchanger is much lower than that of the anion exchanger. The minimum pH values are different for different reaction times. For example, the minimum pH is observed at 0.08 hours 3:3, 0.5 hours – 1:5, 6 hours – 2:4.

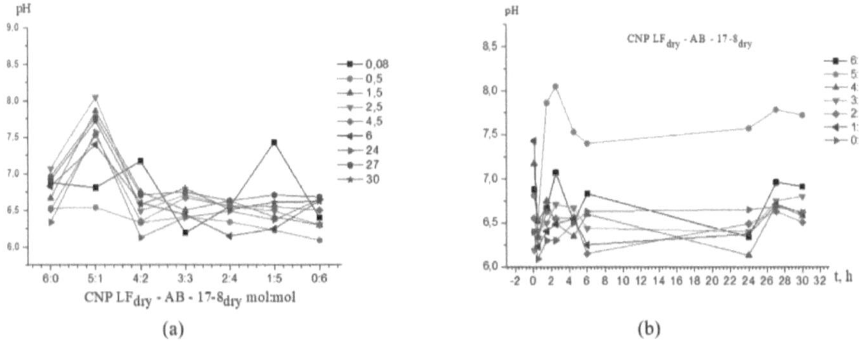

FIGURE 7.4 Time dependence of pH values of aqueous solutions in the presence of CNPLF$_{dry}$–AB-17-8$_{dry}$ interpolymer systems.

Depending on the pH of time (Figure 7.4(b)), the pH value of 5: 1 differs significantly from others and indicates a low content of protons in $CNPLF_{dry}$–AB-17-8_{dry}.

The next study was the control of electrochemical properties of interpolymer systems consisting of $CNPLF_{dry}$-AB-17-8_{sw} (Figure 7.5). Curved dependencies show that their forms resemble interpolymer systems with swollen ions (Figure 7.1). The maximum electrical conductivity is observed in the ratio of cationexchanger:anion exchanger 5:1. Minimum points are found in the ratios of 4:2 and 2:4. Distinctive drawing of this system is the difference between the maximum values of interpolymer systems and the minimum values that are significantly higher than in the case of $CNPLF_{sw}$–AV-17-8_{sw}.

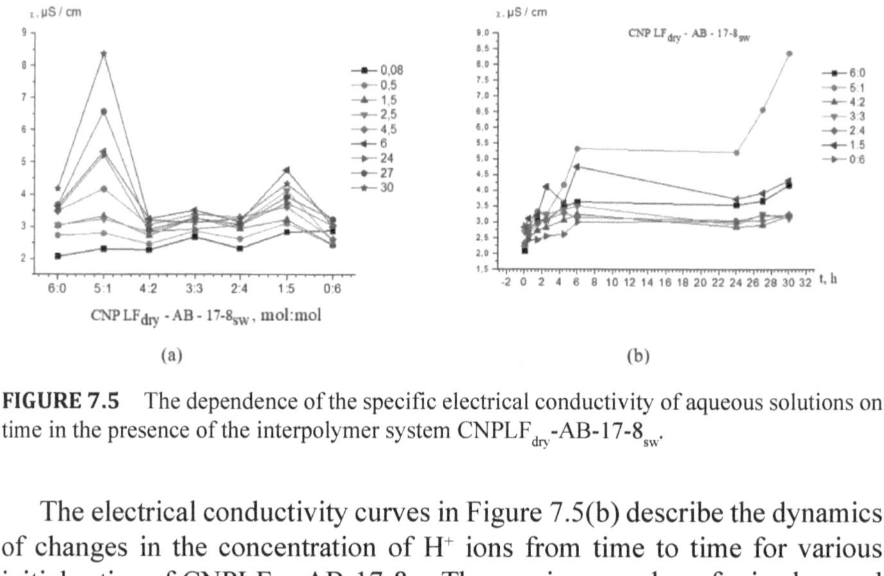

FIGURE 7.5 The dependence of the specific electrical conductivity of aqueous solutions on time in the presence of the interpolymer system $CNPLF_{dry}$-AB-17-8_{sw}.

The electrical conductivity curves in Figure 7.5(b) describe the dynamics of changes in the concentration of H^+ ions from time to time for various initial ratios of $CNPLF_{dry}$-AB-17-8_{sw}. The maximum value of χ is observed for the ratio of 5:1 and for 0:6. The remaining ratios show a slight increase in electrical conductivity.

The change in the pH of interpolymer systems $CNPLF_{dry}$-AB-17-8_{sw} is shown in Figure 7.6. For all ratios, a decrease in the pH of the aqueous medium is observed, except for 0.5 hours. Results in a 5:1 ratio are significantly different from others. At the maximum point there are points beyond 0.08; 1.5; 4.5 and 6 hours. Further, as the interaction time increases, the pH value decreases, indicating an increase in the proton content in the solution.

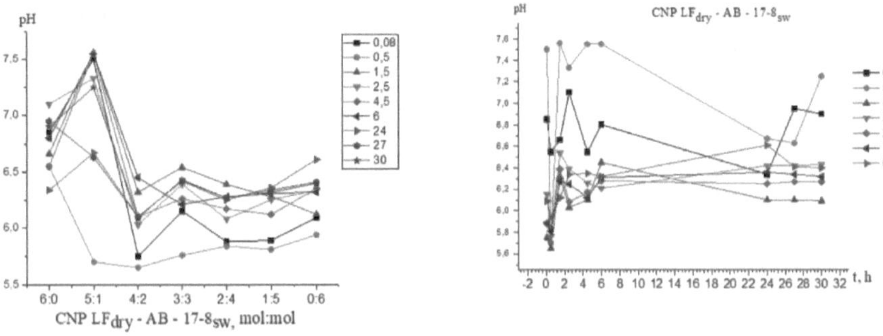

(a) (b)

FIGURE 7.6 The dependence of the pH of aqueous solutions on time in the presence of the interpolymer system $CNPLF_{dry}$-AB-17-8_{sw}.

The results of Figure 7.6(b) are consistent with those of Figure 7.6(a). The highest pH values over time are observed for a ratio of 5:1. Further, several values are observed of the initial cation exchanger $CNPLF_{dry}$-AB-17-8_{sw}. Moreover, it is necessary to note the appearance of the transition region at 2.5 hours. Such a transition is most likely due to the conformational transition of the internodal links of the ion exchangers.

The results obtained for the system $CNPLF_{sw}$–AB-17-8_{dry} (Figure 7.7) differ significantly from the previously considered systems. In the new system, at ratios of 3:3 and 2:4, regions of maximum appear. At the same time, the region of the maximum at 5:1 was preserved for this system. The values of the maximum electrical conductivity are observed during the interaction of 30 hours. The minimum χ is reflected at 0.08 hours.

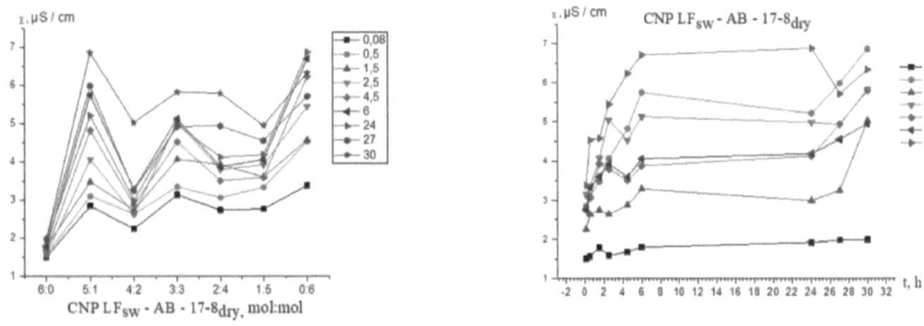

(a) (b)

FIGURE 7.7 The dependence of the specific electrical conductivity of aqueous solutions on time in the presence of the interpolymer system $CNPLF_{sw}$–AB-17-8_{dry}.

In Figure 7.7(b), the maximum value for χ is observed for a ratio of 0:6. A noticeable decrease in χ is also observed at a ratio of 5:1. The minimum value of χ is typical for a ratio of 4:2, which coincides with the data in Figure 7.7(a). The pH dependence observed for Figure 7.8(a) on the ratio is practically identical to the previously obtained results. The difference is observed in the order of the curves in the region of maximum and minimum.

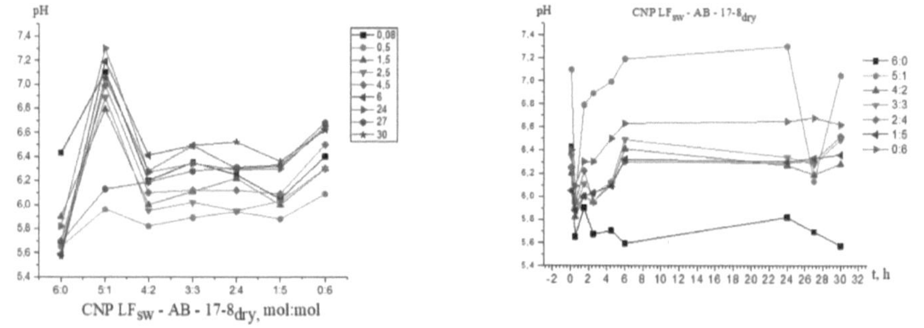

(a) (b)

FIGURE 7.8 The dependence of the pH of aqueous solutions on time in the presence of the interpolymer system CNPLF$_{sw}$–AB-17-8$_{dry}$.

In Figure 7.8(b), the points at the 5:1 ratio are very different from the others. In 6 hours, the H^+ concentration reaches a minimum and practically does not change until 24 hours, then sharply decreases, indicating an increase in concentration at 27 hours. Thus, the obtained results show that the electrochemical behavior of cross-linked polyelectrolytes constituting an interpolymer pair is affected by the initial state, structures, and structure of the studied ion exchangers.

7.3.2 INFLUENCE OF THE INITIAL STATE ON X OF INTERPOLYMER SYSTEMS

In the system of cation exchanger$_{dry}$-anion exchanger$_{dry}$, a maximum of electrical conductivity appears in the region where the cation exchanger predominates (5: 1, CNPLF$_{dry}$–AB-17-8$_{dry}$, Figure 7.3). At 4:2, a minimum of electrical conductivity is observed. In the right region from the minimum point, the electrical conductivity of the system constantly increases and their values exceed the χ data previously observed at a ratio of 5:1.

When studying interpolymer systems consisting of a dry cation exchanger and a swollen anion exchanger, we observe a sharp change in the electrochemical behavior of the systems (Figure 7.5). Conductivity values at 5:1 (cation exchanger:anion exchanger) are 1.5 times higher than data recorded for systems with dry ion exchangers. In the right region, the data on χ drop sharply (almost 2 times). In contrast to the previous system, a maximum region appears at a ratio of cation exchanger:anion exchanger 1:5. Thus, these results show that a change in the initial state of one component dramatically affects the electrochemical behavior of interpolymer systems, which is reflected in the subsequent conformational properties of ion exchangers.

The next system under study differs from the previous one in that the initial state of the ion exchangers in the interpolymer system is in the reverse order: the cation exchanger in the swollen, anion exchanger in the dry state (Figure 7.7). An external review of the figure shows that the measurement results and the course of the curves differ sharply from the previously studied 2 systems. In the first, there are 2 areas of maximum and 2 areas of minimum. In addition to the previously detected points at a maximum of 5: 1, there is another relationship with the points at a maximum of 3: 3 (cation: anion exchange resin). The values of electrical conductivity at 5:1 are slightly reduced, at the point of the second maximum their values are much lower than at a ratio of 5:1. On the right side of the curves, when the dry anion exchange resin predominates in the interpolymer system, the values of χ sharply increase and their values are comparable with the results obtained for the maximum point at 5:1.

The last figure characterizes the changes in the electrochemical behavior of interpolymer systems consisting of swollen ion exchangers (Figure 7.1). The following differences in the behavior of the interpolymer system are observed: (i) the values of χ at the maximum point will decrease significantly from previous systems; (ii) the second maximum has disappeared. The minimum point is observed at a ratio of 2:4 (cation exchanger:anion exchanger). The χ values drop significantly on the right side of the curves.

7.4 CONCLUSION

The obtained experimental results on χ for interpolymer systems consisting of dry and swollen ion exchangers indicate that changes in the initial state of one component change the electrochemical behavior of ion exchangers and interpolymer systems. It follows from the obtained results and conclusions that when studying the physicochemical and sorption properties of ion exchangers in interpolymer systems, it is necessary to take into account the initial state of the components.

KEYWORDS

- **activation**
- **electrochemical properties**
- **hydrometallurgy**
- **industrial ion exchangers**
- **interpolymer system**
- **interpolymer system**
- **lewatit CNPLF**
- **remote interaction**

REFERENCES

1. Beaugeard, V., Muller, J., Graillot, A., Ding, X., Robin, J. J., & Monge, S., (2020). Acidic polymeric sorbents for the removal of metallic pollution in water: A review. *React. Funct. Polym., 152*, 104599. https://doi.org/10.1016/j.reactfunctpolym.2020.104599.
2. Sui, N., & Huang, K., (2019). Separation of rare earths using solvent extraction consisting of three phases. *Hydrometallurgy, 188*, 112–122. https://doi.org/10.1016/j.hydromet.2019.06.012.
3. Botelho, Jr. A. B., Pinheiro, É. F., Espinosa, D. C. R., Tenório, J. A. S., & Baltazar, M. D. P. G., (2021). Adsorption of lanthanum and cerium on chelating ion exchange resins: Kinetic and thermodynamic studies. *Sep. Sci. Technol., 2021*. In press.
4. Martin, D. M., Jalaff, L. D., Garcia, M. A., & Faccini, M., (2021). Selective recovery of europium and yttrium ions with cyanex 272-polyacrylonitrile nanofibers. *Nanomaterials, 9*, 1648.
5. Abu Elgoud, E. M., Ismail, Z. H., Ahmad, M. I., El-Nadi, Y. A., Abdelwahab, S. M., Aly, H. F., Abu, E. E. M., Ismail, Z., Ahmad, M. I., et al., (2019). Sorption of lanthanum(III) and neodymium(III) from concentrated phosphoric acid by strongly acidic cation exchange resin (SQS-6). *Russ. J. Appl. Chem., 92*, 1581–1592.
6. Kolodynska, D., Hubicki, Z., & Fila, D., (2019). Recovery of rare earth elements from acidic solutions using macroporous ion exchangers. *Sep. Sci. Technol., 54*, 2059–2076.
7. Allahkarami, E., & Rezai, B., (2019). Removal of cerium from different aqueous solutions using different adsorbents: A review. *Process. Saf. Environ. Prot., 124*, 345–362.

8. Jörissen, J., (1996). Ion exchange membranes as solid polymer electrolytes (spe) in electro-organic syntheses without supporting electrolytes. *Electrochim. Acta, 41*(4), 553–562. https://doi.org/10.1016/0013-4686(95)00342-8.
9. Pehlivan, E., & Altun, T., (2007). Ion-exchange of Pb^{2+}, Cu^{2+}, Zn^{2+}, Cd^{2+}, and Ni^{2+} ions from aqueous solution by Lewatit CNP 80. *Journal of Hazardous Materials, 140*(1, 2), 299–307. doi: 10.1016/j.jhazmat.2006.09.011.
10. Kurian, M., (2020). Cerium oxide-based materials for water treatment: A review. *J. Environ. Chem. Eng., 8*(5), 104439. https://doi.org/10.1016/j.jece.2020.104439.
11. Petrovm G., Zotovam I., Nikitinam T., & Fokinam S., (2021). Sorption recovery of platinum metals from production solutions of sulfate-chloride leaching of chromite wastes. *Metals, 11*, 569. https://doi.org/10.3390/met11040569.
12. Jumadilov, T., Khimersen, K., Malimbayeva, Z., & Kondaurov, R., (2021). Effective sorption of europium ions by interpolymer system based on industrial ion-exchanger resins amberlite IR120 and AB-17-8. *Materials, 14*(14), 3837. https://doi.org/10.3390/ma14143837.
13. Altshuler, H. N., Ostapova, E. V., Malyshenko, N. V., & Altshuler, O. H., (2017). Sorption of nicotinic and isonicotinic acids by the strongly basic anion exchanger AB-17-8. *Russian Chem. Bul., 66*, 1854–1859. https://doi.org/10.1007/s11172-017-1957-7.
14. Jumadilov, T., Totkhuskyzy, B., Malimbayeva, Z., Kondaurov, R., Imangazy, A., Khimersen, K., & Grazulevicius, J., (2021). Impact of neodymium and scandium ionic radii on sorption dynamics of amberlite IR120 and AB-17-8 remote interaction. *Materials, 14*, 5402. https://doi.org/10.3390/ma14185402.

CHAPTER 8

Thermostable Composition Materials on the Basis of Imide and Anhydride of 4-Sulfoisophthalic Acid and Ed-20

F. M. MAMEDALIYEVA, E. T. ASLANOVA, and B. A. MAMEDOV

Institute of Polymer Materials, Azerbaijan National Academy of Sciences, Sumgait, Azerbaijan

ABSTRACT

In this chapter, the methods of the preparation and formulation of thermally curing compositions using ED-20 epoxide resin and imide of 4-sulfoisophthalic acid, as well as anhydride of 4-sulfoisophthalic acid are presented. The thermal properties of the obtained polymer compositions have been investigated by DTA and TGA methods. The developed composition materials have thermal stability, heat resistance, and also high values of the thermogravimetric index (TGI). The thermostable compositions can be used in the electric-technical industry for coatings of the copper wires of mounting wires, electro-technical contacts, and also in potting compounds.

8.1 INTRODUCTION

The epoxide resins are universal, self-adhesive material used for the manufacture of compounds, composites, and also for pouring various surfaces and glue applying. The polymer materials on the basis of epoxide resins have such a complex of strength, heat-physical, dielectric, adhesion, moisture protection and other indices, which no group of high-molecular compounds

Advanced Polymer Structures: Chemistry for Engineering Applications. Omar Mukbaniani, DSc, Tamara Tatrishvili, PhD & Marc Jean M. Abadie, DSc (Eds.)

has. These indices of epoxypolymers, in combination with the technologically convenient aggregate state of epoxide compositions, stipulated their widespread use in industry, transport, construction, agriculture, and in various areas of the national economy due to low shrinkage during curing, high adhesion to various materials, chemical resistance, good physical-mechanical and excellent dielectric properties [1, 2].

The epoxide compositions based on ED-20 resin are most widely used as coatings, lacquers, and glues. However, the film-forming coatings made of unmodified ED-20 are characterized by low physical-mechanical and thermal indices [3, 4]. The low heat resistance and impact resistance, and also the absence of elasticity limit the use of ED-20 as anti-corrosion and electro-insulation coatings [5]. For elimination of these lacks, the modifiers containing various reactive functional groups (hydroxyl, carboxyl, carbonyl, etc.), which favor the improvement of the exploitation indices of epoxide coatings are introduced into the composition of the epoxide composition.

The epoxide resins containing reactive epoxide groups are cured with many low-molecular compounds, oligomers, and polymers. Among the substances often used for this purpose – primary aliphatic and aromatic di-and polyamines (polyethylene polyamine, hexamethylenediamine, diethylenetriamine, m-phenyldiamine, etc.), dicyandiamide, anhydrides of di- and tetracarboxylic acid, tertiary amines, their complexes, etc. The aliphatic diamines are able to cure epoxide resins in the cold, and curing with aromatic diamines is carried out at higher temperatures. Along with carboxylic acids, the anhydrides of dicarboxylic acid are used as hardeners of epoxide resins.

The industrial aliphatic and aromatic diamines used for curing of epoxide resins are toxic and volatile. The composition materials on their basis have low viability, and their use as hardeners is accompanied by considerable self-heating, which hinders the preparation of large-sized products. The aromatic diamines containing amide, imide, and sulfo groups in its composition are less toxic. The epoxide resins cured with such hardeners are self-extinguishing and have higher heat and thermal resistance [6]. Considering the above-stated one, it was of interest to obtain effective and low-toxic hardeners for epoxide resins.

8.2 EXPERIMENTAL PART

8.2.1 SYNTHESIS OF IMIDE OF 4-SULFOISOPHTHALIC ACID

18.6 g (0.1 mol) of m-xylene-2-sulfoamide and 44.5 ml of sulfuric acid (H_2SO_4) was loaded into flask and in stirring, 102 g of potassium chromate was added. After completion of the reaction, the stirring was continued for another 1–1.5 h. The resulting reaction mass was poured into a glass with ice water and filtered on Shot filter. Then it was dissolved in a soda solution to an alkaline medium and acidified with hydrochloric acid, dried first in air, then in a vacuum at 75–80°C. The product yield was 75%.

8.2.2 SYNTHESIS OF ANHYDRIDE OF 4-SULFOISOPHTHALIC ACID

About 7.5 g (0.03 mol) of the dry monoammonium salt of 4-sulfoisophthalic acid was added to 50 ml of chlorosulfonic acid. The salt was immediately dissolved and the temperature of the reaction mixture spontaneously rose to 45–50°C. Then the temperature was raised to 110°C and heated for 1.5 hours. At the end of the experiment, the white needle crystals fell out. The reaction mass was cooled to room temperature, and the precipitate was filtered out, several times treated with dry chloroform. It was dried under vacuum at 95–100°C to a constant mass. Yield – 5.5 g (80.4%).

8.3 RESULTS AND DISCUSSION

With the aim of preparation of effective and low-toxic hardeners for epoxide resins, we have synthesized the imide of 4-sulfoisophthalic acid (1); and anhydride of 4-sulfoisophthalic acid (2):

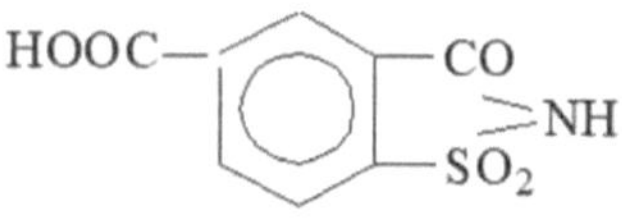

(1)

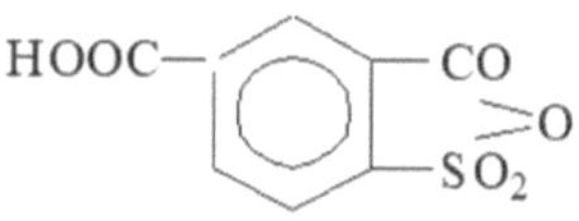

(2)

It was known that the aromatic tri- and tetracarboxylic acids and their various derivatives are widely used as the initial monomers for synthesis of thermostable polymers [7, 8]. We have developed the methods of synthesis of imide of saccharin derivatives of 4-sulfoisophthalic acid. By oxidation of m-xylene-4-sulfamide obtained on the basis of m-xylene, with alkali metal bichromate, an imide of 4-sulfoisophthalic (saccharin-5-carboxylic) acid was isolated with good yields on the scheme:

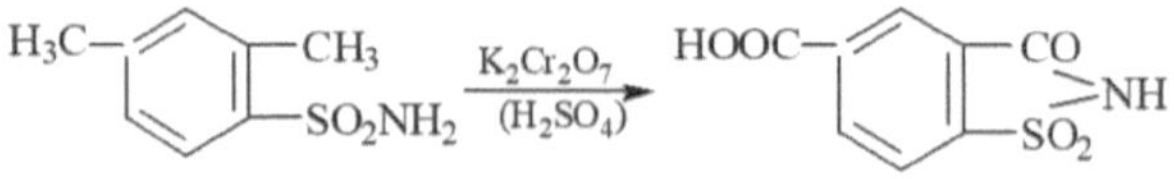

For the synthesis of monoanhydride of 4-sulfoisophthalic acid, the monoammonium salt of 4-sulfoisophthalic acid was first obtained in 10% hydrochloric acid solution, followed by heating of the obtained ammonium salt with chlorosulfonic acid on the scheme:

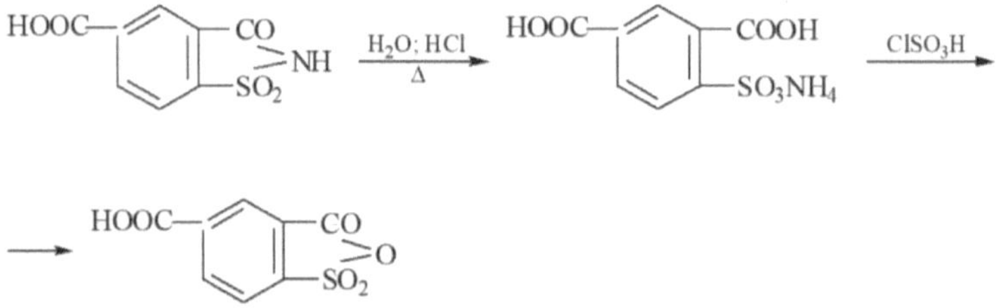

The imide and anhydride of 4-sulfoisophthalic acid have been used as a hardener of epoxide resin ED-20 (Table 8.1). The optimal quantity and curing regime for all hardeners have been found by methods of thermogravimetric and differential-thermal analysis on TGA and DTA curves [9]. According to the results of analysis, the optimal quantity of hardener-compounds (1) and (2) is 80% of the stoichiometric value. It should be noted that the compounds (1) and (2) cure the epoxide resin ED-20 at high temperature regime: 100°C/1 h + 140°C/2 h + 160°C/2 h and 160°C/1 h + 220°C/2 h + 250°C/2 h, respectively, in this case a degree of curing reaches 90% and 85%. Some kinetic parameters of curing for epoxide compounds on the basis of resin ED-20 cured with compounds (1) and (2) are presented in Table 8.2.

TABLE 8.1 Basic Characteristics of Imide (1); and Anhydride (2) of 4-Sulfoisophthalic Acid

Structural Formula	Yield (%)	M.p. (°C)	Elemental Analysis (%) Found (Calculated) (%)			
			C	H	N	S
(1) 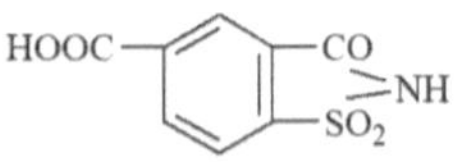	40	280	42.11 (42.29)	2.11 (2.21)	6.09 (6.16)	13.96 (14.11)
(2) 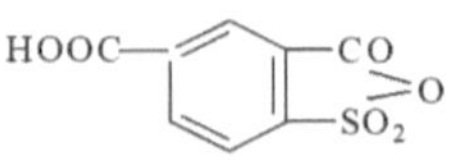	85	260–262	47.05 (42.10)	1.96 (1.75)	–	15.68 (14.03)

TABLE 8.2 Some Kinetic Parameters of Curing for Epoxide Compounds on the Basis of Resin ED-20 Cured with Compounds (1) and (2)

System	$E_{act.}$ Curing (kJ/mol)	Reaction Order	Degree of Curing (%)
ED-20 + Comp. (1) + UP 606/2	97.31	1.7	95
ED-20 + Comp. (2) + UP 606/2	89.11	1.69	95

For decrease of the curing temperature, a widespread accelerator UP 606/2 (2.4.6-tridimethylaminomethyl phenol) has been tested. It has been detected that UP 606/2 accelerator is technological and economical (1 mass p.) and decreases the curing temperature of the composition with the compound (1) to 120°C, and with the compound (2) to 150°C.

The developed composition materials have thermal stability, heat resistance, as well as thermogravimetric indices (TGI), allowing to make a conclusion about the temperature, at which the composition material does not lose its exploitation properties for 20,000 h. The data obtained during the study process are shown in Table 8.3.

TABLE 8.3 Thermogravimetric Indices and Heat-Resistance of the Developed Composition Materials

System	TGI (°C)	Vicat Heat Resistance (C)	Thermal Stability (°C)
ED-20 + Comp. (1) + UP 606/2	135	145	240
ED-20 + Comp. (2) + UP 606/2	130	140	230

The analysis of the physical-mechanical properties of epoxide compositions cured with compounds (1) and (2), before and after thermal aging showed that they are characterized by good physical-mechanical characteristics, which almost do not change after thermal aging at 120°C for 20,000 hours. The obtained results are presented in Table 8.4.

TABLE 8.4 Some Physical-Mechanical Properties of Cured Epoxide Compositions on the Basis of ED-20 Before and After Thermal Aging

Index	Hardener		After Thermal Aging at 120°C/20,000 h		PEPA	MA
	Comp. 1	Comp. 2	Comp. 1	Comp. 2		
Breaking stress (MPa):						
at tension	75	67	42	38	65	70

TABLE 8.4 *(Continued)*

Index	Hardener		After Thermal Aging at 120°C/20,000 h		PEPA	MA
	Comp. 1	Comp. 2	Comp. 1	Comp. 2		
at compression	120	130	100	88	110	120
at bending	102	99	100	90	80	90
Specific elongation (%)	2.2	1.5	1.0	1.0	2.0	1.5

Thus, the synthesized new imide of 4-sulfoisophthalic acid (1); and anhydride (2) allows to expand the assortment of hardeners for epoxide resins and to obtain new composition materials with improved exploitation properties [10].

8.4 CONCLUSION

For the preparation of the effective and low-toxic hardeners of epoxide resins, the imide and anhydride of 4-sulfoisophthalic acid have been synthesized. These compounds have been used as a hardener of epoxide resin ED-20. The thermal properties of the obtained polymer compositions are determined. It has been established that they possess thermal stability, heat resistance, as well as high values of the thermogravimetric index (TGI).

The obtained compositions can be used in the electro-technical industry for coatings of the copper wires of mounting wires, electro-technical contacts, and also in potting compounds.

KEYWORDS

- **4-sulfoisophthalic acid**
- **DTA and TGA methods**
- **ED-20 epoxide**
- **epoxide resins**
- **epoxypolymers**
- **tetracarboxylic acids**
- **thermogravimetric index**

REFERENCES

1. Vorobyev, A., (2003). Epoxide resins. *Components and technologies,* p. 8.
2. Prokopchuk, N. R., Yu. Klyuev, A., Kozlov, N. G., & Latyshevich, I. A., (2016). *Use of Epoxide Resins in Thermo-Cured Compositions* (Vol. 4, pp. 87–99). Trudy BGTU.
3. Gulay, O. I., (2001). In: Gulay, O. I., & Serednitskiy, Ya. A., (eds.), *Properties of Composition Materials on the Basis of Organosilicon Lacquer KO-921 with Structured Epoxide Resin ED-20* (Vol. 12, pp. 21–23). Plasticheskiye massy.
4. Sumenkova, O. D., (2001). In: Sumenkova, O. D., Osipchik, V. S., Lebedeva, E. D., Kononova, O. A., (eds.), *Influence of Fillers on Curing Processes and Properties of ED-20* (Vol. 12, pp. 35–37). Plasticheskiye massy.
5. Amirova, L. M., Sayfutdinov, A. F., Maksumova, A. F., & Amirov, R. R., (2001). *Russian J. Appl. Chem., 74*(11), 1881, 1882.
6. Aslanov, T. A., & Demyannik, N. Ya., (1991). *Abstracts of Conference "Polymer-91"* (p. 122). Baku.
7. Bessonov, M. I., Koton, M. M., Kudryavtsev, V. V., & Layus, L. A., (1988). *Polyimides – Class of Thermostable Polymers* (p. 328). M.: Nauka.
8. Irzhak, V. I., & Rozenberg, B. A., (1979). *Lattice Polymers* (p. 248). M.: Nauka.
9. Kurenkov, V. F., (1990). *Workshop on Chemistry and Physics of Polymers* (p. 299). M.: Nauka.
10. Dickinson, P. R., & Sung, C. S. P., (1992). Kinetics and mechanisms of thermal imidization studies by UV–visible and fluorescence spectroscopic techniques. *Macromolecules, 25*(14), 3758–3768.

CHAPTER 9

Synthesis and Use of Oligonaphthylamines in Making of Heat-Resistant Electro-Conductive Rubbers

G. N. ABASZADE, R. A. AKHMEDOVA, CH. O. ISMAILOVA, and B. A. MAMEDOV

Institute of Polymer Materials, Azerbaijan National Academy of Sciences, Sumgait, Azerbaijan

ABSTRACT

By oxidative polycondensation of 1-naphthylamine in the presence of hydrogen peroxide, the soluble and meltable polyfunctional polyconjugated oligomers, consisting of aminonaphthylene links and showing the thermostable, semiconductive, and paramagnetic properties, and also high reactivity in the reactions characteristic for aromatic amine groups have been obtained. The composition and structure of the synthesized samples of oligonaphthylamine (ONA) have been established by the methods of elemental, chemical, and IR spectral analyzes, and the molecular weight (MW) parameters – by gel-penetrating chromatography ($\bar{M}_w$ = 1100 ÷ 1610 and $\bar{M}_n$ = 670 ÷ 940).

The synthesized ONAs have been used as an active additive for the preparation of rubber mixtures on the basis of butyl (BR) rubber. ONA samples show the high electrical conductivity, and their joint use with electro-conductive carbon black allows to obtain the resins with $\sigma_v \sim 10^{-8} \div 10^{-6}$ Om^{-1} cm^{-1}. The content growth of ONA from 22.5 to 45.0 m.p. (from mass of BR) instead of carbon black leads to an increase of specific electro-conductivity of the obtained rubbers. The percolation effect is reached at a content of ~23.6 m.p. of ONA for rubbers obtained from BR. ONA can act

Advanced Polymer Structures: Chemistry for Engineering Applications. Omar Mukbaniani, DSc, Tamara Tatrishvili, PhD & Marc Jean M. Abadie, DSc (Eds.)

as an antioxidant in the composition of composites, thereby increasing the period of their effective exploitation.

9.1 INTRODUCTION

Various types of electro-conductive or antistatic rubbers are widely used in various fields of technique and industry [1]. The preparation of such rubbers is possible in the use of electro-conductive fillers, since thermoplasts, thermoelastoplasts, and rubbers are good electrical insulators. The metal powders, various marks of carbon black and powdered graphite are usually used as such fillers [2, 3]. At the same time, the use of electro-conductive fillers of organic nature is more advisable. Such additives give to rubbers, besides electrical conductivity, also valuable physical-mechanical properties due to the good compatibility of the components.

For the achievement of high electrical conductivity, the particles of electro-conductive fillers must form continuous electro-conductive structures. In the use of metal powders, the polymer composition materials do not have the necessary elastic properties, since the polymers are incompatible with metal particles. As a result, it occurs a rapid destruction of the electro-conductive structures due to the deformation of the products during exploitation. Due to the good compatibility of the components, the use of electro-conductive additives of an organic nature, especially oligomer and polymer compounds, allows to obtain electro-conductive composition materials with good exploitation properties [4–7].

The polyfunctional aromatic polyconjugated oligomers attract attention as electro-conductive additives. It should also be noted that such oligomers also have thermal- and radiation resistance, paramagnetism, semiconductivity, stabilizing and antistatic activity, solubility, meltability, and high reactivity in various chemical conversions [5–10].

Taking this into account, we have synthesized the polyfunctional oligomers of 1-naphthylamine, including reactive amine groups in each elementary unit. Oligonaphthylamines (ONAs) have been synthesized by the oxidative polycondensation of 1-naphthylamine in the presence of hydrogen peroxide.

9.2 EXPERIMENTAL PART

The synthesis of 1-naphthylamine (ONA) oligomers was carried out in a three-neck flask with a volume of 250 ml, equipped with a thermometer, a mechanical stirrer and a reflux condenser. About 14.2 g (0.1 mol) of naphthol, 34 g of 30% solution of hydrogen peroxide (0.3 mol of H_2O_2) and 85 ml of distilled water were placed in the flask. The oxidative polycondensation reaction was carried out at 363 K for 4 h at intensive stirring of the reaction mixture. The synthesized ONA samples have been purified from residue of monomer with washing of hot distilled water and dried at 373 K in vacuum cupboard (13.3 Pa) to constant mass.

The IR spectra of ONA were taken for thin films of 1-naphthylamine oligomers applied on NaCl monocrystals on Specord M-80 spectrometer Aglient Cary 630 of firm "Aglient Technologies." The UV spectra of ONA samples were taken in ethanol using UV-Specord M-80 of Aglient Cary 630 FTIR of firm "Aglient Technologies."

The molecular weights (MW) and molecular weight distribution (MWD) of oligomers was determined on gel-chromatograph "Waters" (refractometric detector) [10]. Three styrene-gel columns with a porosity of 200, 500, and 1,000 Å were used. Eluent – tetrahydrofuran. Eluent feeding rate – 1.0–1.1 ml/min. The sample was injected for one minute as 0.2% solution of ONA in tetrahydrofuran. For the calculation, it was used the calibration dependence, which is described by the equation V_R = 30.2–3.8 lgM, where V_R is the retention volume, ml; M – molecular weight. The average molecular weights were calculated on formulas:

$$\bar{M}_w = \Sigma W_i M_i \text{ and } \bar{M}_n = \frac{1}{\Sigma W_i / M_i}$$

where; W_1 is the mass portion of fractions with molecular weight M_i (it was determined as a ratio of the area of the ith part of the chromatogram to all area).

The electrical measurements have been carried out on generally accepted method at direct current by means of 3–16 amplifier, and at alternating current by means of R-571 bridge – at low frequencies and Q meter – at high frequencies ($5\times10^4 - 3\times10^7$ Hz). A rubber mixture on the basis of butyl rubber (BR) and synthesized oligonaphtholamine were made using laboratory rollers. The rubber was firstly plasticized on the rollers, and then the necessary components, including the oligomer compound, were added in a certain sequence and mixed. Depending on the peculiarities of the components of the mixture, the temperature of the rollers was adjusted in the front roller in the range of 305–315 K, in the final roller in the range of 345–348 K.

Following the sequence of introduction of the components, the preparation of the rubber mixture was carried out as follows: (1) BR – 0; stearin tech. – 3 min; captax – 5 min, thiuram – 7 min, zinc oxide + oligomer – 10 min; sulfur – 15 min, total mixing time – 20 min, vulcanization temperature – 313–323 K. (2) Then, by means of hydraulic press PG-63, the obtained mixtures were pressed in special press-molds as plates with a thickness of 1.5–2.0 mm. Using standard knives, the samples of the necessary forms and sizes were cut from the obtained plates for determination of the physical-mechanical (tensile strength, specific elongation, and residual deformation), electrical and other properties.

The strength and deformation properties of the samples were determined by means of tearing machine RM-250.

The electrical resistance of the sample section (R) in Ohm was calculated on the formula:

$$R = \frac{V}{I}$$

where; V is the voltage value in the sample area between voltage electrodes, measured by an electrometer; I is the current passing through the sample.

Specific volume electrical resistance – ρ_v (Ohm cmм) was calculated on formula:

$$\rho_v = \frac{R_{cp} \cdot h \cdot b}{1}$$

where; R_{av} is the average arithmetic value of the electrical resistance of the sample (Ohm); 'h' is the sample thickness (cm); 'b' is the sample width (cm); and 'l' is the distance between voltage electrodes (cm).

Specific volume electro-conductivity (σ_v), $Ohm^{-1}\ cm^{-1}$, was calculated on formula:

$$\sigma_V = \frac{1}{p_V}$$

9.3 RESULTS AND DISCUSSION

The samples of oligonaphtylamine (ONA) were synthesized by the oxidative polycondensation reaction of naphthols in the presence of hydrogen peroxide. It has been established that 1-naphthylamine in the presence of aqueous solution of H_2O_2 in the temperature range of 343÷368 K undergoes the oxidative polycondensation reaction and forms the oligomer products with yield ~25÷85% (on monomer) depending on synthesis conditions. It is seen from Table 9.1 that with temperature growth from 343 K to 368 K the yield of oligomer products is increased from ~ 25% to 85%. A growth of the reaction duration to 5 h and the ratio of H_2O_2: NA (mol) in the range of 1.0÷3.0 also shows a noticeable positive influence on the yield of oligomer products. However, the basic mass of oligomer products is formed for 2–3 h. According to the results, the optimal condition of carrying out of the process is as follows: $[NA]_o$ = 1.5 mol/l, $[H_2O_2]_o$ = 3.0 mol/l, T=363 K, τ=4 h. Under optimal conditions, the yield of oligomer products reaches up to 93.5%.

The results of elemental analysis of ONA samples obtained in the various conditions of 1-naphtylamine are practically identical. However, the values of molecular-weight indices of ONA samples synthesized under various conditions are differed; with growth of the reaction temperature and the oxidizer content, these indices are noticeable (Table 9.2).

The IR spectra of ONA samples, synthesized under various conditions are not practically differed. In these spectra, the absorption bands of >NH groups (3200 ÷ 3440 cm^{-1} intensive and wide-valence vibrations, 1,286 cm^{-1} – deformation vibrations), of naphthalene ring (1,592, 1,572, 1,512, and 1,458 cm^{-1}), isolated aromatic CH groups (786 cm^{-1} – out-of-plane deformation vibrations) have been fixed. Thus, on the basis of the results of elemental and IR spectroscopic analyzes, it can be concluded that the macromolecules of the synthesized samples consist mainly of the following structural units.

TABLE 9.1 Conditions of Carrying Out and Yields of Oligomer Products of the Oxidative Polycondensation Reaction of 1-Naphtylamine in the Presence of H_2O_2

$[NA]_o$ (mol/l)	$[H_2O_2]_o$ (mol/l)	T (K)	τ (h)	Oligomer Yield (%)
1.5	1.5	343	4	25.1
1.5	1.5	353	4	54.3
1.5	1.5	363	4	71.5
1.5	1.5	368	4	84.9
1.5	1.5	363	1	26.2
1.5	1.5	363	2	50.3
1.5	1.5	363	3	65.2
1.5	1.5	363	5	74.2
1.5	1.5	363	8	76.3
1.0	1.5	363	4	81.8
1.0	2.0	363	4	84.5
1.0	3.0	363	4	85.0

TABLE 9.2 Some Indices of ONA Samples Synthesized Under Various Conditions (τ =5 h)

$[NA]_0$ (mol/l)	$[H_2O_2]_0$ (mol/l)	T (K)	Elemental Composition			Average Molecular Weight	
			C	H	N	$\bar{M}_w$	$\bar{M}_n$
1.0	1.5	343	85.11	4.97	9.92	1,100	670
1.0	1.5	353	84.58	5.12	10.30	1,300	740
1.0	2.0	363	84.91	4.87	10.22	1,460	780
1.0	3.0	363	84.52	4.75	10.73	1,610	940

It is known from the scientific literature that the aniline oligomers include the following structural units:

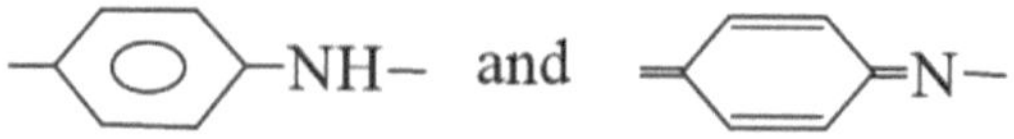

Similarly, the naphthylamine links in ONA macromolecules can exist in amine and quinoimine structural forms, the ratio of which is predetermined by the condition of the synthesis of ONA samples:

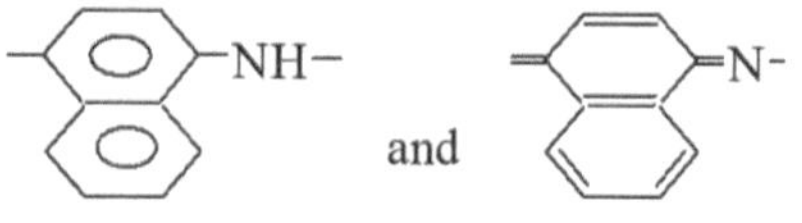

The UV spectra of the synthesized ONA samples confirm the proposed structure; a wide intensive peak of 210 ÷ 235 nm with a maximum at 220 nm, and also less intensive absorption bands with maxima at 295 and 335 nm have been fixed. The first of them is the E-band ($\pi \rightarrow \pi^*$ excitation, and the second one characterizes $n \rightarrow \pi^*$ transitions of the unshared electrons of the nitrogen atom. The last peak indicates the availability of a system of polyconjugated bonds.

The synthesized ONA samples – black powders, are well dissolved in polar organic solvents (CH_3OH, C_2H_5OH, CH_3COCH_3, DMFA, etc.). They are melted under load at 373–398 K, depending on the synthesis conditions. Their composition and structure have been established by the methods of elemental, chemical, and IR spectral analysis and molecular-weight indices – by a method of gel-permeating chromatography (Table 9.1).

ONA exhibits paramagnetic and semiconductor properties. The EPR spectra of the solid powdered samples of ONA have a singlet form ($g = 2.0023 \pm 0.0002$, $\Delta H = 0.512$ MTl), the concentration of paramagnetic centers in their composition is $(1.5 \div 4.2) \times 10^{17}$ spin/cm^3 depending on the synthesis condition. Thus, ONA can be attributed to the high-resistance polymer semiconductors; the values of the specific volume electrical conductivity of the ONA samples are $\sigma = \sim 10^{-12} \div 10^{-13}$ Ohm^{-1} cm^{-1}.

Consequently, these oligomers are of interest as an electro-conductive additive, but can also act as an antioxidant in the composition of composites and thereby increase their heat- and thermal stability, and also a lifetime of their effective exploitation.

The synthesized ONA samples have been used as an active electro-conductive additive for the preparation of rubber mixtures on the basis of rubbers, for example, BR. In this case, the rubber mixtures on the basis of BR have been prepared according to the standard receipt of ingredients with the only difference that instead of carbon black (partially) the ONA samples (from 20.0 to 45 m.p. per 100 m.p. of rubber) have been used (Table 9.3).

It has been established that the introduction of ONA instead of carbon black into the composition of the rubber mixtures leads to an increase of the ultimate strength, relative elongation and to a decrease of the modulus of elasticity of the obtained rubbers. For example, for rubbers obtained by vulcanization of a mixture on the basis of BR, including 20 mass p. of ONA instead of carbon black, the ultimate strength is increased to 22.6÷25.3 MPa, the specific elongation reaches 593.6÷670.2%, and the modulus at elongation by 200% is decreased from 9.4÷9.8 to 8.1÷8.4 MPa.

TABLE 9.3 Compositions of Rubber Mixtures and Physical-Mechanical and Electrical Indices of the Obtained Vulcanizates on the Basis of Butyl Rubber

SL. No.	Ingredients	Mixtures (mass, g)			
		1	2	3	4
1.	Butyl rubber	100.0	100.0	100.0	100.0
2.	Technical stearin	3.0	3.0	3.0	3.0
3.	Captax	0.65	0.65	0.65	0.65
4.	DFH	1.3	1.3	1.3	1.3
5.	Zinc oxide	5.0	5.0	5.0	5.0
6.	Carbon black	50.0	30.0	15.0	5.0
7.	ONA	0	20.0	35.0	45.0
8.	Sulfur	2.0	2.0	2.0	2.0

Mixtures	Vulcanization Time (min.)	Properties: Tensile Strength (MPa)	Specific Elongation (%)	Residual Deformation (%)	Elongation Module (MPa) 200%	Elongation Module (MPa) 300%	$\sigma \times 10^8$ (Ohm^{-1} cm^{-1})
1.	40	19.5	710.0	4.2	9.4	14.1	–
	60	18.3	700.0	4.5	9.0	13.2	–
	80	19.1	663.3	4.3	9.8	14.3	–
2.	40	22.6	670.2	7.5	8.1	12.0	0.18
	60	25.0	662.7	7.0	7.9	11.4	0.35
	80	25.3	593.6	6.6	8.4	12.4	0.42
3.	40	22.0	683.7	9.7	6.2	10.2	2.4
	60	22.9	652.4	11.5	6.7	10.6	6.6
	80	23.7	633.2	12.6	7.1	11.7	8.2
4.	40	25.3	675.1	11.3	6.0	8.6	28
	60	27.6	654.4	12.9	6.3	9.8	45
	80	29.8	588.2	14.2	6.8	10.3	63

Along with this, the thermal stability and lifetime of the obtained rubbers are increased, which has been apparently connected with the structural peculiarity of ONA; the condensed aromatic structural fragments in the aromatic polyconjugation chain stipulate the high thermal stability, and an availability of amine groups in the naphthalene rings – antioxidant activity. Since ONA samples exhibit high electrical conductivity, their joint use with electro-conductive carbon black allows to obtain the rubbers with a specific volume conductivity 10^{-8}–10^{-6} Ohm^{-1} cm^{-1}. The growth of ONA content from 20 to 45.0 m.p. (from mass of rubber) instead of carbon black leads to an increase of specific electro-conductivity of the obtained rubbers. The percolation effect is reached at a content ~23.6 m.p.

The reinforcing properties of ONA in the composition of the rubber composite have been probably stipulated by the optimal combination of such indices as small particle size, low density and good compatibility of components and participation of amine groups in the formation of the spatial grid.

Thus, the accumulation of the static electric charges on the surface of rubber-technical products with the use of the developed rubbers during their exploitation is minimized.

9.4 CONCLUSIONS

1. By oxidative polycondensation of 1-naphthylamine, the polyfunctional polyconjugated soluble and meltable oligomers, including the corresponding naphthylamine links with high reactivity in the reactions characteristic for aromatic amine groups have been obtained.
2. It has been shown that in partial substitution of carbon black by an oligomer of 1-naphthylamine in the composition of the vulcanizate from BR, the obtained rubbers exhibit high heat-physical, physical-mechanical, and electrical properties.

KEYWORDS

- **aminonaphthylene**
- **butyl rubber**
- **molecular weights**
- **oligonaphthylamine**
- **oxidative polycondensation**
- **rubber mixtures**

REFERENCES

1. Herbert, N., (2002). Polymers, electrically conducting. In: *Ullman's Encyclopedia of Industrial Chemistry (*p. 429).
2. Kaverinskiy, V. S., & Smekhov, F. M., (1990). *Electrical Properties of Paint and Varnish Materials and Coatings* (p. 158). M.: Khimiya.
3. Valipour, A. Ya., Modhaddam, P. N., & Mamedov, B. A., (2012). *Life Sci. J., 9*(4), 409–421.
4. Valipour, A. Ya., Modhaddam, P. N., & Mamedov, B. A., (2012). *Archives De Science Journal, 65*(7), 14–20. Switzerland, Geneva.
5. Pud, A., Ogurtsov, N., Korzhenko, A., & Shapoval, Q., (2003). *Prog. Polym. Sci., 28,* 1701–1758.
6. Ragimov, A. V., Mamedov, B. A., & Gasanova, S. G., (1997). *Polymer International, 43*(4), 343–348.
7. Mammadov, B. A., Ahmadova, R. A., Mashayeva, S. S., et al., (2014). *European Applied Sciences,* 147–153.
8. Tefera, M., Geto, A., Tessema, M., Admassie, S., et al., (2016). *J. (Elsevier) Chemistry, 210,* 156–162.
9. Food Bhandari, H., Srivastav, R., Choudhary, V., & Dhawan, S. K., (2010). *J. Thin Solid Films, 519*(3), 1031–1039.
10. Doğan, F., Şirin, K., Kolcu, F., & Kaya, İ., (2018). *Journal of Electrostatics, 94,* 85–93.

CHAPTER 10

Regularities of Epoxidation of the Cotton Oil Under Conditions of Conjugated Oxidation with Hydrogen Peroxide in the Presence of Propane Acid and Chlorinated Cationite Ku-2×8

M. SH. GURBANOV, T. I. ALKHAZOV, S. A. RZAYEVA, and A. A. SALIMOVA

Institute of Polymer Materials, Azerbaijan National Academy of Sciences, Sumgait, Azerbaijan

ABSTRACT

The regularities of epoxidation of the cotton oil (CO) in the presence of H_2O_2, propionic acid, and chlorinated cationite KU-2×8 as a catalyst have been investigated. It has been shown that this process involves the oxidation reaction of propane acid (PA) to peroxypropane acid and interaction of CO with peroxypropane acid. The reaction has a first-order concentration of PA and H_2O_2, and the chlorinated cationite accelerates only the formation of peroxypropane acid. The influence of temperature, mixing intensity of the reaction mixture, quantity of the catalyst, and ratio of the reacting components on the epoxidation reaction rate and quality of the purposeful product has been studied. Some kinetic and activation parameters of the epoxidation reaction of the CO have been determined.

Advanced Polymer Structures: Chemistry for Engineering Applications. Omar Mukbaniani, DSc, Tamara Tatrishvili, PhD & Marc Jean M. Abadie, DSc (Eds.)

10.1 INTRODUCTION

The epoxidized unsaturated oils are valuable products in the paint and varnish industry and are used as a plasticizer-stabilizer for polyvinyl chloride [1]. The epoxidation of unsaturated compounds, including higher unsaturated acids is carried out by peroxyacids in a solution of organic acids [2–5]. It has been shown [3–5] that this process can be carried out by organic peroxyacid at the time of its formation from the corresponding carboxylic acid and hydrogen peroxide in the presence of inorganic (H_2SO_4, HNO_3, H_3PO_4) and organic (n-toluene sulfoacid, trichloroethane acid) acids, salts of transition metal [cerium (IV) and zirconium (IV) sulfates], KU-2 cationite and Mn (IV) complexes. The strong water-soluble acids show the high activity in the formation of peroxyacids, but the carrying out of conjugated oxidation of carboxylic acids and epoxidation of unsaturated compounds in the presence of such catalysts is ineffective due to the difficulty of isolation of the catalyst from the end products of the reaction. The epoxidation of unsaturated compounds with peroxyacids is also inadvisable due to the explosive nature of the preparation process, isolation of peroxyacids and two-stage nature of the process. The most advantageous is the use of ion-exchange catalysts both in the synthesis reactions of peroxyacids [6, 7], and for carrying out of the epoxidation reaction of unsaturated compounds with peroxyacids at the time of their formation [8, 9].

In this chapter, the results of investigation of regularities of the epoxidation reaction of the cotton oil (CO) by peroxypropane acid at the time of its formation in the presence of chlorinated cationite KU-2×8 (CCU) have been presented.

10.2 EXPERIMENTAL PART

The experiments were carried out in a three-neck flask with a volume of 250 ml, provided with a hydraulic seal, mechanical stirrer and reflux condenser. The certain quantities of propane acid (PA), hydrogen peroxide, CCU, and a solution of CO in toluene were placed in the flask at a given temperature. Then the stirrer was activated, and the reaction was followed the reaction course by changing the content of hydrogen peroxide and peroxypropionic acid (PPA) by titration. After consumption of necessary quantity of hydrogen peroxide, the reaction was completed and the reaction mass was transferred to the separating funnel, where the lower water layer was separated from the upper organic layer.

The content of H_2O_2, PPA, and PA was determined in the water layer. A solvent was added to the organic layer and washed with warm distilled water until complete absence of carboxylic acid and hydrogen peroxide in it. Toluene was distilled from the washed solution of epoxidized oil in toluene (at 323–328 K and 13.3 kPa) and the purposeful product was isolated. Then, the physical-chemical indices of epoxidized cotton oil (ECO) were determined.

CCU with a chlorine content 9.5 mass % and an exchange capacity of 3.54 mg-eqv g^{-1} used as the catalyst were obtained by chlorination of KU-2×8 cationite in the presence of $HCl + H_2O_2$ system [10], the efficiency and stability of which in the synthesis of peroxyacids are shown in work [11].

During carrying out of the experiments, 30% aqueous solution of H_2O_2, PA of the "pure" mark and CO with an iodine number 95.0 were used. The content of the epoxide groups and iodine number was determined according to the method described in work [12], and the content of H_2O_2 and PPA – on methodology [13].

10.3 RESULTS AND DISCUSSION

The conjugated oxidation of CO in the presence of PA, H_2O_2, and CCU is a sequential, two-stage process. In the first stage, PPA is formed from PA and H_2O_2 in the presence of CCU, in the second stage of the CO process, PPA is oxidized on its formation. Although the first stage of the process is reversible [14], however, in view of the consumption of peroxyacid for epoxidation on formation of epoxy product, all process becomes irreversible. The best results on epoxidation can be achieved on condition that the second stage rate of the process is equal to or greater than the first stage rate [14], since in this case the accumulation of peroxyacid is excluded. Consequently, the change of the reaction parameters should be carried out in such ranges, in which the optimal behavior of the first and second stages, and also the high quality of the purposeful product is simultaneously provided.

With this purpose, the influence of diffusion factors, changes of temperature, quantity of CCU and molar ratio of the reacting components on the epoxidation reaction rate and quality of the purposeful product has been studied.

It has been previously established in investigation of the formation of peroxypropane acid in the H_2O_2 + PC + CCU system that the mixing intensity higher 90 rev.·min^{-1} provides the reaction behavior in the kinetic field, and a size of grains of CCU catalyst does not essentially influence on the course of the reaction, and the pore-diffusion factors also do not play a determining role [15, 16]. A similar dependence is observed in carrying out of the epoxidation reaction of the CO in the presence of H_2O_2 and CCU. Due to the availability of diffusion factors during carrying out of the reaction without mixing, the peroxypropane acid accumulates in the system. In this case, the peroxyacid almost does not contact with CO, and therefore an accumulation of the epoxy product does not occur. It has been established that the best condition for carrying out of the epoxidation reaction is the mixing intensity in the interval 180–200 rev min^{-1}. During carrying out of the reaction without mixing, the yield of epoxy products from theory is 8.1% and increases to 97% at mixing rate 180–200 rev min^{-1}. The mixing rate essentially influences also on quality of the epoxy product and process rate. The initial rate of H_2O_2 consumption increases from 0.51 to 1.23×10^{-3} mol l^{-1} s^{-1}, PPA accumulation rate is decreased from 4.41 to 1.43×10^{-4} mol l^{-1} s^{-1} (Table 10.1). Taking into account these results, further investigation was carried out at mixing intensity 180–200 rev min^{-1}.

TABLE 10.1 Influence of Mixing Intensity on the Quality of the Purposeful Product, Consumption Rate of H_2O_2 (W_1) and Accumulation of PPA (W_2)*

Mixing Intensity (rev min^{-1})	Iodine Number	Epoxide Number	Yield from Theory (%)	τ (h)	W_1×10^{-3} (mol l^{-1} s^{-1})	W_2×10^{-4} (mol l^{-1} s^{-1})
0	95.5	0.81	8.1	7.5	0.51	4.41
60–70	2.2	8.1	81	5.5	0.72	3.0
110–120	1.2	8.5	85	4.5	1.0	2.20
180–200	0.15	9.7	97	3.5	1.2	1.81
290–300	2.2	8.1	81	4.0	1.23	1.43

*Molar ratio CO:H_2O_2:PA = 1:2:2; T = 333 K, a quantity of the catalyst – 10% from mass of PA and H_2O_2.

Note:

1. The qualitative indices of the end product have been determined at reaction duration for 8 h.

2. Some process parameter values: A = 5.01×10^5 l mol^{-1} min^{-1}; E = 56.5 kJ mol^{-1}; ΔH^0 = 51.0 kJ mol^{-1}; $\Delta S^{\#}$ = –150.5 kJ mol^{-1} deg^{-1}; $\Delta G^{\#}$ = 46.1 kJ mol^{-1}.

The influence of temperature on the epoxidation reaction behavior of the CO in the conjugated system was studied in the range of 313–343 K (Figure 10.1). With an increase of temperature from 313 to 343 K, a regular increase of H_2O_2 consumption rate is observed. For example, at 343 K, half of the quantity of H_2O_2 necessary for the epoxidation of CO is consumed for 30 min., while at 313 K, 150 min. is required. It can be seen from curves (Figure 10.1(b)) that a ratio of the formation rate and consumption of PPA is noticeably changed at all studied temperatures for 30–60 min., and in the beginning, PPA formation rate prevails over its consumption rate. At the end of this period, the concentration of peroxyacid is slightly changed, and the complex character of the kinetic curves obtained at various temperatures has been stipulated with the difference in the activation energy values of the formation of peroxyacid and its consumption in the epoxidation reaction.

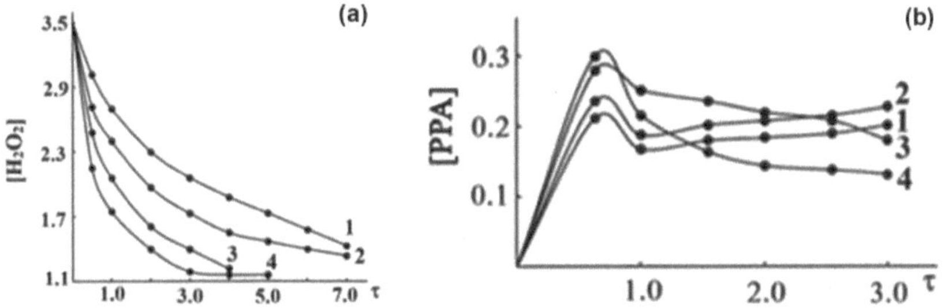

FIGURE 10.1 Kinetic curves of consumption of H_2O_2 (mol l^{-1}) (a); and accumulation of PPA (mol l^{-1}) (b) at various temperatures in the epoxidation reaction of the cotton oil in the conjugated system in the presence of CCU.

Note: Quantity of CCU – 10% from mass of H_2O_2 and PA, molar ratio $[CO]_0:[H_2O_2]_0:[C_2H_5OOH]_0 = 1:2:2$. τ is the time (h); also for Figures 10.2 and 10.3. T(K): (1) 313; (2) 323; (3) 333; and (4) 343.

It has been established that the highest yield of the epoxy product (~97% from theory) is provided at 323–333 K. Although the high temperature leads to a decrease of the process duration, however, in this case the epoxide number (91% from theory at 343 K) of the purposeful product is also decreased, which has been connected with the partial opening of the epoxide ring in its composition.

The influence of a quantity of the catalyst on the yield of the purposeful product was investigated at 333 K with the change of a quantity of CCU from 5 to 20% from mass of PA and H_2O_2. It has been shown that the most optimal catalyst content is about 10%. In this case, the yield of the epoxidized product is 98% (the process duration is 3.5 hours). At the catalyst content of 5%, the reaction is completed for 6.5 h, but due to long contact with PA, it occurs the partial opening of the epoxide ring, which leads to a decrease of the epoxy product yield to 72%. With further increase of the quantity of the catalyst, the process duration is decreased in 3 times, but in this case, the opening of the epoxide ring of the purposeful product due to the high content of sulfoacidic groups of the catalyst also occurs. The epoxidation reaction rate grows from 0.67×10^{-4} to 2.55×10^{-4} mol l^{-1} s^{-1} with an increase of a quantity of CCU from 5 to 20%. The effective consumption of H_2O_2 is provided in the presence of 10% of the catalyst (Figure 10.2(a)). In the presence of 20% of the catalyst, a considerable accumulation of PPA in the reaction medium is observed (Figure 10.2(b)). In view of the fact that CCU catalyzes only the stage of formation peroxyacid and does not influence on rate of the epoxidation stage, an increase of rate of the first stage of the process due to an increase of a quantity of the catalyst leads to an increase of the concentration of PPA in the reaction mixture.

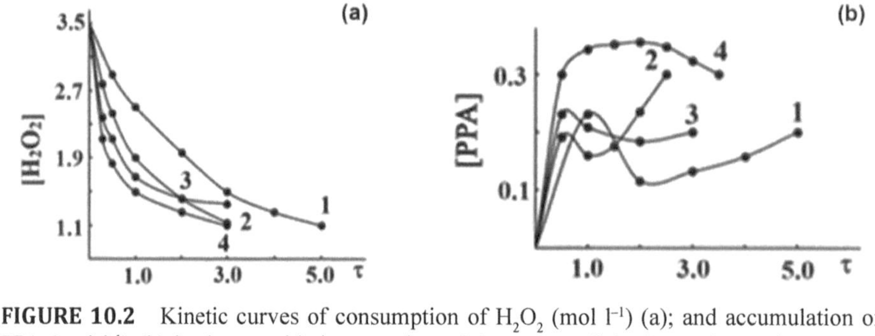

FIGURE 10.2 Kinetic curves of consumption of H_2O_2 (mol l^{-1}) (a); and accumulation of PPA (mol l^{-1}) (b) in the epoxidation reaction of the cotton oil in the conjugated system at various quantities of CCU.

Note: $[CO]_0:[H_2O_2]_0:[PA]_0 = 2:2:1$; τ – 333 K. Quantity of CCU (from mass of H_2O_2 and PA): (1) 5; (2) 10; (3) 15; and (4) 20.

The efficiency of this process depends mainly on the degree of use of H_2O_2. Due to this, the influence of a quantity of H_2O_2 on the yield of epoxidized CO was studied. The investigations showed that during carrying out of the reaction in the presence of an equimolar quantity of hydrogen peroxide (per CO), the process duration is considerably increased (Figure 10.3(a)). In twofold excess of H_2O_2, the epoxidation is completed for 4.0 hours. However, in this case, almost half of H_2O_2 taken in the reaction remains in the system and its isolation from the reaction zone is almost impossible. The highest yield of the purposeful product (97%) is provided precisely in twofold excess of H_2O_2. The carrying out of the epoxidation reaction with the participation of stoichiometric quantity of hydrogen peroxide and oil favors the complete consumption of H_2O_2, although in this case, the reaction duration is increased to 7 h, and the yield of the purposeful product is decreased from 97.0 to 71.0%. At the same time, the character of curves of PPA accumulation is also considerably changed (Figure 10.3(b)), the content of which in the reaction mass reaches a maximum (0.19 mol l^{-1}) in twofold excess of H_2O_2. The initial consumption rate of H_2O_2 is decreased with an increase of a quantity of the latter one from 1.0 to 2.0 mol, the accumulation rate of PPA is increased from 3.66×10^{-3} to 5.00×10^{-3} mol l^{-1} min^{-1}.

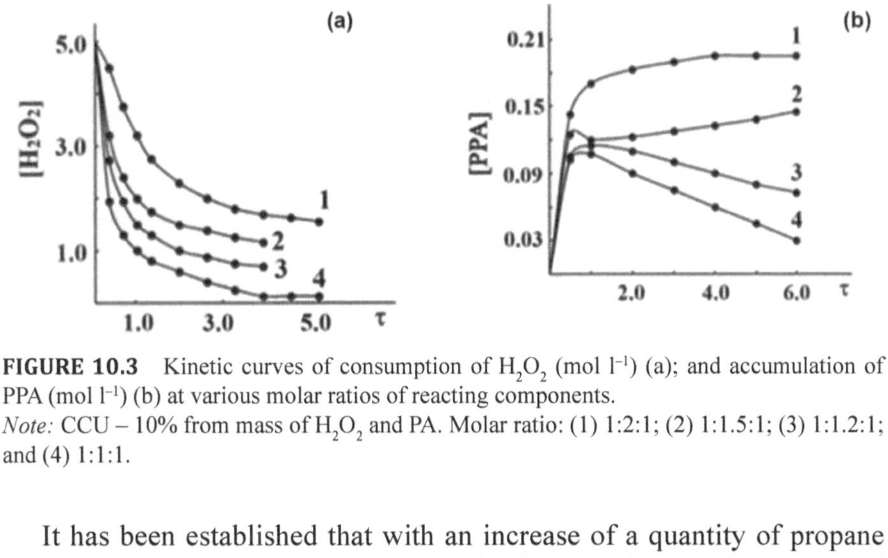

FIGURE 10.3 Kinetic curves of consumption of H_2O_2 (mol l^{-1}) (a); and accumulation of PPA (mol l^{-1}) (b) at various molar ratios of reacting components.
Note: CCU – 10% from mass of H_2O_2 and PA. Molar ratio: (1) 1:2:1; (2) 1:1.5:1; (3) 1:1.2:1; and (4) 1:1:1.

It has been established that with an increase of a quantity of propane acid (PA) from 0.1 to 0.6 mol, the epoxidation reaction rate increases and the process duration is decreased from 7.0 to 2.0 h. It also follows from the data in Table 10.2 that an increase of the molar ratio of C_2H_5COOH/H_2O_2 to 2.0 leads to a growth of the epoxide number and product yield. Further, the yield of the epoxidized product is decreased from 91.0 to 32.0%, which has been connected with the opening of the epoxide ring in excess of C_2H_5COOH.

TABLE 10.2 Influence of a Quantity of H_2O_2 (Tests No. 1–4) and C_2H_5COOH (Tests No. 5–7) on Yield and Quality of Epoxidized Cotton Oil*

Number of Test	Quantity of Reagents (mol)			τ (h)	Epoxide Number	Iodine Number	Yield from Theory (%)
	PA	CO	H_2O_2				
1.	2.0	1.0	1.0	7.0	7.0	2.1	70
2.	2.0	1.0	1.2	6.5	8.1	1.5	81
3.	2.0	1.0	1.5	5.0	8.7	1.2	87
4.	2.0	1.0	2.0	3.5	9.7	0.4	97
5.	1.0	1.0	2.0	3.0	9.0	1.1	90
6.	3.0	1.0	2.0	3.0	8.1	0.7	81
7.	6.0	1.0	2.0	3.0	3.1	0.2	31

*T = 333 K; quantity of CCU – 10% from mass of PA + H_2O_2.

In Table 10.1, some kinetic parameters of the conjugated oxidation of CO with H_2O_2 in the presence of PA and CCU are also presented. It can be seen that the process is characterized by sufficiently high values of the initial rate and the reaction rate constants of the first stage, noticeably low value of the activation energy and sufficiently high negative value of the activation entropy.

Thus, it can be concluded on the basis of the obtained results that the epoxidation of CO with hydrogen peroxide in the presence of PA and CCU is a complex process and involves a number of sequentially and in parallel proceeding stages. In the first stage of the process, PA is oxidized with hydrogen peroxide to PPA:

$$H_2O_2 + C_2H_5COOH \underset{\kappa_2}{\overset{\kappa_1,\ CCU}{\rightleftharpoons}} C_2H_5COOOH + H_2O \tag{1}$$

The stage of PPA formation proceeds in the aqueous phase (a.ph.), and the epoxidation stage – in the organic phase (org.ph.), consequently, PPA is distributed between these phases:

$$[C_2H_5COOOH]_{a.ph.} \rightleftharpoons [C_2H_5COOOH]_{org{\cdot}ph} \tag{2}$$

$$\bar{\gamma}_1 = \frac{[\tilde{N}_2H_5COOOH]_{org.ph.}}{[\tilde{N}_2H_5COOOH]_{a.ph.}} \tag{3}$$

Similarly, PA distribution occurs in the aqueous and organic phases:

$$[C_2H_5COOH]_{a.ph.} \rightleftarrows [C_2H_5COOH]_{org.ph.} \tag{4}$$

$$\gamma_2 = \frac{[C_2H_5COOH]_{orh.ph.}}{[\tilde{N}_2H_5COOH]_{a.ph.}} \tag{5}$$

CO epoxidation on double bond proceeds mainly in the organic phase:

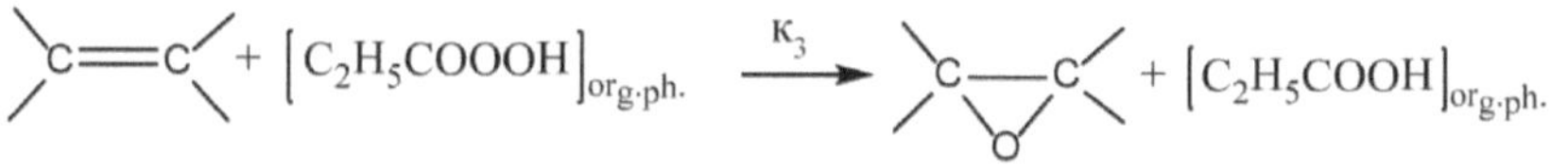

(6)

Along with these main stages, an occurrence of the side reaction is possible. PA soluble in the organic phase favors partial opening of epoxide cycle with the formation of ester:

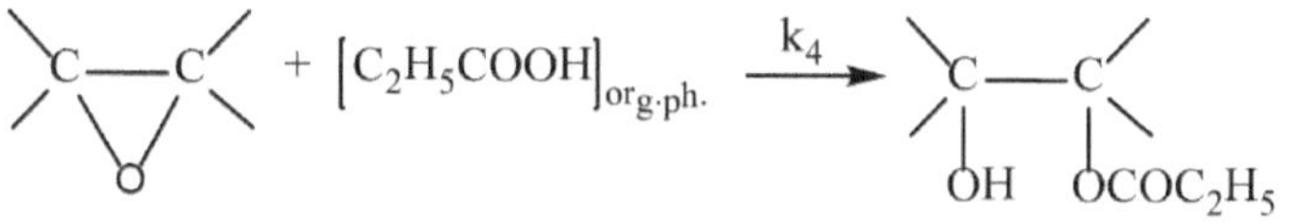

(7)

Then the mathematical description of the epoxidation process is a system of the following differential equations:

$$-\frac{d[H_2O_2]}{dt} = \{k[C_2H_5COOH][H_2O_2] - [H_2O_2] - k_2[C_2H_5COOOH]\}[CCU] \tag{8}$$

$$-\frac{d[CO]}{dt} = k_3[CO][C_2H_5COOOH]_{org.ph.} \tag{9}$$

$$\frac{d[ECO]}{dt} = k_3[CO][C_2H_5COOOH]_{a.ph.}\gamma_1 - k_4[ECO][C_2H_5COOH]_{a.ph.}\gamma_2 \tag{10}$$

$$\frac{d\left[Est.CO\right]}{dt} = \kappa_4\left[ECO\right]\left[C_2H_5COOH\right]_{a.ph.}\gamma_2 \quad (11)$$

The bond between the concentrations of substances in the reaction medium is established using the material balance:

$$[H_2O_2]_t = [H_2O_2]_o - [H_2O_2]_{consump.} \quad (12)$$

$$[CO]_t = [CO]_o - [ECO] - [Est.CO] \quad (13)$$

$$[C_2H_5COOH]_t = [C_2H_5COOH]_o - [C_2H_5COOOH] - [Est.CO] \quad (14)$$

where; $[H_2O_2]_o$ and $[H_2O_2]_t$ are the initial and current concentrations of H_2O_2; $[CO]_o$ and $[CO]_t$ are the initial and current concentrations of CO (or double bond of CO); [ECO] and [Est.CO] are the current concentrations of epoxidized CO and ester, respectively; $[C_2H_5COOH]_o$ and $[C_2H_5COOH]_t$ are the initial and current concentrations of PA.

The distribution coefficients of PA and PPA in the aqueous and organic phases at various temperatures are determined experimentally.

10.4 CONCLUSIONS

1. It has been established that the epoxidation of cotton oil (CO) in the presence of chlorinated cationite KU-2×8 in the conjugated system with H_2O_2 proceeds with sufficiently high rate.
2. In the presence of chlorinated cationite KU-2×8, the mixing intensity higher than 180 rev min^{-1} provides the reaction in the kinetic field. With the process temperature rise from 313 to 343 K, the epoxidation rate increases, but the quality of the purposeful product is deteriorated.
3. It has been shown that the chlorinated cationite KU-2×8 accelerates only the stages of formation of peroxypropane acid and does not influence on the oxidation reaction rate of the double bond of the CO and the opening of the epoxide ring of the purposeful product, which leads to an increase of the selectivity and efficiency of the process, and also the quality of the purposeful product.

KEYWORDS

- **chlorinated cationite**
- **cotton oil**
- **epoxidation**
- **epoxidized cotton oil**
- **hydrogen peroxide**
- **peroxy propionic acid**
- **peroxyacid**
- **propane acid**

REFERENCES

1. Blagonravova, A. A., & Nepomnyzshiy, A. I., (1970). *Lacquer Epoxy Resins* (p. 230). M.: Khimiya.
2. Chalabiyev, Ch. A., Guseinov, M. M., & Salakhov, M. S., (1978). ZhPKh, *51*(9), 2036–2041.
3. Akhmedyanova, R. A., Turmanov, R. A., & Kochnev, A. M., (2015). *Influence of Nature of Vegetable Oils on Their Epoxidation Process with Hydrogen Peroxide in the Presence of Peroxophosphowolframate Catalytic System, 17*(18), 25–28. Herald of technological University.
4. Kozlov, Yu. N., Mandelli, D., Voytiski, K. V., & Shilpin, G. B., (2004). ZhPKh, *78*(3), 433–457.
5. Yermakov, O. A., & Yanova, N. A., (1975). *Kinetics and Catalysis, 16*(4), 1100–1106.
6. Sadykhov, O. A., Guseinov, M. M., & Chalabiyev, Ch. A., (1984). ZhOrKh, *20*(8), 1638–1645.
7. Chalabiyev, Ch. A., Sadykhov, O. A., & Guseinov, M. M., (1983). *Kinetics and Catalysis*, *24*(6), 1400–1405.
8. Kuptsevich, O. Ya., & Peak, Z. G., (1996). ZhOrKh, *32*(6), 817–819.
9. Krylov, O. V., (2002). *Kinetics and Catalysis, 43*(2), 310–316.
10. A.c. 1728250 USSR, ICI^4 C08F8/20. Method of preparation of oxidizer-resistant cationite.
11. A.c. 1685932 USSR, ICI^4 C07C409/24. Method of preparation of aliphatic percarboxylic acids.
12. Sorokin, M. F., & Lyulyushko, K. A., (1971). *Workshop on the Chemistry and Technology of Film-Forming Substances* (p. 263). M.: Khimiya.
13. Eugeniuze, M., Kornelia, M., & Marlene, K., (2015). Technological aspects of chemoenymatic epoxidation of fatty acids, fatty acid esters and vegetable oils: A review. *Molecules, 20*, 21481–21493.
14. Gurbanov, M. Sh., Mamedov, B. A., & Chalabiyev, Ch. A., (2005). *Chemistry and Petrochemistry, 1*, 10–19.

15. Gurbanov, M. Sh., Chalabiyev, Ch. A., Mamedov, B. A., & Efendiyev, A. A., (2005). ZhPKh, *78*(10), 1707–1711.
16. Gurbanov, M. Sh., & Mamedov, B. A., (2009). ZhPKh, *82*(8), 1384–1388.

CHAPTER 11

Polyethylene Terephthalate as a Reducing Agent in High-Temperature Oxidation-Reduction Reactions

L. SHAMANAURI[1] and D. DZANASHVILI[2]

[1]*R. Dvali Institute of Machine Mechanics, 10, Mindeli Str., Tbilisi*

[2]*R. Agladze Institute of Inorganic Chemistry and Electrochemistry, Ivane Javakhishvili Tbilisi State University, Tbilisi*

ABSTRACT

In this chapter, a new technology with the use of secondary polymer-polyethylene terephthalate (PETP) for removing metallic lead from lead oxide residues is proposed instead of expensive traditional reducers (coke, natural gas, etc.). In the reaction medium carbonates Na_2CO_3 and K_2CO_3 smelts are used instead of traditional chlorides, after melting of which the harmful products for health are created. In our technology, the recovery process is going on at much lower temperatures (750–800°C), in the result of which a fairly high quality (98%) metal lead is obtained. This technological method is much more environmentally friendly, safe, and economically viable compared with traditional technologies.

11.1 INTRODUCTION

The rapid development of the world automotive industry has led to a significant growth trend in the production of lead-acid starter batteries. Lead, and

Advanced Polymer Structures: Chemistry for Engineering Applications. Omar Mukbaniani, DSc, Tamara Tatrishvili, PhD & Marc Jean M. Abadie, DSc (Eds.)

its alloys are also indispensable products for many industries. Therefore, the demand for it is growing every year around the world.

Lead production with the processing of raw materials is associated with significant technical, economic, and environmental problems (one ton of lead requires about 1,000 tons of ore processing). In the last decades, the lead content in the ore has decreased by 20–50%, which makes the processing of new ores even more unprofitable.

Thus, due to the depletion of lead natural resources, its uptake into recycled raw materials is becoming increasingly relevant. At the same time, the constant updating of motor batteries leads to the accumulation of lead and its compounds in the environment. Due to its high toxicity, lead scrap poses major environmental problems, necessitating the development of new, environmentally safe, innovative technologies for recycling lead.

Such productions are much more profitable due to low energy costs and other technological advantages: high lead content in raw materials, less conventional energy consumption per unit of finished product, relatively small volume of slag, dust, and gases generated during processing, etc.

Also major environmental problems are caused by polyethylene terephthalate waste used as packaging material and a carrier for alcohol and soft drinks, the number of which is constantly increases in the world, whose disposal is of great importance for improving the ecological condition of the environment. Currently, only a small fraction of the residual polymers are recycled and returned to production.

The aim of the work was to develop a new technology for the production of metallic lead using secondary raw materials – lead oxide and polyethylene terephthalate obtained from batteries. This would make it possible to replace the expensive regenerative coke with cheap used polyethylene terephthalate, which would have a significant economic and environmental impact.

In the literature, data on the use of carbonates as fluxes in lead recovery technology are scarce. Their use is known at relatively low temperatures of 750–800°C [1].

One of the main problems in lead production is the lead transferred to the slag, which depends on the solubility of lead oxide in the flux. By some data its concentration in the slag changes in the range 10–35% [2, 3].

For ascertainment of the amount of the lead recovered to the slag, we studied the solubility of lead oxide in the melt carbonates with different content at 750–800°C. The solubility of the lead oxide in the melt carbonates also has an important meaning for the process stability and increasing lead output.

It is known from the literature that the solubility of oxides in general, including lead oxide, depends on both the melting temperature and the melting composition [4, 5]. In particular, the concentrations of cations in the melt, which have a high charge density and significantly change the base-acid properties of the melt, are of great importance [6]. Such ions include lithium (Li +), magnesium (Mg^{2+}), and calcium (Ca^{2+}).

The aim of the experiments was to determine the dependence of the solubility of lead oxide (PbO) on the melting temperature and composition in double and triple carbonate alloys – K_2CO_3-Na_2CO_3, K_2CO_3-Na_2CO_3-$CaCO_3$ [7].

It is known that the melting temperatures of Na_2CO_3 and K_2CO_3 of the individual carbonates are 854°C and 891°C, respectively, while the decomposition temperatures are equal to 1,000 and 1,200°C. The decomposition temperature of $CaCO_3$ is relatively low – 840°C [8, 9].

Due to the fact that the melting temperature of the double and triple carbonates eutectic is 700–720°C, as the working temperature was selected 750–800°C. At this temperature, the evaporation of both carbonates and lead oxide is minimal and therefore the target product-metal lead losses are small [10].

11.2 EXPERIMENTAL SECTION

The experiments were conducted in a shaft electric furnace with a quartz screen embedded. The screen had a refractory ceramic roof with three miles in it – one to carry the weight of the lead, the other to react with the air, and the third to remove the exhaust gases from the reaction area. The construction of the roof also allowed for visual observation of the slope.

The experiments were conducted under a special vent umbrella. The temperature in the reaction area was measured with a chromium-alumel thermocouple. Carbonate salts were decomposed in corundum pans.

Carbonate salts in the amount of 50–50 g was used for K_2CO_3-Na_2CO_3 smelting and 45–45–10 g for Na_2CO_3-K_2CO_3-$CaCO_3$ smelting for the experiment.

Accordingly, the weighed salts were loaded into corundum pans placed in a shaft furnace, where the temperature was 850°C. After melting the salts the temperature was reduced to 750 or 800°C, depending on the conditions of the experiment. In the carbonate melt we had lead oxide in the form of one-gram weights, after a delay of 20–25 minutes the sample was taken for analysis.

To determine the solubility of lead recovered in melts of carbonate mixtures of different compositions, analysis of solid powders obtained by atomic absorption spectroscopy was performed after cooling the melts, for which nitric acid solutions of medium samples (approximately 0.5 g) of these powders were pre-prepared. The analysis was performed on the tool Analyst 200 (PerkinElmer).

11.3 RESULTS AND DISCUSSION

The experimental results are presented in Figure 11.1.

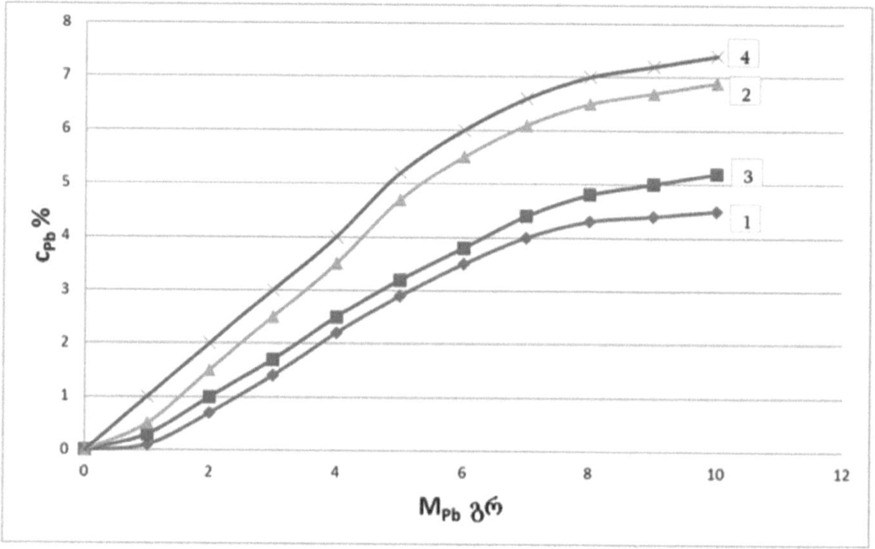

FIGURE 11.1 Dependence of lead oxide solubility on temperature and fracture composition (fracture weight MPbO = 100 g).

Note: K_2CO_3-Na_2CO3-$CaCO_3$: (1) 750°C; (3) 800°C; and K_2CO_3-$NaCO_3$: (2) 750°C; (4) 800°C.

From Figure 11.1, it is seen that the solubility of lead oxide in the flood of K and Ca at 750°C is maximum 8%, and in K_2CO_3-Na_2CO_3-$CaCO_3$ no more than 5%. As it was expected at 800°C these numbers grow insignificantly.

Decreased solubility of lead oxide in the addition of calcium carbonate to alkali metal alloys may be explained by complex physicochemical processes due to the high specific charge density of calcium ions compared with potassium and sodium ions [11]. Lead ion is likely to be bound by calcium (Ca^{2+}) ions at the expense of the formation of various compounds. Such may be calcium seals $Ca[(PbCO_3)]$, $Ca_2[(PbCO_3)]_2$, it is possible to form other mixed complexes, the solubility of which differs from the solubility of lead oxide [12]. Thermal decomposition of calcium carbonate also produces calcium rust with a melting point of 2,880°C and poorly soluble carbonate melt. The production of these substances significantly increases the melting temperature of the flux, which is undesirable for the stability of the low-temperature lead recovery process.

It is known [1, 13] that, in the case of lead oxide reduction by coke, it is pre-thermally sintered so that the reducing gases (CO, H_2) generated during coke combustion can easily penetrate between the lead oxide granules and achieve a high recovery rate.

Instead of reducing lead oxide with coke, we used secondary polymer polyethylene terephthalate in the experiments, and instead of sintering lead oxide, it was briquetted with its polyethylene terephthalate, for the following reasons:

i. Lead oxide is briquetted at low temperature (200–240°C), under minimum pressure and therefore, energy consumption is negligible in terms of unit output;
ii. The lead oxide and bonding polymer in the briquette are homogeneously distributed, which leads to a high probability of contact of the regenerating and regenerating metal in the technological process;
iii. The regenerative gases and air generated in the technological process will pass smoothly between the briquettes, which leads to a high percentage of lead recovery.

To achieve a homogeneous distribution of the reducing agent and lead oxide in the briquette, a polymer shredding technology has been developed, which involves pre-thawing the polymer at 270°C, mechanical grinding and separation. Briquetting was done by means of hydraulic pressure.

Briquettes in which the concentration of polyethylene terephthalate varied from 5 to 50% were prepared to select the optimal concentrations of the reducing and reducing metal oxide. Lead oxide was also recovered in the shaft furnace.

To recover the lead, the briquettes were placed in carbonates permeated in a corundum tub and air was blown through a compressor. The recovery process lasted for 15–20 minutes, then the corundum crucible was removed from the furnace, cooled, and the recovered lead was separated from the flue.

The obtained lead samples were analyzed to determine the quality of the lead. The results of the experiments are given in Table 11.1.

TABLE 11.1 Recovering of the Lead Oxide by PETP*

Sample N	Content of PETP (%)	Content of Lead in the Oxide Briquette (%)	Content of Lead in the Briquette (%)	Weight of Metallic Lead (gr.)	Output of the Lead (%)
1.	10	90	25.06	21.18	87.8
2.	15	85	23.67	21.35	90.2
3.	20	80	22.27	21.54	94.9
4.	25	75	20.88	19.85	95.1
5.	30	70	19.49	18.51	95.0
6.	35	65	18.09	17.18	95.0
7.	40	60	16.70	15.41	92.3
8.	45	55	15.30	13.94	91.1
9.	50	50	13.92	12.53	90.0

*The weight of briquette M = 200 gr; T = 750°C.

As shown in Table 11.1, the ratio of reducing and reducing lead oxide is possible over a wide range, although the optimal ratio is 25% polyethylene terephthalate and 75% lead oxide. The percentage yield of lead was reduced at lower or higher concentrations of regenerative polyethylene terephthalate. The reason for this is the lack of a restorer in the first case, and the excess of a restorer in the second case. As the combustion duration and temperature increase, the evaporation rate of lead oxide, metallic lead, and flux increases, and the yield of the resulting product-reduced lead decreases.

After the experiments, the flux was weighed and analyzed to determine the amount of evaporated salts and the concentration of lead oxide dissolved in the flux, the results of which are given in Table 11.2.

TABLE 11.2 Lead Oxide Content in Lead Slag After Recovery of the Oxide with Polyethylene Terephthalate in the Melt K_2CO_3–Na_2CO_3, M = 200 g, T = 750°C

Sample N	Content of Lead Oxide in the Briquette (%)	Content of PETP (%)	Content of Lead Oxide in the Briquette (%)	Weigh of Slag (gr.)
1.	10	90	3.41	190
2.	15	85	3.18	189
3.	20	80	2.25	192

TABLE 11.2 *(Continued)*

Sample N	Content of Lead Oxide in the Briquette (%)	Content of PETP (%)	Content of Lead Oxide in the Briquette (%)	Weigh of Slag (gr.)
4.	25	75	1.98	191
5.	30	70	2.00	188
6.	35	65	1.95	185
7.	40	60	1.95	186
8.	45	55	1.8	185
9.	50	50	1.7	183

From Table 11.2, it is seen that the content of the lead oxide in the slag after reduction of the oxide with polyethylene terephthalate, the percentage is the lowest in the following ratio of reducing and reducing lead oxide: 25% polyethylene terephthalate and 75% lead oxide, which also has the highest lead yield.

During the experiments, the quality of the recovered lead was checked by spectral analysis using the method of atomic absolute spectroscopy, according to which it can be said that the quality of the recovered lead is quite high and the percentage of pure lead in it is 97.01% to 97.9%, which is quite high.

Experiments have confirmed the possibility of using carbonate alloys both individually and in mixed form to reduce lead oxide, although potassium and sodium carbonate alloys should be used in flux to conduct the process at relatively low temperatures (750–800°C). Carbonate smelting, which performs the function of flux, at the same time shields lead ions from oxygen, reduces the evaporation of lead and its compounds, which is important both for improving the environmental condition and for the health of production personnel.

11.4 CONCLUSIONS

Studies have shown that the use of polyethylene terephthalate as a restorative agent in the absence of metallic lead has a number of advantages over coke recovery:

1. Lead dioxide in other known technologies must be converted into granules before recovery so that the carbon dioxide obtained by

burning coke can easily penetrate between the granules (fragments). The yield of metallic lead does not even exceed 85–87%; In our case, permeable polyethylene terephthalate is used for briquetting of lead dioxide, which is in direct contact and evenly distributed between the lead dioxide particles, which ensures maximum contact of the gases released during the combustion of the reducing reagent with the lead oxide particles rising from the lead (92–95% from the theoretical output).

2. The recovery temperature of lead with coke ranges from 1,100–1,200°C, which leads to intensive evaporation of lead oxide and metallic lead, causes environmental pollution, and the process is less profitable economically. The operating temperature of our experiments did not exceed 800–850°C, which leads to less evaporation of lead and its compounds, and the technology is both more environmentally friendly and economically more profitable than traditional technologies.
3. When recycling lead by traditional methods, flus waste containing 8–10% lead is obtained, which is difficult to recycle, and the vaporized and filtered lead compounds are almost never used for recycling, creating even more significant environmental problems.

Using flue carbonates for lead recovery, if lead and its compounds are trapped, it will be possible to re-briquette them with the permeable regenerator and return them to production, which will further reduce lead and carbonate losses while the technology becomes environmentally safe and economically viable.

KEYWORDS

- **carbonates smelts**
- **lead oxide**
- **lead oxide**
- **metallic lead**
- **physicochemical processes**
- **polyethylene terephthalate (PETP)**
- **recovery reactions**

REFERENCES

1. Morachevski, A. G., Weisgant, Z. I., Ugolkov, V. L., et al., (2006). Establishment processes during the processing of active mass of fractures of pig batteries. Ж. *Approx. Chemistry, T. 79*(2C), 242–250.
2. Meiler, H. W., Nolum, J. B., & Richardson, F. D., (1966). The activity of lead oxide in blast furnace slag. *Trans. Journ., 756*, 121, 122.
3. Barbin, N. M., & Barbina, T. M., (2017). Lead oxide solubility in molten NaOH and NaOH and $NaCO_3$. *Physics and Chemistry of Lequides, 55*(6).
4. Qingfeng, L., Flemming, B., Irina, P., & Niels, B., (1999). Complex formation during dissolution of metal oxides in molten alkali carbonates. *Journal of the Electrochemical Society, 146*(7), 2449–2454.
5. Li, Q., Flemming, B., Irina, P., & Niels, J. B., (1999). *Complex Formation During Dissolution of Metal Oxides in Molten Alkali Carbonates.* Materials Science Group, Department of Chemistry, Technical University of Denmark, 2800 Lyngby, Denmark.
6. Efimov, A. I., (1983). *Properties of Inorganic Compounds.* Lawyer, Chemistry.
7. Posipajko, V. P., (1977). Diagrams of floating saline system 4\2. *Metallurgy*, 304.
8. Takeshi, H., & Katsuyasu, S., (2007). Effect of sodium carbonate on reduction of lead sulfate by carbon solid. *Kagaku Kogaku Renbunshu, 33*(6), 622. doi: 10 1252.
9. Knunianz, I. L., et al., (1995). *Chemical Encyclopedia* (Vol. 4, p. 639). M.: Soviet Encyclopedia.
10. Shamanauri, L., & Dzanashvili, D. (2007). *Dictionary of Chemistry 21 Chemistry and Chemical Technology*, 258.
11. Sergeev, D., Yazhenskikh, E., Kobertz, D., & Müller, M., (2019). *Vaporization Behavior of Na_2CO_3 and K_2CO_3*, 125.
12. Yunjian, M., & Keqiang, Q., (2015). Recovery of lead from lead paste in spent lead acid battery by hydrometallurgical desulfurization and vacuum thermal reduction. *Waste Manag., J. Wasman*, doi: 10.1016.
13. Loskutov, F. M., (1995). *Complex Formation During Dissolution of Metal Oxides in Molten Alkali Carbonates* (p. 528). Metalurgy pig-M. Metalurgy.

CHAPTER 12

Development of Technology for the Production of Geopolymer Binders Based on Thermally Modified Clay Rocks

ELENA SHAPAKIDZE,[1] MARINA AVALIANI,[2] MARINE NADIRASHVILI,[1] VERA MAISURADZE,[1] IOSEB GEJADZE,[1] and TAMAR PETRIASHVILI[1]

[1]*Ivane Javakhishvili Tbilisi State University, Al. Tvalchrelidze Caucasian Institute of Mineral Resources, Tbilisi, Georgia*

[2]*Ivane Javakhishvili Tbilisi State University, R. Agladze Institute of Inorganic Chemistry and Electrochemistry, Tbilisi, Georgia*

ABSTRACT

The high energy intensity of cement production and CO_2 emissions into the atmosphere, in recent years, raised the question of its alternative, which can be a geopolymer binder.

Geopolymer binders are considered a promising basis for creating a technological platform for the resource-saving building materials industry. These materials have a number of valuable properties and in the near future may serve as an alternative to modern binders based on Portland cement.

The main raw materials for the production of geopolymers are natural and artificial aluminosilicate materials (kaolin and ordinary clay rocks, feldspars, slags, fly ash and other), the hardening of which is activated by low-modulus alkali metal silicates.

Advanced Polymer Structures: Chemistry for Engineering Applications. Omar Mukbaniani, DSc, Tamara Tatrishvili, PhD & Marc Jean M. Abadie, DSc (Eds.)

The chapter is devoted to the development of technology for the production of geopolymer binders based on thermally modified clay rocks of Georgia, such as shale, mudstone, and low-melting clay.

12.1 INTRODUCTION

Geopolymer binders are a new direction in the creation of technologies for the production of energy-efficient building materials.

Geopolymer materials are considered as binding systems obtained on the basis of finely ground aluminosilicate materials, which are mixed with alkalis or solutions of their salts (as a rule, hydroxides, silicates or carbonates of sodium or potassium) with an alkaline reaction. After dissolution of aluminosilicates in alkalis, they recondense and form an amorphous three-dimensional framework structure. That is, geopolymer is a three-dimensional aluminosilicate mineral polymer [1]. Geopolymers are also considered nanomaterials.

Geopolymer binders, in comparison with Portland cement, are characterized by environmental friendliness, durability, and low carbon dioxide emission during their production. When 1 ton of geopolymer binder is obtained, 0.18 ton of CO_2 is released into the atmosphere, which is 5 times less than in the production of Portland cement [2]. Therefore, geopolymers are considered as an alternative to Portland cement.

At the initial stage of obtaining geopolymer binders, metakaolin was used as an aluminosilicate material – a product of heat treatment of kaolin clays at temperatures of 750–850°C. When heated, the dehydration of kaolinite occurs according to the following scheme: $Al_2O_3\ 2SiO_2\ 2H_2O \rightarrow Al_2O_3\ 2SiO_2 + 2H_2O$. Heat treatment increases the amount of the amorphous phase, which increases the reactivity of aluminosilicates.

Alkaline activation of metakaolin makes it possible to obtain a geopolymer binder of high strength and network structure. In the presence of $Ca(OH)_2$, the reaction proceeds according to a different scheme: the shape of the gel and the shape of the network change [3].

The widespread use of metakaolin in the production of building materials is hindered by the fact that the world's reserves of kaolin and kaolinite clays are limited, and in many countries, including Georgia, they are not mined.

Recently, thermal modification of polymineral clays and shales has been widely used to obtain metakaolin [4–6].

Earlier, we have developed a regime for thermal modification of shale in order to obtain metakaolin, which is one of the main components for the synthesis of geopolymer [7–10].

The aim of this work is to develop a technology for producing geopolymer binders based on thermally modified clay rocks of Georgia, for which local rocks were used: clay shale, mudstone, and low-melting clay.

12.2 EXPERIMENTAL METHODS AND MATERIALS

12.2.1 MATERIALS

For the study were used clay rocks widespread in Georgia: clay shale from Kvareli, mudstone from Teleti, clay from Gardabani, blast furnace slags from the Rustavi Metallurgical Plant.

An alkaline activator was used as a mixing liquid: NaOH, Na_2CO_3, or $Na_2O(SiO_2)n$, or a mixture of these.

12.2.2 METHODS

The materials were studied by methods of chemical, X-ray, microscopic, differential thermal and electron microscopic analyzes.

The mineral composition of clays was determined using an Optika B-383POL polarization microscope (Italy).

A NETZSCH derivatograph with STA-2500 REGULUS thermogravimetric and differential thermal analyzer (TG/DTA) was used for thermogravimetric analysis. Samples were heated to 1,000°C, in a ceramic crucible, heating rate 10°C/min. Reference substance α-Al_2O_3.

The X-ray phase analysis was carried out using a Dron-4.0 diffractometer ("Burevestnik," St. Petersburg, Russia) with a Cu-anode and a Ni-filter. U = 35 kv. I = 20 mA. Intensity – 2 degrees/min. λ = 1.54178 Å.

Scanning electronic microscope (SEM) measurements were performed on a JEOL scanning electronic microscope JSM-6510LV (well-appointed by energy-disperse X-max No. 20 micro-X-ray spectral analyzer produced by Oxford Instruments). SEM measurements were carried out by means of reflected (BES) as well as secondary (SEI) electrons at an accelerating voltage (at 20 kV). The working distance was approximately 15 mm. Micrographs have been taken at the diverse enlargements.

12.2.3 OBJECTS OF STUDY

The object of study was geopolymer binders synthesized on the basis of thermally modified clay rocks.

12.3 RESULTS AND DISCUSSION

Table 12.1 shows the chemical compositions of clay rocks.

According to the results of microscopic analysis, the structure of clay rocks is as follows:

1. **Shale:** Pelitic structure, with poorly pronounced slate elements. With a multiple increase it can be seen that the main components of the pelitic structure are chlorite and hydromica minerals, iron hydroxides (hematite) and organic compounds. The coarse fraction is represented by quartz-feldspathic minerals (Figures 12.1).

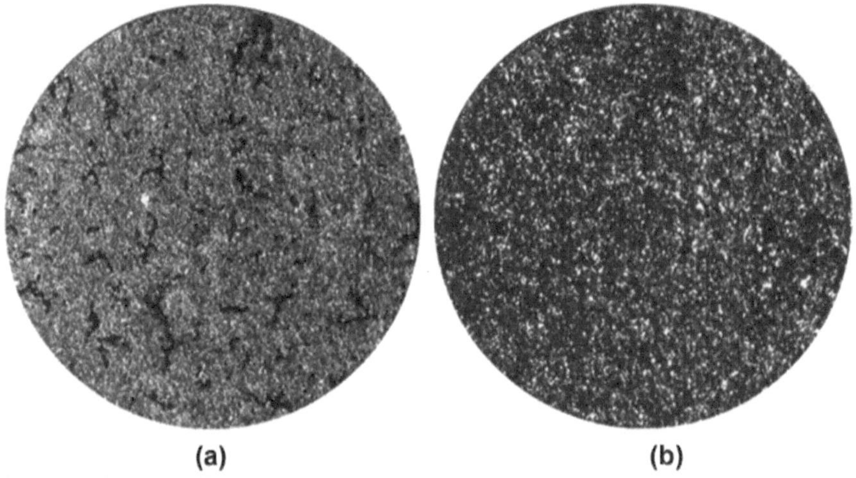

FIGURE 12.1 Micrographs of shale, magnification 100x.

2. **Mudstone:** Siltstone structure – mineralogy is represented mainly by quartz, plagioclase, calcite, glauconite, a small amount of pyroxene and biotite. Most of them are quartz-feldspar minerals (Figure 12.2).

TABLE 12.1 Chemical Compositions of Clay Rocks (wt.%)

Material	L.O.I.	SiO_2	TiO_2	Al_2O_3	Fe_2O_3	FeO	Mn_2O_3	CaO	MgO	SO_3	Na_2O	K_2O
Shale	4.50	59.95	0.89	17.30	3.45	3.65	0.59	1.53	2.43	0.30	2.20	2.20
Mudstone	7.01	47.19	–	15.90	13.36	–	0.10	6.30	4.10	1.39	2.86	1.30
Clay	10.60	52.84	–	15.07	6.47	–	–	7.06	2.49	1.36	1.19	2.17

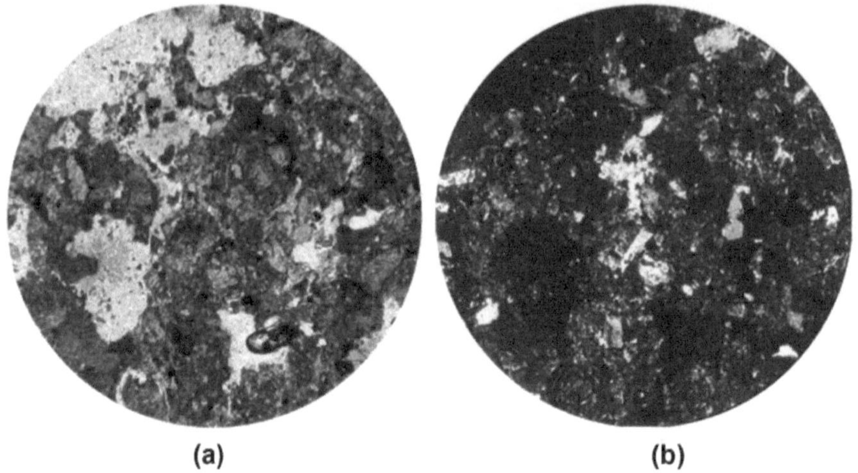

FIGURE 12.2 Micrographs of mudstone, magnification 100x.

3. **Clay:** It is carbonized, the integrity of the structure is broken. Most of the residual structure consists of pelitic carbonate-micaceous minerals, partly of quartz, feldspar, amphibole, and particles of pyroxene grains and fragments. Iron oxides and ore minerals are noted in single amphibole and pyroxene grains (Figure 12.3).

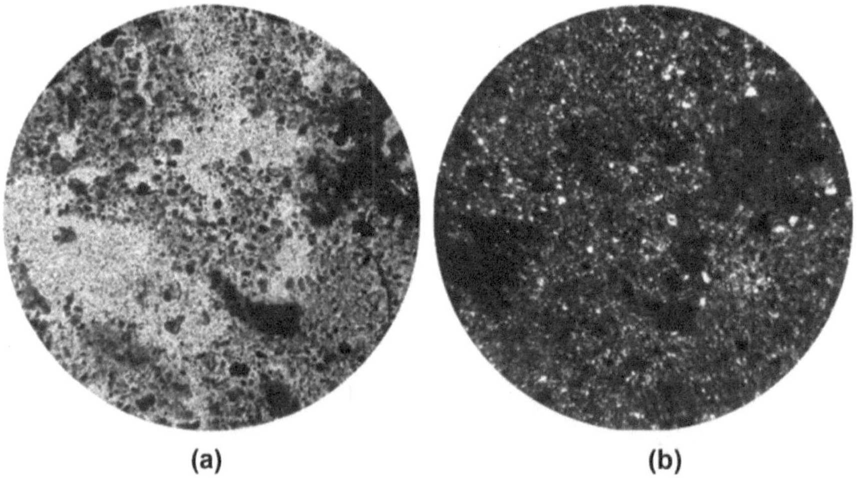

FIGURE 12.3 Micrographs of clay, magnification 100x.

Clay minerals (14.66–14.96, 7.14, 4.25, 3.66, 2.86, 2.327 Å) are noted on the X-ray diffraction patterns (Figure 12.4); quartz (3.34 Å); feldspar (3.87 Å), calcium carbonate (3.03 Å).

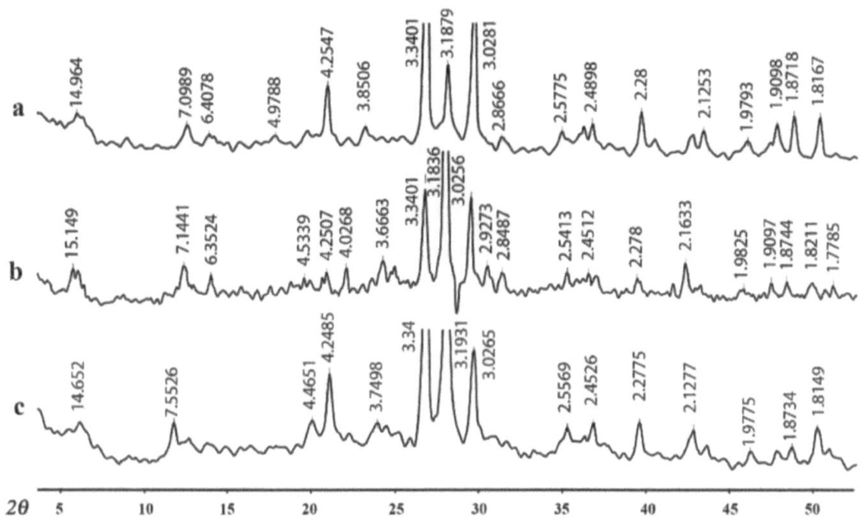

FIGURE 12.4 X-ray diffraction patterns of clay rocks: (a) shale; (b) mudstone; and (c) clay.

Clay materials were thermally modified for synthesizing the geopolymer binder.

According to the test results (Figure 12.5(a–c)), it is obvious that clay rocks thermally modified at 700°C have a rather high pozzolanic activity.

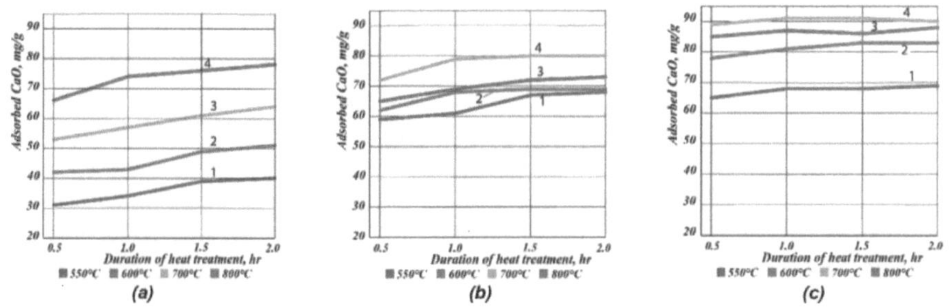

FIGURE 12.5 The kinetics of CaO absorption by heat-treated clay rocks from a saturated solution: (a) shale; (b) mudstone; and (c) clay.

12.3.1 STUDY OF PHYSICAL-MECHANICAL PROPERTIES OF OBTAINED GEOPOLYMER BINDERS

Heat treatment (modification) of clay materials was carried out in a muffle furnace at a heating rate of 250–300°C per hour and at an exposure time of 1 hour at a maximum temperature of 700°C. The samples were cooled naturally at room temperature.

For the preparation of geopolymer compositions, granulated blast furnace slag and modified clay rocks were used, which were ground in a laboratory ball mill to a specific surface area of 8,000–9,000 cm^2/g. As an alkaline activator, we used: sodium alkali – NaOH, sodium carbonate – Na_2CO_3, or liquid glass (LG) – $Na_2O(SiO_2)n$, or a combination of them. The alkaline activator was added in an amount that obtain the dough of normal density.

Samples of 20 × 20 × 20 mm were formed, which were removed from the mold 2 days after forming. Some samples were stored in air, some in water, and some in air-wet conditions. The curing temperature was 20°C for 28 days, after which the samples were tested for mechanical strength.

After being removed from the mold, part of the samples was subjected to heat treatment according to the mode: 80C for 20 hours and after cooling, they were tested for mechanical strength.

As shown by the test results (Table 12.2), after heat treatment (80°C for 20 hours), the strength of geopolymer binders significantly increases compared to binders that have hardened under normal conditions. Research has made it possible to develop different compositions of geopolymer binders [9, 10].

TABLE 12.2 Compositions of Geopolymer Binders and the Results of Physical and Mechanical Testing

SL. No.	Components (%)		Alkaline Component (Dry Matter above 100%) (%)	Compressive Strength After 28 Days of Curing Depending on Curing Conditions (kg/cm^2)			Compressive Strength After Heat Treatment (kg/cm^2)
				Air	Water	Air-Wet	
1.	Slag (80)	*Shale (20)	NaOH (10)	410	452	440	690
2.	Slag (80)	*Shale (20)	Na_2CO_3 (10)	210	245	240	537
3.	Slag (80)	*Shale (20)	$Na_2O(SiO_2)n$ (10)	187	334	212	488
4.	Slag (80)	*Mudstone (20)	NaOH (10)	469	480	418	695
5.	Slag (80)	*Mudstone (20)	Na_2CO_3 (10)	335	420	390	685

TABLE 12.2 *(Continued)*

SL. No.	Components (%)		Alkaline Component (Dry Matter above 100%) (%)	Compressive Strength After 28 Days of Curing Depending on Curing Conditions (kg/cm²)			Compressive Strength After Heat Treatment (kg/cm²)
				Air	Water	Air-Wet	
6.	Slag (80)	*Mudstone (20)	NaOH (2.5) + $Na_2O(SiO_2)$ n (7)	536	472	450	856
7.	Slag (80)	*Clay (20)	NaOH (10)	460	510	478	630
8.	Slag (80)	*Clay (20)	Na_2CO_3 (10)	175	223	217	575
9.	Slag (80)	*Clay (20)	$Na_2O(SiO_2)$n (10)	75	88	85	150
10.	Slag (80)	*Clay (20)	NaOH (4) + Na_2CO_3 (6)	215	254	230	266
11.	Slag (80)	*Clay (20)	NaOH (4) + $Na_2O(SiO_2)$n (10)	850	940	935	1,025
12.	Slag (80)	*Clay (20)	Na_2CO_3 (4) + $Na_2O(SiO_2)$n (10)	112	145	156	320

12.3.2 STUDY OF MORPHOLOGY OF GEOPOLYMER BINDERS

Figures 12.6 and 12.7 show the micrographs obtained using SEM measurements by means of reflected (BEC) as well as secondary (SEI) electrons for geopolymer binders No. 6 and No. 11 (Table 12.2), which had the best strength indicators. As can be seen in the figures, heat treatment promotes compaction of the material, which has a positive effect on its mechanical strength.

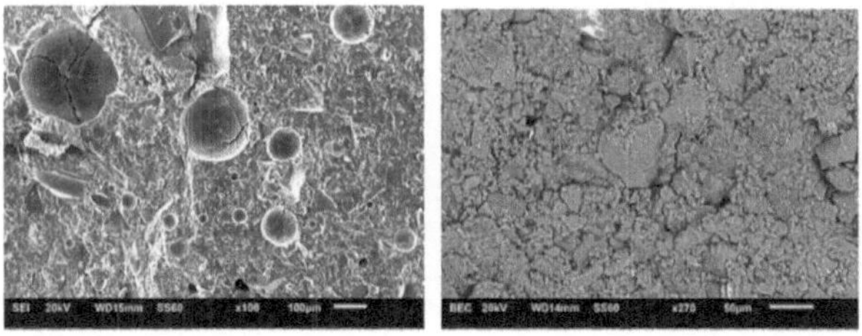

(a) (b)

FIGURE 12.6 Micrographs of geopolymer binder No. 6: (a) after 28 days of curing under normal conditions (SEI, enlargement 100); (b) after 2 days of curing and heat treatment (BEC, enlargement 270).

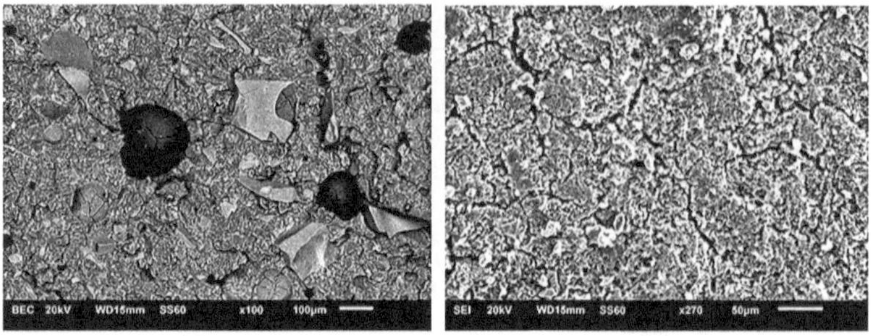

(a) (b)

FIGURE 12.7 Micrographs of geopolymer binder No. 11: (a) after 28 days of curing under normal conditions (BEC, enlargement 100); (b) after 2 days of curing and heat treatment (SEI, enlargement 270).

12.4 CONCLUSION

Thus, the possibility of obtaining geopolymer binders with high physical and mechanical characteristics based on thermally modified local clay rocks, granulated blast furnace slag and an alkaline solution of various compositions has been proved.

ACKNOWLEDGMENTS

The authors are grateful to the Shota Rustaveli National Science Foundation of Georgia, with the financial support of which this work was carried out [grant number FR-18-783].

KEYWORDS

- **clay shale**
- **geopolymer binder**
- **geopolymer materials**
- **low-melting clay**
- **mudstone**
- **thermal modification**

REFERENCES

1. Davidovits, J., (1988). Soft mineralurgy and geopolymers. *Proceeding of Geopolymer 88 International Conference* (pp. 49–56). The Université de Technologie, Compiègne, France.
2. Davidovits, J., (1993). Carbon-dioxide greenhouse-warming: What future for Portland cement. *Proceedings, Emerging Technologies Symposium on Cement and Concretes in the Global Environment.* Portland Cement Association, Chicago, Illinois.
3. Granizo, M. L., Alonso, S., Blanco-Varela, M. T., & Palomo, A., (2002). Alkaline activation of metakaolin: Effect of calcium hydroxide in the products of reaction. *J. Amer. Ceram. Soc., 85*(1), 225–231. doi: 10.1111/j.1151-2916.2002.tb00070.x.
4. Castello, L. R., Hemandes, H. J. F., Scrivener, K. L., & Antonic, M., (2011). Evolution of calcined clay solid as supplementary cementitious materials. *Proceedings of a XII International Congress of Chemistry of Cement* (p. 117). Madrid: Instituio de ciencias de la Construction Eduardo Torroja.
5. Gaifullin, A. R., Rakhimov, R. Z., & Rakhimova, N. R., (2015). The influence of clay additives in Portland cement on the compressive strength of the cement stone. *Mag. Civil Eng., 7*, 66–73.
6. Rakhimov, R. Z., Rakhimova, N. R., & Gaifullin, A. R., (2017). Influence of the addition of dispersed fine polymineral calcined clays on the properties of Portland cement paste. *Advances in Cement Research, 29*(1), 21–32. https://doi.org/10.1680/jadcr.16.00060.
7. Shapakidze, E., Nadirashvili, M., Maisuradze, V., Gejadze, I., Petriashvili, T., Avaliani, M., Todradze, G., & Khuchua, E., (2018). Development of geopolymeric binding materials based on the calcined shales. *Ceram. Adv. Technol., 20*(2), 31–38.
8. Shapakidze, E., Nadirashvili, M., Maisuradze, V., Gejadze, I., Petriashvili, T., Avaliani, M., & Todradze, G., (2019). Elaboration of optimal mode for heat treatment of shales for obtaining metakaolin. *Eur. Chem. Bull., 8*(1), 31–33. doi: 10.17628/ecb.2019.8.31-33.
9. Shapakidze, E., Avaliani, M., Nadirashvili, M., Maisuradze, V., Gejadze, I., & Petriashvili, T., (2020). Obtaining of geopolymer binders based on thermally modified clay rocks of Georgia. *Nano Studies/ European Chemical Bulletin, 20*(1), 43–52. doi: 10.13140/RG.2.2.18756.17281.
10. Shapakidze, E., Avaliani, M., Nadirashvili, M., Maisuradze, V., Gejadze, I., & Petriashvili, T., (2020). *Composite Materials Engineering: Modeling and Technology: Materials Science* (pp. 351–358). Apple Academic Press, USA. PART III, Chapter 25. doi: 10.13140/RG.2.2.28081.12644.

CHAPTER 13

Influence of Synthesis Ways and Conditions on Phase Formation and Superconductivity Properties of Tl-Based HTS

METSKHVARISHVILI IOSEB,[1,2] LOBZHANIDZE TEA,[3] DGEBUADZE GURAM,[1] BENDELIANI BEZHAN,[1] METSKHVARISHVILI MAGDA,[4] GIORGANASHVILI GIORGI,[1] GIORGADZE KRISTINE,[3] and GABUNIA VAKHTANG[1,5]

[1]*Laboratory of Cryogenic Technique and Technologies, Ilia Vekua Sukhumi Institute of Physics and Technology, Tbilisi, Georgia*

[2]*Department of Microprocessor and Measurement Systems, Faculty of Informatics and Control Systems, Georgian Technical University, Tbilisi, Georgia*

[3]*Department of Chemistry, Faculty of Exact and Natural Sciences, Ivane Javakhishvili Tbilisi State University, Tbilisi, Georgia*

[4]*Department of Engineering Physics, Faculty of Informatics and Control Systems, Georgian Technical University, Tbilisi, Georgia*

[5]*Petre Melikishvili Institute of Physical and Organic Chemistry, Ivane Javakhishvili Tbilisi State University, Tbilisi, Georgia*

ABSTRACT

The chapter presents the comparative analysis of the sol-gel (SG) and solid-state reaction (SSR) routes to synthesize thallium-based superconductors.

Advanced Polymer Structures: Chemistry for Engineering Applications. Omar Mukbaniani, DSc, Tamara Tatrishvili, PhD & Marc Jean M. Abadie, DSc (Eds.)

Samples were prepared by a two-step method and by sealed quartz tube technique in ambient pressure. We could conclude that using wet chemistry offers some advantages compared to the classical solid-state ceramics processing, significantly better chemical homogeneity, and higher reactivity of the precursor powder.

13.1 INTRODUCTION

High-temperature superconducting oxides have several constituent ions. For example, three for Y-1223 superconductors, four for Bi-1223 and Tl-1223 superconductors, and sometimes more than four. Therefore, in solid solutions, amplify the problems solid-state reaction (SSR) and co-precipitation methods. Therefore, ceramic superconductors with the same overall composition often exhibit different properties, reflecting small phase purity and compositional homogeneity differences. The sol-gel (SG) process has an advantage over the other methods to achieve homogeneous mixing of the component cations on the atomic scale and form bulk superconducting materials from gels.

Thallium-and mercury-based superconductors are generating considerable interest, as these systems set records for the transition temperature to the superconducting state. Tl-1223 and Hg-1223 had transition temperatures $T_c \approx 115$ K and $T_c \approx 133$ K when prepared under ambient conditions. The critical temperature can reach 133.5 K and 164 K when synthesized under high pressures, respectively [1–4]. The formula unit and crystal structure of the $TlBa_2Ca_{n-1}Cu_nO_{2n+2+\delta}$ system are similar to those of the $HgBa_2Ca_{n-1}Cu_nO_{2n+2+\delta}$ system, where (*n*) is the number of adjacent Cu-O layers. Mercury- and thallium are very toxic and volatile at high temperatures. More important for this family, to achieve a high purity superconductivity phase critically depends on the precursor and synthesis conditions used. Obtaining precursors of high purity and reactivity in the various papers was solved by different methods [5–11].

Loureiro et al. investigated the importance [5] of the average copper valence in the precursor for synthesized homologous series of Hg-1223 superconductors under high pressure. Calcium oxide and barium nitrates were used to avoid eliminating residual carbonate in the precursor. SSR methods prepared precursors. They showed that average copper valence is a crucial dependence on the oxygen intake of non-stoichiometric phases during precursor synthesis at a specific temperature, time, and cooling process treatment. S. Lee and coauthors [6] proposed a freeze-drying method for synthesizing highly homogeneous precursors. First, materials have used nitrates. As a result, 75% of Hg-1223 superconducting fazes and $T_c \approx 133$ K were obtained from Hg-1223 samples prepared from freeze-dried precursors. Metal-organic chemical vapor deposition methods are presented in articles [7, 8]. Precursor films have been derived by employing $Ba(hfa)_2mep$, $Ca(hfa)_2tet$, and $Cu(dpm)_2$ – as metal sources. In this case, the final synthesis necessarily requires thallium fluorine to nucleate the Tl-1223 phase. They obtained high transport properties for samples with very short transition temperatures. Sin et al. [9] simultaneously studied the influence of rhenium addition and the in-situ gelation process using acrylamide monomers to synthesize of precursors. The authors conclude that the rhenium-based precursor is much more stable against moisture and carbonation. The urea combustion method for synthesizing precursors is presented by T.M. Mendonca and coauthors [10]. The results showed that the single-phase 99% vol. of the Hg-1223 phase in urea samples was observed. Brylewski et al. [11] reported the SG method for synthesized $Ba_2Ca_2Cu_3O_x$ precursors. The starting materials used nitrate and nitrate tetrahydrate, and ethylenediaminetetraacetic acid as the complexing agent. Subsequently, 89.1% of the high volume fraction (Hg, Pb)-1223 phase was observed. The authors think that the EDTA chelating agent may be a reason for this.

Since achieving a high purity superconductivity phase critically depends on the precursor properties, the present study is devoted to synthesizing the precursors by two different methods. It is must note that for both SG and SSR ways, oxides, and carbonate-containing materials were used as starting materials. For the SG route, poly(vinyl alcohol)/poly(vinyl acetate) was used as the complexing agent [12, 13]. The present work also investigated the precursor's properties dependence on heat treatments. A characterization of precursors obtained by the SG and SSR approaches on the superconducting parameters of $TlBa_2Ca_2Cu_3Dy_xO_{8+\delta}$ is investigated, analyzing the results of their X-ray diffraction, FTIR analysis, ac magnetic susceptibility comparatively, and transport critical current densities.

The phase method was used to study the real parts $-4\pi\chi'$ of the linear susceptibility. Errors in the determination of χ' at frequencies higher than 1 kHz do not exceed 1% when $-4\pi\chi > 0.1$. For $4\pi\chi < 0.1$, the errors are increased in proportion to diminishing the magnitude of susceptibility and frequency. For the measurements of intergranular critical current densities, we used the method of high harmonics. The error of measurement of high harmonics was approximately 2% when the measured signal was less than 0.2 μV and no more than 0.5% when the signal was higher. The measurements were mainly performed at $h = 1$ Oe, f=20 kHz, and $H = 0$. The Earth's magnetic field was shielded to less than 10^{-3} Oe by the use of Permalloy screens [14, 15].

13.2 EXPERIMENTAL PART

For synthesized $TlBa_2Ca_2Cu_3O_{8+\delta}$ samples, we used the two-step method. In the first stage, a Tl-free precursor was prepared before proceeding to the second stage, where Tl_2O_3 was added to the precursor before final sintering. We note that for both methods, starting materials utilized powder materials $BaCO_3$ (99.0% Oxford Chem Serve), $CaCO_3$ (99.98% Oxford Chem Serve), CuO (99.999% Sigma-Aldrich), and Tl_2O_3 (99.99% Sigma-Aldrich).

Ba:Ca:Cu=2:2:3 multiphase ceramic precursors were prepared by two methods: the SG method and the SSR method:

1. **Sol-Gel Method (SG):** The initial reactants were dissolved separately, $BaCO_3$ and $CaCO_3$ in acetic acid and CuO in nitric acid. When complete dissolution was achieved, all the solutions were mixed, and poly(vinyl alcohol)/poly(vinyl acetate) ($[-CH_2CHOH-]_n$/ $[CH_2CH(O_2CCH_3)]_n$, (Sigma-Aldrich) was added as the complexing agent. The solution was stirred slowly, continuously up to 80°C, and then they obtained green gel was dried slowly, at a temperature rate of 1°C per minute up to 300°C. The resulting black powder was then ground in an agate mortar. Then, the powder was calcined at 900°C in the air with a heating rate of 2°C/min for 12 h, with two intermediate grindings.
   The resulting powders were ground, separated into six parts, and pressed in the form of a disc. To eliminate CO_2 from the precursors, each pellet was individually annealed in a tube-type furnace at temperatures of 700°C, 800°C, 900°C, 915°C, 930°C, and 945°C under a flowing oxygen partial pressure of 0.5 bar for 12 h. For all synthesis temperatures, the heating rate was 1°C/min. First, the samples were heated until the temperature of the planned synthesis and then turned on the oxygen and kept at this temperature for 12 h. After the synthesis was completed, we turned off the oxygen, and then the samples cooled to room temperature inside the furnace.
2. **Solid-State Reaction Method (SSR):** The materials $BaCO_3$, $CaCO_3$, and CuO were mixed in the stoichiometric ratio Ba:Ca:Cu=2:2:3, and then ground carefully in an agate mortar. The resulting powder

mixture was calcined in an alumina crucible in the air in a muffled furnace, with four intermediate grindings at 900°C for 60 h. Then, as described above, the resulting powders were separated into six parts, pressed, and annealed under flowing O_2 at various temperatures.

In the second step, both $Ba_2Ca_2Cu_3O_x$ precursors prepared by the SG and SSR methods separately were mixed with Tl_2O_3 according to the composition $TlBa_2Ca_2Cu_3O_{8+\delta}$. After final grinding, the powder was pressed into a disc-shaped pellet 6 mm in diameter, and 3 mm thick, by using a hydraulic press under a pressure of 400 MPa. The samples were wrapped in platinum foil and then individually put into quartz tubes, and quartz tubes were evacuated up to 10^{-3} Torr and sealed. Thereafter, a quartz tube was inserted into a programmed muffle furnace. The temperature of the furnace was raised at a rate of 25°C/min up to 900°C and held at this temperature for 8 h. After the synthesis completed, the furnace was quickly cooled to room temperature.

X-ray powder diffraction (XRD) patterns were obtained on a Dron-3+PC diffractometer with CuKα *radiation.* The Fourier transformed IR of the samples was taken in the region 400–4,000 cm^{-1} on a Cary 600 series FTIR spectrometer using the KBr disc technique. Scanning resolution 0.5 cm^{-1}. The samples were pulverized into a fine powder and then mixed with potassium bromide powder using a weight ratio of 1:100. The IR absorption spectra were measured immediately after preparing the discs.

13.3 RESULTS AND DISCUSSION

As described above, achieving a higher purity superconductivity phase critically depends on the precursor used. To obtain precursors with optimal properties, the best method is thermal annealing at the oxygen partial pressure, especially if precursors will be prepared from oxide-containing carbonates. The synthesis of precursors in oxygen pressure provides the elimination of the carbonates in samples and the cation homogeneity and the oxygen content. The absorption spectra of the precursors, annealed at various temperatures, are shown in Figure 13.1(a and b) for the range of 400–4,000 cm^{-1}. The signature of the $_3$-triply degenerated stretching mode of carbonate at ~1,460–1,360 cm^{-1} is observed as SSR and for SG-precursors annealed at 700–900°C temperatures. The ν_2-doubly degenerated stretching modes of carbonate (870–860 cm^{-1}) are observed only in precursors that have been synthesized at 700C and 800°C [16–18]. In contrast, for the SG-precursor annealed at 915°C, the remaining CO_3^{2-} impurity species are not observed and for the SSR-precursor annealed at 945°C.

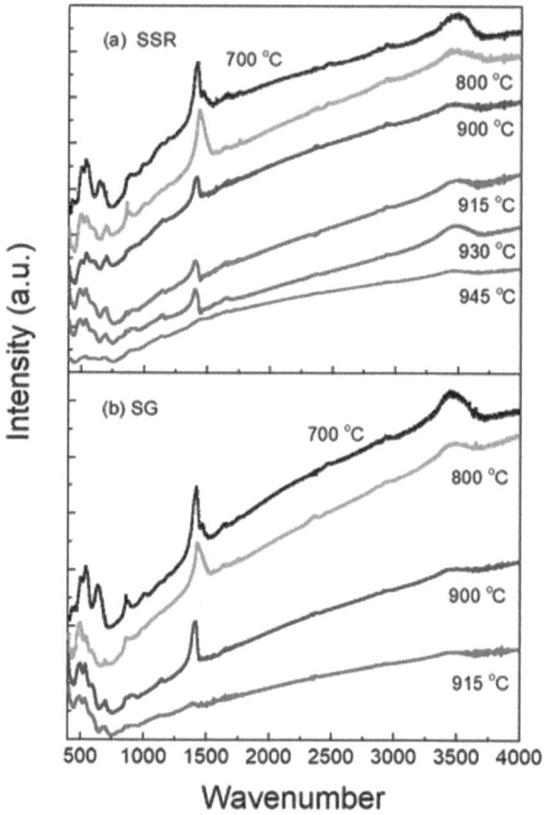

FIGURE 13.1 IR analysis of $Ba_2Ca_2Cu_3O_y$ after annealing at a flowing oxygen partial pressure of 0.5 bar: for the SSR-precursor at 700°C–945°C (a) and for the SG-precursor at 700°C–915°C (b).

The X-ray diffraction patterns of the $Ba_2Ca_2Cu_3O_y$ precursor powder, prepared by the SG method at 800, 900°C and 915°C temperatures are plotted in Figure 13.2(a–c). In contrast, the IR results for $BaCO_3$ XRD measurements $BaCO_3$ are fixed only at 800°C. In samples annealed at 900°C and above, no carbonate CO_3^{2-} is observed, such as SG and for SSR precursors. A sample practically consists of only two phases of $BaCuO_2$ and Ca_2CuO_3, and only small amounts of $Ba_2Cu_3O_5$ and unreacted CaO phases are identified. High purity precursors were obtained for SG and SSR at 915°C and

945°C, respectively. The precursor consists of only two phases of $BaCuO_2$ and Ca_2CuO_3 [19–21]. From the XRD measurement results, we can conclude that the sensitivity of X-ray diffraction to fixing carbonate is not sufficient.

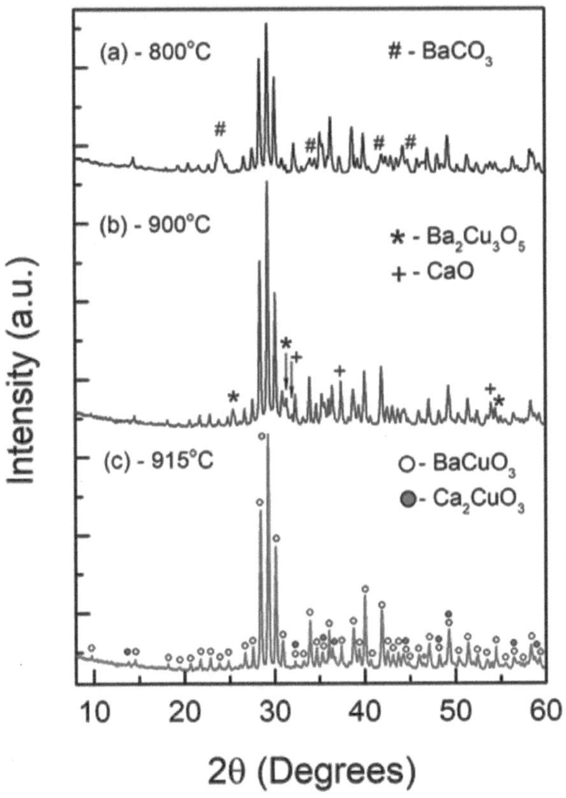

FIGURE 13.2 XRD patterns of the $Ba_2Ca_2Cu_3O_y$ precursor, prepared by the SG method at 800, 900°C, and 915°C temperatures.

Figure 13.3(a and b) presents a comparison of the XRD patterns of the superconducting samples with the nominal composition $TlBa_2Ca_2Cu_3O_y$, prepared on the best precursor powders of both methods: SSR synthesized at 945°C and SG synthesized at 915°C. From the XRD results, one will notice that small amounts of impurities such as Tl-1212 and $BaCuO_2$, which are usually present in the preparation of the Tl-1223 phase, have appeared. Both prepared samples are nearly a single phase, and have a tetragonal structure with lattice parameters for the SSR method a=3.846 (Å), and c=15.908 (Å) and SG method a=3.847 (Å), and c=15.920 (Å).

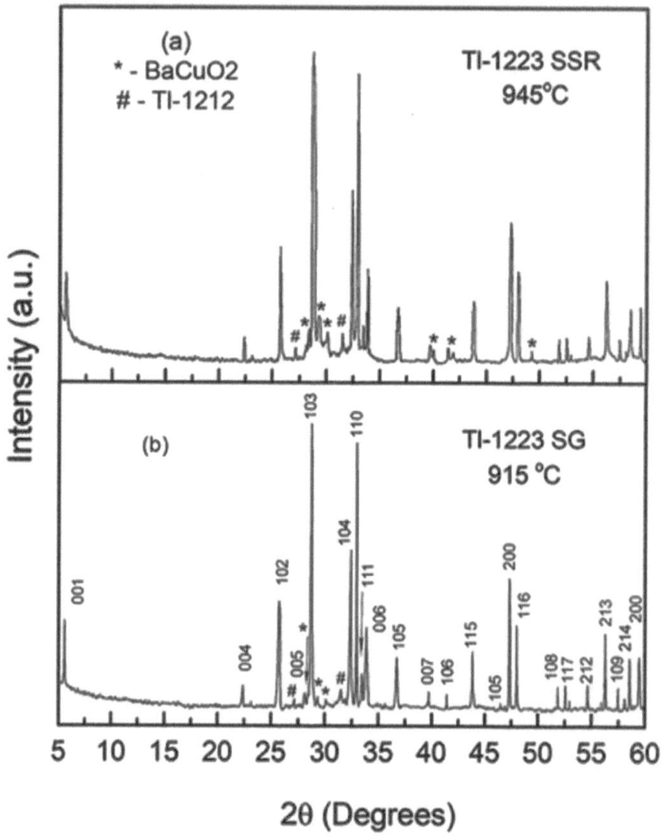

FIGURE 13.3 XRD patterns of the $TlBa_2Ca_2Cu_3O_{8+\delta}$ samples, prepared on the best precursor powders of both methods: (a) SSR synthesized at 945C; and (b) SG synthesized at 915°C.

The temperature dependencies of the susceptibility versus temperature for the Tl-1223 samples prepared on the based precursors that are synthesized by the SSR and SG methods are presented in Figure 13.4(a and b). The samples synthesized with precursors prepared using the SSR method, which gated thermal treatments at a temperature 900°C, 915°C, 930°C, and 945°C diamagnetic onset are at Tc≈114 K, 117 K, 118 K, and 120 K, respectively. The full diamagnetic state was observed at T_c≈78 K (915°C), T_c≈88 K (930°C), and 92 K (945°C). Unlike the SSR in the SG samples, the diamagnetic onset temperatures of the superconducting transition for the SG samples synthesized at 900°C and 915°C are T_c≈117 K and 120 K, respectively. The full diamagnetic states are T_c≈90 K and 102 K, respectively.

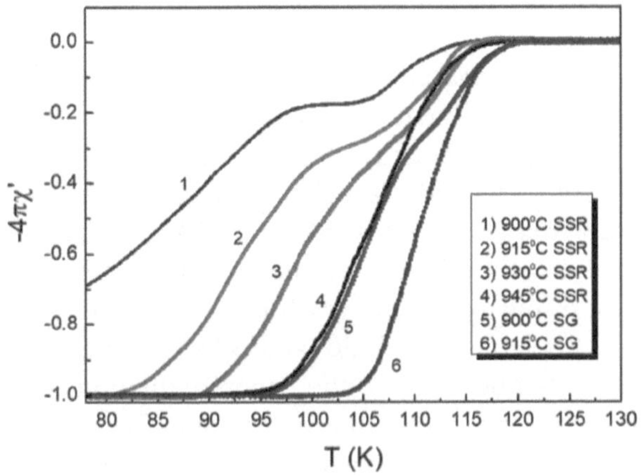

FIGURE 13.4 Temperature dependences of the real-$4\pi\chi$' part of *ac* susceptibility for the SG and SSR samples.

Figure 13.5 presents the dependence of the intergrain critical current density J_c of the Tl-1223 samples synthesized with precursors prepared at various temperatures, as for the SSR and SG methods. The maximal value of J_c at 78 K for SSR (945C) is 128 A/cm^2, and the SG sample (915°C) is 174 A/cm^2. This is because, although the transition temperatures for SSR (945°C) and SG (915°C) are the same, complete diamagnetic states are observed at 92 K and 102 K, respectively.

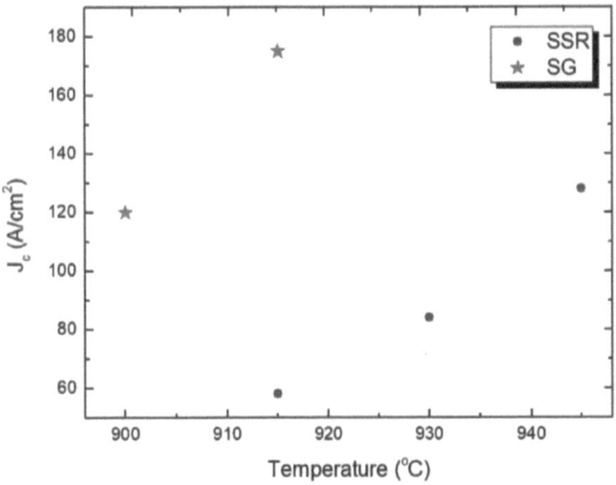

FIGURE 13.5 Dependencies of the J_c transport critical current densities versus precursor synthesis temperature, for the SSR and SG samples.

13.4 CONCLUSION

We have reported the preparation of Tl-1223 high-temperature superconductors by the sealed quartz tube technique. For the synthesis samples, we used the two-step method. In the first stage, multiphase precursors were prepared in the second stage where Tl_2O_3 was added. Precursors were synthesized by two methods, SG and SSR methods, and examined the influence of heat treatment on precursors. These results showed that to obtain high-purity precursors for SG methods, sufficiently heat treatments were performed at 915C and 945°C for SSR methods. The diamagnetic onset temperature of the superconducting transition for the samples prepared by precursors SSR at 945C and SG at 915°C is approximately $T_c \approx 120$ K and full diamagnetic state observed at $T \approx 92$ K and $T \approx 102$ K, respectively. The value of J_c at 78 K for SSR (945°C) is 128 A/cm^2 in comparison to the SG (915°C) sample and is equal to 174 A/cm^2. Although the transition temperatures for both samples are the same, the full diamagnetic state is observed at various temperatures. As a result, we could conclude that the SG method was demonstrated as a successful alternative to the SSR method, allowing faster production of the precursors without any carbonate contamination.

ACKNOWLEDGMENTS

This work was supported by Shota Rustaveli National Science Foundation (SRNSF), grant number: FR/261/6-260/14, Project title: Sol-gel Methods and Polymerization for synthesized Tl-based Polycrystalline Superconductors.

KEYWORDS

- **phase formation**
- **sol-gel process**
- **solid-state reaction methods**
- **superconductivity properties**
- **Tl-1223 HTS**
- **X-ray powder diffraction**

REFERENCES

1. Jasim, K. A., (2013). The effect of cadmium substitution on the superconducting properties of $Tl_{1-x}Cd_xBa_2Ca_2Cu_3O_{9-\delta}$ compound. *J. Supercond. Nov. Magn., 26*, 549–552.
2. Wu, J. Z., (2013). What have we learnt from the highest-Tc superconducting Hg-based cuprates? *Physica C., 493*, 96–99.
3. Iyo, A., Tanaka, Y., Ishiura, Y., Tokumoto, M., Tokiwa, K., Watanabe, T., & Ihara, H., (2001). Study on enhancement of Tc (≥130 K) in $TlBa_2Ca_2Cu_3O_y$ superconductors. *Supercond. Sci. Technol., 14*, 504–510.
4. Yamamoto, A., Takeshita, N., Terakura, C., & Tokura, Y., (2015). High pressure effects revisited for the cuprate superconductor family with highest critical temperature. *Nat. Commun., 6*, 8990.
5. Loureiro, S. M., Stott, C., Philip, L., Gorius, M. F., Perroux, M., Floch, S. L., Capponi, J. J., et al., (1996). The importance of the precursor in high-pressure synthesis of Hg-based superconductors. *Physica C., 272*, 94–100.
6. Lee, S., Shlyakhtin, O. A., Mun, M. O., Bae, M. K., & Lee, S., (1995). A freeze-drying approach to the preparation of $HgBa_2Ca_2Cu_3O_{8+x}$ superconductor. *Supercond. Sci. Technol., 8*, 60–64.
7. McNeely, R. J., Belot, J. A., Hinds, B. J., Marks, T. J., Schindler, J. L., Chudzik, M. P., Kannewurf, C. R., et al., (1997). Efficient route to $TlBa_2Ca_2Cu_3O_{9+x}$ thin films by metal-organic chemical vapor deposition using TlF as a thallination source. *Appl. Phys. Lett., 71*, 1243–1245.
8. McNeely, R. J., Belot, J. A., Schindler, J. L., Chudzik, M. P., Kannewurf, C. R., Zhang, X. F., Miller, D. J., & Marks, T. J., (1998). Novel metal-organic chemical vapor deposition/T1F annealing route to thin films of $Tl_1Ba_2Ca_2Cu_3O_{9+x}$. *J. Supercond., 11*, 133–134.
9. Sin, A., Odier, P., & Nuez-Regueiro, M., (2000). Sol–gel processing of precursor for high-T_c superconductors: Influence of rhenium on the synthesis of $Ba_2Ca_2Cu_3O_x$. *Physica C., 330*, 9–18.
10. Mendonca, T. M., Tavares, P. B., Correia, J. G., Lopes, A. M. L., Darie, C., & Araujo, J. P., (2011). The urea combustion method in the preparation of precursors for high-T_C single phase $HgBa_2Ca_2Cu_3O_{8+d}$ superconductors. *Physica C., 471*, 1643–1646.
11. Brylewski, T., Przybylski, K., Morawski, A., Gajda, D., Cetner, T., & Chmis, J., (2016). Soft-chemistry synthesis of $Ba_2Ca_2Cu_3O_x$ precursor and characterization of high-T_c $Hg_{0.8}Pb_{0.2}Ba_2Ca_2Cu_3O_{8+\delta}$ superconductor. *J. Adv. Cera., 5*, 185–196.
12. Geisari, N., & Kalnins, M., (2016). Poly(vinyl alcohol)– poly(vinyl acetate) composite films from water systems: Formation, strength-deformation characteristics, fracture. *IOP Conf. Ser.: Mater. Sci. Eng., 111*, 012009.
13. Zeng, R., Lu, L., & Dou, S. X., (2008). Significant enhancement of the superconducting properties of MgB_2 by polyvinyl alcohol additives. *Supercond. Sci. Technol., 21*, 085003.
14. Metskhvarishvili, I. R., Dgebuadze, G. N., Lobzhanidze, T. E., Bendeliani, B. G., Metskhvarishvili, M. R., Giorgadze, K. P., & Gabunia, V. M., (2020). Influence of dysprosium addition on the phase formation and transport properties of Hg-1223 superconductor. *J. Supercond. Nov. Magn., 33*, 3401–3405.
15. Metskhvarishvili, I. R., Dgebuadze, G. N., Bendeliani, B. G., Lobzhanidze, T. E., Metskhvarishvili, M. R., & Mumladze, G. N., (2014). Low ac field response of Bi-based superconductors with addition of antimony oxide. *J. Phys. Conf. Ser., 507*, 012032.

16. Baranauskas, A., Jasaitis, D., & Kareiva, A., (2002). Characterization of sol-gel process in the Y-Ba-Cu-O acetate-tartrate system using IR spectroscopy. *Vib. Spectrosc., 28*, 263–275.
17. Devi, P. S., & Maiti, H. S., (1994). A novel autoignited combustion process for the synthesis of Bi-Pb-Sr-Ca-Cu-O superconductors with a $T_c(0)$ of 125 K. *J. Solid State Chem., 109*, 35–42.
18. Gotor, F. J., Odier, P., Gervais, M., Choisnet, J., & Monod, P. H., (1993). Synthesis of $YBa_2Cu_3O_{7-x}$ by sol-gel route. *Physica C., 218*, 429–436.
19. Wong-Ng, W., McMurdie, H., Paretzkin, B., Zhang, Y., Davis, K., Hubbard, C., Dragoo, A. L., & Stewart, J. M., (1987). Reference x-ray diffraction powder patterns of fifteen ceramic phases. *Powder Diffr., 2*, 257–265.
20. Wong-Ng, W., & Cook, L., (1994). A review of the crystallography and crystal chemistry of compounds in the BaO-CuOx system. *Powder Diffr., 9*, 280–289.
21. Whiter, J. D., & Roth, R. S., (1997). *Phase Diagrams for High-Tc Superconductors* (p. 170). Amer. Ceramic. Society, Westerville, Ohio, (USA).

CHAPTER 14

Deep Learning Applications in Predicting Polymer Properties

LELA MIRTSKHULAVA

Department of Computer Science, Ivane Javakhishvili Tbilisi State University, Tbilisi, Georgia

ABSTRACT

A main fundamental goal in polymer development is to design new polymers with specific properties, but this process is budget unfriendly and takes too much time, even years. The goal of the given methodology is to automate this process using machine learning (ML) and deep learning (DL) algorithms to predict the polymer properties. We are using graph convolutional neural networks (GCNN) to represent polymer molecules as graphs. In the given chapter, our idea is to train DL algorithms on a huge amount of datasets to identify unrecognized patterns and propose new materials. Materials informatics is capable of considerably improving the process of the development of new materials including polymers. Polymer informatics is an interdisciplinary field using artificial intelligence (AI), ML, and DL tools for new polymers development and design.

14.1 INTRODUCTION

Briefly described, a polymer is a material that consists of macromolecules that compose many repeating subunits. Both natural and synthetic polymers play crucial roles in everyday life thanks to their broad spectrum of properties [1, 2]. Polymers count over 500 billion dollars in goods shipment and

Advanced Polymer Structures: Chemistry for Engineering Applications. Omar Mukbaniani, DSc, Tamara Tatrishvili, PhD & Marc Jean M. Abadie, DSc (Eds.)

this amount is expected to grow at twice. The polymer community would be well served by organizing and exploiting experimentally and computationally generated data. The problem of nomenclature is exacerbated in sequenced defined polymers especially when it comes to a synthetic polymer [3, 4].

A synthetic polymer is not a single entity. Individual polymer samples can be described by distributions which with complicated monomeric structures leads to nonstandard naming conventions. Moreover, commercial trade names complicate matters. For instance, polystyrene can be described by 1,800 different names. Even the International Union of Pure and Applied Chemistry (IUPAC) naming conventions and Chemical Abstracts Service Registry Numbers created its own numbering system [5–7].

The emerging field known as polymer informatics is a new direction and subfield of materials informatics deeply gaining acceptance among the classes of materials. In our case, we focus on polymer science with a series of challenges using artificial intelligence (AI) techniques such as machine learning (ML) and deep learning (DL) approaches to design new materials. We analyze the challenges faced in polymer science through a huge amount of datasets and give new research methods to solve the issues through computational algorithms which can reduce the time and cost of developing new types of polymers. Previously, we applied neural networks for medical applications and cybersecurity issues [9–14].

DL or deep neural network is a subfield of an AI that mimics the functions and workings of the human brain in processing data and creating patterns for use in decision making. The word "deep" in DL means the use of multiple layers in artificial neural networks (ANN). Early research works showed that DL is concerned with an unbounded number of layers of bounded size, which permits practical and optimized implementation retaining theoretical universality.

A huge amount of scientific data together with computational and algorithmic advances can reduce the time and cost of developing new polymers. The emerging field named "Materials informatics" has gained recognition in a number of classes of materials, including oxides and metals [16–35].

In materials discovery, we encounter four paradigms:

- Original paradigm is the essential process of experimentation;
- The second paradigm implies physics-based models;
- The third paradigm of computer simulations includes integrated computational materials engineering (ICME); and

- The fourth paradigm should be used in conjunction with the prior paradigms.

14.2 POLYMER (ADDITION POLYMER) STRUCTURE AND PROPERTIES

Polymers can be referred to as materials built from repeating chemical chains. Polymer structures may consist of different components started with a chain of backbones (links bonded chemically). A chemical bond holds atoms together in any molecule. Any atom forming two or more chemical bonds can form a chain. An addition polymer is a type of polymer that is formed between monomers by chain addition reactions containing a double bond. Molecules of ethene can enter the polymerization process with each other forming a new polymer with the name the polyethylene under the right conditions [16] (Table 14.1).

TABLE 14.1 Some Important Addition Polymers Derived from Ethylene

Polymer	**Repeating Unit**
Polyethylene	$-[CH_2-CH_2]-$
Polypropylene	$-[CH_2-CH(CH_3)]-$
Poly(vinyl chloride)	$-[CH_2-CH(Cl)]-$
Polyacrylonitrile	$-[CH_2-CH(C{\equiv}N)]-$
Poly(vinyl acetate)	$-[CH_2-CH(O-C(=O)-CH_3)]-$
Polystyrene	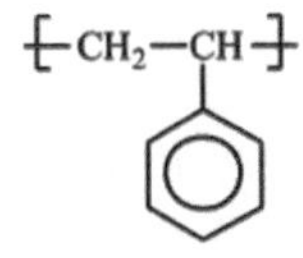

TABLE 14.1 *(Continued)*

Polymer	Repeating Unit
Poly(methyl methacrylate)	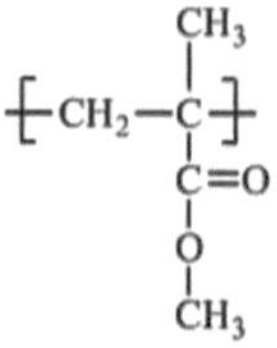
Poly(vinylidene chloride)	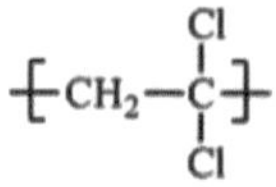

14.3 METHODS

Polymer informatics is an interdisciplinary domain using AI, ML, and DL tools in polymers development, design, and discovery (Figure 14.1). Mainly there are methods to solve inverse issues according to the polymer. Polymer informatics is a particular case of Materials Informatics. There are important challenges related to polymeric systems and their properties. We focus on the discussion of AI approaches for designing new materials.

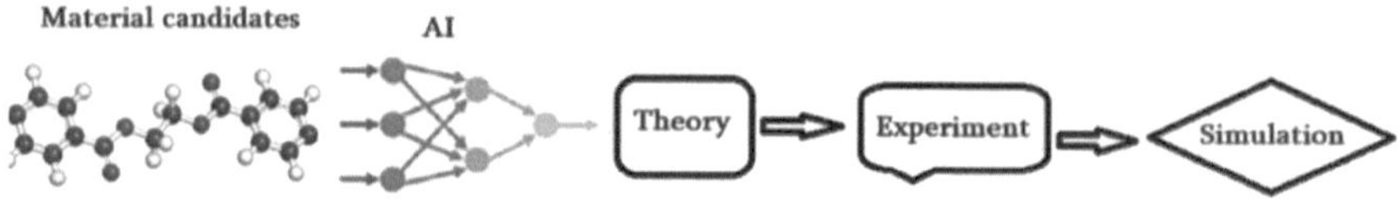

FIGURE 14.1 Material informatics framework.

n-polymer development, a deep understanding of the relationship between structure and properties is required for innovating new polymers. Density functional theory (DFT) suits well the polymer properties estimation process. Through small molecules are generated dataset. We can use a subset from this dataset where molecules represent graphs, atoms represent vertices, and bonds can be represented as edges.

14.4 HOW TO DERIVE POLYMER PROPERTIES USING DEEP NEURAL NETWORKS?

Deep Neural Network applications allowed access to a wideband chemical space towards small molecule's predictive synthesis. Through DFT calculations has been enabled bulk polymers property prediction on a set of repeat units for a fingerprint provision. After that, the final polymer properties have been derived using ML and DL algorithms. Using direct morphological information for fingerprinting instead of using DFT would be a great advance in the discovery and development of the new material. To use this methodology a strong understanding of the material descriptors influencing the properties of the polymers is required.

Our idea in the given chapter is to train ML algorithms on a huge amount of datasets for identifying patterns unrecognized before and propose new candidate materials. Materials informatics is capable of considerably improving new materials development including polymers (Figures 14.2 and 14.3).

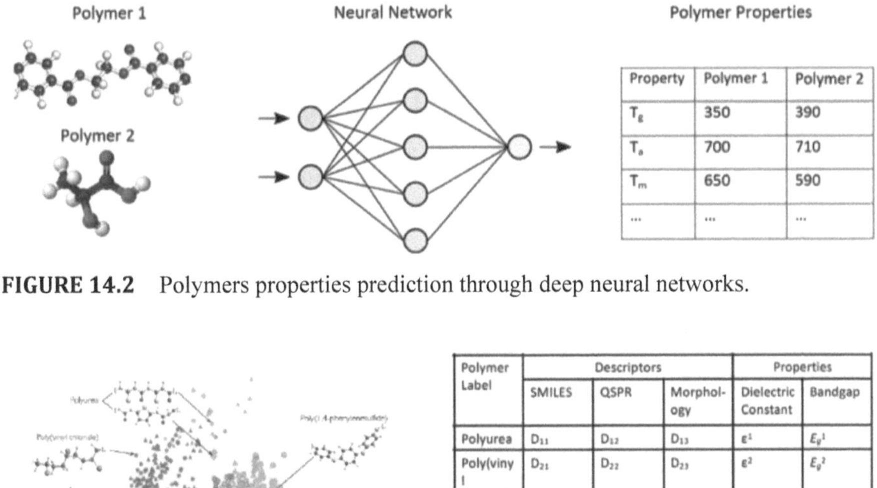

FIGURE 14.2 Polymers properties prediction through deep neural networks.

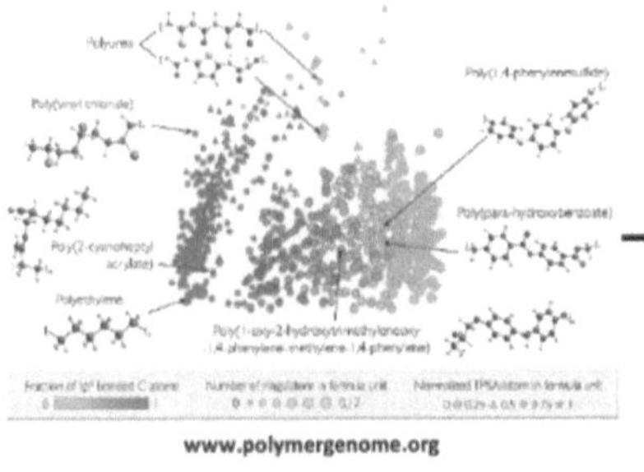

Polymer Label	Descriptors			Properties	
	SMILES	QSPR	Morphology	Dielectric Constant	Bandgap
Polyurea	D_{11}	D_{12}	D_{13}	ε^1	E_g^1
Poly(vinyl chloride)	D_{21}	D_{22}	D_{23}	ε^2	E_g^2
Poly(2-cyano heptyl acrylate)	D_{31}	D_{32}	D_{33}	ε^3	E_g^3
......	.....	.....	.....	.....	.....
C6H4-C4H2S-C4H2S-C4H2S-	D_{n1}	D_{n2}	D_{n3}	ε^n	E_g^n

FIGURE 14.3 Traditional methodology.

14.5 EXPERIMENTAL RESULTS

Polymer Name	**Poly(vinylbenzene) poly(styrene) Poly(1-phenylethylene)** PS
SMILES	**[*]CC([*])c1ccccc1** Polystyrenes Polyvinyls
Similar Polymers	**Poly[1-(2-methylphenyl)ethylene]** **Poly{1-[2-(methoxymethyl)phenyl]ethylene}**

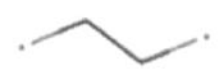

Electronic Properties

Bandgap (Bulk)	6.7 ± 0.5 eV
Bandgap (Single Chain)	6.8 ± 0.3 eV
Ionization Energy	7.0 ± 0.4 eV
Electron Affinity	0.4 ± 0.3 eV
Electron Injection Barrier wrt Al	4.1 ± 0.3 eV
Electrical Conductivity	Select dopant

Dielectric & Optical Properties

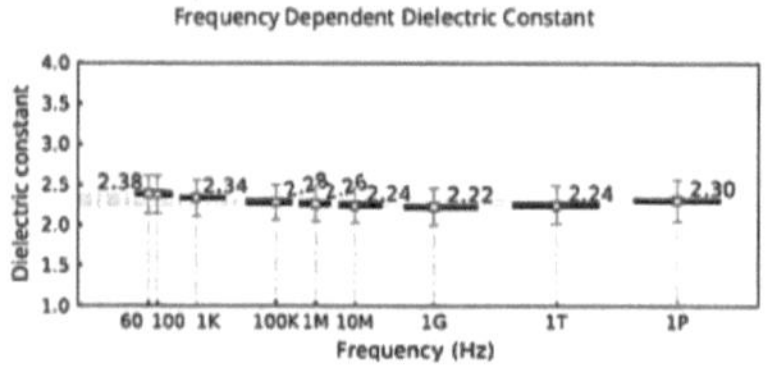

Dielectric Constant (Crystal)	2.8 ± 0.5
Refractive Index (Bulk Resin)	1.49 ± 0.05
Refractive Index (Crystal)	1.6 ± 0.1

Thermal Properties

Glass Transition Temperature	218 ± 26 K
Melting Temperature	331 ± 38 K
Thermal Decomposition Temperature	642 ± 49 K

Physical & Thermodynamic Properties

Density	0.9 g/cc
Atomization Energy	-5.1 eV/atom
Specific Heat	2 J/gK
Cohesive Energy Density	65.8 ± 12.1 Cal/cc
Fractional Free Volume	0.30 ± 0.05

Mechanical Properties

Tensile Strength	< 50.8 MPa
Young's Modulus	1044.04 ± 786.55 MPa

Other Properties

Tendency to Crystallize	40 ± 25 %
Limiting Oxygen Index	18 ± 7 %

Solubility Properties

Solvent	Acetic acid Dichloromethane Chloroform Nitrobenzene Benzene Chlorobenzene THF 1,4-dioxane NMP M-cresol Toluene Ethyl acetate DMF N-butanol DMAc Acetonitrile
Non-solvent	Water Methanol Nhexane Diethyl ether Ethanol Isopropanol DMSO Acetone
Hildebrand Solubility Parameter	17.0 ± 0.9 $MPa^{1/2}$

Permeability Properties

Gas	He	H_2	CO_2	N_2	O_2	CH_4
Permeability (Barrer)	4.1	9.1	10.8	4.2e-1	1.5	3.2
log(Permeability)	0.6 ± 0.2	1.0 ± 0.3	1.0 ± 0.2	-0.4 ± 0.2	0.2 ± 0.2	0.5 ± 0.2

Solvents are collected based on a comparison between ML predicted Hildebrand solubility parameter of given polymer and *experimentally obtained Hildebrand solubility parameter (δ) of solvents. * δ2 = δd2 + δp2 + δh2 (Figure 14.4).

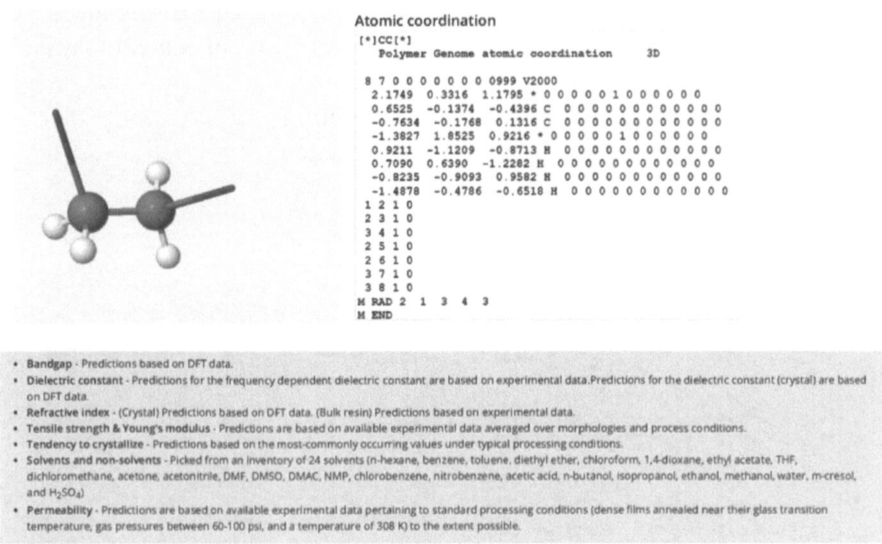

FIGURE 14.4 3D structure of monomer has optimized by UFF force field. Purple wires are added to show two endpoints of the monomer.

14.6 GRAPH NEURAL NETWORK (GNN) APPLICATIONS FOR REPRESENTING THE POLYMER MOLECULES

Our second approach includes graph neural networks (GNNs). GNN is a class of DL capable of inferring data represented by graphs. GNNs are neural networks capable of performing prediction on different levels: node (vertices), edge, and graph. A graph is the fundamental part of GNN. A graph represents a data structure comprising two components: vertices (V) and edges (E). We can define graph G as G = (V, E). A graph can represent molecules taking them directly as input [15, 22, 26].

GCNN are capable of predicting the polymer's dielectric constant and energy bandgap. They can achieve significant agreement using DFT. GNN outperforms ML algorithms. GCNN relies only on the polymers' morphological data and substitutes the hand-crafted descriptors offering high accuracy in fast predictions.

GCNNs represent molecules as the graphs where atoms are represented as nodes and bonds as edges. Features can be considered as atom type, charge, bond type, etc. A trained GCNN model can be applied to any molecule (Figure 14.5 and Table 14.2).

Main steps in training procedure:

i. Define the molecule graph G = (A, B, X, Y), where A is the set of atoms, B is the set of bonds, X represents the atom context matrix, Y represents the bond context matrix.
ii. Learn knowledge from the graph structure.
iii. Learn to predict the products of the reaction.

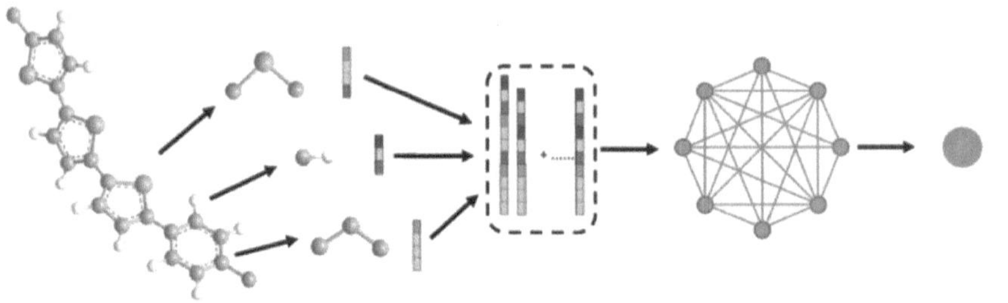

FIGURE 14.5 GCNN methodology for polymer properties prediction.

TABLE 14.2 Comparing the GCNN and the Hand-Crafted Methodologies

Hand-Crafted Feature Classes	GCNN Generated Feature Classes
Structural heterogeneity	Group number
Chemical ordering	Period number
Maximum packing efficiency	Electronegativity
Stoichiometry	Covalent radius
Element property	Valence electrons
Valence orbital	First ionization energy
Ionic property	Electron affinity
	(s, p, d, f) Block
	Atomic volume
	Atomic distance

14.7 CONCLUSION

In the given work, we have used ML algorithms to predict bandgap, dielectric constant, and other properties from a large dataset of polymers. Our results showed that GCNN can offer a very effective methodology for accurate and automated prediction of polymer properties based on the atomic and morphological character of polymer. The polymer scientists will benefit by following the actions to organize, exploit computationally and experimentally generated data.

KEYWORDS

- **artificial intelligence**
- **deep learning**
- **graph neural networks**
- **machine learning**
- **polymer deep learning**
- **polymer informatics**

REFERENCES

1. Fried, J. R., (2014). *Polymer Science and Technology* (3rd edn.).
2. Kuenneth, C., Rajan, A. C., Tran, H., Chen, L., Kim, C., & Ramprasad, R., (2021). *Polymer Informatics with Multi-Task Learning, Patterns 2*, 100238.
3. Dima, A., Bhaskarla, S., Becker, C., Brady, M., Campbell, C., Dessauw, P., Hanisch, R., et al., (2016). Informatics infrastructure for the materials genome initiative. *JOM, 68*, 2053–2064.
4. https://www.sciencenewsforstudents.org/article/explainer-what-are-polymers (accessed on 02 January 2022).
5. https://en.wikipedia.org/wiki/Polymer (accessed on 02 January 2022).
6. Debra, J. A., & Juan, J. De. P., (2017). Polymer informatics: Opportunities and challenges. *ACS Macro Letters, 6*(10), 1078–1082. doi: 10.1021/acsmacrolett.7b00228.
7. https://chem.libretexts.org/Bookshelves/Introductory_Chemistry/Book%3A_Introductory_Chemistry_(CK-12)/25%3A_Organic_Chemistry/25.19%3A_Polymerization_-_Addition_Polymers (accessed on 02 January 2022).
8. https://www.mathworks.com/help/deeplearning/ref/trainnetwork.html (accessed on 02 January 2022).
9. Al-Dabass, D., & Mirtskhulava, L., (2021). Chapter 1 – application of dynamical systems based deep learning algorithms to model emergent characteristics for healthcare diagnostics. *Handbook of Computational Intelligence in Biomedical Engineering and Healthcare* (pp. 1–29). Academic Press, ISBN 9780128222607. https://doi.org/10.1016/B978-0-12-822260-7.00009-1.
10. Shishniashvili, E., Mamisashvili, L., & Mirtskhulava, L., (2020). Enhancing IoT security using multi-layer feedforward neural network with tree parity machine elements. *International Journal of Simulation Systems, Science & Technology. Special Issue: Conference Proceedings*. UKSim2020.

11. Mirtskhulava, L., Gulua, N., & Meshveliani, N., (2019). IoT security analysis using neural key exchange protocol. GESJ: Computer science and telecommunications. *Reviewed Electronic Scientific Journal, 2*(57).
12. Mirtskhulava, L., (2018). Monitoring brain attacks using AI diagnosis. *AI Med. Journal.* CA, USA. Featured Article https://chem.libretexts.org/Bookshelves/Introductory_Chemistry/Introductory_Chemistry_(CK-12)/25%3A_Organic_Chemistry/25.19%3A_Polymerization_-_Addition_Polymers (accessed on 02 January 2022).
13. Karam, J., Al-Majeed, S., Yalung, C. N., & Mirtskhulava, L., (2016). Neural network for recognition of brain wave signals. *International Journal of Enhanced Research in Science, Technology & Engineering, 5*(10). ISSN: 2319-7463.
14. Mirtskhulava, L., Wong, J., Pearce, G., & Al-Majeed, S., (2015). Artificial neural network model in stroke diagnosis. *Conference: 2015 17th UKSIM-AMSS International Conference on Modelling and Simulation.* At Cambridge, UK. doi: 10.1109/UKSim.2015.33.
15. Reddy, M. C., Sivaramakrishna, L., & VaradaReddy, A., (2012). The use of an agricultural waste material, jujuba seeds for the removal of anionic dye(Congo red) from aqueous medium. *J. Hazard Mater., 203*, 118–127. https://ai-med.io/neurology/monitoring-brain-attacks-ai-diagnosis/ (accessed on 02 January 2022).
16. Perego, M. G., (2021). *Free-Radical Polymerization Toolbox.zip* https://www.mathworks.com/matlabcentral/fileexchange/64340-free-radical-polymerization-toolbox-zip MATLAB Central File Exchange.
17. https://dmol.pub/dl/gnn.html (accessed on 02 January 2022).
18. https://www.polysciences.com/german/catalog-products/polymers/amine-functional-polymers (accessed on 02 January 2022).
19. https://www.chemicalbook.com/ProductChemicalPropertiesCB8854425_EN.htm (accessed on 02 January 2022).
20. Mueller, T., Kusne, A. G., & Ramprasad, R., (2016). *Reviews in Computational Chemistry* (pp. 186–273). John Wiley & Sons, Inc: Hoboken, NJ. https://pubmed.ncbi.nlm.nih.gov/22209325/ (accessed on 02 January 2022).
21. Miquelard-Garnier, G., & Roland, S., (2016). Beware of the Flory parameter to characterize polymer-polymer interactions: A critical reexamination of the experimental literature. *Eur. Polym. J., 84,* 111–124.
22. https://graphneural.network/#installation (accessed on 02 January 2022).
23. https://en.wikipedia.org/wiki/Methyl_methacrylate (accessed on 02 January 2022).
24. http://www.odbornecasopisy.cz/en/post/polymer-synthesis-gets-a-jolt-of-caffeine--2903 (accessed on 02 January 2022).
25. https://hal.archives-ouvertes.fr/hal-02271688v1 (accessed on 02 January 2022).
26. https://towardsdatascience.com/a-gentle-introduction-to-graph-neural-network-basics-deepwalk-and-graphsage-db5d540d50b3 (accessed on 02 January 2022).
27. https://www.sciencedirect.com/science/article/pii/0166218X88900121 (accessed on 02 January 2022).
28. https://www.nature.com/articles/pj20136 (accessed on 02 January 2022).
29. https://en.wikipedia.org/wiki/Myoglobin (accessed on 02 January 2022).
30. https://www.sciencenewsforstudents.org/article/explainer-what-are-polymers (accessed on 02 January 2022).
31. http://quantum-machine.org/datasets/ (accessed on 02 January 2022).
32. https://en.wikipedia.org/wiki/Radical_polymerization (accessed on 02 January 2022).

33. https://webbook.nist.gov/cgi/cbook.cgi?ID=58-08-2 (accessed on 02 January 2022).
34. https://molview.org/ (accessed on 02 January 2022).
35. https://www.sciencedirect.com/science/article/pii/B9780128027349000196 (accessed on 02 January 2022).

CHAPTER 15

Fractal in Modern Science, Fractalization Theory

SHAHRIAR GHAMMAMY,[1] MEHDI GHAMAMI,[2] FARZAD HAGHIGHI,[1]
SEYED HAMED HOSSEINI,[1] GOLNOOSH MIVEHCHI,[1]
POYA SOROSHIAN,[1] and SEYEDEH SABA TABATABAEI[1]

[1]*Department of Chemistry, Faculty of Science, Imam Khomeini International University, Ghazvin, Iran*

[2]*Department of Mechanical Engineering, Isfahan University of Technology, Isfahan, Iran*

ABSTRACT

In this research, the subject of self-similarity and fractal has been addressed. Self-similarity is one of the newest and most interesting properties observed in various fields of science. According to this phenomenon, many natural phenomena can be interpreted, and then self-similarity of natural phenomena can be extracted. The dimension of self-similarity is an non-integer number that indicates the fractal properties of the phenomenon. Different methods of extracting of the self-similarity dimension have been expressed and using one of these methods, the dimension of metal oxide nanoparticles (NPs) has been studied. The used self-similarity dimensions method is independent of the type of material and can be applied to any nanomaterial. The properties of the nanomaterial can be described based on its fractal dimension number.

Advanced Polymer Structures: Chemistry for Engineering Applications. Omar Mukbaniani, DSc, Tamara Tatrishvili, PhD & Marc Jean M. Abadie, DSc (Eds.)

15.1 INTRODUCTION

Benoit Mandelbrot or Mandelbrot was born on November 20, 1924, in Warsaw, Poland, in a Lithuanian Jewish family. His father was a second-hand clothes seller and his mother was a doctor. He learned mathematics from his two uncles and immigrated to France with his family in 1936. This migration made him more interested in mathematics. The outbreak of World War II led to his lack of university education, which gave him the advantage of not looking at the phenomenon of existence through the eyes of a mathematician or academic scientist and led him to develop very interesting methods of geometry in mathematics.

The very similarity found in nature can be considered the original home of fractals. The most beautiful forms are found in nature and natural behaviors after the riot. Nature always refers to beauty in every moment. What we have found in nature in all ages has emphasized beauty, but the order that nature creates seeks disorder. No forest can be found where the trees have grown in regular rows by nature but looking at a dense forest in this chaos. It is never seen in the layout. Gaston Maurice Julia, an Algerian mathematician, wrote his famous paper on rational functions and the repetition of polynomials in mathematics in 1918 under the supervision of the French mathematician Emile Peykar.

He was a doctoral student in mathematics and a World War I veteran. Unfortunately, what he had gained in mathematics was forgotten after the war, until the Polish Mandelbrot continued his work. Mandelbrot, an employee of the IBM programming company at Harvard University in the United States, was fascinated by the complex and irregular structure of natural phenomena and, with the help and inspiration of Julia, created a mathematical rhythm pattern. The term fractal was introduced to mathematics in 1976 by Mandelbrot. When researching the length of the England coast, Mandelbrot found that when measured on a large scale, it was longer than when it was smaller.

Fractal is derived from the Latin word (fractus) or (fractum) meaning broken, which indicates one of its main identifiers – divisible; a type of geometric structure in which each small part has a pattern of the whole structure and can be considered a version of the structure itself on a small scale. He found that similar patterns seem to be repeated at different scales, for example the main trunk and the largest branches of a tree create a specific pattern for the branches, but when we focus on a particular large branch, the smaller branches remain the same level of the details that are revealed from the branch pattern [1].

Fractal geometry is one of the new branches of mathematics that has shown unparalleled flexibility and capability in interpreting and simulating various forms of nature. Using fractal geometry, mathematicians, and researchers were presented with a clear horizon for recounting the behavior of seemingly uneven and chaotic functions and sets. Today, they use fractal laws in such ways: the process of describing many diseases of medical science in explaining natural phenomena of the earth, controlling the process of scaling and distribution of fluids in chemical engineering and predicting many phenomena in other fields of sciences.

Fractal dimension analysis using imaging processing methods is a powerful tool to obtain the morphological information of nanomaterials that are being used today. Many people are fascinated by images called fractals. Fractal geometry is beyond the ordinary people's perception of mathematics, which they find complex and tedious formulas. This geometry combines mathematics with art and shows that equations are nothing but a set of numbers. What makes fractals even more interesting is that they are the best mathematical descriptions available for many natural phenomena, such as beaches, mountains, or parts of living things [2].

Although fractal geometry is closely related to technology and computers, some people worked on fractals long before the invention of the computer. Two of the most important properties of fractals are their self-similarity and their non-integer dimension. If you look closely at the leaves, you will notice that the shape of each small leaf (part of the larger leaf) is similar to the whole leaf, and it can be said that one leaf is similar to the larger branch itself. The same is true for fractals: you can zoom in and out over and over again to see the same shape after each step. Figure 15.1 shows a self-similarity pattern in plant leaves.

FIGURE 15.1 Pattern of self-similarity in plant leaves.

Classical geometry deals with objects of correct dimensions: zero-dimensional points, one-dimensional lines, two-dimensional plane curves and shapes such as squares and circles, and three-dimensional objects such as cubes and spheres. However, many natural phenomena can be better described by using dimensions between two arithmetic numbers. So, while one dimension is a straight line, depending on how closely the space is twisted and curved, the fractal curve will be between one and two.

The more the fractal pattern fills the flat plate, its dimension gets closer to two. Likewise, the "mountain fractal view" reaches dimensions between two and three. Thus, the dimension of a fractal scene, which consists of a large hill covered with small and small hills, approaches two. But if the mountain contains a rugged surface consisting of a large number of medium hills, the next will have nearly three.

There are different types of fractals, two of the most popular of which are: (i) complex number fractals; and (ii) iterated function system or IFS.

15.2 FRACTAL EXAMPLE

15.2.1 MANDELBROT SET

The Mandelbrot Set is a collection of points on a mixed screen. To build the Mandelbrot set, we must use an algorithm based on the recursive formula:

$$Z_n = Z_{2n-1} + C \qquad (1)$$

We divide the points on the mixed screen into two categories:

i. Points within the Mandelbrot set; and
ii. Outside Mandelbrot set.

Figure 15.2 shows part of the mixed page. The points of the Mandelbrot sets are marked in black.

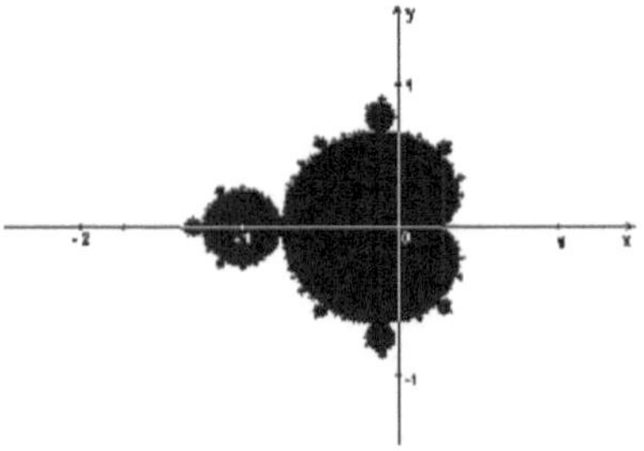

FIGURE 15.2 The Mandelbrot set.

15.2.2 JULIA SET

Julia sets are closely related to the Mandelbrot sets, and the repetitive function used to produce them is the same process as the Mandelbrot sets. The only difference is how you use this formula. To draw an image of the Mandelbrot series, we always repeat the formula for each point C on the composite plane, starting with $Z_0 = 0$. If we want to create an image of a Julia set, in all the steps of its formation, C must be constant, while the value of Z_0 is variable. The value of C determines the shape of the Julia set. In other words, each point on the mixed page corresponds to a specific Julia set.

IFS fractals are created based on simple plane transformations, such as scaling, shifting, and axis rotation. Creating an IFS fractal involves the following steps:

- Define a set of pages transformations;
- Draw an initial pattern on the page (any pattern);
- Convert the initial pattern using the transformations defined in the first step;
- Convert new image (a combination of original and converted patterns) using the same set of conversions;
- Repeat the fourth step as many times as possible (in theory, this method can be repeated over and over again).

The most famous IFS fractals are the Sierpiński Triangle and the Koch Snowflake.

15.2.3 SIERPIŃSKI TRIANGLE

The Sierpiński Triangle is a fractal that is formed by connecting the midpoints of each side of an equilateral triangle. We have to do the repetitions over and over again. Figure 15.3 shows the first four steps of making the Sierpiński triangle. Figure 15.3 shows the developmental stages of the Sierpiński triangle.

Using this example, we can prove that the dimension of fractals is not a numerical integer. We must first understand how the "dimension" of an object behaves as its linear dimension increases. In one dimension, we can consider a line segment. If the linear dimension of the line segment doubles, the length (characteristic size) of the line also doubles. In two dimensions, if the linear dimensions of a square are doubled, the size of the feature which is the area, increases by a factor of four. Also in three dimensions, if the linear dimension of a box doubles, its volume increases by a factor of 8.

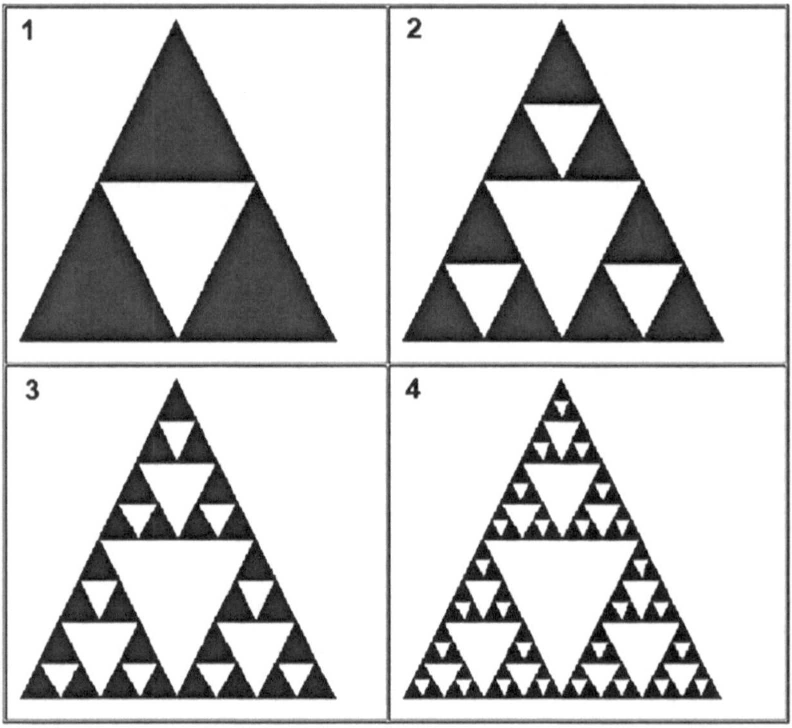

FIGURE 15.3 The Sierpiński triangle.

This relationship between the D dimension, the linear scale L, and the result of increasing the size S can be generalized and written as follows:

$$S = L \cdot D \quad (2)$$

By rewriting this formula, depending on how the size is changed as a function of linear scaling, we get a description for the dimension:

$$D = \log(S)/\log(L)$$

The result of these calculations shows that the fractal dimension is non-integer.

15.2.4 KOCH SNOWFLAKE

To make a Koch Snowflake, we have to start with an equilateral triangle with side length, for example 1. In the middle of each side, we will add a new equilateral triangle with one-third side and repeat this process indefinitely. The length of the borders or the environment is infinite. 3.4/3.4/3.4/3.4/3...

However, the area is less than the area of a circumferential circle around the main triangle. This means that an infinitely long line surrounds a finite surface. The final snowy structure of Koch resembles a shoreline.

Figure 15.4 shows the four steps of Koch snowflake formation:

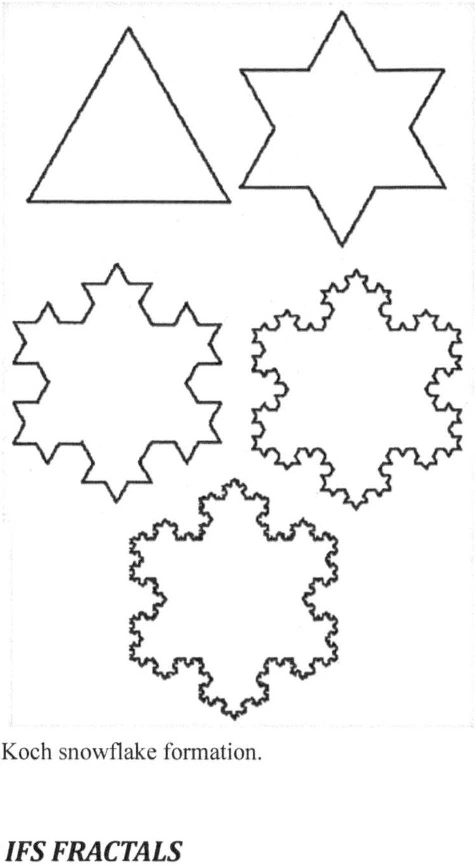

FIGURE 15.4 The Koch snowflake formation.

15.2.5 OTHER IFS FRACTALS

Fern leaves and spirals that are IFS fractals (Figure 15.5).

FIGURE 15.5 Simulated fractals in the software environment.

15.2.6 APPLICATION OF FRACTALS

- Astrophysics;
- Biology;
- Computer graphics;
- Fractal dimension analysis using imaging analysis method is a powerful tool to obtain the morphological information of nanomaterials that are used today.

15.3 FRACTAL CALCULATIONS

15.3.1 CALCULATION OF FRACTAL DIMENSION

Dimension in fractal geometry is one of the criteria in fractal geometry which includes packing, dimension, box count dimension, Hausdorff dimension and modified box count dimension. Due to the simplicity of the box count dimension, this method is commonly used in calculating fractal dimensions. Whit Using of MATLAB and SPSS software, image analysis can be achieved. In MATLAB, an image is stored in three dimensions, the first and second dimensions of which are to determine the coordinates of points. Therefore, to calculate the fractal dimension of the first and second dimensions of the images in MATLAB software is required. You can also use SPSS and MATLAB software to obtain the fractal dimensions of different images, compare them with each other, and from these features can in some cases lead to a specific result or process (such as the formation of a cancerous mass and how it can spread). Therefore, the use of fractal geometry can be used in the diagnosis of various disease [3].

Using the box-counting method, the fractal dimension can be obtained the following formula:

$$D = \lim_{r \to 0} \frac{\log Nr}{\log(1/r)}$$

D In the formula equals to the fractal dimension, Nr equals to the number of squares that are part of the fractal dimension. By looking at the above formula, a logarithmic diagram can be drawn in which the horizontal and vertical axes are like this formula:

$$\log Nr$$

$$\log\left(\frac{1}{r}\right)$$

As mentioned, there are several methods for calculating fractal dimensions as follow:

- Self-similar dimension;
- Fractal dimension of mass;
- The fractal dimension of the spar sphere;
- Euclidean fractal dimension;
- Network fractal dimension;
- Fractal reduction analysis;
- Fractal dimension of box counting.

These seven methods are used to calculate fractal dimensions, each of which has its own approximation and has its own computational complexity. In general, the box counting method has the advantage of simplicity of calculations and is usually used to calculate fractal dimensions [4].

For the fractal dimension of the box count, we assume that F is a non-zero and finite subset digits and (NR) assume that set with a maximum dimension, (r) that can cover (f). (RN) the least number of fractal.

$$Df = \lim_{r \to 0} \frac{\log Nr(F)}{-\log r}$$

The circle of cubes in the coordinates of *Rn* indicates the cube as follows:

If $[m_1 r, (m_1 + 1)r] \times \cdots \times [m_n r, (m_n + 1)r]$,
then $m_n, \ldots, m_1$ are integers.
Finally, $N_r \sqrt{n}(F) \leq N_r'(F)$.
If $r\sqrt{n} < 1$,
then $\log N_r \sqrt{n}(F) / \log(r\sqrt{n}) \leq \log N_r'(F)/(-\log \sqrt{n} - \log r)$.
If we limit $\dim_B F = \lim_{r \to 0}(\log N_r'(F) / -\log r)$,
then $\overline{\dim}_B F \leq \overline{\lim}_{r \to 0}(\log N_r'(F) / -\log r)$.
Finally, $N_r'(F) \leq 3^n N_r(F)$.

Using SPSS software, when a number of images are available, the average dimensions, middle, maximum, minimum, amplitude, harmonic mean and distortion can be obtained and compared. In statistical science, distortion means the degree of asymmetry of the probability distribution, if it is a measure of the presence or absence of asymmetry in the probability distribution graph. So whatever the numerical distortion is small and tends to zero. The graph has more symmetry, and the larger the numbers, the more asymmetric the graph has.

15.4 FRACTAL CALCULATIONS IN MATLAB

MATLAB is a software for numerical calculations and a fourth-generation programming language. This software was initially developed based on the C programming language. The word MATLAB means both numerical calculations and the corresponding programming language itself. All data is stored in MATLAB as a matrix. Images are stored in MATLAB in the form of a three-dimensional matrix, the first and second dimensions of which are used to determine the coordinates of points and the third dimension to determine the color of points. In MATLAB, images are defined as two, three or four-dimensional matrices. Image quality depends on two parameters: dimensional accuracy and depth accuracy when shooting and storing the image. Depth accuracy refers to the number of bits that are allocated from computer memory to any point (pixel) of the image. MATLAB powerful software has made it possible to perform calculations as well as image processing. Using the programs given to this software, fractal dimension calculations of any metal oxide nanoparticle or any nano-drug are possible, as well as histogram diagrams [5].

One of the magnifying instruments that uses electrons instead of light rays is a scanning electron microscope (SEM). This microscope is a study tool in nanotechnology that can produce images of objects as small as 10 nanometers by electron bombardment. Using this tool, preparing a sample for study, requires information about the features of the mechanism, its components and its function. SEM allowed researchers to study larger samples more easily and more clearly. The bombardment of the sample causes electrons to be released from the sample to the positively charged plate, where these electrons are converted into signals. The movement of the beam on the sample provides a set of signals on which the microscope can display an image of the surface of the sample on a computer screen [6]. This microscope provides the following information about the sample:

i. **Sample Topography:** Surface properties.
ii. **Morphology:** The shape, size, and placement of particles on the surface of the body.
iii. **Composition:** The components that make up the sample.

In this device, a larger image of the sample can be created with the help of electrons (instead of light). Figure 15.6 is an example of a SEM image of a metal oxide nanoparticle:

FIGURE 15.6 The scanning electron microscope image of a metal oxide nanoparticle.

Using image processing in MATLAB software, the graph of the image analysis results extracted from the SEM of this metal oxide nanoparticle was extracted as in Figure 15.7.

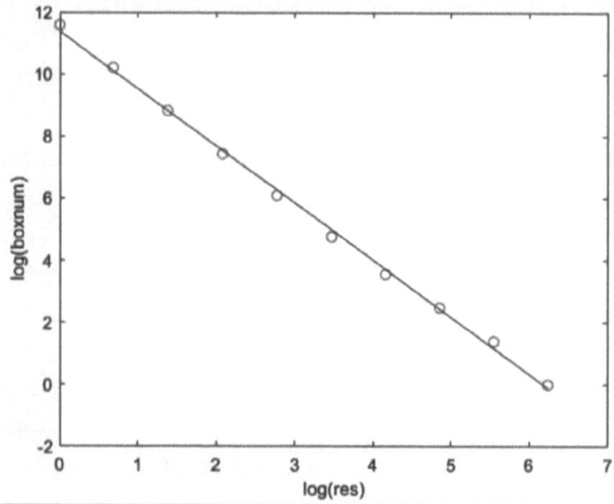

FIGURE 15.7 The image processing of the SEM image.

15.5 CALCULATION WITH SPSS

SPSS is short for statistical package for the social science, this magical software is used by a wide range of scientists and researchers for hard and complex statistical data analysis. It was created in 1968 by SPSS and was later acquired by IBM in 2009. Whenever someone want to do in-depth statistical analysis most scientist choose SPSS software and they know it as the best tool. So the most research institution prefer to use SPSS for analysis and extracting data. Because it has a higher level of simplicity and accuracy for researching projects.

To expand definition this software is used for data managing, advanced analysis, multivariate analysis, business intelligence, etc., so it has been used by a wide range of market researcher, health researcher, survey companies, education researchers, marketing organization, and many related companies to gain useful and effective data and investigate statistical data.

Statistics that include basic software:

1. **Descriptive Statistic:** Checkered tabulation, frequencies, descriptions, exploration, and relative description statistics.

2. **Bivariate Statistics:** Medians, t-test, dispersion analysis, correlation, non-parametric tests.
3. **Predicting Numerical Outputs:** Linear return.
4. **Predicting for Group Identification:** Factor analysis, cluster analysis, separator.

So, we can divide statistics into two different category:

i. **Descriptive Statistics:** These are methods that is used to summarize large amount of data. Some of description are used in every conversation, for example if we try to give information about average income we are using descriptive data.
ii. **Inferential Statistics:** These are the methods which we infer results from the collected data. Inferential Statistics helps us to ask question like "is there a difference?" or "is there a relationship" and answer to them by powerful knowledge of mathematics.

15.5.1 DATA ANALYSIS BY USING SPSS

There are three basic steps in analysis using SPSS. First, we have to enter the raw data and save them in a file. In the second step, we must select the required analysis and after that specify them. Third, we have to check the output and investigate final result.

15.5.2 CORRELATION COEFFICIENT AND REGRESSION COEFFICIENT

One of the types of descriptive research (non-experimental) is correlation research. In this type of research, the relationship between variables is the basis of the purpose of the research analyzed.

Correlation research can be divided into three categories according to purpose:

i. Two-variable correlation study;
ii. Regression analysis; and
iii. Correlation matrix analysis or covariance.

In bivariate correlation analysis, the aim is to investigate the relationship between two variable in research. In regression analysis the purpose of predicting changes in one or more dependent variables to predict the output (whit respect to changes in independent variables).

Some study of two-variable correlation is investigated in a table that is called correlation matrix or covariance. Some of the research in which correlation matrix or variance is investigated is analysis factor model structural equation. In factor analysis, the goal is to summarize a set of data or achieve concealed variables.

15.5.2.1 BIVARIATE CORRELATION ANALYSIS

In this kind of analysis, the goal is to determine the degree of coordination of changes in two variables. In this regard for this aim in term of size scale variable, we provide appropriate indicator. Since in most studies, the correlation of two variables of the distance scale by the default two-variable normal distribution is used to measure the variable, so the correlation coefficient is calculated. In such analysis, the Pearson torque is correlation coefficient, or Pearson correlation coefficient.

15.5.2.2 REGRESSION ANALYSIS

In such a study that uses regression analysis, the final goal is usually predicting one or more output variables from one or more predictor variables. We use regression model if the aim is to predict an output variable from several predictor variables. We use a multivariate regression model, If the goal is to simultaneously predict several output variables from predictor variables. In multiple regression analysis, the aim is to find variables predicted to anticipate changes in the output variable. Entering predictive variables in regression analysis is done in different ways. We will discuss three basic methods here:

i. Simultaneous method;
ii. Step-by-step method; and
iii. Hierarchical method;

In the simultaneous method, all predictors are entered into the analysis together. The step-by-step method based on a predictor variable with the highest zero-correlation coefficient with the output variable is entered into the analysis. After that, other variable predicting is entered in term of the fractional correlation coefficient, partial and semi partial in this method. New variable semi-segregated or segregated correlation coefficient, all variables already entered in the equation as the last input variable is reviewed and it has lost its significance with the introduction of a new variable, the equation it comes out. And finally in the hierarchical method, the order of entry of variables into the analysis is based on a theoretical or experimental framework by researcher. The decision which is adopted before the start of the analysis is based on the following principle:

- Relationship between cause and effect;
- The relationship of variable in previous research;
- Project structure;

15.5.2.3 ANALYSIS OF COVARIANCE OR CORRELATION MATRIX

In this research, the researcher wants to summarize the correlation of variables into more limited factors. When the correlation of a set of variables slows down or determined the underlying properties of a data, the factor analysis method is used. In other words scientists wants to test a specific model in term of the relationship of the variables. For both purposes, it is necessary to analyze the covariance matrix of measured variables.

1. **Factor Analysis:** The covariance matrix can be analyzed in factor analysis with two different goals: "exploratory goal" and "confirmatory goal." If the researcher does not have a hypothesis about the number of attribute factor, is called exploratory factor analysis (EFA) and if a hypothesis exist, it is called confirmatory factor analysis (CFA).
2. **Structural Equation Model:** In such study that aims to test a specific model of the relationship between variables, the analysis of structural equation models is used. The data in the form of covariance or correlation matrices and a set regression equation are compiled between variables.

15.6 CONCLUSION

1. Working with MATLAB Software: In MATLAB, images are defined as two-, three-, or four-dimensional matrices. Images qualities depends on two parameters, one is dimensional accuracy and the other is depth accuracy when saving images.
2. The Meaning of Depth Accuracy is the number of bits that are allocated from the computer memory to each point (pixel) of the image.

You consider that we give the SEM image of a nano-amiodarone to MATLAB software according to the function we have defined for the software. Matrix diagrams and histograms can be easily obtained: (fractal dimension of this nanomaterial 1.9138). According to the obtained diagrams, the spatial value of each point on the diagrams can be found in the software. The graph obtained from MATLAB Log (res) is an independent variable. This factor represents 1/r or the inverse of the pixel length.

(boxnum) Log is a dependent variable that indicates the number of pixels. Obtaining the spatial value of these points shows the relationship between the measurement length and the number of pixels This software uses the same rulers for all images. The difference for each image is in the (boxnum) Log, which results in the existing fractal pattern of each image. Despite the chaotic structure that the images have, referring to these diagrams takes a certain pattern order and shows. To achieve the correlation coefficient between the points on the graphs, we enter the table data into SPSS software.

Using Pearson correlation coefficient, we obtain the correlation coefficient of these points. After calculating the correlation coefficient of the points, we analyze the data in the table using linear equation regression coefficient. We get the slope of the FD line Which has a slight difference in the dimension calculated by MATLAB software, the width is also obtained from the beginning. So the max and min numbers and the data uncertainty are obtained. Now we can find the linear equation for each nanomaterial:

$$\text{Log (boxnum)} = x\ \text{Log(res)} + x$$

To obtain a frequency distribution, the hist histogram command can be given to MATLAB and its histogram diagram can be obtained. In this work, the fractal dimensions of the effective materials used in the synthesis of nanomaterials are calculated, and the results are entered into SPSS for statistical analysis. The purpose of this study is to prove the fractal behavior in materials (nanomaterials, composites, polymers, etc.). Recently, researchers have concluded that some properties of drugs and other materials can be justified and interpreted to the extent of their stability in the body using fractal dimension properties.

Using MATLAB software, the degree of homogeneity or heterogeneity of nanomaterials is checked using SEM images. This method is effective in product grading in some steel and food companies. Some hypotheses in the field of fractal dimension measurement were investigated in this study. The results show differences in fractal dimensions in different magnifications and slices of images. These differences are clearly visible in images such as cauliflower and ferns. Therefore, the hypothesis of fractal dimensions being the same in all sections of a sample was rejected. We named this kind of theory fractalization theory and started the study of universe phenomena by this type of view and theory. Some of our data was published in different fields [8–11].

KEYWORDS

- **geometry**
- **image processing**
- **iterated function system**
- **Mandelbrot set**
- **nanoparticle**
- **scanning electron microscope**
- **Sierpiński triangle**

REFERENCES

1. Kopelman, R., (1986). Rate processes on fractals: Theory, simulation and experiment. *Journal of Statistical Physics, 42*, 185–200.
2. Sander, L. M., (1987). *Fractal Growth* (Vol. 256, pp. 82–88). Scientific American.
3. Witten, T. A., & Sander, L. M., (1983). Diffusion limited aggregation. *Physics Review B, 27*, 5686–5697.
4. Liebovitch, L. S., & Toth, T., (1989). A fast algorithm to determine fractal dimensions by box counting. *Physics Letters A, 141*, 386–390.
5. Tabanfar, Z., et al., (2016). Brain tumor detection using electroencephalgram linear and non-linear features. *Iranian Journal of Biomedical Engineering, 10*(3), 211–221.
6. Sarkar, N., & Chaudhuri, B. B., (1992). An efficient approach to estimate fractal dimension of textural images. *Pattern Recognit., 23*, 1035–1041.

7. Maus, S., & Dimri, V. P., (1994). Fractal properties of potential fields caused by fractal sources. *Geophys. Res. Lett., 21*, 891–894.
8. Kazemi, K. A., Ghamami, S., & Bahrami, Z., (2018). Fractal properties and morphological investigation of nano hydrochlorothiazide is used to treat hypertension. *BMC Pharmacology and Toxicology, 19*, 1–9.
9. Kazemi, K. A., Ghamami, S., & Bahrami, Z., (2019). Fractal properties and morphological investigation of nano-amiodarone using image processing. *Signal, Image and Video Processing, 13*, 281–287.
10. Lashgari, A., Ghamami, S., Shahbazkhany, S., Salgado-Morán, G., & Glossman-Mitnik, D., (2015). Fractal dimension calculation of a manganese-chromium bimetallic nanocomposite using image processing. *Journal of Nanomaterials, 2015*, 2015–2024. doi: 10.1155/2015/384835.
11. Lashgari, A., Ghamami, S., Bahrami, Z., Shomossi, F., Salgado-Morán, G., & Glossman-Mitnik, D., (2015). Morphological investigation and fractal properties of realgar nanoparticles. *Journal of Nanomaterials, 2015*, 8. Article ID 130698. doi: 10.1155/2015/130698.

PART III

Composites and Nanostructures

CHAPTER 16

Study of Possible Negative Impact of a New Wood Composite Containing Triethoxysilylated Styrene on a Living System in Experiment

G. NAKHUTSRISHVILI,[1] M. BERULAVA,[2] E. TAVDISHVILI,[1]
LEVAN LONDARIDZE,[1] and D. DZIDZIGURI[1]

[1]*Ivane Javakhishvili Tbilisi State University, Faculty of Exact and Natural Sciences, Tbilisi, Georgia*

[2]*Sukhumi State University, Tbilisi, Georgia*

ABSTRACT

After the negative impact of phenol-formaldehyde resin-based building tiles on human health have been identified, the problem has become especially acute, and the need arose to create new composite materials, which are environmentally friendly and safe for health.

The aim of the present study was to investigate the potential adverse effects of a new, triethoxysilylated styrene based woody composite on the morphofunctional activity of various tissues of adult white mice.

The research has shown that the new wood composite, which uses triethoxysilylated styrene to bind wood, causes no changes in the total peripheral blood leukocyte count of adult white mice. Also, no negative effect of this composite on the cytoarchitectonics and proliferative activity of various tissues were observed.

Advanced Polymer Structures: Chemistry for Engineering Applications. Omar Mukbaniani, DSc, Tamara Tatrishvili, PhD & Marc Jean M. Abadie, DSc (Eds.)

The new wood composite does not change the morphofunctional activity of various organs of white mice, indicating that the triethoxysilylated styrene used to bind wood shavings have no a cytotoxic impact on living organisms.

16.1 INTRODUCTION

Nowadays, the production of the wood-polymer composites (WPC), which are widely demanded in industry, construction, and everyday life, is of particular importance. Consequently, the number of environmentally friendly, new binder compounds and the research in the field of wood composites has also increased. It has become very important and demanding to create composites that are easy to process, environmentally friendly and inexpensive [1–3]. In this regard, thermosetting substances, thermoplastics, and wood composites based on biopolymers possess good mechanical, thermal, and water-absorbing properties. The simplicity of technological processes for the production of tiles based on thermoactive resins, the physical and mechanical properties of these materials have led to the creation of a large number of tile around the world. At the same time, in the last decade, there has been an increased interest in tile regarding environmental safety, as the latter has been questioned by scientists. The main issue is whether everyday furniture or building materials made from phenol-formaldehyde resins are first class carcinogens and harmful to the body [4–10].

These materials nowadays have a number of shortcomings, the most important of which is that emissions from these materials contain cancerogenic harmful formaldehyde for the human organism. World Health Organization experts it was shown that small concentrations of this substance in the human body cause the following undesirable symptoms [11, 12]. One of the technical disadvantages of particleboard is the instability to moisture (water absorption is about 16–20%). In USA the new regulation in production of materials using phenol-formaldehyde resins, about emission of formaldehyde from wood composites materials was planned from 2013 [13].

The negative effect of wood shavings and monoterpenes on the health of workers in furniture factories was also investigated [2]. According to the literature, the building material polystyrene, used as a binder for the ceiling, does not have a harmful effect and complies with the American standard (ASTM) [14–16].

With interesting properties are characterized the previously obtained composite materials on the basis of renewable plant raw materials were as a binder phenethoxysiloxanes PhES-50, PhES-80, liquid glass (LG), colophony, and wood glue are used [17–20]. The properties of composites materials using different additives are much higher than the characteristics of today produced particleboard, strength on a bend is in the range 14–16 MPa, density 0.8–1.0 g/cm^3, water absorption is in the range 2.0–8.0% and temperature constancy on Vickat is in the range 110–130°C. On some characteristics, the received materials exceed to known analogs, which indicates the advantages of these composites In comparison with the existing ones.

From this point of view, the new composite material, in which triethoxysilylated styrene [21, 22] is used as a binder is considered environmentally friendly and safe for health. Thought, at the same time, the possible negative effect of this composite on a living organism has not been established.

The aim of the present study was the investigation of the potential adverse effects of the new composite material obtained during hot pressing and in situ-polymerization of triethoxy (vinylphenethyl)silane (triethoxysilylated styrene) with pine sawdust.

16.2 EXPERIMENTAL PART

16.2.1 MATERIALS AND METHODS

The study was conducted on white adult mice (25–30 g). A wood composite based on styrene silicate and various tissues of adult mice (blood, liver, and brain) were used as research material. Changes in the histoarchitectonics of the studied tissues were assessed by microscopic observation of paraffin embedded tissue sections. Goriaev camera was used to count the total number of leukocytes in the blood. Duration of experiments was 90 days.

16.2.2 PREPARATION OF THE MATERIAL FOR THE STUDY UNDER THE LIGHT MICROSCOPE

In order to examine the tissues (liver) under a light microscope, the materials were fixed in the 4% formaldehyde solution prepared on the Na/K phosphate buffer. After the fixation, 5% of EDTA solution was used to decalcify the femur of adolescent rats. Dehydration of the material took place in an increasing range of alcohols of different concentrations. Tissues were embedded in a wax-paraffin mixture. About 5–7 μm thick slices were stained with hematoxylin-eosin. Tissue samples were studied under a light microscope (Zeiss Primo Star, Germany). To estimate the mitotic index (‰), at least 5,000 cells were counted.

Determination of the total number of leukocytes in peripheral blood 20 μl of blood taken from the tail of animals anesthetized with diethyl ether were dissolved in transferred into 400 μl of 3% acetic acid and kept at room temperature for 20 min. A certain volume of the sample was placed in a Goryaev chamber, the number of leukocytes was counted according to a standard protocol using a light microscope (at low magnification, lens 20, eyepiece 10).

16.2.3 STATISTICAL ANALYSIS

The data are expressed as mean ± SD. Students' t-test was used for comparison among the different groups. $P<0.05$ was considered statistically significant.

16.3 RESULTS AND DISCUSSION

To assess the effectiveness of this composite, experimental animals were divided into two groups:

i. **Control Group:** Intact animals.
ii. **Experimental Group:** Animals kept in cages covered with the test composite for 90 days.

The experiment lasted 90 days. The research material (peripheral blood) was periodically taken from the tails of animals, and the total number of leukocytes was counted in the Goryaev chamber. Figure 16.1 shows the curvature of the change in the total number of leukocytes in the peripheral blood of animals in the control and experimental groups.

Figure 16.1 represents a new composite of triethoxysilylatedstyrene + wood causes no change in the total number of leukocytes in the blood of white mice.

In parallel, changes in the histoarchitectonics of tissues of animals in the control and experimental groups were studied. Figure 16.2 shows micrographs of the histoarchitectonics of the liver of animals of both groups. Figure 16.2(a) and (b) shows the characteristic histoarchitectonics of intact liver tissue. The large binuclear hepatocytes are well distinguished in radial direction to the central vein, with sinusoidal capillaries and a visible space of Disse, between the rods. Single mitotic figures are observed in the liver parenchyma (Figure 16.2(a) and (b)). It should be noted that morphological and functional changes in the liver tissue of the animals of the experimental group were not observed.

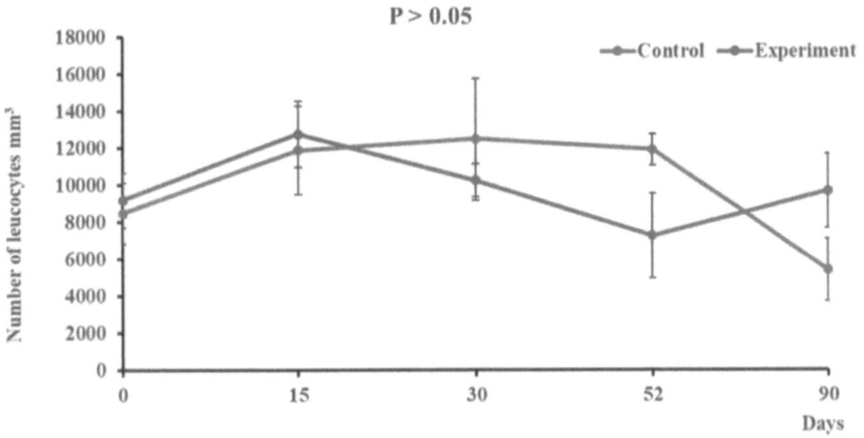

FIGURE 16.1 The effect of wood composite containing silicate styrene on the total number of leukocytes in the peripheral blood of adult mice.

Changes in the mitotic index of hepatocytes were evaluated to determine the possible negative effect of the composite on the proliferative activity of liver cells in animals of both groups. Studies have shown that the composite based on styrene silicate does not affect the mitotic index of animals in the experimental group. In particular, the mitotic index of the liver of experimental animals (0.2 ± 1.4‰) remained similar to the corresponding indicator of control animals (0.4 ± 0.7‰). Thus, the results show that the tiles used in our experiment containing styrene silicate do not have a negative effect on the histoarchitectonics and proliferative activity of these tissues.

Similar results are described using the example of wood composites in which organic and organosilicon polymers are used as binders [14–16].

In parallel, the effect of the composite on changes in histoarchitectonics and proliferative activity of the brain tissue were studied in both groups of animals. No morphofunctional changes in the brain tissue of the experimental group were observed. The same results were obtained when studying the liver tissue.

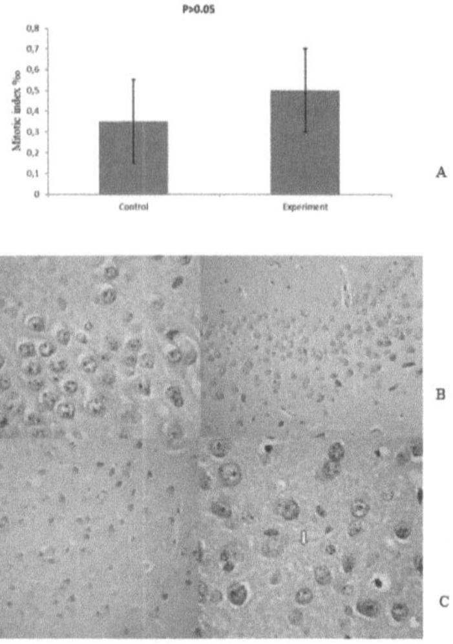

FIGURE 16.2 Study of the effect of triethoxysilylatedstyrene on liver histoarchitecture and proliferative activity in adult white mice. (A) mitotic index ($p > 0.04$); (B) and (C) mitotic figures (90X7, H&E).

There was no significant difference between the values of the mitotic index of the brain in the control (0.35 ± 0.2‰) and experimental (0.5 ± 0.2‰) groups (Figure 16.3).

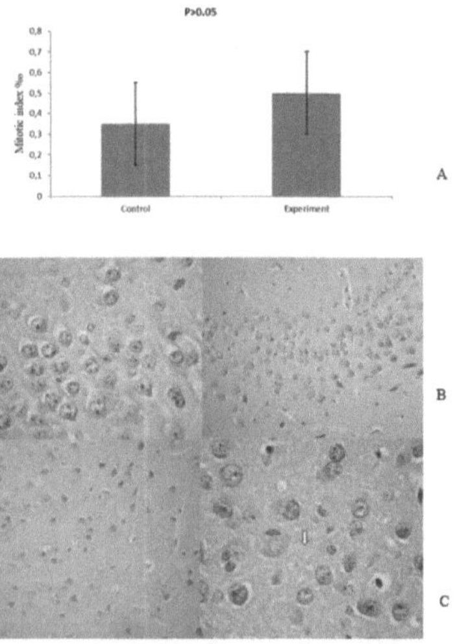

FIGURE 16.3 Study of the effect of styrene silicate on the histoarchitecture and proliferative activity of adult white mice in the brain. (A) mitotic index ($p > 0.04$); (B) and (C) mitotic figures (90X7, H&E).

16.4 CONCLUSION

Thus, the tiles containing triethoxysilylated styrene as a binder used in our experiment did not have a negative effect on the histoarchitectonics of these tissues, the no negative impact on various tissues white adult mice.

ACKNOWLEDGMENTS

We would like to thank Prof. Omar Mukbaniani for the provided materials and cooperation.

KEYWORDS

- **leukocytes**
- **mitotic index**
- **thermoactive resins**
- **triethoxysilylated styrene**
- **white adult mice**
- **wooden sawdust tile**
- **wood-polymer composites**

REFERENCES

1. Papadopoulos, A. N., (2019). *Advances in Wood Composites*. Laboratory of Wood Chemistry and Technology, Department of Forestry and Natural International Hellenic University, GR-661 00 Drama, Greece; antpap@for.ihu.gr.
2. Håkan, L., Katja, H., Ing-Liss, B., Mats, H., & Rask-Andersenc, A., (2017). *Respiratory Symptoms and Lung Function in Relation to Wood Dust and Monoterpene Exposure in the Wood Pellet Industry, 122*(2), 78–84.
3. Mohammad, L., Khan, Z. R., Sunil, K. S., & Gupta, M. K., (2020). A state-of-the-art review on particulate wood polymer composites: A state-of-the-art review on particulate wood polymer composites: Processing, properties and applications. *Polymer Testing, 89*, 106721.
4. Qiang, Y., Dongyang, W., Januar, G., & Stuart, B., (2008). *J. of Thermoplastic Composite Materials, 21*, 195.
5. Chia-Huang, L., Tung-Lin, W., Yong-Long, C., & Jyh-Horng, W., (2010). *Holzforschung, 64*, 699–704.
6. Haihong, J., & Pascal, K. D., (2004). *J. of Vinyl & Additive Technology, 10*(2), 59.

7. Poletto, M., Dettenborn, J., Zeni, M., & Zattera, A. J., (2011). *Waste Management, 31*, 779.
8. Adjovi, E. C., Olodo, E. T., Niang, F., Guitard, D., Foudjet, A., & Kamdem, D. P., (2013). *International Journal of Scientific & Engineering Research, 4*(4), 344.
9. Cagla, K. S., Emel, Y., & Iskender, Y., (2016). *Polymer, 99*, 580.
10. Shafeen, P., & Nagarajan, N. M., (2015). *International Journal of Innovative Science, Engineering & Technology, 2*(8), 645.
11. https://www.cancer.gov/about-cancer/causes-prevention/risk/substances/formaldehyde/formaldehyde-fact-sheet (accessed on 02 January 2022).
12. https://www.ncbi.nlm.nih.gov/pmc/articles/PMC2855181/ (accessed on 02 January 2022).
13. http://www.govtrack.us/congress/bill.xpd?bill=s111-1660 (accessed on 02 January 2022).
14. Goliadze, M., Bedineishvili, N., Berulava, M., Markarashvili, E., Mukbaniani, O., & Dzidziguri, D., (2019). New wood composites influence of the morpho-functional activity of the different tissues of adult and old white mice. *PolyChar 26 World Forum on Advanced Material.*
15. Goliadze, M., Bedineishvili, N., Berulava, M., & Dzidziguri, D., (2019). Comparative study on impact of the new compositions on the histoarchitectonics of the white mice organs. In: *6th International Symposium on Polymers and Advanced Materials.* Batumi.
16. Goliadze, M., Berulava, M., Razmazashvili, M., & Dzidziguri, D., (2017). The wood compositions, having different technical features and their influence on the biological systems. *Fifth Caucasian International Symposium on Polymers and Advanced Materials*. Tbilisi.
17. Mukbaniani, O., Aneli, J., Buzaladze, G., Markarashvili, E., & Tatrishvili, T., (2016). Composites on the basis of straw with some organic and inorganic binders. *Oxidation Communications, 39*(3-II), 2763–2777.
18. Mukbaniani, O., Aneli, J., Buzaladze, G., Tatrishvili, T., & Markarashvili, E., (2017). Biocomposite on the basis of leaves. *Oxid. Commun., 40*(I-II), 430–440.
19. Mukbaniani, O., Brostow, W., Hagg, L. H. E., Aneli, J., Tatrishvili, T., Markarashvili, E., Dzidziguri, D., & Buzaladze, G., (2018). *Pure and Applied Chemistry, 90*(6), 1001–1009.
20. Mukbaniani, O., Aneli, J., Buzaladze, G., Tatrishvili, T., & Markarashvili, E., (2015). *Wood Biorefinery Conference* (pp. 467–473). NWBC 2015, Helsinki, Finland.
21. Mukbaniani, O., Aneli, J., Londaridze, L., Markarashvili, E., & Tatrishvili, T., (2021). Ecologically friendly polymer composites on the base of leafs. *Oxidation Comm.* In press.
22. Mukbaniani, O., Brostow, W., Aneli, J., Londaridze, L., Tatrishvili, T., & Markarashvili, E., (2021). Composites on the basis of wood sawdust and triethoxysilylated styrene with low water absorption. *Oxid. Comm.* In press.

CHAPTER 17

Recent Advances in Elastic Polymer Composites for Neutron Shielding Applications

JOBIN JOY,[1,2,3] K. M. PRAVEEN,[3] HANNA J. MARIA,[2] JÓZEF T. HAPONIUK,[1] and SABU THOMAS[2]

[1]*Department of Polymer Technology, Faculty of Chemistry, Gdańsk University of Technology, Gdańsk, Poland*

[2]*School of Energy Materials, Mahatma Gandhi University, Kerala, India*

[3]*Department of Mechanical Engineering, Muthoot Institute of Science and Technology, Kerala, India*

ABSTRACT

Ionizing radiation, which includes particles such as alpha, beta, and neutron particles, and also electromagnetic radiation such as X-rays and Gamma rays, are emitted in a nuclear reaction due to radioactive decay. If not efficiently shielded, these radiations will cause considerable damage to living cells, causing cancer, tumors, etc. So, shielding these radiations is the utmost priority as far as safety is concerned. Concrete, lead, aluminum, etc., are usually used as shielding materials, but for complex geometries and uneven surfaces, it is complicated to use these materials due to their rigid nature. The development of flexible elastomeric composites for shielding applications has been the main area of focus for the research community for a long time. The main objective of this chapter is to review the latest development made in elastomeric composites for neutron shielding applications.

Advanced Polymer Structures: Chemistry for Engineering Applications. Omar Mukbaniani, DSc, Tamara Tatrishvili, PhD & Marc Jean M. Abadie, DSc (Eds.)

17.1 INTRODUCTION

Radiations are generally classified as ionizing and non-ionizing types. Ionizing radiations are more energetic than non-ionizing, due to which it interferes with the molecular structure of a material or living tissues and is very harmful if not efficiently utilized. Like alpha, beta, gamma, and X-rays, neutron radiations are also ionizing in nature. As the name suggests neutron is a chargeless particle [1], which external electric or magnetic fields cannot influence; hence they travel at a great speed, with high penetrating power (as shown in Figure 17.1), and directly interact with the nuclei of an atom. Sometimes these interactions will cause the emission of secondary radiations like γ rays, protons, and alpha particles. So shielding materials must also attenuate these secondary radiations as far as neutron shielding is concerned [2].

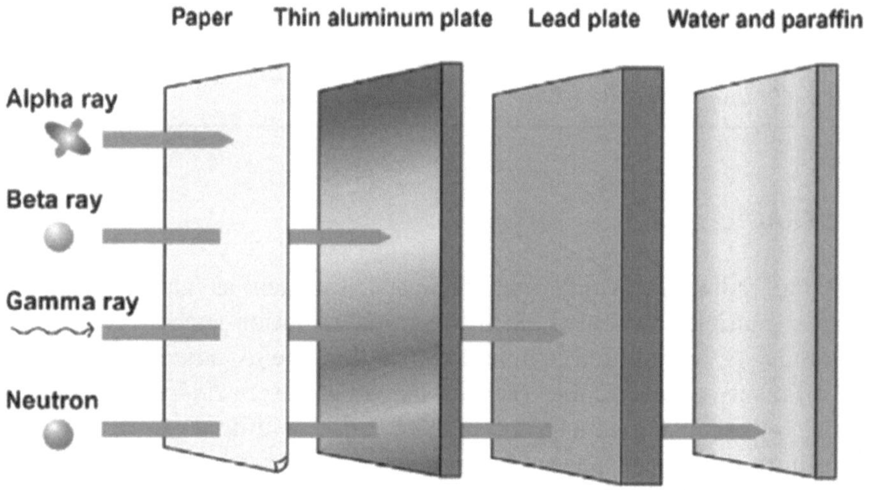

FIGURE 17.1 Penetration strength of radiations.
Source: Reprinted with permission from Ref. [3]; ©2015 Elsevier.

Neutron sources are used in a wide variety of applications like research reactors, nuclear Icebreakers, aircraft Carriers, submarines, etc. [4]. With the developments in neutron science and technology, neutron sources are extensively used in applications like neutron radiography, neutron capture therapy, neutron activation analysis, etc. [5]. Apart from many benefits, there are also risks to the safety of workers, the public, and the environment involved with these applications, which must be assessed and controlled if required [6]. The basic principle of a neutron shielding mechanism is a two-step process, as shown in Figure 17.2. The first step is the deceleration of the incoming fast neutrons to thermal neutrons (low atomic weight and hydrogen-rich materials are used for that, e.g., water, paraffin, concrete, etc.), and second step is the absorption of thermal neutrons (materials with high neutron absorption cross-section like boron and its compounds are used to achieve this).

Conventional shielding materials like concrete, lead, etc., are heavy and with poor flexibility. Elastomeric composite materials, on the other hand, are lightweight, highly flexible, and with good material properties [7].

The interaction between a neutron and the nucleus is expressed in terms microscopic cross-section is used for understanding the interaction process, but in actual practice, where measurements are done on thick samples, the term macroscopic cross-section (symbol: Σ_t and unit: cm^{-1}) is used [8]. Neutrons are used in a wide variety of applications like density gauges, moisture gauges, activation analysis, research purposes (physics, medicine), a trigger for nuclear weapons, etc. The general classification of neutron sources is – alpha neutron sources, gamma neutron sources, spontaneous fission neutron sources, fission reactors, accelerators, etc. Therefore, the chances of radiation leakage from these sources would be very high [9].

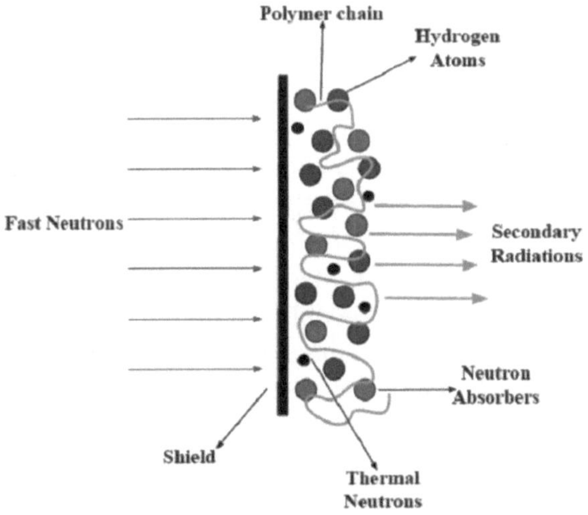

FIGURE 17.2 Neutron shielding mechanism.

Thus, various safety measures like ALARA (as low as reasonably achievable) must be implemented to ensure a safe working environment. ALARA includes three critical factors – TIME (minimizing the time of exposure), DISTANCE (doubling the distance between the source and the personnel), and SHIELDING (using appropriate radiation-absorbing materials) [10]. Usage of elastomeric shielding materials decreases the chances for radiation leakage. These materials are also more cost economic and less toxic than other conventional shielding materials. From this chapter, we will review the latest developments in elastomeric composites in neutron shielding applications.

17.2 ELASTOMERS

An elastomer is a flexible polymeric material with high impact resistance, good toughness and low stiffness. Elastomers also have high elasticity, viscoelastic properties and glass transition temperature (T_g) below room temperature. Elastomers are chemically or physically crosslinked polymers with weak intermolecular forces, characterized by a high degree of elongation, which helps them regain their original shape on removal of external force. These qualities make elastomers a good choice for industrial purposes like adhesives, hybrid composites, radiation shielding fabrics, etc. [7].

17.2.1 TYPES OF ELASTOMERS USED IN NEUTRON SHIELDING

The benefits of elastomers as neutron shielding materials are evident when flexibility is of prime concern for applications, like shielding of curved and irregular surfaces, radiation shielding clothing, space crafts, etc. Some standard elastomers which were greatly used are natural rubber (NR), silicon rubber, EPDM rubber, LDPE/HDPE rubber, styrene-butadiene rubber, etc. [11]. In a study conducted by T. Özdemir et al. [12] EPDM is used as a matrix material because of its flexible nature and high hydrogen content, which improves the neutron shielding capability. Depending upon the type of neutron interactions, sometimes the secondary radiations like gamma, proton, etc., are also emitted. So, the composite material prepared must also be capable of shielding these secondary radiations also [12]. High-density polyethylene (HDPE) is used in the study along with the other filler materials to shield not only neutrons, also gamma absorption was taken into account. From results, it is clear that by adding h-BN and Gadolinium(III) oxide nanoparticles (NPs), the absorption of both neutron and gamma fluxes shows an enhancement of 200–280%, 14–52% [13].

17.2.2 PROCESSING OF ELASTOMERIC COMPOSITES

The processing of an elastomer is basically a four-step process:

- **Step 1:** Mastication (Here the elastomer molecules are broken by deformation or shearing to obtain an optimum level of viscosity).
- **Step 2:** Mixing (Here filler materials are added immediately after mastication).
- **Step 3:** Shaping (Here the mixed compound is shaped by using extrusion or molding).
- **Step 4:** Curing (also known as Vulcanization) imparts cross linking to the prepared material).

In laboratories, the mastication process is usually done on a 2-roll mill or on a laboratory scale internal mixer. The process of shaping or curing is traditionally done on a compression molding machine [14, 15] (Figure 17.3).

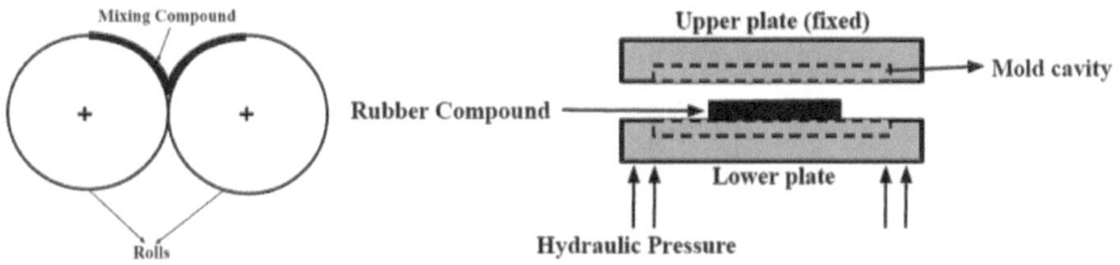

FIGURE 17.3 Two roll milling and compression molding.

17.2.3 NEUTRON ABSORBING FILLERS USED IN ELASTOMERIC COMPOSITES

The properties, which are expected to be attained by using filler materials, can be physical, chemical, or mechanical and practically, it is not possible to improve all the properties in one go. Hence, a compromise has to be made to reach an optimum level. For neutron shielding, filler materials with a high neutron absorption cross-section are usually used in an elastomeric matrix [16]. Boron carbide (B_4C), boric acid, boron oxide, hexagonal boron nitride (hBN), etc., are commonly used filler materials along with crosslinking agents, accelerators, activators, retarders to synthesize the elastomeric composite (Table 17.1).

TABLE 17.1 A Summary of the Neutron Absorbing Fillers Used in the Processing of Elastomeric Composites

SL. No.	Matrix	Filler Materials	Shielding	References
1.	Silicon resin	Bismuth (Bi) (for gamma radiation) and boron-10 (10B) (for neutron shielding)	Gamma and neutron	[17]
2.	Clay-polyethylene	Recycled low density polyethylene (LDPE)	Neutron	[18]
3.	Ethylene-vinyl acetate (EVA)	Boron carbide (B_4C)	Neutron, flame retardant	[19]
4.	High-density polyethylene (HDPE)	Cadmium oxide (CdO) nanoparticles	Neutron	[20]
5.	High-density polyethylene (HDPE)	micro-sized and nano-sized CdO	Neutron	[21]
6.	Silicone rubber (polydimethylsiloxane)	Bismuth (III) oxide and hexagonal boron nitride (hBN)	Neutron and X/gamma ray	[22]
7.	Low-density polyethylene (LDPE)	Boron carbide (B_4C)	Neutron	[23]
8.	Silicone rubber	Boron carbide, hollow beads, and zinc borate	Neutron	[24]
9.	Ultra-high molecular weight polyethylene (UHMWPE)	Boron carbide (B_4C)	Neutron	[25]
10.	EPDM rubber (ethylene propylene diene monomer)	Boron trioxide	Neutron	[12]
11.	Thermoplastic natural rubber (TPNR)	Boron carbide (B_4C)	Neutron	[26]
12.	Natural rubber latex (NRL)	Boron carbide (B_4C)	Neutron	[27]

17.2.4 NEUTRON INTERACTION WITH ELASTOMERIC COMPOSITES

There are three modes of interaction between a neutron and an atomic nucleus:

1. **Elastic Scattering:** Here, kinetic energy is conserved, i.e., the total kinetic energy before and after collision remains the same.
2. **Inelastic Scattering:** Here, the total kinetic energy after collision is less than the total kinetic energy before. Here the nucleus is left in an excited state.
3. **Absorption:** This is also known as radiative capture. Here the neutron is absorbed by the nucleus and becomes excited by forming the next higher isotope. It emits gamma radiation on de-excitation. This type of interaction is ideal for neutron shielding applications (Figure 17.4).

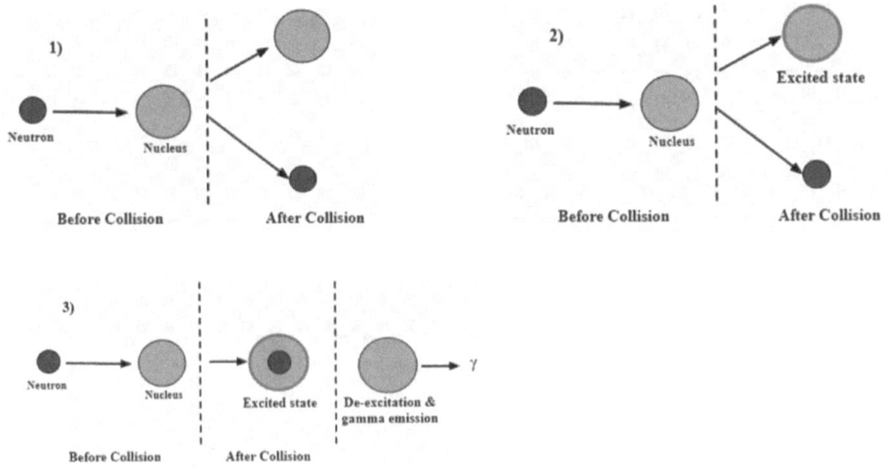

FIGURE 17.4 (1) Elastic scattering; (2) inelastic scattering; and (3) absorption.

17.3 CHARACTERIZATION OF NEUTRON SHIELDING ELASTOMERIC COMPOSITES

In this section of the chapter, different characterization studies of neutron shielding elastomeric composites were discussed. Boron-based filler materials and a matrix with high hydrogen content are ideal for neutron shielding composites. In a study by A. Güngör et al. [28] hBN as a neutron absorbing filler and EPDM as a matrix were used to synthesize a thermal neutron shielding elastomeric composite. For thermal neutron attenuation tests, Pu-Be radiation source with an activity of 2 Ci was used. From mechanical tests, it was clear that with the increase in hBN content it shows elastic-like mechanical properties (as shown in Table 17.2) but a decrease of mechanical properties like tensile strength and toughness. But from neutron attenuation tests, it is clear that the addition of hBN particles in the EPDM matrix improved the shielding efficiency (as shown in Figure 17.5) (up to 61.5% neutron radiation absorption was observed) [28].

TABLE 17.2 Results Obtained from Mechanical Testing

	Tensile Strength (MPa)	Elongation at Break (%)	Elastic Module (MPa)	Toughness (J)	Hardness (Shore A)
hBN0	8.9±0.2	438.8±25.1	7.1±0.3	13.7±0.3	60±1
hBN20	7.6±0.4	265.1±32.6	14.2±0.5	7.6±0.2	71±1
hBN40	4.2±0.3	118.1±11.1	101.3±3.8	2.9±0.2	87±1

Source: Reprinted with permission from Ref. (28); ©2019 Elsevier.

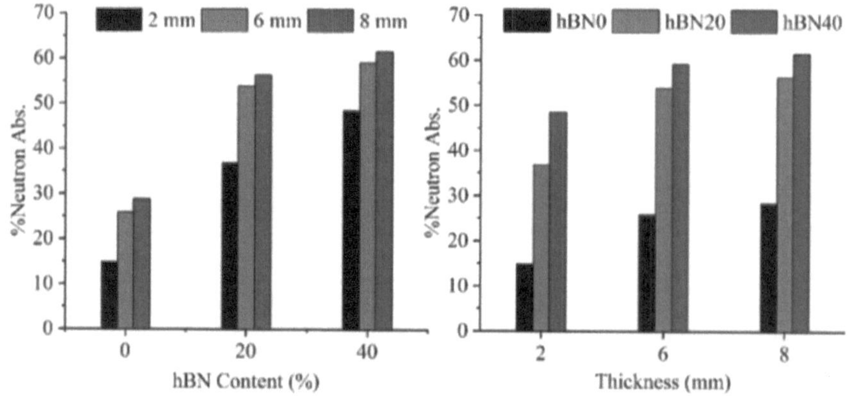

FIGURE 17.5 Neutron attenuation results of the samples with varying hBN content and sample thickness.
Source: Reprinted with permission from Ref. [28]; ©2019 Elsevier.

In the work by Aysegül Canel et al. [29] low cost and durable epoxy resin was used as a matrix. Different samples were prepared with different filler metals like Fe (Iron), Bi (Bismuth), Ta (Tantalum) and WC (Tungsten

Carbide). An epoxy is used with its hardener in the ratio of 3:1 (Epoxy: Hardener). The metal filler percentage were varied from 10% to 40%. Simulation codes GEANT4 and FLUKA were applied for simulation by using sources of 4.5 MeV 241Am-Be for neutrons and 1.25 MeV 60Co for gamma. Although all samples showed good simulated shielding performances the best shielding efficiency was obtained for Tantalum-Epoxy sample. Neutron effective removal cross-sections, Σ (unit – cm^{-1}) and gamma linear attenuation coefficients, μ (unit – cm^{-1}) were estimated to determine the shielding efficiency. For neutron shielding Ta40 is the best shield with 0.677 cm^{-1} effective removal cross section and also the Ta40 is the best gamma-ray shield with 0.43 cm^{-1} linear attenuation coefficient [29] (Figure 17.6).

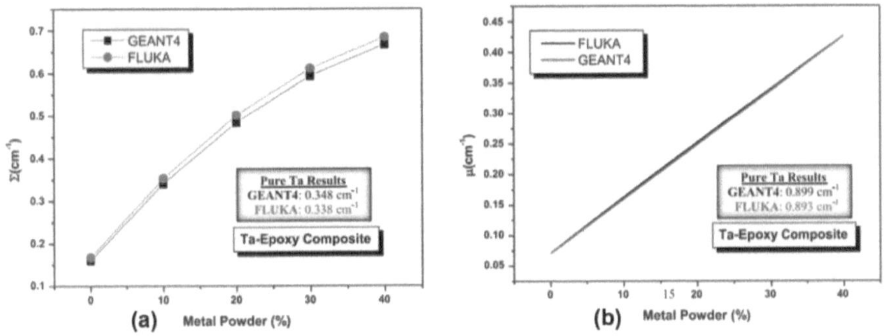

FIGURE 17.6 (a) Neutron effective removal cross-sections of Ta-epoxy composite; and (b) gamma linear attenuation coefficients cross-sections of Ta-epoxy composite.
Source: Reprinted with permission from Ref. [29]; ©2019 Elsevier.

Surface modified B_4C powder was embedded in the LDPE matrix by Suna Avcıoğlu et al. [23] The shielding composites were prepared with filler ratios 0.6, 1, and 1.7 wt.%. Due to surface modification the filler particle size included nano-sized (20 nm) primary particles to submicron (500 nm) particles (as shown in the High-resolution field emission scanning electron microscopy (FESEM) image, Figure 17.7), which in turn dramatically improved the shielding efficiency up to 39%. The neutron shielding efficiency testing was done as a function of composite thickness and B_4C loading. Improved shielding efficiency was obtained by increasing the amount of B_4C loading and increasing plate thickness [23] (Figure 17.8).

FIGURE 17.7 FESEM image of B_4C powder.
Source: Reprinted with permission from Ref. [23]; ©2020 Elsevier.

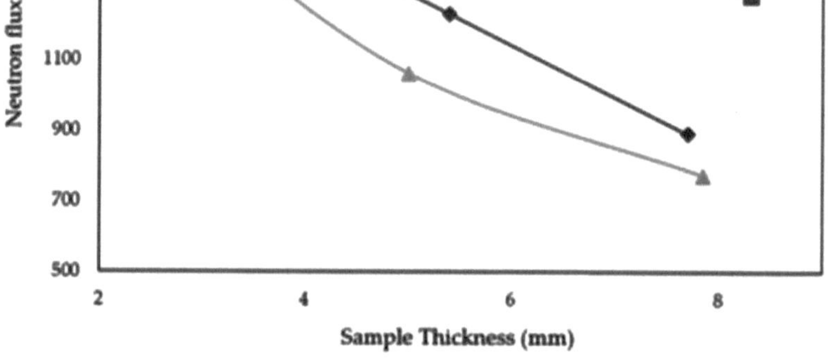

FIGURE 17.8 Neutron shielding test results.
Source: Reprinted with permission from Ref. [23]; ©2020 Elsevier.

Neutron shielding properties for both the fast and thermal neutrons were estimated for high density polyethylene (HDPE) reinforced with micro-sized and nano-sized cadmium oxide (CdO) (weight fractions of 10% and 40%) composites by Ahmed M. El-Khatib et al. [21] From this study it is clear that for fast neutrons, nano-CdO/HDPE shows better shielding properties, whereas for thermal neutrons, micro-CdO/HDPE shows better results. Also, the mechanical properties of composites were greatly influenced by particle addition percentage and size. Figure 17.9 shows the comparison between the transmitted fraction of fast neutrons for both nano-cadmium and micro-cadmium. It was found that the nano-cadmium composites show a lesser value for transmitted fraction which means a better shielding efficiency (due to the better dispersion of nano-cadmium in the matrix) in comparison to micro-Cadmium.

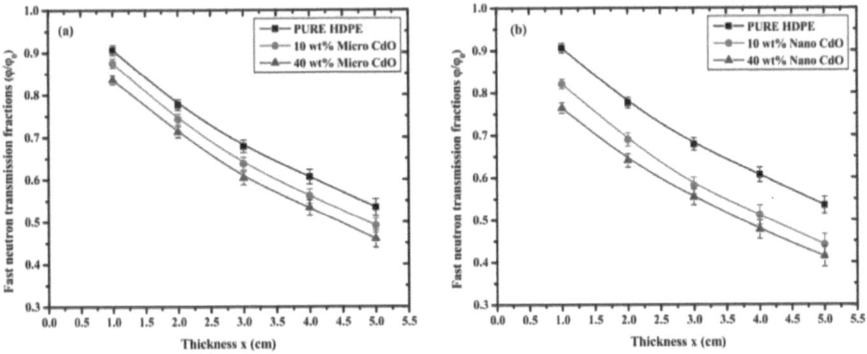

FIGURE 17.9 Fast neutron transmission fractions for micro- and nano-CdO versus sample thickness.
Source: Reprinted with permission from Ref. [21]; ©2020 Elsevier.

But in the case of thermal neutron shielding, the composites (as shown in Figure 17.10) containing micro-CdO show better shielding efficiency, as the micro-CdO particles are larger, agglomerated, and not evenly distributed, which increases the probability of neutron interaction within the composites containing micro-CdO particles [21].

In the study by Oussama Mehelli et al. [25] a hybrid elastomeric composite was prepared with ultra-high molecular weight polyethylene (UHMWPE) fibers, epoxy, and boron carbide (B_4C) particles. In this study, the prepared composites are referred to as Spectra/Epoxy/X B_4C (where; 'X' shows the amount of B_4C particles). The neutron transmission ratio (I/I_0) was measured through different sample thicknesses to measure neutron shielding performance. Here I/I_0 is expressed as:

$$I/I_0 = e^{-\Sigma x} \text{ or } -\text{Ln}\,(I/I_0) = \Sigma x \quad (1)$$

where; 'I_0' is the incident neutron beam flux; 'I' is the transmitted neutron beam flux; 'X' is the shielding thickness (in cm); and 'Σ' is the macroscopic cross-section (cm^{-1}) of a specific material for neutron shielding.

Also the screening or shielding ratio (S) is measured as:

$$S = (I_0 - I/I_0) \quad (2)$$

The macroscopic cross-section (Σ) was determined from Eqn. (1) by varying the sample thickness and measuring the radiation dose as shown in Figure 17.11.

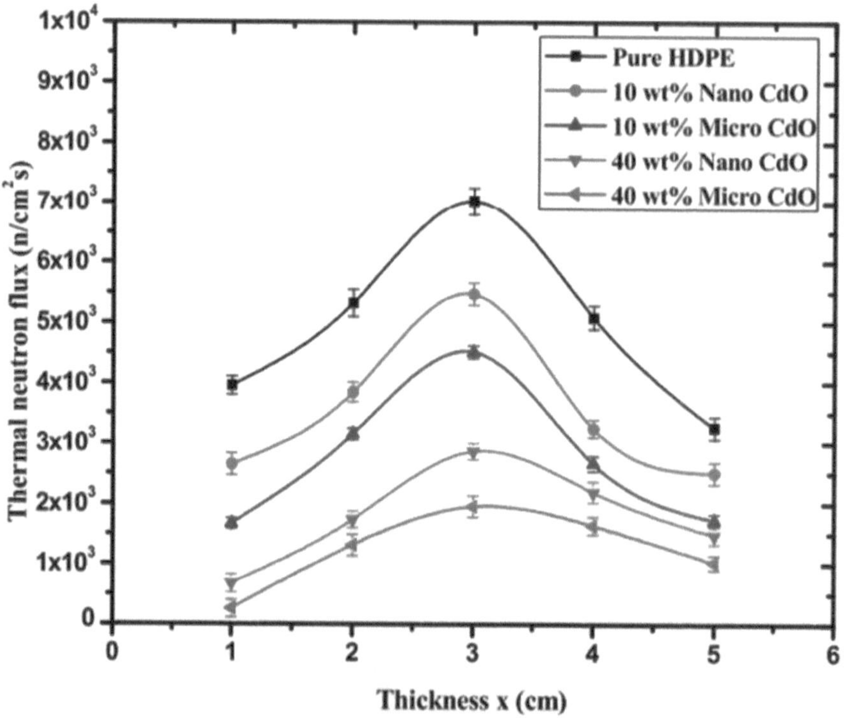

FIGURE 17.10 Thermal neutron flux for micro- and nano-CdO versus sample thickness. *Source*: Reprinted with permission from Ref. [21]; ©2020 Elsevier.

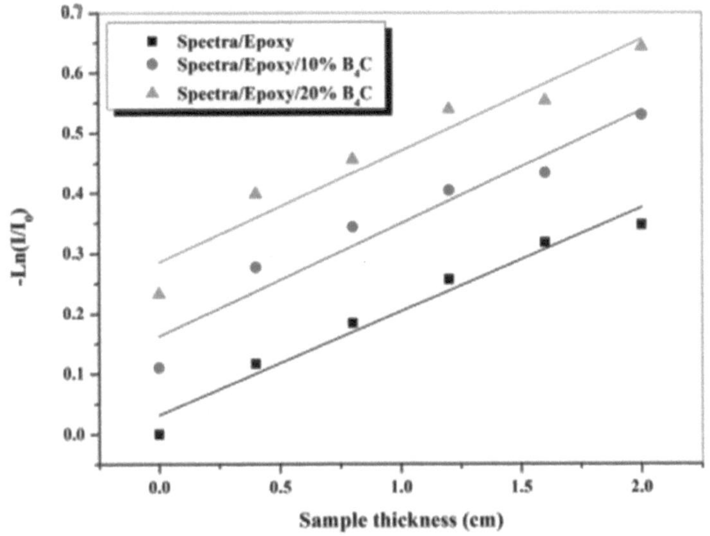

FIGURE 17.11 ln (I/I_0) as a function of the thickness.
Source: Reprinted with permission from Ref. [25]; ©2021 Elsevier.

With the help of Eqns. (2) and (3), for each composite, macroscopic cross-section (Σ) and the screening ratio (S) were determined as shown in Figure 17.12.

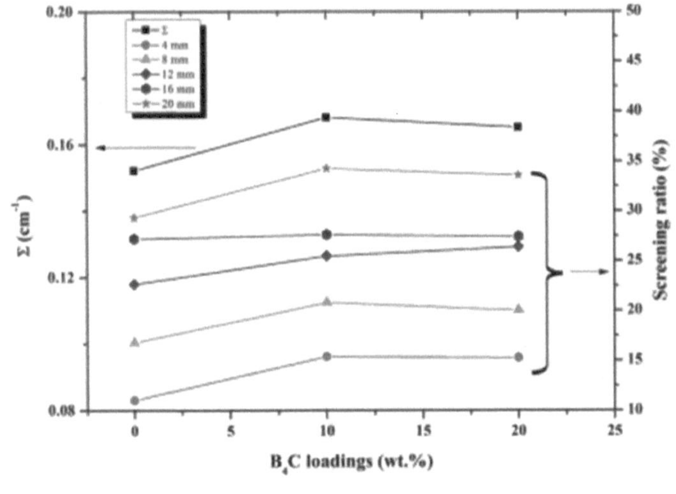

FIGURE 17.12 Macroscopic cross-section (Σ).
Source: Reprinted with permission from Ref. [25]; ©2021 Elsevier.

The estimated values of neutron shielding parameters are shown in Table 17.3. The Spectra/Epoxy/10 B4C displayed an improved Σ value of 0.188 cm^{-1}, which in turn shows the better shielding efficiency of the sample [25].

TABLE 17.3 Neutron Shielding Data of the Prepared Composites

Neutron Shielding Data of the Studied Materials*							
Sample	**S4 (%)**	**S8 (%)**	**S12 (%)**	**S16 (%)**	**S20 (%)**	**Σ (1/cm)**	**Λ (cm)**
Spectra/epoxy	11.021	16.785	22.649	27.199	29.323	0.168	4.126
Spectra/epoxy/10% B_4C	15.369	20.829	25.48	27.604	34.277	**0.188**	3.687
Spectra/epoxy/20% B_4C	15.268	20.02	26.39	27.401	33.569	0.185	3.747

*S4, S8, S12, S16, and S20 are the screening ratio for the samples thickness 4, 8, 12, 16, and 20 mm, respectively.

Source: Reprinted with permission from Ref. [25]; ©2021 Elsevier.

An attempt has been made to synthesize an elastomeric composite with ultra-high molecular weight polyethylene (UHMWPE) as a matrix with different contents of WO_3 for investigating both the neutron and proton shielding by Sayyed et al. [31]. Figure 17.13(a)) shows the SEM image of composite with 5 wt.% WO_3, which are uniformly distributed in the UHMWPE matrix with very low tendency to agglomeration. Additionally, Figure 17.13(b)) reveals that the neutron shielding improves by absorbing the dose released from the neutron source by 9.0585%, 11.3950%, 15.6651%, 26.9175%, and 27.4423% UHMWPE with 1 to 5% of WO_3 [31].

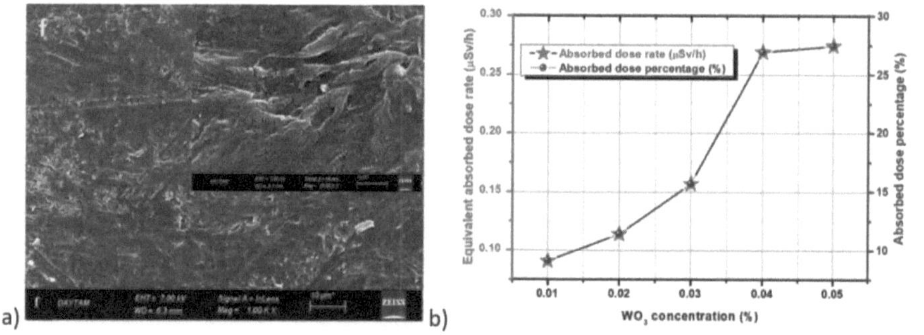

FIGURE 17.13 (a) SEM results of UHMWPE with 5 wt.% WO_3; and (b) equivalent dose rate for fast neutron radiation for all polymer samples.
Source: Reprinted with permission from Refs. [25, 31]; ©2021 Elsevier.

17.4 CHALLENGES IN NEUTRON SHIELDING

Some of the essential challenges encountered while preparing composites for neutron shielding are discussed here. In the study conducted by Tonguç Özdemir et al. [15]. It was observed that the increase of hBN (hexagonal boron nitride) content in the silicone rubber (poly-dimethyl siloxane) increases the elastic modulus and decreases the values of maximum stress, strain value at the break, energy to break. The decrease in the properties occurs because of discontinuities in the matrix (due to an increase in hBN content) [15].

Pin Gong et al. [32] conducted a study in which methyl vinyl silicone rubber (VMQ) was used as a matrix and filler materials like B_4C, PbO, and benzophenone (BP) were added. Results show that with an increase in the filler content, the tensile strength and elongation of the composites decreased due to agglomerations. Also, the tear strength of the composites first increased and then dropped [32].

For use as potential flexible neutron shielding materials, natural rubber (NR) and wood/NR composites were filled with boron oxide (B_2O_3) or boric acid (H_3BO_3). For NR and wood/NR composites with H_3BO_3 as fillers, the scorch and cure times increased with increasing H_3BO_3 content. On the other hand, for NR and wood/NR composites with B_2O_3 the scorch times were relatively constant but cure times gradually decreased up to 50 phr in NR composites and 40 phr in NR/wood composites, and again increased on increasing B_2O_3 content. In case of tensile strength and elongation at break, both the values will decrease on increasing boron oxide (B_2O_3) and boric acid (H_3BO_3) contents [14].

From the above studies, it is clear that the filler materials used for neutron absorption in the elastomeric matrix may cause an improvement or degradation of the physio-mechanical properties, so in most of the cases a compromise has to be made while synthesizing the composites.

17.5 APPLICATIONS OF NEUTRON SHIELDING ELASTOMERS

One of the most important applications of neutron shielding elastomers comes as a protective clothing for protection from nuclear radiation. In the study by Samir Ushah El-Kameesy et al. [34] emergency personal wearing clothes were fabricated from borated silicone rubber composite.

Neutron shielding elastomers play a crucial part in shielding fast radiations from medical linear accelerators in radiation therapy. In a study by

Younes Afkham et al. [35] a silicon resin matrix was filled with nanoparticles (NPs) of Fe_3O_4 and B_4C to design the shielding elastomers. This composite was intended to shield the fast neutrons produced during radiation therapy from 18 MeV photon beam of a Varian 2100 C/D linac (linear accelerators).

Storage of radioactive wastes like spent nuclear fuel, etc., is of prime concern for both the normal functioning of the nuclear power plant and the safety of the workers. The half-lives of the radioactive isotopes from the nuclear wastes normally range up to thousands of years, and also the nuclear waste generates alpha, beta, gamma particles and numerous neutrons. Hence the storage casks for storing the nuclear wastes should be made of an effective elastomeric composite to shield not only neutrons but also the secondary radiations [36].

For aerospace applications, an effective shielding elastomer must have a high electron density for enhancing the electromagnetic interaction, must be capable of producing fewer secondary radiations in space, and should be light in weight. Elastomeric composites with these material properties will prove to be effective and economical [33].

Another application of neutron shielding elastomers comes in the field of nuclear-based techniques for explosives detection by using multilayer perceptron models (MLPs) coupled with pulsed fast thermal neutron activation (PFTNA) technique [30].

17.6 CONCLUSION

This chapter aims to give the reader a basic idea about some essential aspects on neutron shielding which are – basic introduction to neutron shielding mechanism, various elastomeric materials used in neutron shielding applications, characterization of neutron shielding effectiveness of elastomeric composites, their challenges and practical applications. The elastomeric composites, which are hydrogenous, and having neutron-absorbing fillers with high neutron absorption cross-section, are very effective as the neutron shielding applications. The other additional properties like flame retardancy, shielding of secondary radiations like gamma, photons, etc., can also be tailored-in by using suitable fillers. An optimum radiation shielding efficiency can only be achieved by compromising with the other physico-mechanical properties as proved by various studies. The conventional shielding materials like concrete, lead, paraffin, etc., are not only rigid, bulky, with high cost but also have no flexibility at all. Due to which the objects with irregular geometries, sharp corners, etc., cannot be shielded effectively, which is a significant drawback. In contrast, the elastomeric composites can be tailored easily by using appropriate filler materials according to requirements.

KEYWORDS

- **elastomeric composites**
- **ethylene-vinyl acetate**
- **field emission scanning electron microscopy**
- **gamma rays**
- **ionizing radiations**
- **X-rays**

REFERENCES

1. Högaasen, H., (1964). *Possible Existence of a Super-Ω* (Vol. 32, pp. 1129, 1130). IL Nuovo Cimento.
2. Nambiar, S., & Yeow, J. T. W., (2012). *Polymer-Composite Materials for Radiation Protection. ACS Appl. Mater. Interfaces, 4*, 5717–5726
3. Issard, H. (2015). *9 – Radiation Protection by Shielding in Packages for Radioactive Materials,* pp. 123–140.
4. Outline, C., & Training, T., (2019). *Neutron Sources*, 289–292.
5. Engineering, C., Mara, U. T., & Alam, S., (2020). *Introduction to Neutron-Shielding Materials*, 1–24.
6. Protection, R., Basic, I., & Standards, S. (2014). *Radiation Protection and Safety of Radiation Sources: International Basic Safety Standards General Safety Requirements Part 3*. IAEA safety standards series, ISSN 1020–525X; no. GSR Part 3, STI/PUB/1578, ISBN 978–92–0–135310–8.
7. Özdemir, T., (2020). Elastomeric micro- and nanocomposites for neutron shielding. *Micro Nanostructured Compos Mater Neutron Shield Appl.*, 125–137.
8. Rinard, P., (1990). Ch.12. *Neutron Interactions with Matter* (pp. 357–377). Los Alamos Tech Reports.
9. *Neutron Sources*, (2010). pp. 1–64. IAEA safety standards series, ISSN 1020–525X; no. GSR Part 3, STI/PUB/1578, ISBN 978–92–0–135310–8.
10. Division, R. S. (2016). Radiation safety and ALARA. *Radio Guided Surgery,* 103–111, https://doi.org/10.1007/978-3-319-26051-8.
11. Engineering, C., Mara, U. T., & Alam, S., (2020). *Introduction to Neutron-Shielding Materials*, 1–23.
12. Özdemir, T., (2017). *Crossmark, 131*(2016), 7–12.

13. Gözde, Ş. İ., Alchekh, A., Keskin, M. A., Baykara, O., Ozkoc, G., Avc, A., et al., (2018). *Physical, Mechanical and Neutron Shielding Properties of h-BN/Gd_2O_3/HDPE Ternary Nanocomposites, 144*(August 2017), 434–443.
14. Ninyong, K., Wimolmala, E., Sombatsompop, N., & Saenboonruang, K., (2017). Potential use of NR and wood/NR composites as thermal neutron shielding materials. *Polym. Test [Internet], 59*, 336–343. Available from: http://dx.doi.org/10.1016/j.polymertesting.2017.02.020.
15. Özdemir, T., & Nur, S., (2018). *Hexagonal Boron Nitride and Polydimethylsiloxane: A Ceramic Rubber Composite Material for Neutron Shielding, 152*, 93–99.
16. Caminos, D., (2020). *Review on Neutron-Absorbing Fillers.* Woodhead Publishing Series, *Composites Science and Engineering*, pp. 25–52. http://doi.org/10.1016/b978-0-12-819459-1.00002-7.
17. Talley, S. J., Robison, T., Long, A. M., Young, S., Brounstein, Z., Lee, K., et al., (2021). Flexible 3D printed silicones for gamma and neutron radiation shielding. *Radiat. Phys. Chem. [Internet]., 188*, 109616. Available from: https://doi.org/10.1016/j.radphyschem.2021.109616.
18. Olukotun, S. F., Gbenu, S. T., Oladejo, O. F., Balogun, F. O., Sayyed, M. I., Tajudin, S. M., et al., (2021). The effect of incorporated recycled low-density polyethylene (LDPE) on the fast neutron shielding behavior (FNSB) of clay matrix using MCNP and PHITS Monte Carlo codes. *Radiat. Phys. Chem., 182.*
19. Chai, F., Wang, G., Liu, F., Jiang, D., Yao, C., Xu, T., et al., (2020). Preparation and properties of flame-retardant neutron shielding material based on EVA polymer reinforced by radiation modification. *Radiat. Phys. Chem. [Internet], 174*, 108984. Available from: https://doi.org/10.1016/j.radphyschem.2020.108984.
20. Arslan, S., & Shahbaz, M., (2020). Investigating the effect of adding CdO nano particles on neutron shielding efficacy of HDPE. *Radiat. Phys. Chem. [Internet], 177*, 109145. Available from: https://doi.org/10.1016/j.radphyschem.2020.109145.
21. El-khatib, A. M., Hamada, M. S., Alabsy, M. T., Mohamed, Y., Abd, M., Badawi, M. S., et al., (2021). Fast and thermal neutrons attenuation through micro-sized and nano-sized CdO reinforced HDPE composites. *Radiat. Phys. Chem. [Internet], 180*, 109245. Available from: https://doi.org/10.1016/j.radphyschem.2020.109245.
22. Yılmaz, S. N., Akbay, İ. K., & Özdemir, T., (2021). A metal-ceramic-rubber composite for hybrid gamma and neutron radiation shielding. *Radiat. Phys. Chem., 180.*
23. Avcıoğlu, S., Buldu, M., Kaya, F., Üstündağ, C. B., Kam, E., Menceloğlu, Y. Z., et al., (2020). Processing and properties of boron carbide (B4C) reinforced LDPE composites for radiation shielding. *Ceram. Int. [Internet], 46*(1), 343–352. Available from: https://doi.org/10.1016/j.ceramint.2019.08.268.
24. Chai, H., Tang, X., Ni, M., Chen, F., Zhang, Y., Chen, D., et al., (2015). Preparation and properties of flexible flame-retardant neutron shielding material based on methyl vinyl silicone rubber. *J. Nucl. Mater. [Internet], 464*, 210–215. Available from: http://dx.doi.org/10.1016/j.jnucmat.2015.04.048.
25. Mehelli, O., Derradji, M., Belgacemi, R., & Abdous, S., (2021). Development of lightweight and highly efficient fast neutrons composites shields based on epoxy, UHMWPE fibres and boron carbide particles. *Radiat. Phys. Chem. [Internet]*, 109510. Available from: https://doi.org/10.1016/j.radphyschem.2021.109510.

26. Zali, N. M., Yazid, H., Harun, M., Rashid, A., & Ahmad, M., (2018). Neutron shielding behavior of thermoplastic natural rubber/boron carbide composites. *Neutron Shielding Behavior of Rubber/Boron Carbide Composites Thermoplastic Natural.*
27. Liao, Y. C., Xu, D. G., & Zhang, P. C., (2018). B4C/NRL flexible films for thermal neutron shielding. *Nucl. Sci. Tech. [Internet], 29*(2), 1–9. Available from: https://doi.org/10.1007/s41365-018-0358-4.
28. Güngör, A., Akbay, I. K., & Özdemir, T., (2019). EPDM rubber with hexagonal boron nitride: A thermal neutron shielding composite. *Radiat. Phys. Chem. [Internet], 165*, 108391. Available from: https://doi.org/10.1016/j.radphyschem.2019.108391.
29. Korkut, H., & Korkut, T., (2019). *Improving Neutron and Gamma Flexible Shielding by Adding Medium-Heavy Metal Powder to Epoxy Based Composite Materials, 158*, 13–16.
30. Hossny, K., Hossny, A. H., Magdi, S., Soliman, A. Y., & Hossny, M., (2020). Detecting shielded explosives by coupling prompt gamma neutron activation analysis and deep neural networks. *Sci. Rep. [Internet],* (0123456789), 1–8. Available from: https://doi.org/10.1038/s41598-020-70537-6.
31. Sayyed, M. I., Taki, M. M., Abdalsalam, A. H., Mhareb, M. H. A., Alajerami, Y. S., Şakar, E., et al., (2021). Fabrication, characterization of neutron and proton shielding investigation of tungsten oxide dispersed-ultra high Mw polyethylene. *Chem. Phys.*, 548.
32. Gong, P., Ni, M., Chai, H., Chen, F., & Tang, X., (2018). Jiangsu key laboratory of nuclear energy equipment materials engineering. *Nucl. Eng. Technol. [Internet].* Available from: https://doi.org/10.1016/j.net.2018.01.005.
33. Chen, S., (2018). *Polymer Based Nanocomposites as Multifunctional Structure for Space Radiation Shielding: A Study of Nanomaterial Fabrications and Evaluations* (Thesis available to public electronically), file:///C:/Users/HP/Downloads/Chen_Siyuan.pdf.
34. El-kameesy, S. U., Kansouh, W. A., Salama, E., El-Mansy, M. K., El-khateeb, S. A., & Megahid, R. M., (2017). *A Developed Material as a Nuclear Radiation Shield for Personal Wearing*, 596–605.
35. Afkham, Y., Mesbahi, A., Alemi, A., Zolfagharpour, F., & Jabbari, N., (2020). *Design and Fabrication of a Nano-Based Neutron Shield for Fast Neutrons from Medical Linear Accelerators in Radiation Therapy*, 1–13.
36. Fu, X., Ji, Z., Lin, W., Yu, Y., & Wu, T., (2021). *The Advancement of Neutron Shielding Materials for the Storage of Spent Nuclear Fuel.* Volume 2021, Article ID 5541047, 13 pp. https://doi.org/10.1155/2021/5541047.

CHAPTER 18

Electrochemical Composite Coatings of Copper-Containing Carbon Materials

TAMAZ MARSAGISHVILI, ZURAB SAMKHARADZE, MANANA GACHECHILADZE, NATELA ANANIASHVILI, and MARINA MATCHAVARIANI

Ivane Javakhishvili Tbilisi State University, R. Agladze Institute of Inorganic Chemistry and Electrochemistry. Mindeli Str. 11, Tbilisi, Georgia

ABSTRACT

One of the dynamically developing areas of modern science is the production and research of new materials with improved performance properties. The chapter is dedicated to obtaining of composite copper-carbon coatings by the electrochemical method. In the second phase, the carbon material is used, obtained by the authors of the chapter in an original way from nectarine kernel. The physical parameters and composition of the carbon material have been determined. Composite coatings of copper-carbon material are obtained from suspensions containing different concentrations of the second phase. It is established that the content of the second phase in the composite coating does not correlate with the amount of carbon material introduced into the suspension, and that the upper layers of the composite coating contain a greater amount of the second phase. The wear of the obtained samples was studied, and it was found that the introduction of a carbonaceous material increases the wear resistance of the composite coating and depends on the number of dispersion phase inserts into the matrix.

Advanced Polymer Structures: Chemistry for Engineering Applications. Omar Mukbaniani, DSc, Tamara Tatrishvili, PhD & Marc Jean M. Abadie, DSc (Eds.)

18.1 INTRODUCTION

Special attention in functional electroplating is paid to the production of electrochemical composition coatings obtained on the basis of various metals. The use of dispersed phase particles of various properties and sizes in the formation of composite electrochemical coatings makes it possible to obtain materials with increased operational properties [1].

The principle of obtaining composite coatings is that dispersed particles of different sizes are deposited from the electrolyte suspension together with the metal, the inclusion of which in the coating significantly increases the properties of the coating (hardness, wear resistance, corrosion resistance).

It is known that the efficiency of composite materials obtained by the electrochemical method is determined by the properties of the dispersed phase [2].

As dispersed phase in the electrolyte can be used solid particles with a size not exceeding 3–5 microns (in some cases tens of microns). Recently, the subject of active research has become carbon materials of different origins (graphene, nanoparticles, technical carbon, fullerene) [3, 4] with unique properties (high electrical conductivity, resistance to high temperatures and aggressive solutions, layered structure, etc.). Due to these properties, they are used in many industries, as well as for the obtaining of solid antifriction electrochemical composite coatings. These types of coatings, by reducing the wear of units and mechanisms, increase their service life and reliability [5].

A technology has been developed for producing a carbon material with a high specific surface area from cellulose-containing, constantly renewable secondary raw materials (hazelnut and walnut shells, nectarine kernel, sawdust, bamboo) using inexpensive reagents. A patent was issued for this method and a positive decision was obtained [6].

18.2 EXPERIMENTAL METHODS AND MATERIALS

18.2.1 MATERIALS

To obtain composite coatings, a carbon material is used, obtained by the above method, from a material containing secondary cellulose – nectarine kernel.

Material milled in a nanomill is a black fine-dispersed (50–100 nm), dry, loose powder with a high specific surface area. Figure 18.1 shows the carbon material we obtained from the nectarine kernel.

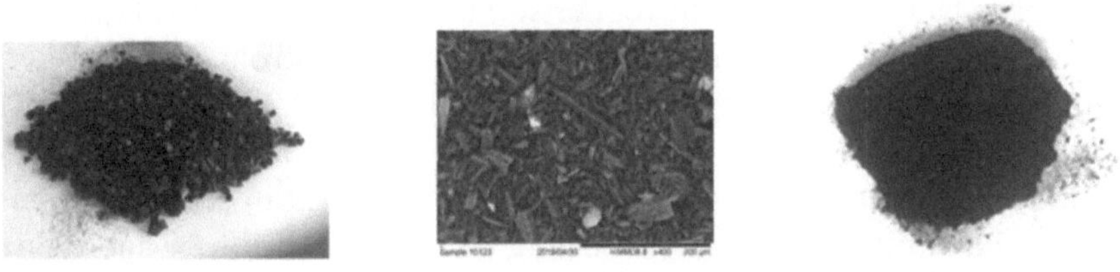

1 2 3

FIGURE 18.1 Images of carbon material obtained from nectarine kernel: (1) original carbon material; (2) carbon material (magnification x400); and (3) sample grounded in a mill.

We measured the composition and some physical characteristics of carbon material, obtaining from the Nectarine Kernel. The measurement results are given in Tables 18.1 and 18.2.

TABLE 18.1 Composition of Carbon Material Obtained from Nectarine Kernel

Sample	Chemical Composition of Samples (%)												
	C	O	Si	Ca	Al	K	Fe	S	Zn	Cr	Ni	Cu	F
Carbon material	91.2–92.5	6.33–7.49	0.07–0.2	0.09–0.13	0.14–0.22	0.16–0.25	0.24–0.34	0.07–0.12	0.09–0.16	0.01–0.16	0.02–0.12	0.03–0.07	0.00–0.03

TABLE 18.2 Physical Characteristics of Carbonaceous Material (Nectarine Kernel)

Sample	Surface Area (m^2/g)	Micropore Area (m^2/g)	Micropore Volume (cm^3/g)	Moisture (%)	Ash Content (%)
Carbon material	520.10	434.67	0.19	9.3	0.2

18.2.2 METHODS

The physical characteristics of the obtained product were measured on a Gemini VII, and the composition was determined under a scanning electron microscope (SEM) (HITACHI TM 300-plus). Ash content of the material [7] and moisture are also determined.

The morphology and composition of the obtained composite coatings were studied using a TESCAN VEGA 3 SEM and an Oxford Instruments, Aztec ONE electrodispersive spectrometer integrated with it.

Composite materials were obtained from suspensions prepared on the basis of a standard solution of copper sulfate (0.8M $CuSO_4$ 5M H_2O + 50 g/l H_2SO_4; pH = 0.35) using a carbon material obtained from nectarine kernel, with second phase concentration (g/l) – 0.04; 0.06; 0.08; 0.1; 0.4; 1.2; 2.0; 5.0.

For good wetting of the material, it was dissolved in a small volume of ethyl alcohol, filtered, and, after washing with electrolyte, the suspension was dispersed (40 kHz, 16 min) to prevent uniform distribution and precipitation.

Composite coatings were obtained on stainless steel cathodes. The anode is copper, the area of which is 5 times the area of the cathode.

The wear resistance of the obtained composite coatings is studied. The wear of the obtained samples was determined by the weight method, at a constant load and at a fixed speed of the grinding disk on a model made by us, similar to the one [5] described in the literature (Figure 18.2).

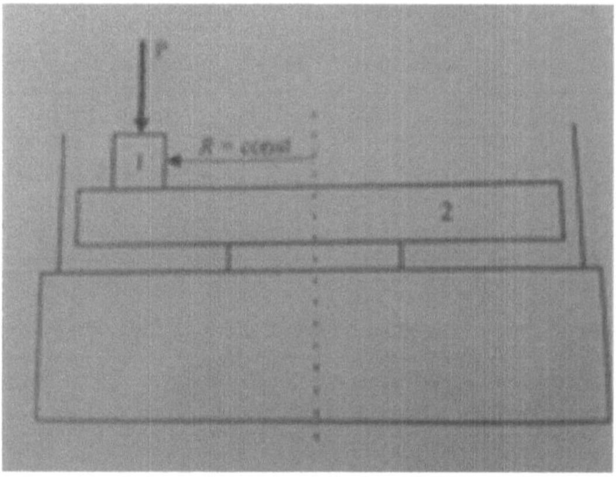

FIGURE 18.2 Scheme of determination of wear resistance. (1) sample; and (2) grinding disk.

18.3 RESULTS AND DISCUSSION

18.3.1 COPPER COMPOSITE COATINGS

Copper composite coatings are obtained from solutions containing carbonaceous material of various concentrations.

Figures 18.3 and 18.4 show the electro-dispersion spectrum of the carbon-containing copper composite coating and an element distribution map in an electrolyte with a content of 0.04 g/l of the second phase.

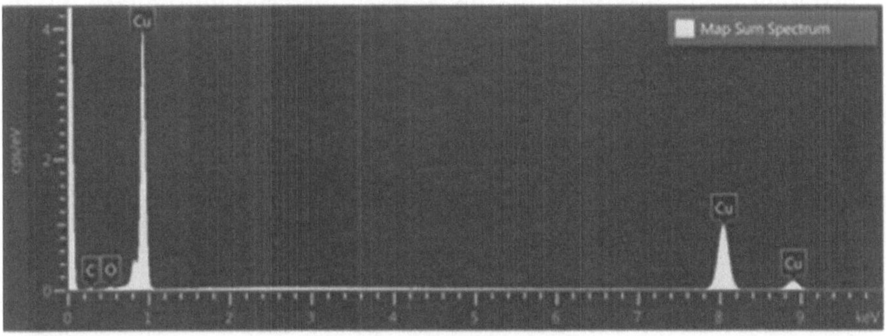

Map Sum Spectrum				
Element	**Line Type**	**Weight Percent**	**Weight Percent Sigma**	**Atomic Percentage**
C	K series	3.65	0.23	16.38
O	K series	0.69	0.08	2.34
Cu	K series	95.66	0.24	81.28
Total		100.00		100.00

FIGURE 18.3 Electro-dispersed spectrum of a carbon-containing copper composite coating in the presence of a carbonaceous material (0.04 g/l) in solution.

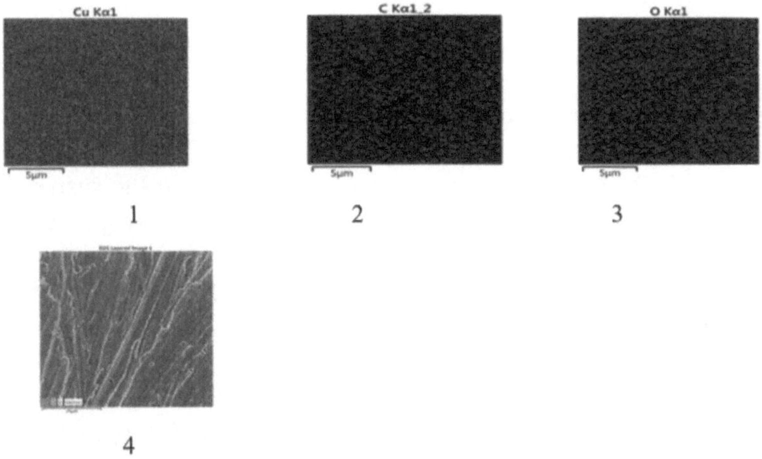

FIGURE 18.4 Distribution map of copper carbon-containing composite coating elements (Cu, O, C) in the presence of carbon material (0.04 g/l) in solution and 3.65% (by weight) in the coating: (1) copper; (2) oxygen; (3) carbon; and (4) averaged energy dispersive spectrum.

According to the element distribution map in the composite material, it can be said that the dispersion phase is uniformly distributed in the upper layer of the coating.

Table 18.3 shows the compositions of copper composite coatings at different concentrations of the carbon material introduced in the suspension.

TABLE 18.3 Average Value of the Second Phase Content in the Coating (weight percent), Depending on the Concentration of Carbonaceous Material in the Suspension (g/l)

SL. No.	Concentration of Carbonaceous Material in the Suspension (g/l)	Carbon Content in Composite Coating (Cu-C), Weight Percent		
		Base Side	Surface	
			Limits	Average
1.	0.04	0.96	3.20–3.65	3.43
2.	0.06	0.76	1.79–3.07	2.63
3.	0.08	1.15	3.77–4.03	3.67
4.	0.1	0.85	3.19–3.20	3.2
5.	0.4	0.93	3.33–3.35	3.34
6.	1.2	0.67	2.56–2.59	2.58
7.	2.0	0.98	3.2–3.76	3.48
8.	5.0	1.05	3.35–3.88	3.62

As can be seen from the table, the content of the second phase in the composite coating fluctuates in the range of 2.58–3.67 (wt.%) and does not correlate with the amount of carbon material inserted in the suspension. It should be noted that the upper layers of the coatings contain a greater amount of dispersed phases at all concentrations of the second phase in the suspension. A study of the morphology of the base side of the coatings showed that it replicated the morphology of the stainless steel base, and its study was less interesting.

Figure 18.5 shows the surface images of the composite material taken under an electron microscope (Evromix ME 2665) "in pure form" and with a different content (weight percent) of carbonaceous material in the coating.

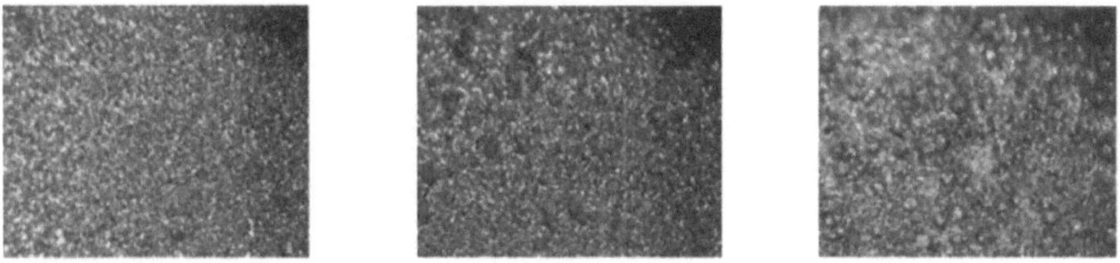

FIGURE 18.5 Images of surfaces of copper composite material (magnification x10) at different content of carbonaceous material in the coating (weight percent): (1) 0.0; (2) 2.63; and (3) 3.48.

As can be seen from the figures, the introduction of carbon material into the composite coatings leads to an increase in grain size.

18.3.2 WEAR OF THE COPPER–CARBON COMPOSITE MATERIAL

To evaluate the mechanical properties of the coatings, Cu-C composite coatings were tested for wear (Figure 18.6). Load P = 200 g (disk radius – R = 12 cm; sample test time 20 min., step 5 min., fixed linear velocity).

It is known from the literature [8, 9] that the inclusion of dispersed particles in the coating causes structural changes in the metal matrix, which affects the properties of the coatings.

At the initial stage of electrolysis, the introduction of the dispersed phase into the coating begins, and the upper layers of the composite coating contain a greater amount of the second phase [10].

Figure 18.6 shows the wear of "clean" and composite coatings at different concentrations of carbonaceous material in the coating during dry friction.

As can be seen from the figure, the wear of the composite material during the first 10 to 15 minutes (the time depends on the content of the second phase in the composite) is less than the wear of the "clean" coating for all samples. As mentioned above, the upper layers contain more of the second phase and the wear of the samples and the time when the wear of the composite is better than that of the "pure" sample is directly proportional to the carbon content of the composite.

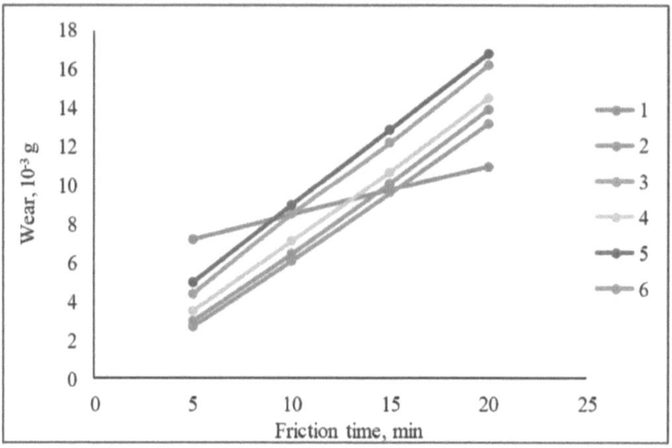

FIGURE 18.6 Wear of composite coatings over time during dry friction at different contents of carbonaceous material (nectarine kernel) in the coating (weight percent): (1) 0.0; (2) 2.58; (3) 2.63; (4) 3.2; (5) 3.43; and (6) 3.62.

Further increase in wear (above 10 to 15 minutes) exceeds the wear of the "clean" sample, which must be due to the fact that the lower layers of the coating contain a small amount of carbon material. Based on the obtained results, it can be concluded that the wear of the composite material depends on the content of the second phase in the coating.

18.4 CONCLUSION

Based on the work done, a conclusion can be made: Carbon material is obtained from secondary cellulose-containing material (nectarine kernel); The physical parameters and composition of the carbon material obtained from nectarine kernel are determined; Optimal conditions for obtaining copper-carbon composite coatings have been experimentally determined; Composite coatings are obtained by electrochemical method, from copper sulfate-based suspensions in the presence of carbonaceous materials of different concentrations; It has been established that the introduction of carbonaceous material into composite coatings causes an increase in grain size; The content of the second phase in the composite coating is 2.58–3.67 (weight percent) and is not correlated with the amount of carbon material inserted in the suspension. The upper layers of the composite coating contain more carbon material.

KEYWORDS

- **carbon material**
- **composite electrochemical cofting (CEC)**
- **composite material**
- **copper composite coatings**
- **electrical conductivity**
- **nickel**

REFERENCES

1. Sayfullin, R. S., (1972). *Combined Electrochemical Coatings and Materials*. M.: Chemistry.
2. Tarasevich, M. R., (1984). *Electrochemistry of Carbonaceous Materials.* Science, Moscow, (in Russian).
3. Mingazova, G. G., Fomina, R. E., & Vodopianova, S. V., (2011). Influence of particles of different nature on the properties of nickel coatings. *Bulletin of KSTU*, *12*, 157–161.
4. Daniel, C. W., Keith, C. K., & Irene, M. C. (1989). *Conversion of Wood Waste into Activated Carbon and its Application.*
5. Kozenkov, O. D., Ptashkina, T. V., Kosilov, A. T., & Zhiliakov, D. G., (2015). Wear resistance of composite electrochemical coatings reinforced with carbon nanomaterials. *Voronezh State Technical University Bulletin, 11*(5), 135–138.
6. Marsagishvili, T. A., Tatishvili, G. D., Ananiashvili, N. Sh., Gachechiladze, M. P., Metreveli, J. A., Matchavariani, M. N., Tskhakaia, E. T., et al., (2021). Method of obtaining sorbents from waste containing plastics and cellulose. *Georgian National Intellectual Property Center.* Patent 2021 7309 B.
7. ASTM D1506 – 15. *Standard Test Methods for Carbon Black-Ash Content*. https://www.astm.org/Standards/D1506.htm (accessed on 02 January 2022).
8. Celukin, V. N., (2009). *Electrochemical Bonding of Composite Coatings Based on Nickel and Media.* Auto reference. Saratov, (in Russian).
9. Kubrak, P. B., Drozdovich, V. B., Zharsky, I. M., & Chaevsky, V. V., (2012). Electrochemical deposition and properties of nickel coatings containing carbon materials. *Electroplating and Surface Treatment,* #2, v. *XX*, 43–49.
10. Vasilenko, E. A., (2013). *Electrochemical Bonding of Composite and Multilayer Coatings Based on Nickel and Nickel-Chromium*. Dissertation.

CHAPTER 19

Composite Materials Based on Noryl and Polyvinyl Chloride

S. S. MASHAYEVA and B. A. MAMEDOV

Institute of Polymer Materials, Azerbaijan National Academy of Sciences, S. Vurgun Str. 124, Az5004, Sumgait, Azerbaijan

ABSTRACT

In order to reduce the cost of Noryl, its compositions with recycled polyvinyl chloride (r-PVC) were obtained, and their physical-mechanical properties were studied depending on their composition. The filler and Noryl were mixed in a ratio of 60:40; 50:50; 40:60, respectively.

It was found that compositions containing 60% Noryl and 40% PVC are more processable in combination with good thermo-physical and physical-mechanical properties.

19.1 INTRODUCTION

At the present time, intensive research is underway to create new polymeric materials and reduce their cost. In this case, the main goal is to preserve the properties of the obtained composite materials as they are in the initial polymers or to improve the corresponding performance characteristics and manufacturability of the obtained polymeric materials. Noryl is one of the most interesting polymeric materials. Noryl is mainly obtained by modifying polyphenylene oxide with polymers such as polystyrene, polyamide, and polypropylene, and is a valuable thermoplastic with high physical-mechanical, impact, and heat resistance, as well as high dielectric properties.

Advanced Polymer Structures: Chemistry for Engineering Applications. Omar Mukbaniani, DSc, Tamara Tatrishvili, PhD & Marc Jean M. Abadie, DSc (Eds.)

Depending on the composition and properties, Noryl may be suitable for a wide variety of applications, such as in electronics, electrical equipment, coating, machinery, etc. [1]. However, one of its main disadvantages is its high cost, which limits its use. In order to reduce the cost of Noryl, its compositions with various polymers and polymer wastes were obtained and studied [2, 3]. The use of polymer waste as a filler in the creation of such compositions is cost-effective and also environmentally beneficial for the disposal of polymer waste.

19.2 EXPERIMENTAL PART

Noryl and filler – r-PVC were used as the object of research. Polymer-polymer compositions were obtained by mixing Noryl and r-PVC in a laboratory mill at a temperature of 160°C. Then, standard samples were prepared on a press at a temperature of 170°C and under a pressure of 10–14 MPa. The filler and Noryl were mixed in a ratio of 60:40; 50:50; 40:60, respectively.

The physical-mechanical properties of the composites were determined by the following methods. The melt flow rate was determined using a CEAST MF-150 capillary rheometer (INSTRON, Italy). The measurement is carried out at 190°C under 5 kg load in accordance with ASTM D1238.

IR spectra were recorded by using an Agilent Cary 630 FTIR spectrometer (Agilent Technologies): ZnSe crystal, wavelength range 600–4,000 cm^{-1}.

The thermal stability of the samples under study was determined on a Q-1500D derivatograph (MOM, Hungary) of the Paulik-Paulik-Erdey system. Sample weighed in 100 mg, channel sensitivity TG-100, DTG-1 mV, DTA-250 μV, T/V-500/5, crucible Pt. Heating rate 5 degrees/min in a stream of air.

19.3 RESULTS AND DISCUSSION

As already noted, Noryl is one of the most valuable materials used in various fields of technology and industry, depending on its composition. Noryl, used in electrical engineering and electronics, must have heat resistance, low flammability, high electrical properties. Noryl used in mechanical engineering has a valuable set of technological and operational properties, a combination of high mechanical and dielectric properties with manufacturability. Noryl, which is used in the manufacture of household appliances, must be easily treated and be resistant to hot water. Despite such valuable properties, the high cost of Noryl limits its use. To reduce the cost of polymer products from Noryl, it is considered expedient to create its composition with other polymers using suitable fillers. Recycled polyvinyl chloride (r-PVC) was chosen as filler. New composite materials are obtained by mixing Noryl and r-PVC in various proportions.

First of all, the structures and compositions of the starting material Noryl and the obtained composite materials based on it were studied by infrared spectroscopy.

Comparison of the IR spectra of the initial polymer matrix and the obtained composites (Figures 19.1 and 19.2) shows that no significant change in the IR spectra is observed in the course of component displacement. Stretching vibrations of C-H groups were observed at 3,055–2,853 cm^{-1}, stretching vibrations of the aromatic ring at 1,600 cm^{-1} and stretching vibrations of the C-Cl bond at 616 cm^{-1}.

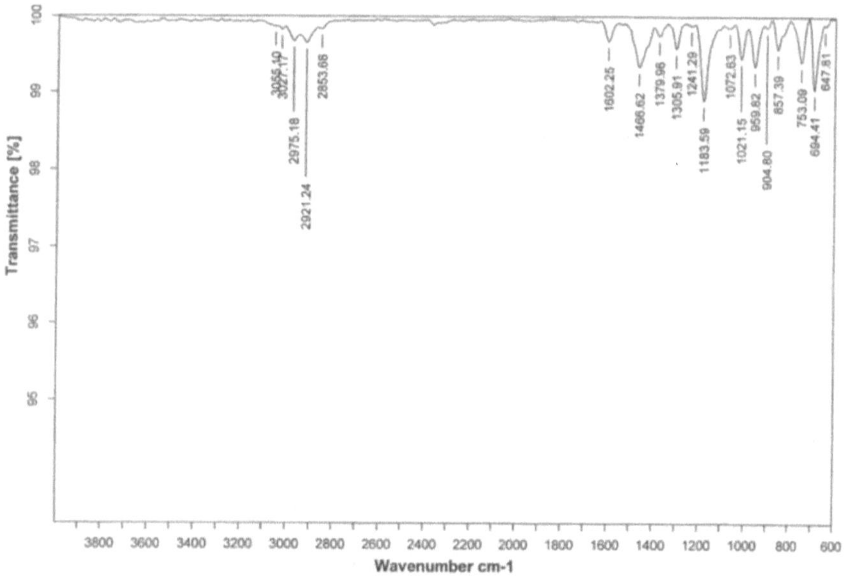

FIGURE 19.1 IR spectrum of Noryl.

Some physical-mechanical properties of composite materials have been studied. The research results are presented in the diagram (Figure 19.3). As can be seen from the diagram, the physical-mechanical properties of the obtained composite materials change depending on the composition of the composite materials. With 40 mass % filler content, the tensile strength is ~ 30 MPa and the elongation at break is ~ 26%.

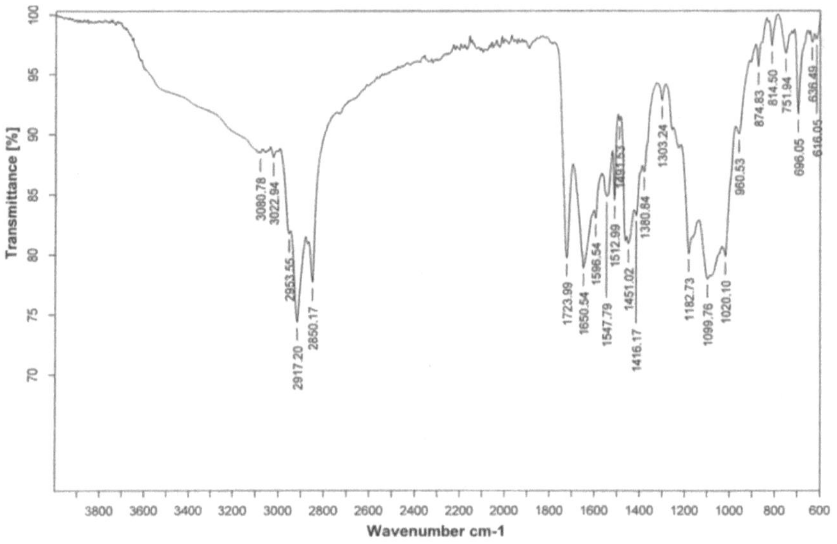

FIGURE 19.2 IR spectrum of the composite based on Noryl and r-PVC.

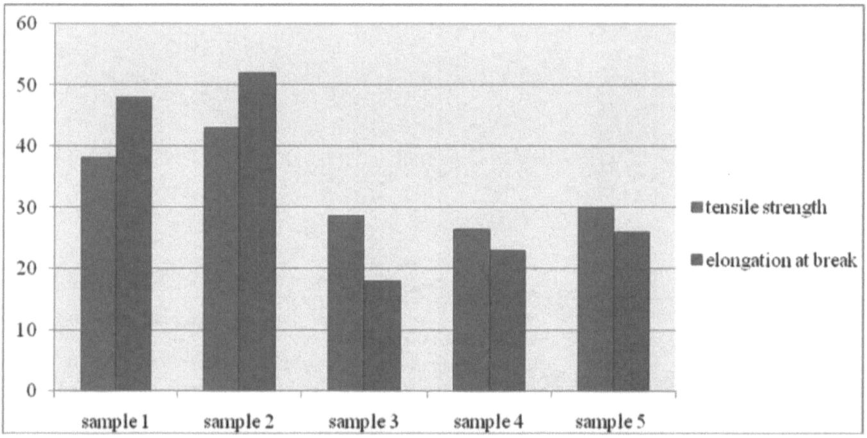

FIGURE 19.3 Variation of tensile strength and elongation at break at different filler content.

In Table 19.1, some properties of composite materials are presented.

TABLE 19.1 Some Properties of Composite Materials

Sample Number	Composition Formulation (mass %)	Melt Flow Rate (g/10 min)	Vicat Softening Point (K)
Sample 1	Noryl (100)	0.323	413
Sample 2	r-PVC (100)	–	413
Sample 3	Noryl:r-PVC (40:60)	0.113	413
Sample 4	Noryl:r-PVC (50:50)	0.132	418
Sample 5	Noryl:r-PVC (60:40)	0.168	418

Thermal resistance was assessed by the temperatures T_{10}, T_{20}, T_{50} at which the samples lost 10, 20, 50% weight, respectively, the temperature of the onset of decomposition T_n and the value of the half-life $\tau_{1/2}$ (Table 19.2).

TABLE 19.2 Some Indicators of Thermal Destruction of the Test Samples

SL. No.	Compos.	$\tau_{1/2}$ (min)	$T_н$ (°C)	T_{10} (°C)	T_{20} (°C)	T_{50} (°C)
1.	Sample 1	73.2	90	325	355	385
2.	Sample 5	70	75	200	215	330

The comparative analysis of thermogravimetric curves showed that, in contrast to Noryl, thermal destruction of the composite material proceeds in two stages. This is undoubtedly due to the presence of polyvinyl chloride in the composition, since PVC exhibits low heat resistance. As is known from [4], when heated above 150°C, the destruction of polyvinyl chloride begins with the release of hydrogen chloride. It is obvious that the first stage of the thermal destruction of the composite corresponds to this precipitation. Analysis of the data corresponding to the change in mass with temperature of the test samples shows that, although the sample begins to decompose at lower temperatures, the mass residues corresponding to a temperature of 500°C have a similar meaning.

The differential thermal analysis curves of the sample showed an endothermic melting peak of the composite corresponding to 210°C. Nevertheless, the temperature range of melting encompasses a rather large region of 175–200°C, which is associated with the defectiveness of the crystal structure of the composite.

19.4 CONCLUSION

Compositions of Noryl with PVC were obtained, and their properties were investigated. It was found that the compositions containing 60 mass% of Noryl and 40 mass% of the filler – r-PVC, show the best physical-mechanical properties in comparison with other compositions (Noryl:PVC (40:60)) and (Noryl:PVC (50:50)). Composite materials can be used in the manufacture of parts for household appliances.

KEYWORDS

- **infrared spectroscopy**
- **noryl**
- **physical-mechanical properties**
- **polymer**
- **polymeric materials**
- **polypropylene**
- **polyvinyl chloride**

REFERENCES

1. Mikhailin, Yu. A., (2006). *Heat-Resistant Polymers and Polymer Materials* (p. 490). St. Petersburg: Publishing house. Profession.
2. Bondaletova, L. I., & Bondaletov, V. G., (2013). *Polymer Composite Materials* (p. 118). Tomsk Polytechnic University Publishing House.
3. Yu, V. P., Korchagina, T. K., & Lobasenko, V. S., (2015). *Diphenyl Oxide Derivatives. Synthesis, Reactions and Applications* (p. 248). Monography Volg. GTU – Volgograd.
4. Iskhakova, A. A., Nosirov, T. D., Parsanov, A. S., & Krasina, I. V., (2019). Properties of polyvinyl chloride. *New Technologies and Materials for Light Industry* (pp. 205–208). Kazan.

CHAPTER 20

Composite Materials on the Basis of Various Binders

OMAR MUKBANIANI,[1,2] TAMARA TATRISHVILI,[1,2] and LEVAN LONDARIDZE[1,2]

[1]*Department of Macromolecular Chemistry, Ivane Javakhishvili University, Ilia Chavchavadze Blvd. 1, Tbilisi, Georgia*

[2]*Institute of Macromolecular Chemistry and Polymeric Materials, Ivane Javakhishvili University, Ilia Chavchavadze Blvd. 13, Tbilisi, Georgia*

ABSTRACT

Ecologically friendly new composite materials with high-technical characteristics are made on the basis of wood sawdust and organic/inorganic binders. These composite materials are obtained on the basis of a new binders phenylethoxysiloxane (PhES-80), liquid glass (LG), polyethylene and colophony (at different pressures and temperatures). The binder used simultaneously acts as both a binder and a reinforcement agent. The surface structure of the new composite materials was studied by means of optical microscopy, scanning electron microscopy, and energy-dispersive X-ray micro-analysis. For composites, tensile strength at bending, impact viscosity, thermogravimetric stability and water absorption coefficient have been examined. Optimal conditions for obtaining new, ecologically friendly composites have been established. The obtained composites are characterized by high mechanical properties, thermal resistance, ecological purity, and low water absorption capacity, which is one order of magnitude smaller than the water absorption of existing particle board.

Advanced Polymer Structures: Chemistry for Engineering Applications. Omar Mukbaniani, DSc, Tamara Tatrishvili, PhD & Marc Jean M. Abadie, DSc (Eds.)

20.1 INTRODUCTION

Agro-industrial waste removal is a serious issue of concern in developing countries. Cellulose is a polysaccharide polymer. This present review [1] explores cellulose history, structure, worldwide production, and extraction of cellulose from agro-waste. A wide spectrum of researches in the arena of properties of cellulose, hemicellulose, and lignin; their degradation; sources and composition of cellulosic and its derivatives from agro-industrial wastes; present status of converting them into value-added products of food and pharmaceutical applications. Cellulose is a tremendous product due to its abundance and characteristic structural properties. The major industrial source of cellulose is vascular plants. The lignocellulosic materials, especially agro-industrial residues, are important as reinforcement products for building construction material industry, in terms of environmental preferences of the modern society. Most paper products generate from wood pulp, while textile fibers are commonly not isolated from woody fibers. The materials based on cellulose and its derivatives have been used for a wide variety of applications, such as food additives, paper manufacturing, pharmaceuticals, or other chemical engineering uses, such as chromatography, paints, and explosives.

Waste is defined as any material, which has not yet been fully utilized, i.e., the leftovers from production and utilization. The waste contains three main constituents: Cellulose, hemicellulose, and lignin, and it can contain various compounds [2].

Cellulose and hemicellulose are carbohydrates that can be broken down by enzymes and acids and then fermented to produce ethanol, renewable electricity, fuels, and biomass-based products [3, 4].

However, waste is an expensive and generally unavoidable result of human activity. It includes plant materials, agricultural, industrial, and municipal wastes, and residues [5].

Food processing wastes food in spillage, spoilage, discarding substandard edible materials, or removing edible food parts in inefficient processing [6]. Waste significantly impacts environmental, economic, and community health [7].

Plants produce about 180 billion tons of cellulose manufacture annually, and it is the largest reservoir of organic carbon on the earth. Cellulose constitutes the most abundant, renewable polymer resource available today worldwide. It has been expected that by photosynthesis, 10^{12} tons are synthesized annually in a rather pure form, for example, in the seed hairs of the cotton plant but mostly are common with lignin and other polysaccharides in the cell wall of woody plants. Cellulose is the structural part of the primary cell wall of green plants, many forms of algae, and the oomycetes. Cellulose is the most common organic compound on the earth.

About 33% of all plant material is cellulose (cotton is 90% and wood is 40–50%). For industrial use, cellulose is mainly obtained from wood pulp and cotton. It is mainly used to manufacture paperboard and paper [8], and it is transformed into a wide variety of derivative products such as cellophane and rayon. Converting cellulose from energy crops into biofuels such as cellulosic ethanol is under exploration as an alternative fuel source.

In recent years great interest in the development of new composites is derived from thermoplastic polymer matrices reinforced with wood filler, because of their environmental and economic benefits [1, 2]. Their renewability, biodegradability, low density, high stiffness and relatively low price are established [3]. Among of these various thermoplastic matrices mainly used in the manufacture of plastic/wood composites was polystyrene [4], which is very popular because of its inapparency, fluidity, and good electrical insulating properties [5].

The use of lignocellulosic fibers has certain disadvantages such as degradation at low temperatures with poor compatibility between the polar lignocellulose filler and the non-polar polymer matrix [6]. Due to the strong intermolecular hydrogen bonding between the lignocellulose fibbers, which tend to agglomerate during mixing with the polymer matrix in the compounding process, resulting composites with low mechanical and thermal properties [7].

The improving of the interface compatibility between thermoplastic polymers and lignocellulosic filler has attracted much attention from researchers [8]. Several modifications of the fiber surface such as reaction with acid compounds [9, 10], alkali treatment [11, 12] and the incorporation of compatibilizer, such as malleated polymer [13, 14] or treatment with coupling agents [15, 16] are reported in the literature.

Among the different coupling agents, functional organosilanes [$RSi(OR')_3$)] are often used. These bifunctional molecules with their alkoxylsilane groups are used to modify the surfaces of natural fibers. After hydrolysis reactions of the fibers, surfaces rich in OH groups can be created. Chemical bonds on the surfaces of the fibers through a siloxane bridges, organofunctional groups bond to the polymer matrix. These groups improve the compatibilization between the fibers and the polymer matrix by the formation of covalent bonds. The silane coupling agents provide bridges between the fibers and the matrix [17–19].

Impregnations using phenethosiloxane, LG is a way, which will be improve several unfavorable features in woods. In this procedure, lightweight, and permeable solid woods will be impregnated with using monomer with low molecular weight (MW), low viscosity and/or high reactivity, which must be capable of fill intra- and/or intercellular spaces in wood and/or chemically bond themselves to certain polysaccharides and lignin present in the wood cell wall.

The aim of this chapter is to obtain bi- or tricomponent composite materials on the basis of sawdust and penylethoxysiloxane, LG, polyethylene, and colophony and to disclose the effects of surface modification on the morphological, thermal and mechanical properties of sawdust reinforced composites, where used binders acts as both a binder and a reinforcement agent.

20.2 EXPERIMENTAL PART

20.2.1 MATERIALS

We have created composites based on dry sawdust on the basis of pine with LG, polyethylene, colophony, as binder and reinforcement agent.

20.2.2 CHARACTERIZATION

1. **Processing:** The composites were prepared by hot pressing of highly dispersed (50 μm) components under pressures up to 15 MPa and temperatures up to 100°C in molds for 15 min. We have created two types of samples: cylindrical (for investigation of water absorption) and rectangular (for mechanical testing).
2. Fourier Transform Infrared Spectra: These were determined with a Varian 660-IR FT-IR Spectrometer. The KBr pellets of samples were prepared by mixing (1.5–2.0) mg of samples, finely grounded, with 200 mg KBr (FTIR grade) in a vibratory ball mixer for 20 s.
   Microstructure of the samples was studied by NMM-800RF/TRF type of optical microscope SEM and EDS observations were conducted. Measurements were performed under a microscope – Tescan Vega 3, LMU, LaB 6 cathode. Maximum accelerating voltage was 30 kV, resolution 2.0 nm. The microscope was also equipped with an energy dispersive spectrometer of X-ray-induced electron beam specimens (EDS, Oxford Systems). EDS was used to analyze the sample compositions.

3. **Bending Testing:** It is also known as flexural testing, was performed on parallelepipeds with the length of 10 cm and the vertical square cross-section of 1 cm^2. Each specimen was placed on two prisms, with the distance of 8.0 cm between the prisms. The indenter was a metal cylinder with the diameter of 10 mm applied from above to the midpoint of the specimen. Bending strength (or flexural strength) is defined as the stress needed to create a breaking point (a crack) in the outer surface of the test specimen [20].
4. **Impact Viscosity Determination:** It is also called *shock viscosity* determination, is a technique applied to soft solids [21, 22] and is essentially a drop impact test. The drop height h is the vertical distance between the upper surface of the tested material (h_1) and the bottom surface of the drop hammer at the end of the impact event (h_2). With the sample mass m and the acceleration g, the work performed by the falling hammer is mg (h_1–h_2), a normalized with respect to the horizontal cross-section of the specimen.
5. **Vicat Softening Depth:** This consists in the determination of the depth of the indentation with respect to the top surface caused by a flat ended indenter with a cross-section of 1 mm^2. The load applied is 10 or 50 N and the cross-section of the indenter end is circular. The term *Vicat hardness* is also in use – really confusing since larger values correspond to lower hardness.
6. **Water Absorption:** It is determined simply as the percentage weight change of the sample after submersion in water. We have performed such measurements after 3 hours and 24 hours exposition to water.

20.3 RESULTS AND DISCUSSIONS

As noted above, the aim of our work is the creation and development of new bi- and tricomponent composites on the basis of sawdust (Pine) and ecologically friendly organo-inorganic binders.

From scientific-technical literature, it is known that the sawdust mainly contain sufficient amounts of starch, cellulose, hemicellulose, and pectic substances, lignin, polyphenols. All this above-mentioned compounds may react with proposed binders. For example phenylethoxysilane (PhES-80) containing ethoxyl group participate to the etherification reaction with a sawdust through the macromolecular and intra-molecular reactions. Processes that occur during the curing are complex and varied. A modern look at an overview of the curing LG itself and in the various homogeneous and heterogeneous systems, the most widely encountered in practice, is presented in a number of reviews [23–26].

Acting as an adhesive or binder LG system goes from liquid to solid in many ways and may be divided into three types:

- the loss of moisture by evaporation at ordinary temperatures;
- the loss of moisture from the system, followed by heating above 100°C; and
- the transition to the solid state by introducing specific reagents, which is called hardeners. Naturally, these three types are used in combination.

In solution, the degree of polymerization of silicate anions is known to depend on two factors – the silica modulus and the solution concentration. Each solution has a distribution of degree of the anion polymerization. Distribution is superimposed on the polymer distribution of anions on the charges, which is also determined by these two factors.

It is possible on the intermediate stages the reactions between of ≡Si-OH band contained LG and cellulose molecule, which are take place with dehydration reactions. These reactions go through obtaining of three-dimension structures. Formation of such structures takes place also at the use of hardener $Na_2SiF_{6,}$ which accelerates the processes.

The binder colophony derives from pine resin, tall oil and stump extractives. It is used naturally or in chemically modified forms: hydrogenated, disproportionated, esterified, polymerized [27]. The colophony is in the structure of different plants main structure containing the isomeric acid rings. The noted structural ring can introduce to reaction both with PhES-50 and PhES-80 or with LG. At presence of colophony it may be realized the donor-acceptor bond with leaves matrix, which is connected with formation of additive intermolecular forces and leads to increasing of material strengthening.

Natural dry wood sawdust was destructed in the mixer like of coffee mill. The fraction of particles with middle sizes near 50 mcm was used in our experiments. The ingredients dispersive powders PhES, LG, PE, colophony, and wood glue with amount of 3–20 wt.% were added to the powder of wood sawdust and carefully mixed in the mixer. The obtained blends were placed to the spatial press forms corresponding to the standards and pressed under various pressures and temperatures.

Composites were measured using the following features: (i) Fourier transition IR spectroscopy (FTIR) study; (ii) limit bending strength; (iii) shock viscosity; (iv) thermo-Vikat method; and (v) water absorption coefficient. Results and analysis are provided below.

From the physical-mechanical characteristics of the composites, the mechanical strength at elongation and shock viscosity were provided with use the standard methods and apparatus. Thermal stability of the materials was established with the use of the well-known method of Vicat. Water absorption was defined by means of weighing of samples before and after the exposition of samples in the distilled water. The results of these measurements are presented in subsections.

20.3.1 FTIR SPECTRAL INVESTIGATIONS

For samples, Fourier transform infrared spectroscopy investigation have been carried out in KBr [28, 29]. The KBr pellets of samples were prepared by mixing (1.5–2.00) mg of samples, finely grounded, with 200 mg KBr (FT-IR grade) in a vibratory ball mixer for 20 s.

In the FTIR spectra of bi- and tricomponent containing composites (Figures 20.1–20.5), one can absorption bands for asymmetric valence oscillation of Si-O-Si bonds with maximum at 1,066 cm^{-1}, characteristic for siloxane bonds in cyclotetrasiloxane fragments well as for etheric C-O-C and C-O-Si bonds this bands are overlaps. In the spectra one can see absorption bands at 1,267, 1,373, 1,429, 1,515, 1,600–1,650, 1,733, 2,800–2,950, 3,363 characteristics for methyl groups, C–H absorption (–/C–/CH_3), CH_2 cellulose – lignin, C=C aromatic, C=C alkene, (C=O ester), C–H methyl, methylene, and phenyl groups, O–H alcohol accordingly [30, 31].

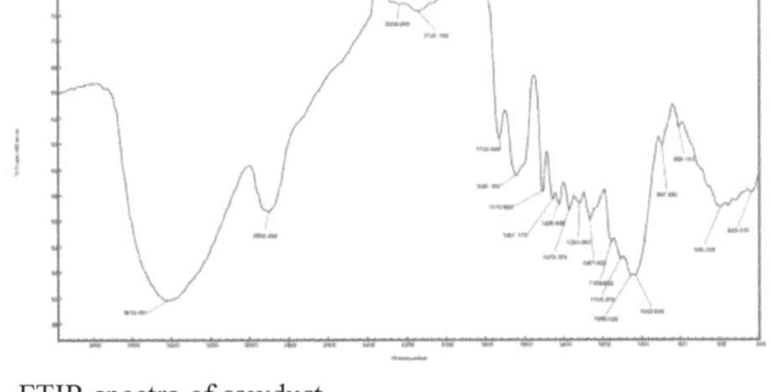

FIGURE 20.1 FTIR spectra of sawdust.

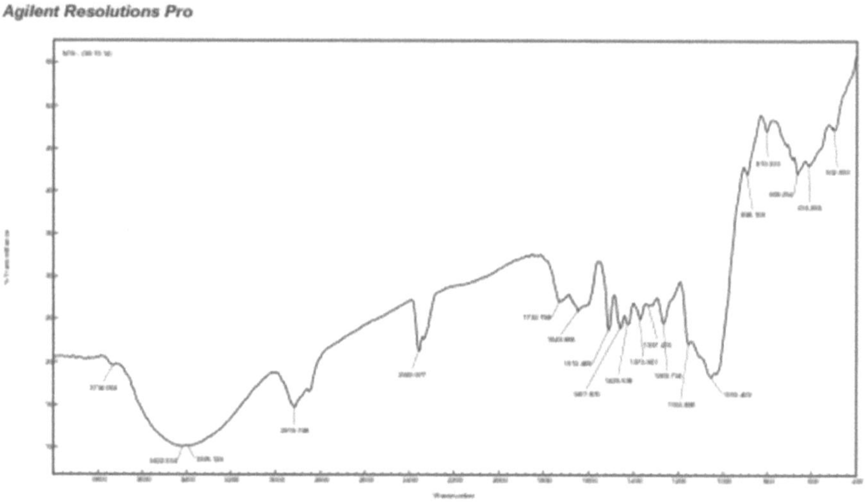

FIGURE 20.2 FTIR spectra of composite 19 PhES-50 (3%) + sawdust (92%) + PE (5%) (Table 20.7) (pressure – 130 kg/cm^2; temperature – 150°C).

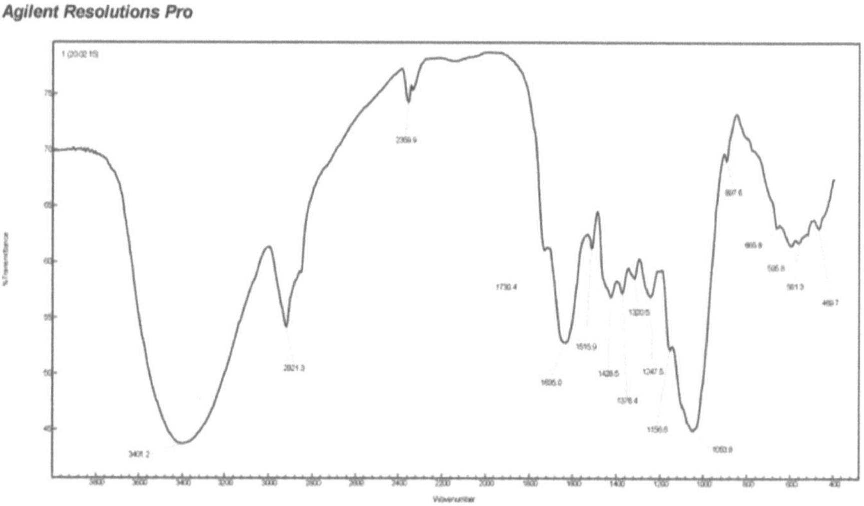

FIGURE 20.3 FTIR spectra of two components composite (1.1) (straw 97% + LG 3%) (Table 20.7).

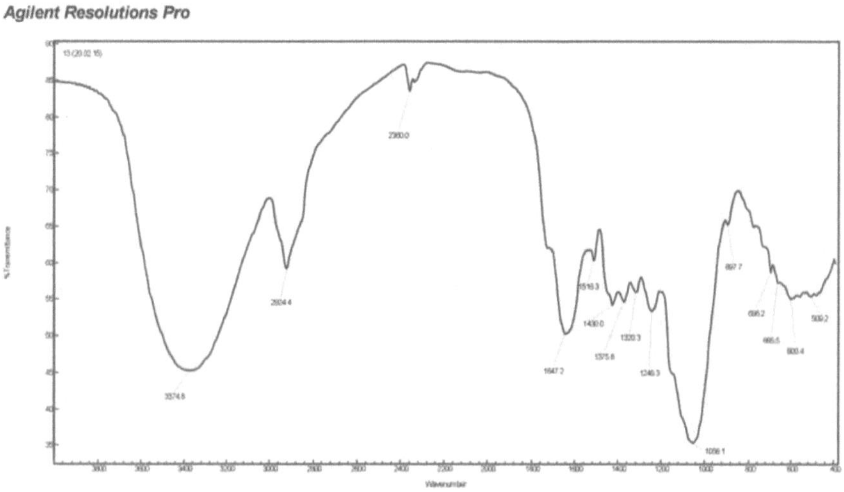

FIGURE 20.4 FTIR spectrum of two components of composite (straw 90% + polyethylene 10%).

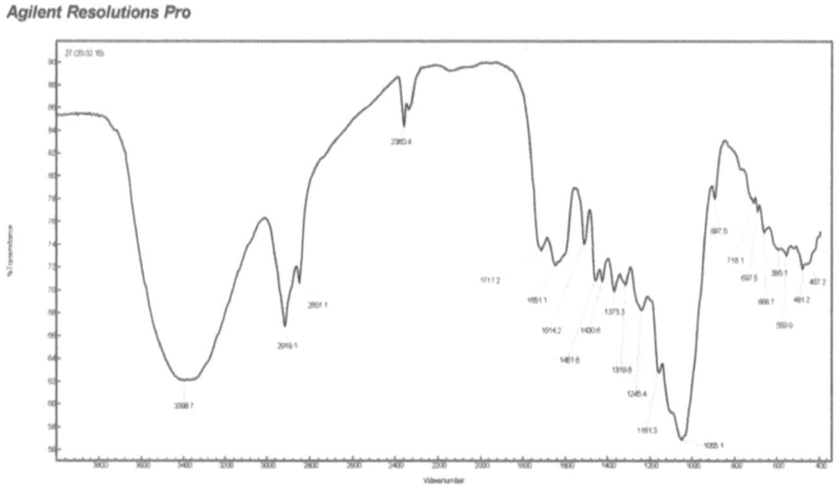

FIGURE 20.5 FTIR spectrum of four-component of composite (straw 85% + PhES-80 – 5% + LG 5% + PE 5%).

20.3.2 OPTICAL MICROSCOPE INVESTIGATIONS OF COMPOSITES

The microstructure of wood composites was studied on the NMM-800RF/ TRF type optical microscope and SEM.

For investigation of the wood material, the sample with 1 cm longevity was prepared. This sample was polished on the sheet 25 mcm SiC paper during 1 h and after this sample was displaced to the sending paper with 5 mcm and the polishing was continued for 1 h. After this procedures, the sample was polished additively by means of byazi, after which the sample was displaced to the optical microscope for investigation of different ranges. It was investigated the samples prepared under pressure 170 kg/cm^2 and at different temperatures. The composite with sawdust (97%) + PhES-80 (3%, at temperature 110, 120, and 125°C and composite with sawdust (90%) + PhES-80 (5% and siliylated styrene (5%)). For samples it was defined both the transverse surface and longitudinal one or the inner part of the sample and the inclusions were observed under conditions with increasing 50, 100, 200, and 500 times (*see* Figures 20.6–20.14).

FIGURE 20.6 Optical microscopic data of transverse surface of composite sawdust (97%) + PhES-80 (3%) obtained at 110°C.

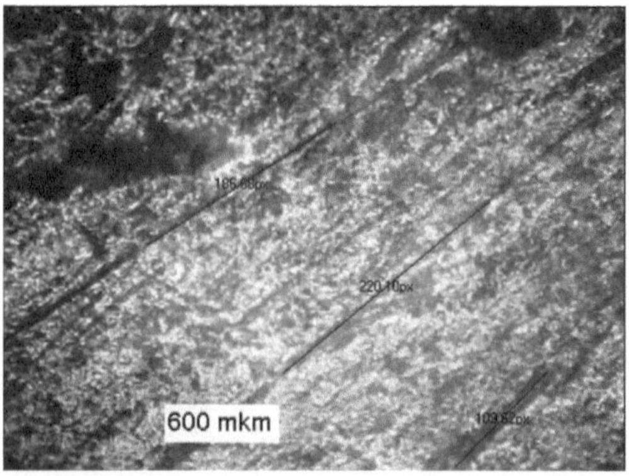

FIGURE 20.7 Optical microscopic data of transverse surface of composite sawdust (97%) + PhES-80 (3%) obtained at 120°C, with cleavage near 600 mcm.

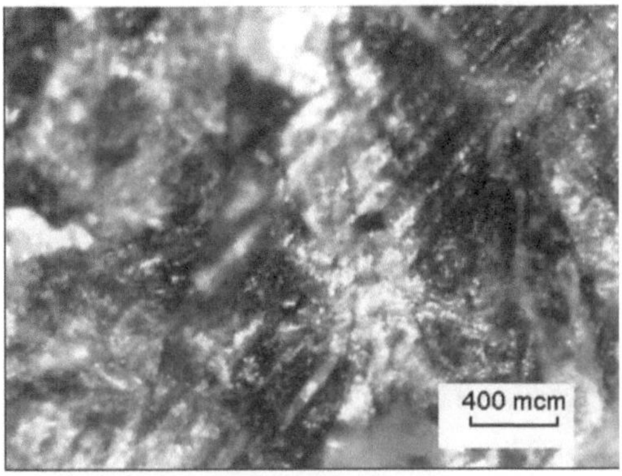

FIGURE 20.8 Optical microscopic data of composite (sawdust – 97%, LG – 3%).

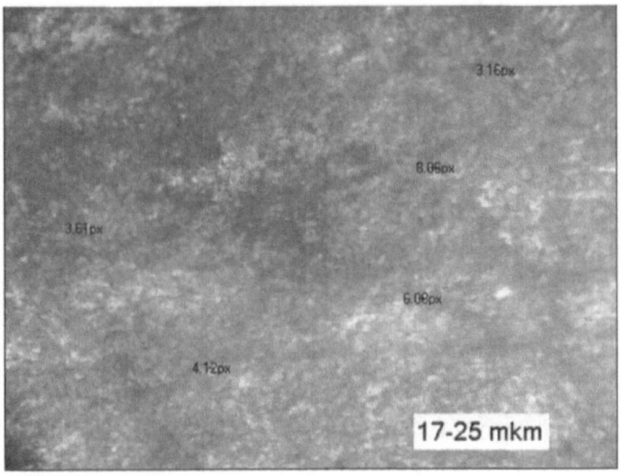

FIGURE 20.9 Optical microscopic data of longitudinal surface of composite sawdust (97%) + PhES-80 (3%) obtained at 110°C.

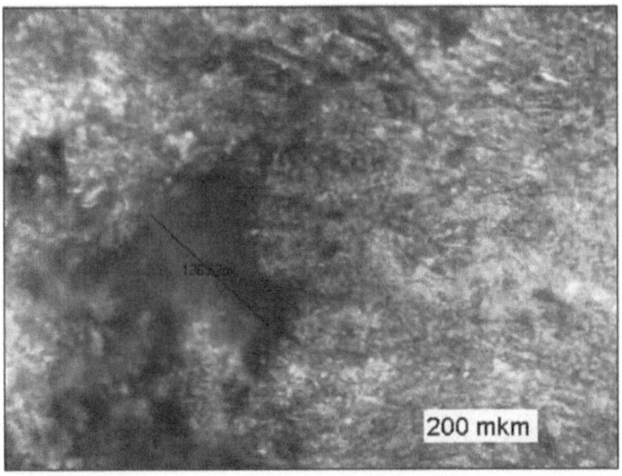

FIGURE 20.10 Optical microscopic data of longitudinal surface of composite sawdust (97%) + PhES-80 (3%) obtained at 110°C, magnification x200, the inclusion size 200 mcm.

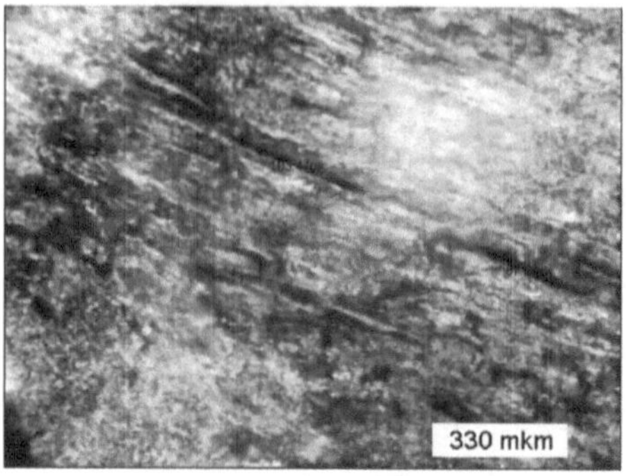

FIGURE 20.11 Optical microscopic data of longitudinal surface of composite sawdust (97%) + PhES-80 (3%) obtained at 110°C, the magnification x200, the cleavage size about 330 mcm.

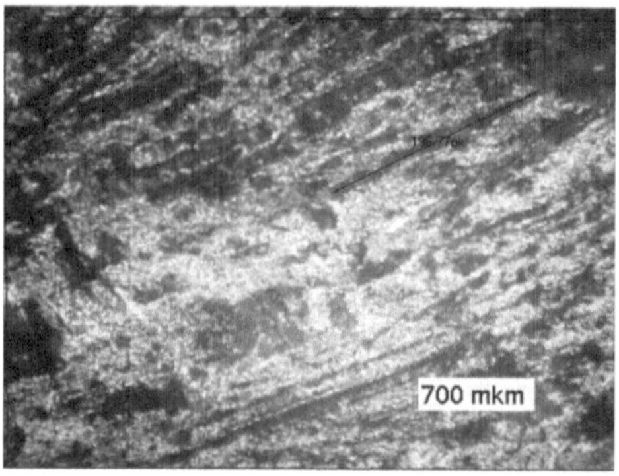

FIGURE 20.12 Optical microscopic data of longitudinal surface of composite sawdust (97%) + PhES-80 (3%) obtained at 120°C, magnification x100, the inclusion size 700 mcm.

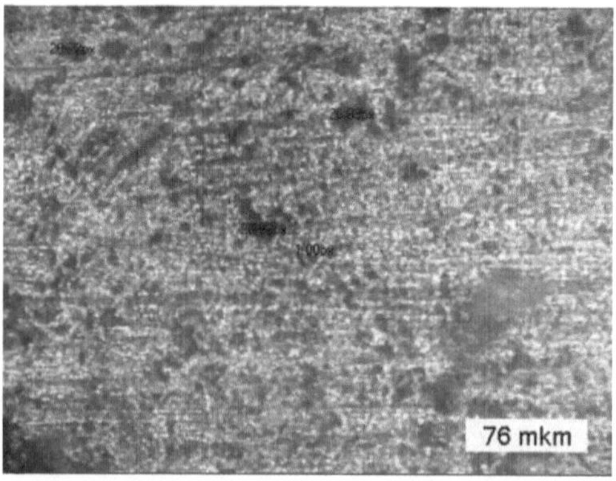

FIGURE 20.13 Optical microscopic data of longitudinal surface of composite sawdust (97%) + PhES-80 (3%) obtained at 125°C, magnification x100, the cleavage size 76 mcm.

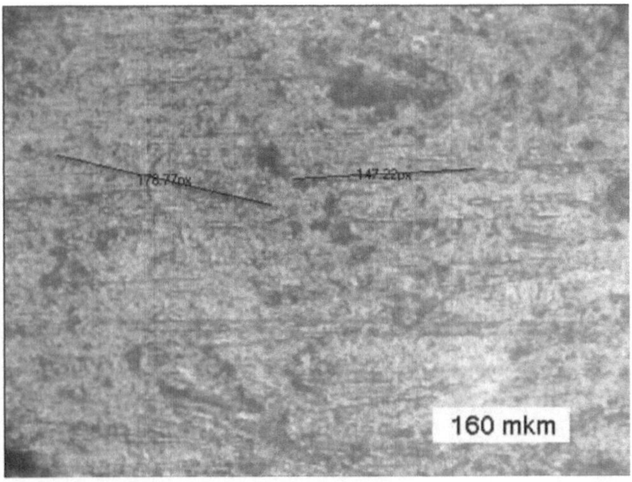

FIGURE 20.14 Optical microscopic data of longitudinal surface of composite sawdust (97%) + PhES-80 (3%) obtained at 125°C, magnification x100, the inclusion size 160 mcm.

From the optical microscope pictures one can see, that the binder rises up to surfaces with the change of the inclusions with sizes in the frames 76–200 mcm and cleavage sizes 160–760. The change of temperature does not influence on the sizes neither on inclusions nor on cleavages.

20.3.3 SCANNING ELECTRON MICROSCOPIC AND ENERGY DISPERSION MICRO-X-RAY ANALYSIS OF COMPOSITE MATERIALS ON THE BASIS OF THE SAWDUST

For composite materials made on the basis of powder like wood sawdust bamboo and different hardeners and additives scanning electron microscopic (SEM) investigations are provided. Besides of energy dispersive X-ray examinations have been carried out in parallel with the micro-spectral (EDS) examinations. SEM and EDS analyzes were conducted with use of the microscope Nikon Eclipse LV 150. The micrograms of SEM were obtained at various (x100–x1000) magnification ratios. Surface analysis, chemical analysis and visualization of the composites obtained on the basis of leaves were studied by method of SEM (Figures 20.15–20.28 and Tables 20.1–20.4).

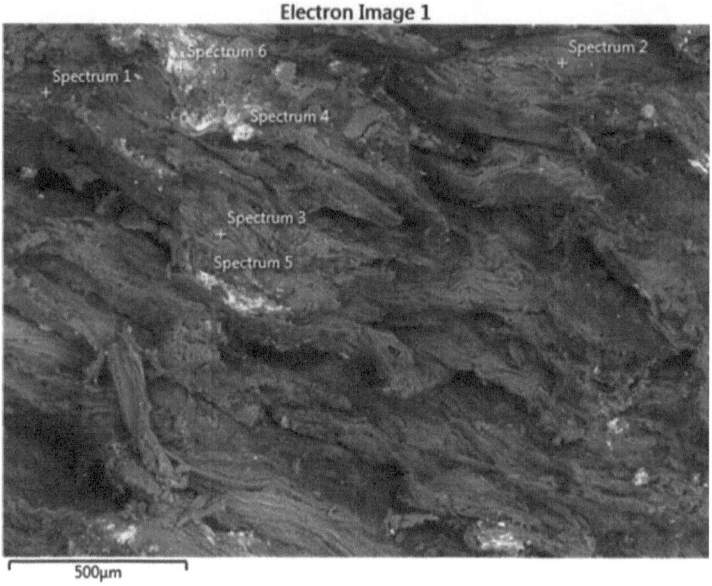

FIGURE 20.15 Scanning electron microscopic micrograms of composite 1.2 (wood sawdust 95% + liquid glass 5%) (Spectrum 2).

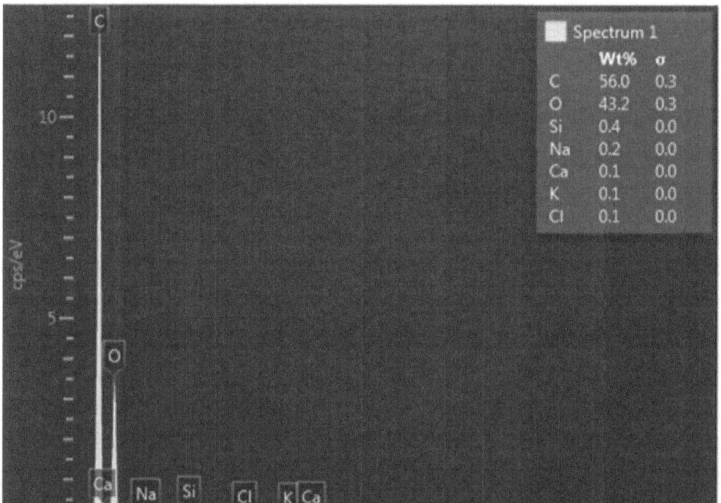

FIGURE 20.16 Energy dispersion micro-X-ray spectral analysis of composite 1.2 (wood sawdust 95% + liquid glass 5%) (Spectrum 1).

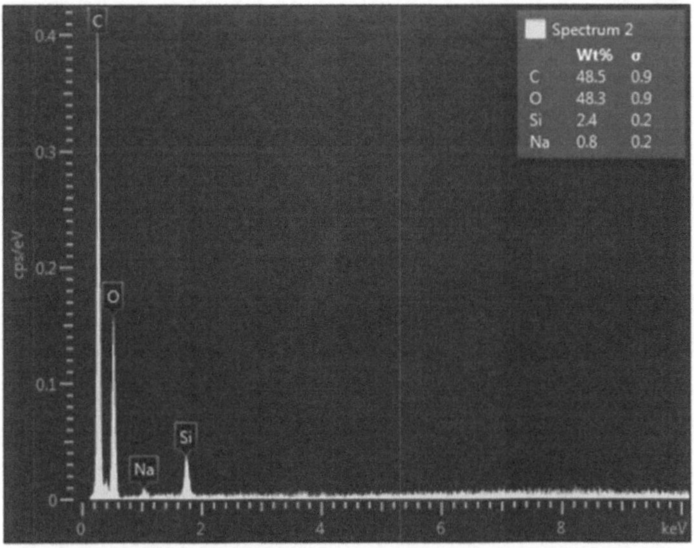

FIGURE 20.17 Energy dispersion micro-X-ray spectral analysis of composite 1.2 (wood sawdust 95% + liquid glass 5%) (Spectrum 2).

TABLE 20.1 Energy Dispersion Micro-X-Ray Spectral Analysis of Composite 1.2 (Wood Sawdust 95% + Liquid Glass 5%)

Result Type	**Weight Percent**					
Spectrum Label	**Spectrum 6**	**Spectrum 1**	**Spectrum 2**	**Spectrum 3**	**Spectrum 4**	**Spectrum 5**
C	46.83	55.96	48.50	50.00	31.35	49.57
O	45.88	43.16	48.33	45.22	40.65	46.70
Na	2.75	0.23	0.79	0.99	0.48	0.93
Al	–	–	–	–	5.65	–
Si	4.54	0.37	2.38	2.31	19.45	2.80
Cl	–	0.07	–	–	–	–
K	–	0.08	–	0.43	0.28	–
Ca	–	0.13	–	1.04	0.45	–
Ti	–	–	–	–	0.41	–
Fe	–	–	–	–	1.30	–
Total	100.00	100.00	100.00	99.99	100.02	100.00

Statistics	C	O	Na	Al	Si	Cl	K	Ca	Ti	Fe
Max	55.96	48.33	2.75	5.65	19.45	0.07	0.43	1.04	0.41	1.30
Min	31.35	40.65	0.23	5.65	0.37	0.07	0.08	0.13	0.41	1.30
Average	47.04	44.99	1.03	–	5.31	–	–	–	–	–
Standard deviation	8.29	2.73	0.89	–	7.05	–	–	–	–	–

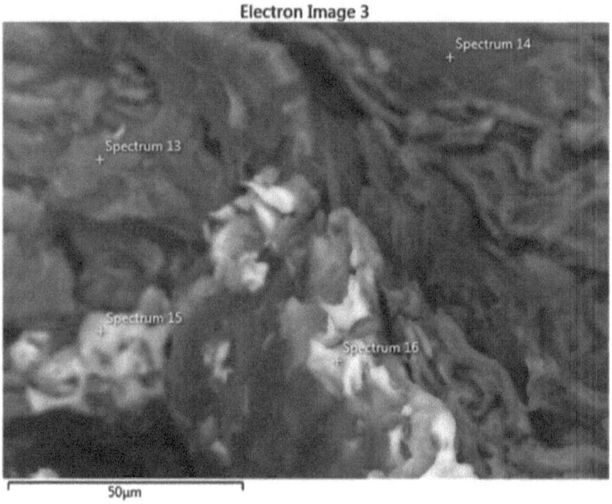

FIGURE 20.18 Scanning electron microscopic micrograms of composite 1.14 (wood sawdust 95% + PE – 5%).

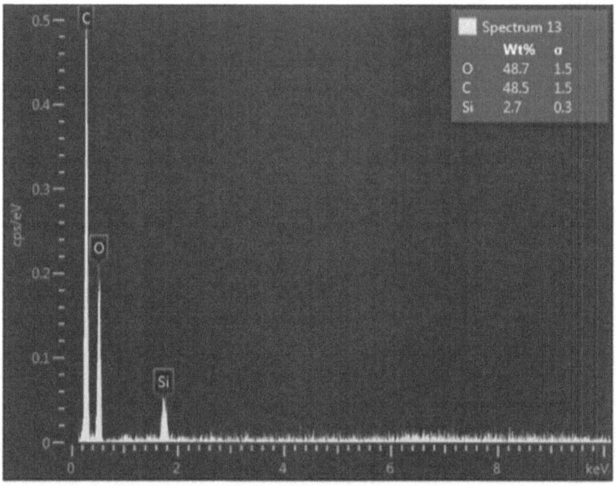

FIGURE 20.19 Energy dispersion micro-X-ray spectral analysis of composite 1.14 (wood sawdust 95% + PE – 5%) (Spectrum 13).

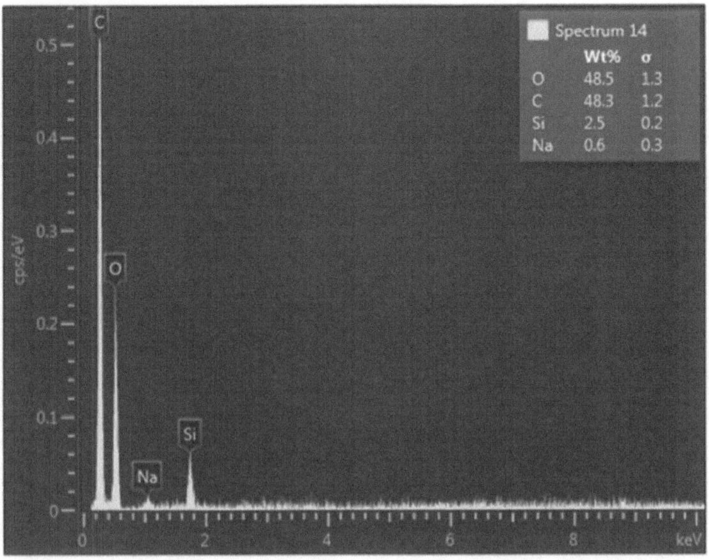

FIGURE 20.20 Energy dispersion micro-X-ray spectral analysis of composite 1.14 (wood sawdust 95% + PE – 5%) (Spectrum 14).

TABLE 20.2 Energy Dispersion Micro-X-Ray Spectral Analysis of Composite 1.14 (Wood Sawdust 95% + PE – 5%)

Result Type	**Weight Percent**			
Spectrum Label	**Spectrum 16**	**Spectrum 13**	**Spectrum 14**	**Spectrum 15**
C	49.71	48.53	48.35	46.82
O	47.39	48.75	48.55	50.51
Na	–	–	0.60	–
Si	2.90	2.72	2.50	2.66
Total	100.00	100.00	100.00	99.99

Statistics	**C**	**O**	**Na**	**Si**
Max	49.71	50.51	0.60	2.90
Min	46.82	47.39	0.60	2.50
Average	48.35	48.80	–	2.70
Standard deviation	1.18	1.29	–	0.17

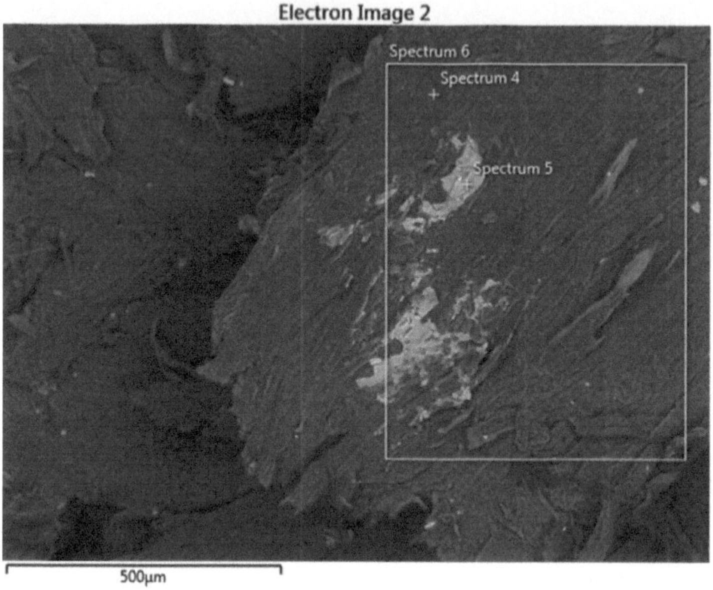

FIGURE 20.21 Scanning electron microscopic micrograms of composite 1.18 (wood sawdust 90% + 10% colophony).

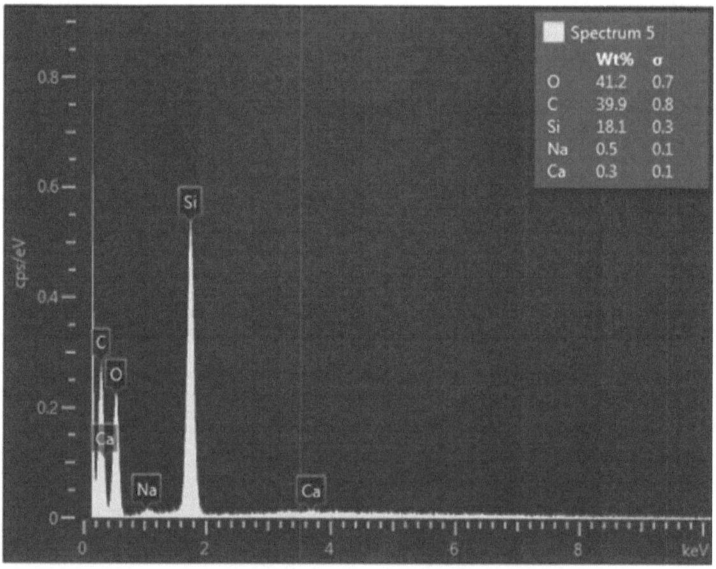

FIGURE 20.22 Energy dispersion micro-X-ray spectral analysis of composite 1.18 (wood sawdust 90% + 10% colophony) (Spectrum 4).

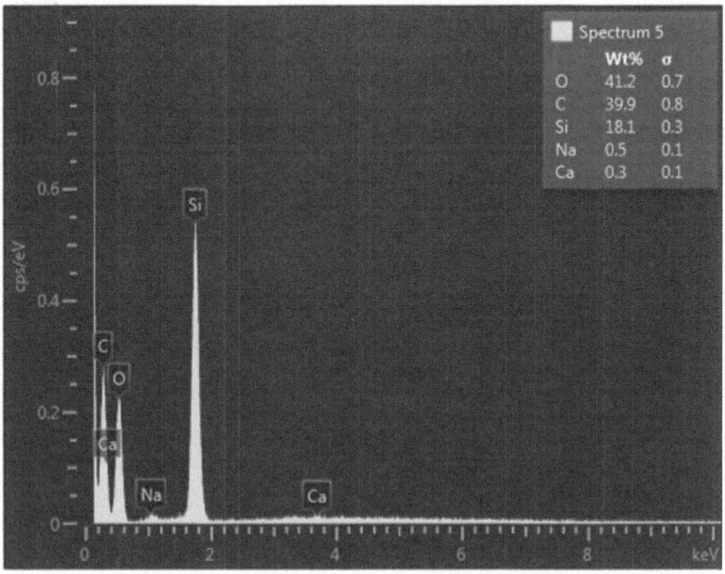

FIGURE 20.23 Energy dispersion micro-X-ray spectral analysis of composite 1.18 (wood sawdust 90% + 10% colophony) (Spectrum 5).

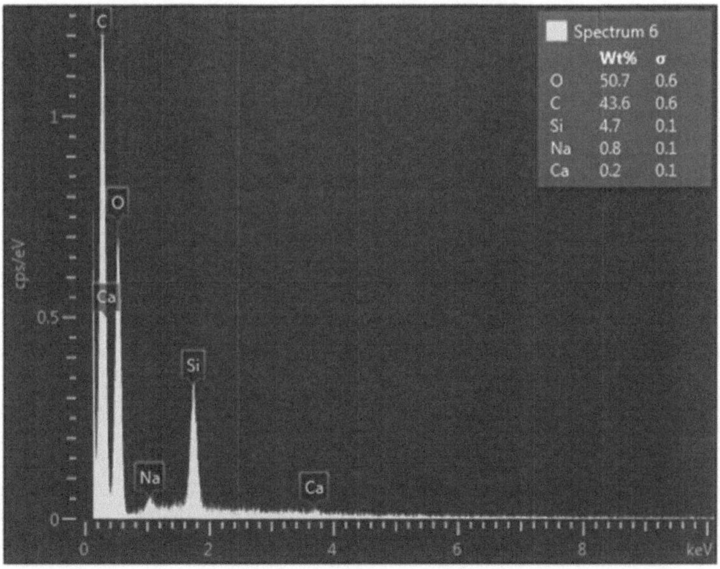

FIGURE 20.24 Energy dispersion micro-X-ray spectral analysis of composite 1.18 (wood sawdust 90% + 10% colophony) (Spectrum 6).

TABLE 20.3 Energy Dispersion Micro-X-Ray Spectral Analysis of Composite 1.18 (Wood Sawdust 90% + 10% Colophony)

Result Type	Weight Percent		
Spectrum Label	**Spectrum 4**	**Spectrum 5**	**Spectrum 6**
C	49.54	38.94	43.61
O	47.20	42.19	50.66
Na	0.85	0.38	0.82
Si	2.41	18.18	4.68
Ca	–	0.31	0.23
Total	100.00	100.00	100.00

Statistics	C	O	Na	Si	Ca
Max	48.64	51.66	0.83	17.08	0.30
Min	40.94	40.19	0.50	3.41	0.24
Average	43.40	45.32	0.70	9.39	
Standard Deviation	5.90	6.79	0.23	7.47	

Sawdust 85% + PhES-50 – 5% + PE – 10%.

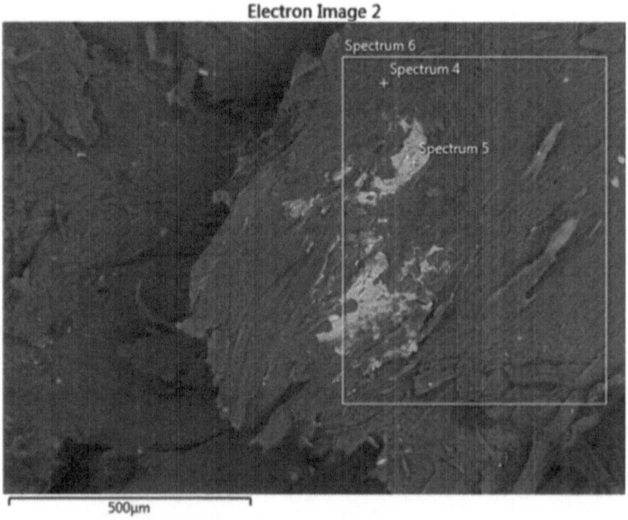

FIGURE 20.25 Scanning electron microscopic micrograms of composite 1.38 (wood sawdust 85% + 10% PhES-50 – 5% + PE – 10%). Sawdust 85% + PhES-50 – 5% + PE – 10%.

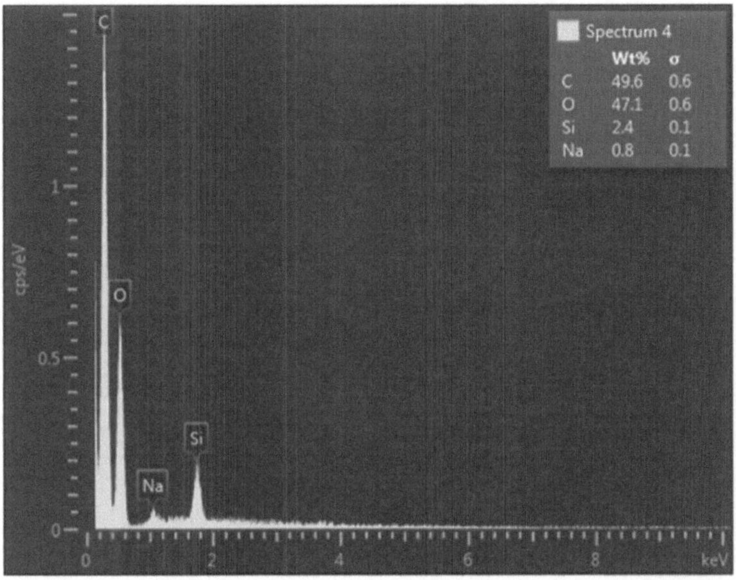

FIGURE 20.26 Energy dispersion micro-X-ray spectral analysis of composite 1.38 (wood sawdust 85% + 10% PhES-50 – 5% + PE – 10%) (Spectrum 4).

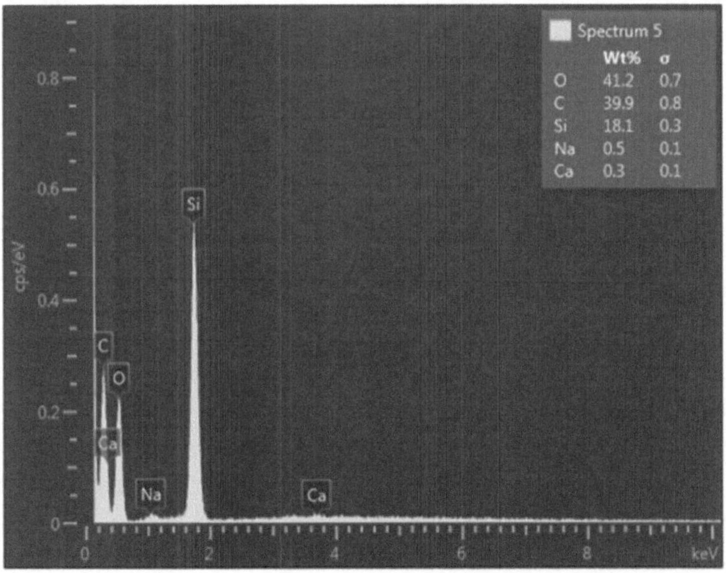

FIGURE 20.27 Energy dispersion micro-X-ray spectral analysis of composite 1.38 (wood sawdust 85% + 10% PhES-50 – 5% + PE – 10%) (Spectrum 5).

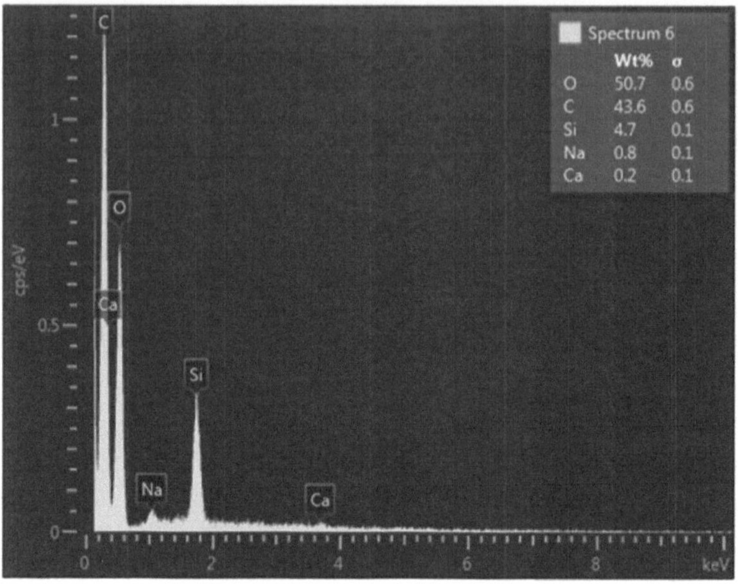

FIGURE 20.28 Energy dispersion micro-X-ray spectral analysis of composite 1.38 (wood sawdust 85% + 10% PhES-50 – 5% + PE – 10%) (Spectrum 6).

TABLE 20.4 Energy Dispersion Micro-X-Ray Spectral Analysis of Composite 1.38 (Wood Sawdust 85% + 10% PhES-50 – 5% + PE – 10%)

Result Type	**Weight Percent**		
Spectrum Label	**Spectrum 4**	**Spectrum 5**	**Spectrum 6**
C	49.64	39.94	43.61
O	47.10	41.19	50.66
Na	0.85	0.48	0.82
Si	2.41	18.08	4.68
Ca	–	0.31	0.23
Total	100.00	100.00	100.00

Statistics	**C**	**O**	**Na**	**Si**	**Ca**
Max	49.64	50.66	0.85	18.08	0.31
Min	39.94	41.19	0.48	2.41	0.23
Average	44.40	46.32	0.72	8.39	–
Standard deviation	4.90	4.79	0.21	8.47	–

High pace of the development of the industry more and more expands the demands on the wood materials, however an increased deficit of the natural wood stipulates the scientific-technical research on the theme of analogical materials.

At this time a great attention attract the materials obtained in result of combination of wood with sub-products formed after it treatment (sawdust, leaves, needless) with use of different binders. This process takes place often as polymerization of polymer substances (for example oligomers) on the surface of dispersed wood product. Besides of there are wide spread the methods of extrusion and hot pressing in the press molds of the high dispersive thermo-plastic polymers and wood products.

The WPC are materials of relatively new generation, in which the role of the binder performs such thermo-plastics polymers as polyethylene, polypropylene, polyvinyl chloride, polystyrene, and others. These materials sometimes are called liquid wood. There are known rather wide assortment of the products made from WPC. Using such methods as extrusion, hot pressing, rotation formatting one obtained such goods as terraces, floor desks, wall panels, roofs coatings, pipes, and so on. WPC are distinguished from analogs by high stability to atmospheric influences, mechanical, and chemical sustainability, water proofing, which allows to use these materials as coatings of washing rooms, sauna, terraces, and docks and so on.

A first production of WPC was prepared in the 90th years of last century. During the first stage were provided the works taking into account of obtaining and the type of wood product. In all cases the main binder was the liquid glass (LG), although in the separate case it was used high dispersive polyethylene of low density (Figures 20.29–20.31).

The composites were obtained with use of following manipulations:

- Weighing of the components with use of analytical balance;
- Dry mixing of the components;
- Loading of the blend to the standard press-forms;
- Heating of blends in the press-forms at definite temperature during fixed time;
- Taking out of the samples from press-forms.

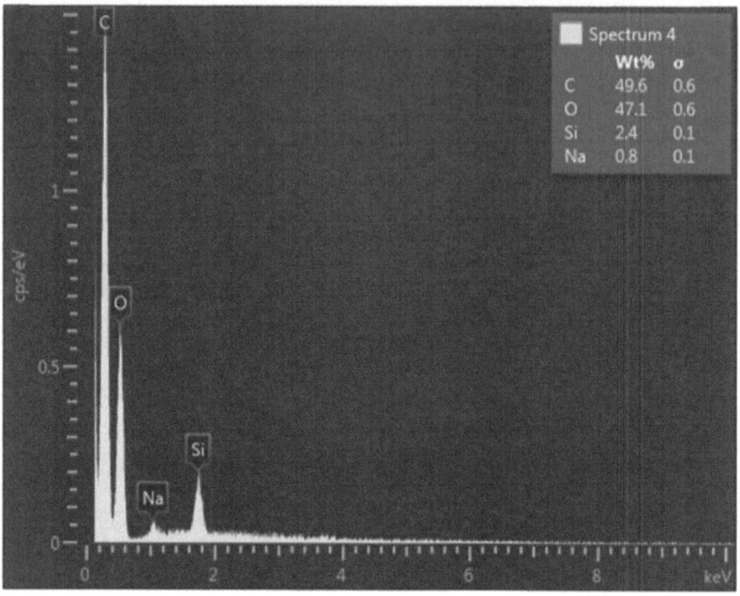

FIGURE 20.29 Energy dispersion micro-X-ray spectral analysis of composite 1.18 (wood sawdust 90% + 10% colophony) (Spectrum 4).

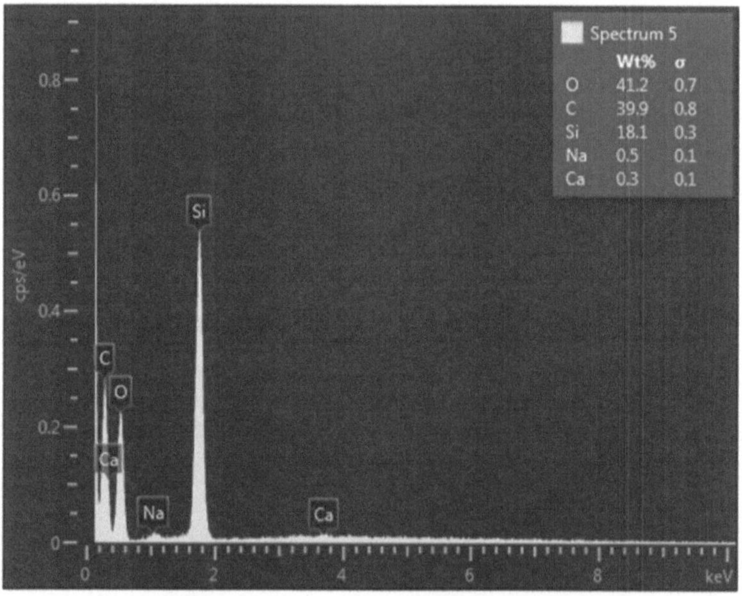

FIGURE 20.30 Energy dispersion micro-X-ray spectral analysis of composite 1.18 (wood sawdust 90% + 10% colophony) (Spectrum 5).

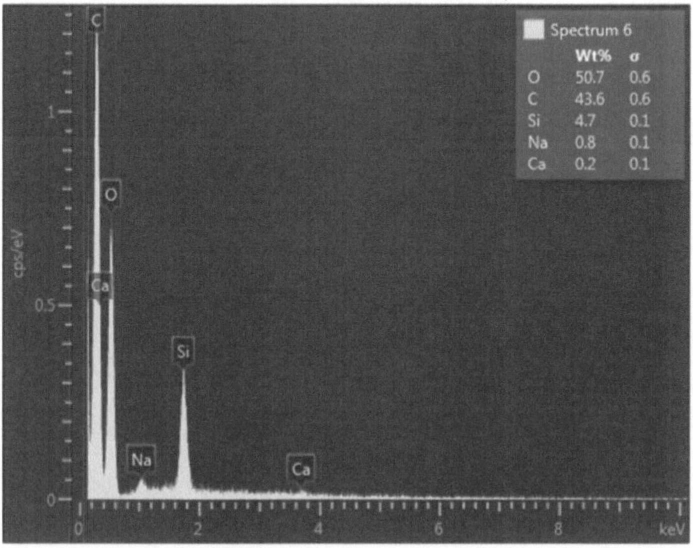

FIGURE 20.31 Energy dispersion micro-X-ray spectral analysis of composite 1.18 (wood sawdust 90% + 10% colophony) (Spectrum 6).

20.3.4 GENERAL PROPERTIES OF COMPOSITES

Following composites on the basis of wood sawdust were obtained and investigated in the project frames: (a) composites with two ingredients (sawdust + binder 1); (b) composites with three ingredients (sawdust + binder 1 + binder 2). The numerical data of the noted parameters for these composites are tabulated in Table 20.5.

The table data allow us to make the following conclusions:

1. The strengthening of the composite containing bamboo powder is 4 times better than analog, which contains bamboo less than 5 wt.%. Although here probably this difference is due to difference in the percent amount of LG, i.e., at more high content of last leads to worse result because of "excessive" wetting of the filler; the composite with high dispersed filler exposes relatively less strengthening in comparison with analog containing more big particles of the same filler, which may be due to armoring function of last filler.
2. The composites containing wood sawdust and additively polyethylene powder show mechanical strengthening higher than that for analog composites without polyethylene. This fact naturally is described by additive amplifier role of the polymer filler.

TABLE 20.5 Some Characteristics of the Composites Obtained on the Basis of the Renewable Raw Materials

SL. No.	Composition (wt.%)	Density (g/cm^3)	Strengthening at Stretching (MPa)	Strengthening at Bending (MPa)	Young Module (Mpa)	Thermal Stability (by method Vica) (°C)	Water Absorption (%)
1.	Bamboo (90%) + LG (10%)	0.9	0.6	4	100	>180	4
2.	Bamboo (95%) + LG (5%)	0.9	2.6	4	52	>180	7
3.	Fine dispersed bamboo (95%) + LG (5%)	0.9	1.6	4	357	>180	6
4.	Wood sawdust + (95%) + LG (5%)	0.9	1.4	4	254	>180	20
5.	Fine dispersed sawdust (95%) + LG (5%)	–	1.9	4	–	>180	19
6.	Wood sawdust (90%) + LG (5%) + polyethylene (5%)	0.9	5.0	6	913	>180	11
7.	Fine dispersed sawdust (90%) + PE (5%) + LG (5%)	0.9	3.8	6	865	>180	9
8.	Wood leaf (95%) + LG (5%)	0.9	1.0	4	–	>180	16
9.	Straw (95%) + LG (5%)	0.9	4.5	4	–	–	20
10.	Pine needles (95%) + LG (5%)	0.9	4	4	832	>150	14
11.	Bamboo (70%) + PhES-80 (30%)	0.7	0.4	0.8	–	–	55
12.	Bamboo (75%) + PhES-80 (25%)	0.7	0.5	1.1	–	–	50
13.	Bamboo (85%) + PhES-80 (15%)	0.8	0.7	2	–	–	38
14.	Cheap board (DSP)	–	2.7	–	–	–	55

3. Comparison of the mechanical properties of the composites containing woods leaves, pine needles and hey shows that composites with first filler are less durable than the ones with other two materials; it is clear that in these cases the needles and hay have thread-like structure and consequently possess the armoring properties.
4. Comparison of the experimental results obtained by us show that the composites on the basis of wood materials waste products show some important mechanical properties and water prove better than ones for widespread construction material-DSP and, what is more important, our composites are pollution free materials.
5. According to the table, apparently, in this case, the binder, as in the case of composite number 1, "excessive wetting" wood filler impairs the mechanical properties of the composite. Preliminary experiments on the composites containing a binder of the type PhES-80 and bamboo have shown that these materials require careful selection of the proportions of the components of the composite materials the above relations are not presented to them provide requirements for important performance characteristics. With this purpose in the future, we will carry out the optimization of composites with ecologically cleaner binder PhES-80.

20.3.4.1 INVESTIGATION OF PHYSICAL-MECHANICAL PROPERTIES

Following composites on the basis of wood sawdust were obtained and investigated in the project frames: (a) composites with two ingredients (sawdust + binder 1); (b) composites with three ingredients (sawdust + binder 1 + binder 2). The numerical data of the noted parameters for these composites are tabulated in Tables 20.6 and 20.7.

The noted properties of the investigated materials were tested in two directions: (i) bending strengthening; and (ii) impact viscosity.

20.3.5 EFFECT OF TECHNOLOGICAL FACTORS ON THE IMPACT VISCOSITY OF THE COMPOSITES BASED ON SAWDUST

It was interesting to define some properties in dependence on some technological factors (temperature, pressure). There were tested the composites obtained above on the impact viscosity. The results of the experimental measurements are presented in Table 20.6.

TABLE 20.6 Dependence of the Value of Impact Viscosity of Composites Containing Sawdust on the Conditions (Temperature, Pressure) of Obtaining

SL. No.	Composite (Mass%)	Technological Parameters		Samples Cross-Section (10^{-5} M^2)	Impact Viscosity[3] (kJ/m^2)
		Pressure (MPa)	Temperature (°C)		
1.	Sawdust (95%) + PhES-80 (5%)	17	90	12	16.3
2.	Sawdust (95%) + PhES-80 (5%)	17	100	12	17.0
3.	Sawdust (95%) + PhES-80 (5%)	17	110	12	19.4
4.	Sawdust (95%) + PhES-80 (5%)	17	120	12	20.8
5.	Sawdust (95%) + PhES-80 (5%)	8	110	11	18.1
6.	Sawdust (95%) + PhES-80 (5%)	10	110	11	17.2
7.	Sawdust (95%) + PhES-80 (5%)	12	110	11	18.8
8.	Sawdust (95%) + PhES-80 (5%)	15	110	12	20.7
9.	Sawdust (95%) + PhES-50 (5%)	8	110	12	16.3
10.	Sawdust (95%) + PhES-50 (5%)	10	110	12	18.9
11.	Sawdust (95%) + PhES-50 (5%)	12	110	9	20.1
12.	PhES-50 (5%) + Sawdust (95%)	15	110	11	22
13.	Sawdust (92%) + PhES-50 (3%) + PE (5%)	8	110	12	21.8
14.	Sawdust (92%) + PhES-50 (3%) + PE (5%)	10	110	13	17.0
15.	Sawdust (92%) + PhES-50 (3%) + PE (5%)	12	110	13	18.3
16.	Sawdust (92%) + PhES-50 (3%) + PE (5%)	15	110	10	13.6
17.	Sawdust (90%) + PhES-50 (5%)+ PE (5%)	13	120	–	–
18.	Sawdust (92%) + PhES-50 (5%)+ PE (5%)	13	130	13	17.7
19.	Sawdust (92%) + PhES-50 (5%)+ PE (5%)	13	150	12	17.7
20.	Sawdust (92%) + PhES-50 (5%)+ PE (5%)	12	120	13	16.9
21.	Sawdust[1] (92%) + PhES-80 (5%) + PE (3%)	12	120	16	19.8
22.	Sawdust[2] (90%) + PhES-50 (3%)[2] + PE (3%) + Vin(OEt)$_3$ (4%)	12	120	14	22.4

[1]Dispersity < 0.5 mm;
[2]Dispersity up to 0.1 mm;
[3]Hummer mass 0.1 kg.

On the basis of Table 20.6, it may be made the following conclusions. Impact viscosity of the composite based on sawdust and PhES-80 depends on the temperature at pressing (pressure is constant). Namely the increasing of temperature in the range 90–120°C leads to increasing of the impact viscosity in the range 16–21 kJ/m, which may be connected to enhancing of the caking at increased temperature (samples 1–4). Analogical process takes place at increasing of technological pressure in the range 8–15 MPa at constant temperature (samples 5–8).

Although this increase in comparison with the first process is lower to some extent. In this case increasing of pressure leads to compressing of ingredients and consequently to increasing of the interaction between them and enhancing of the physical and chemical contacts. In the same conditions the composites with PhES-50 the mechanical strengthening increases in the interval 16–22 kJ/M^2.

It must be noted that the variation of temperature and pressure in the case of PE does not lead to any essential change of mechanical strengthening. This fact indicates on the existence of some additive factors. Here it may be supposed that at temperatures higher than 150°C the softening of PE reaches to level when the part of PE is leaked from press forms, which is equal to decreasing of concentration of PE.

Comparison of the samples 21 and 22 show that the higher is dispersive the sawdust, the higher is its mechanical properties. This fact is described in terms of increasing of the contact surface between ingredients.

Bending strength of the tested compositions was taken in two modes: during increase the temperature at fixed pressure by using working molds appropriate standards, in a wide temperature range. The temperature in the forms was varied in the range 80–170°C, while the pressure at intervals of 10–15 MPa. Samples were prepared for 10 minutes.

Experimental data of the bending strength is given in Table 20.7. Data of Table 20.7 allow us to make the following conclusions.

The bending strength of the composite, which was obtained from 95% wood Sawdust (average grain size is less than 1 mm) and 5% binder PhES-80, generally depending on the technological parameters – temperature and pressure. The strength of the composites, which are derived in the temperature range 90–120°C and at a fixed pressure 17 MPa monotonously increases between 4.5 MPa and 21 MPa within. The bending strength composites that are obtained at the pressure interval 8–15 MPa at the constant temperature (110°C) conditions, monotonously decreases with an increase of technological pressure (8 MPa 22 MPa. Clearly, monolithic high-strength material is obtained at optimal values of the parameters of the technological conditions. According to our data, the best results are obtained in the case when the temperature of the composite 120°C and the pressure of 17 MPa, or when the temperature is 110°C and a pressure 8 MPa.

Composites, which are received in the same mode, where PhES-50 used instead PhES-80, the influence of the technological regime on the impact bending strength is less noticeable. This option will vary in the range 12–15 MPa (when composite making regime is 110°C temperature and 8–15 MPa pressure) and 14–19 MPa (a composite making regime is the same, but 3% of the amount taken PhES-50).

Composites containing polyethylene (5%) generally exhibit higher mechanical properties and depending on the mode of getting less than it's not-containing analogs. It is obvious that in this case the polyethylene as a binder performs an additional role. It should be noted that composite containing 3% of the polyethylene gives a better properties to composites than 5% by weight, which indicates that the elastic phase which is obtained by polyethylene with the percentage reduction it reduced and thus the strength of the composite increases. For example, wood composite, which includes PhES-80 (3%) and polyethylene (3%) demonstrated a 33 MPa equal bending strength limit. Also distinguished by high strength composite, which additionally contains a binder vinyltriethoxysilane (29 MPa) and dispersed sawdust. This result can be explained by the last phase separator surface (sawdust-binder) and raising the heterogeneous bonds with increasing concentration. The experiments show that the composites making technological conditions significantly affect the microstructure of the composites.

Further, it was provided the systematic investigation of the impact viscosity for composites based on sawdust and different binders at their separate and combined application in composites. The experimental data are presented in Table 20.7.

20.3.5.1 STRENGTHENING ON BENDING

The samples 1.9–1.12 present the composites with PhES-80 and sawdust (particles sizes up to 1 mm), the impact viscosity of which increases with increasing of temperature (90–120°C) at constant pressure (17 MPa). This fact shows that these conditions cause rather high mechanical properties and give the possibility to suggest that here it takes place some heterogenic reactions between active groups of the ingredients (PhES-80 and sawdust), which lead to increasing of the chemical bonds concentration and consequently – mechanical properties. Although this parameter at 110C and 8–15 MPa this parameter for obtained composites changes only up to 15 MPa.

TABLE 20.7 Mechanical Properties of the Composites based on Sawdust and Different Binder

SL. No.	Composite	Mass (g)	Volume (cm^3)	Density (g/cm^3)	Mass in Water 3 h Exposition (g)
1.	Sawdust 97% + LG 3%	150	130	3.6	22.0
2.	Sawdust 95% + LG 5%	150	130	6.2	25.0
3.	Sawdust 90% + LG 10%	150	130	4.0	19.0
4.	Sawdust 85% + LG 15%	150	130	7.1	27.0
5.	Sawdust 97% + PhES-50 – 3%	250	130	12.2	23.3
6.	Sawdust 95% + PhES-50 – 5%	250	130	3.6	2.1
7.	Sawdust 90% + PhES-50 – 10%	150	130	9.5	30.0
8.	Sawdust 85% + PhES-50 – 15%	150	130	9.2	20.0
9.	Sawdust 97% + PhES-80 – 3%	150	130	11.4	32.0
10.	Sawdust 95% + PhES-80 – 5%	150	130	8.0	25.0
11.	Sawdust 90% + PhES-80 – 10%	150	130	13.4	35.2
12.	Sawdust 85% + PhES-80 – 15%	150	130	9.0	25.0
13.	Sawdust 90% + PE – 10%	150	130	10.0	35.0
14.	Sawdust 95% + PE – 5%	150	130	16.3	39.03
15.	Sawdust 85% + PE – 15%	150	130	17.0	32.5
16.	Sawdust 80% + PE – 20%	150	130	15.6	49.6
17.	Sawdust 95% + Colophony – 5%	150	130	19.9	52.4
18.	Sawdust 90% + Colophony – 10%	150	130	11.8	21.0
19.	Sawdust 85% + Colophony – 15%	150	130	11.0	30.0
20.	Sawdust 80% + Colophony – 20%	150	130	23.3	46.3
21.	Sawdust 95% + 5% Wood glue	150	130	16.5	36.2
22.	Sawdust 90% + 10% Wood glue	150	130	20.4	35
23.	Sawdust 85% + 15% Wood glue	150	130	40.0	46.0
24.	Sawdust 80% + 20% Wood glue	150	130	23.9	63.5
25.	Sawdust 94% + PhES-50 – 3% + LG 3% + Hardener	150	130	16.7	33.3
26.	Sawdust 90% + PhES-50 – 5% + LG 5% + Hardener	150	130	17.2	26.0

TABLE 20.7 *(Continued)*

SL. No.	Composite	Mass (g)	Volume (cm^3)	Density (g/cm^3)	Mass in Water 3 h Exposition (g)
27.	Sawdust 80% + PhES-50 – 10% + LG 10%	150	130	18.6	40.7
28.	Sawdust 80% + PhES-50 – 15% + LG 5%	150	130	18.6	42.6
29.	Sawdust 80% + PhES-50 – 5% + LG 15%	150	130	16.9	17.3
30.	Sawdust 94% + PhES-80 – 3% + LG 3% + Hardener	150	130	17.1	41.2
31.	Sawdust 90% + PhES-80 – 5% + LG 5%	150	130	20.2	40.2
32.	Sawdust 91% + PhES-50 – 3% + LG 3% + PE – 3%	150	130	24.5	37.4
33.	Sawdust 85% + PhES-50 – 5% + LG 5% + PE – 5%	150	130	15.0	38.6
34.	Sawdust 91% + PhES-80 – 3% + LG 3% + PE – 3%	150	130	15.3	37.3
35.	Sawdust 85% + PhES-80 – 5% + LG 5% + PE – 5%	150	130	18.3	3.08
36.	Sawdust 94% + PhES-50 – 3% + PE – 3%	150	130	16.7	43.0
37.	Sawdust 90% + PhES-50 – 5% + PE – 5%	150	130	14.0	41.5
38.	Sawdust 85% + PhES-50 – 5% + PE – 10%	150	130	20.0	40.0
39.	Sawdust 94% + PhES-80 – 3% + PE – 3%	150	130	17.3	44.0
40.	Sawdust 90% + PhES-80 – 5% + PE – 5%	150	130	13.0	46.0
41.	Sawdust 85% + PhES-80 – 5% + PE – 10%	150	130	18.0	41.6
42.	Sawdust 85% + PhES-50 – 5% + PVA 10%	150	130	17.3	38.0
43.	Sawdust 88% + PhES-503% + LG3% + PE – 3% + $VinSi(OEt)_3$ – 3%	150	130	15.0	45.0
44.	Sawdust 80% + PhES-505% + LG 5% + PE – 5% + $VinSi(OEt)_3$ – 5%	150	130	19.0	37.5

TABLE 20.7 *(Continued)*

SL. No.	Composite	Mass (g)	Volume (cm^3)	Density (g/cm^3)	Mass in Water 3 h Exposition (g)
45.	Sawdust 88% + PhES-80 – 3% + LG3% + PE – 3% + VinSi(OEt)$_3$ – 3%	150	130	13.8	35.0
46.	Sawdust 80% + PhES-80 – 5% + LG 5% + PE – 5% + VinSi(OEt)$_3$ – 5%	150	130	16.7	46.0
47.	Sawdust 90% + PhES-80 – 5% + PE – 5% Colophony	150	130	13.0	42.0
48.	Sawdust 94% + LG3% + PE – 3%	150	130	17.1	50.0
49.	Sawdust 90% + LG 5% + PE – 5%	150	130	23.0	45.0
50.	Sawdust 94% + LG 4% + PhES – 802%	150	130	9.9	24.4
51.	Sawdust 94% + LG 2% + PhES – 804%	150	130	7.6	34.2
52.	Sawdust 85% + LG 5% + PE – 10%	150	130	15.6	54.0
53.	Sawdust 85% + LG 10% + PE – 5%	150	130	15.2	47.9
54.	Sawdust 85% + LG 7.5% + PE – 7.5%	150	130	14.4	57.6
55.	Sawdust 85% + PhES-50 – 7.5% + PE – 7.5%	150	130	8.8	44.4
56.	Sawdust 85% + PhES-50 – 10% + PE – 5%	150	130	9.2	36.7

The composites obtained in the analogical regime, where PhES-50 is used instead of PhES-80 the effect of the technological conditions does not observe. This parameter change in the range 12–15 MPa (when the sample is obtained at 110°C and pressure 8–15 MPa) and 14–19 (when the regime of obtaining of composites is same and PhES-50 amount is 3%).

The strengthening for composite obtained on the basis of blend of sawdust (95%) with particles sizes less than 1 mm and 5% PhES-80 generally depends on the technological parameters – temperature and pressure. Namely the strengthening of composites obtained at temperatures 90–120°C and pressure 17 MPa monotonically increases in the range 4.5–21 MPa. The strengthening of composites obtained at 8–15 MPa and constant temperature (110°C) monotonically decreases with increasing of technological pressure. It is clear that the monolithic material may be obtained only at optimal technological parameters. In our case best results are obtained under conditions of 120°C and pressure 17 MPa or when temperature is equal to 110°C and pressure – 8 MPa.

It was interesting to define – how effects the thermal aging of composites on their thermal stability. On the Figure 20.33 there are presented the curves corresponding to composite Sawdust (95%) + LG (5%) before (2) and after (1) aging. One can see that after aging thermal stability of the composite decreases to some extent because of the formation of some structural defects (caverns, empties) and removal of the light fraction of the composite in the result of prolonged heating.

However in case of composite Sawdust (90%) + LG (5%) + PhES-80 (5%) effect of aging is opposite – the thermal stability of the initial composite is lower than that after aging (Figure 20.32). Obviously in this case some additive reactions with participating of LG and PhES-80 take place, in result of which some more stable products form in composite.

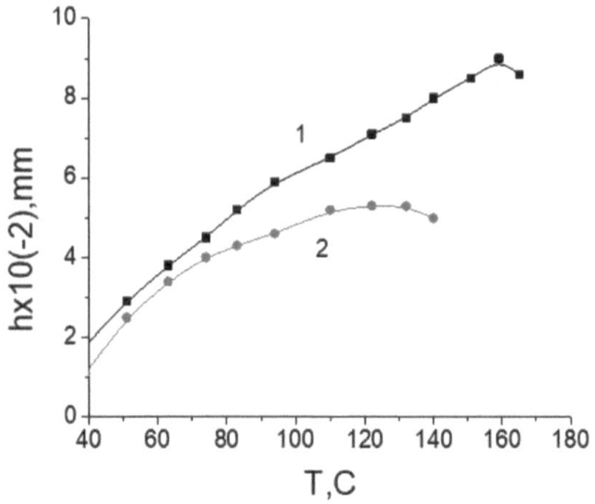

FIGURE 20.32 The thermal stability of the composite sawdust (90%) + LG (5%) + PhES-80 (5%) obtained in result of hot pressing at 130°C and 13 MPa before (1); and after aging (2).

This dependence is characteristic for such composites, obtained in the noted conditions. In the next investigations we attempt to obtain the composites with more thermal stability by way of inserting 2 or even 3 binders. However, after these experiments only some cases we obtained more or less stable composites. For example, with rather high thermal stability is characterized the composite Sawdust (95%) + LG (5%) (Figure 1.34), while in case of composite Sawdust (90%) + PE (3%)+ PhES-80 (3%) + vinyltrietoxisilane (4%) (Figure 20.34) is sufficiently low, which indicates on bed wettability of the components in this composite and increasing of "soft phase."

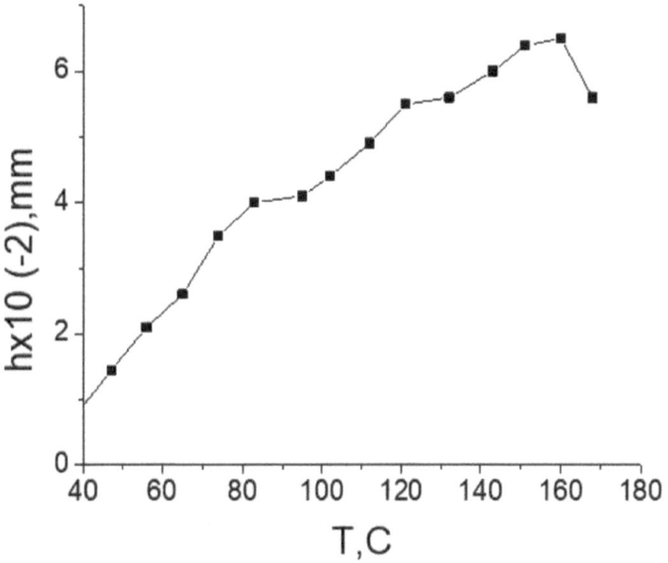

FIGURE 20.33 Thermal stability of the composite sawdust (95%) + LG (5%) obtained at 130°C and 13 MPa.

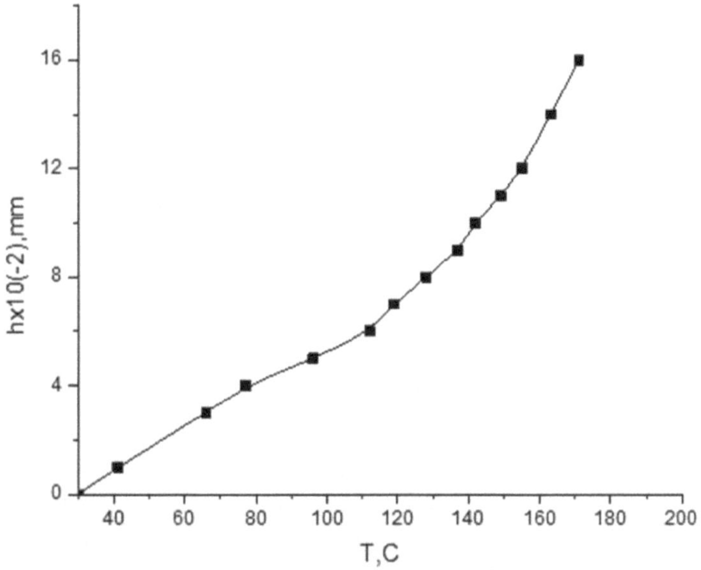

FIGURE 20.34 Thermal stability of composite sawdust (90%) + PE (3%) + PhES-80 (3%) + vinyltriethoxysilane (4%).

In case of composites: (1) Sawdust (85%) + LG (5%) + PE (5%) + PhES-80 (5%); (2) PE (5%) + PhES-50 (5%); and (3) Sawdust (85%) + LG (5%) + PhES-50 (5%) thermal stability little differs one from others. Probably the microstructure of these composites differs little from each other.

The composites containing only one type of binder – LG at different contents show the increasing thermal stability at increasing of concentration of this binder (Figure 20.35).

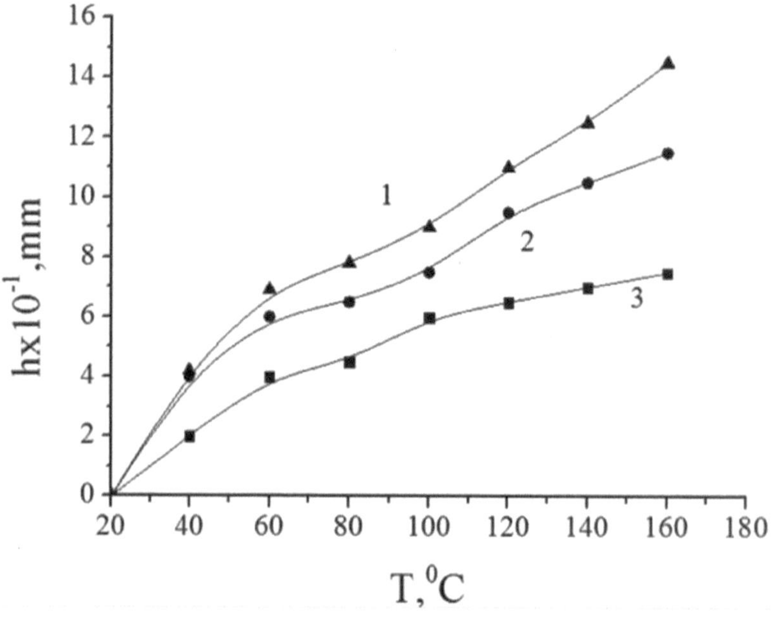

FIGURE 20.35 Temperature dependence of softening of composites based on sawdust with 3 (1); 5 (2); and 10% (3) of LG.

The composites containing two types of PhES (fraction 50 and 80) show rather high stability, although they differ little from each other (Figure 20.36).

With more high thermal stability are characterized the composites: sawdust +3% PhES 80 + 3% PE (1); and 5% PhES 80 + 10% PE (2). Obviously more high content of PE in composites leads to more high value of thermal stability of composite (Figure 20.37). In the last case probably, the heterogenic structure creates with rather high combination of convenience and high interaction between phases.

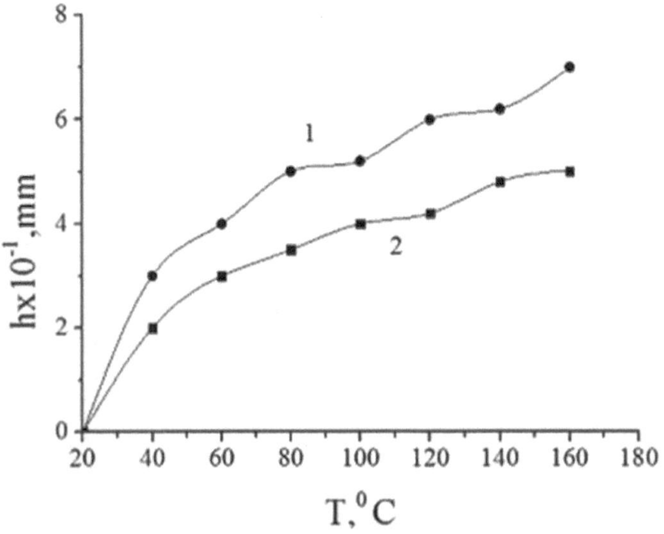

FIGURE 20.36 Temperature dependence of softening of composites based on sawdust with PhES-50 (1); and PhES-80 (2).

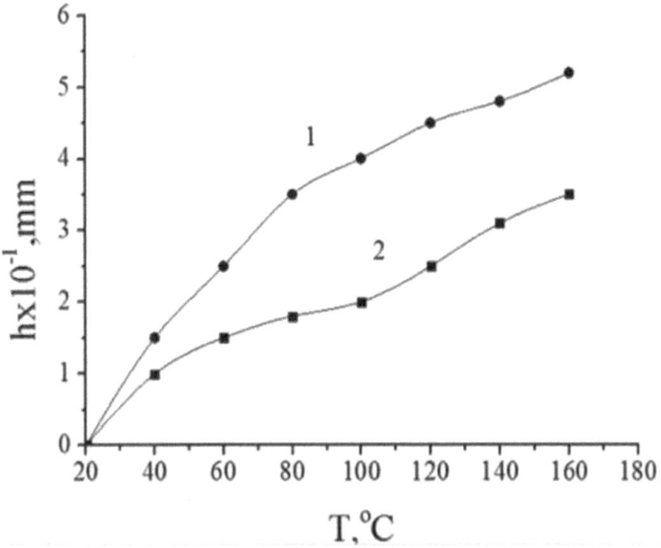

FIGURE 20.37 Temperature dependence of softening of composites based on sawdust with 3% of PhES-80 + 3% PE (1); and 5% PhES-80 + 10% PE (2).

20.3.6 INVESTIGATIONS OF THE WATER ABSORPTION OF COMPOSITES BASED ON SAWDUST

Water absorption is one of the most important exploitation characteristics for wood composites, because they are usually used under conditions of wet atmosphere. The composites based on sawdust and different binders were tested on water absorption.

Water absorption of samples was prepared as the known standard method. The initial weight of each specimen was taken before immersing into the water by electronic balance with an accuracy of 10^{-4}.

Samples were then immersed in collected rain water of pH value 5.6 at room temperature of 31°C. After a specific period of 24 hours interval the samples were taken out and wiped with tissue paper to remove the moisture. Then the weight of the specimens was taken and tabulated. The readings were taken for 21 days for an interval of 24 hours. The (water absorption) moisture content, M(t) absorbed by each specimen is calculated from its weight before, w(0) and after, w(t) absorption by using the following equation:

$$M(t) = 100\left[\frac{W_t - W_0}{W_0}\right]$$

The results of corresponding experiments are presented on Table 20.9. Table 20.9 suggest that water absorption greatly is depended on the composition of compounds, as well as making its technological conditions. For example, water absorption of composite, consisting from PhES-50 (5%)+ PE (5%) + Sawdust (90%) is only 0.06%, in other at that time for other composites, water absorption is about 1–2%. This fact shows that the composite contains relatively low concentrations of micro-emptiness spaces in the composite mass. It is expected existence of various separate emptiness in spaces.

It should be noted, that the value of the water absorption of composite differs not only in its absolute sense, but rather a sign. Negative values correspond to composites, from which water has driven the reaction products. It should be noted that the value of the water absorption of composite differs not only in its absolute sense, but rather a sign. Negative values correspond to composites, from which water has driven the reaction products. The positive water absorption characterize composites, which are relatively less washed up these products and at the same time production of material diffused water is completely not driven out of the composite, which is depends on the micro-emptiness number, on their size, and including the Channels among them. The bending strength (compression) test showed that the positive water absorption of samples has a high resistance compared with samples with negative water absorption. It shows that in the composites with negative water absorptions more excreted the substances that could participate in the realization of internal bonds in the material.

Thus, in the second quarter of the plan obtained by the analysis of the results of the testing of the materials can be made to the basic conclusion that the Sawdust's wood composites based on physical properties of these composites are the development of recipes and technological conditions for their acceptance. We hope that the study materials as a result of a systematic study of the structural analysis of some additional methods will allow a deeper understanding of the nature of the differences in the properties of materials.

Thus, the results of the analysis of the testing materials, can be made to the basic conclusion that the wood sawdust composites physical-mechanical as well as water absorption mainly depends on the compositions (recipes) of materials, and technological conditions (temperature, pressure) for their acceptance. It must be noted that not all composites appear stable in the medium of water. Some of them were destructed after exposition of them in water for 24 hours (in accordance with standard). Such composites are based on LG (1.1–1.3) and PhES-50 (1.5, 1.8). It is characteristic that the composites containing relatively low concentrations of the water are not stable in water medium. However, the same composite with 15% LG (1.4) is rather stable in comparison with analogs with less content of LG. Don't destructed but have rather low water stability the composites based on sawdust and PhES-es. Probably these microstructures contained micro-empties and cavities, which are communicated each to other. The water diffusion through these channels leads to effective water sorption by these composites. Essentially from composites noted above the ones contenting the PE distinguish very essentially (compare several percentage for composites contenting PE (1.13, 1.15, 1.16) with 43–78% for composites with PhES-50 or PhES-80 (1.6, 1.7). Obviously low water absorption of composites with PE is due to high hydrophobicity of the composites contenting PE because of good distribution of PE between particles of sawdust. With high hydrophobicity are characterized the composites contenting colophony and wood glue. The level of water absorption of composites containing the last three binders has nearly one and same order, because of all they are good hydrophobic substances. To some extent good hydrophobicity show also the composites containing vinyltriethoxysilane and some composites with a combination of PhES and PE 1.41, 1.46, 1.55, 1.56). Optimization of the recapture and technological conditions – the most real ways to reach to high hydrophobicity of wood composition materials (Table 20.8).

20.4 CONCLUSIONS FOR COMPOSITES ON THE BASIS OF SAWDUST

1. Here is presented a new composite material on the basis of wood sawdust and such binders as LG, phenylethoxysisilane of two types PhES-50 and PhES-80 (first molecular mass is less than for second one) in directions of obtaining and investigation of composite structure and such physical properties, as mechanical strengthening (at bending and impact viscosity), thermal stable properties and water absorption.
2. The structural investigations were provided with the use of optical and scanning electron microscope (SEM) for define of the morphology of the composites and the character of distribution of the ingredients in the composite body with use of the picture of optical and SEMs. Energy dispersive X-Ray micro-analysis was used as the method for establish the elemental analysis of the researched materials. From mechanical properties there were studied the strengthening at bending and the impact viscosity. The methods of Vica and thermal gravimetry were used for investigation of the thermal stability of the composites. The water absorption of composites was studied with the use of standard methodic.
3. It was shown generally that the mechanical properties of the composites essentially depend on the conditions of obtaining of the composites. For example, the strengthening for composites containing of sawdust with 5% increases with increasing technological temperature in the range 90–120°C at constant (17 MPa) pressure. Besides of this parameter for the same composite but obtaining at constant temperature (110°C) also increases when the technological pressure increases up to 15 MPa. It is shown that the mechanical properties of the composites is the higher the higher is degree of dispersion of the sawdust particles.
4. Mechanical strengthening of the composites depends essentially on the recapture factor too. It was observed that the dependence

TABLE 20.8 Water Absorption of Samples at Various Temperatures and Pressure

SL. No.	Composite	Mass (g)	Volume (cm^3)	Density (g/cm^3)	Mass in Water 3 h Exposition (g)	Water Absorption, 3 h Exposition (%)	Mass in Water 24 h Exposition (g)	Water Absorption, 24 h Exposition (%)
1.	PhES-80 (5%) + Sawdust (90%), P = 120 kg/cm^2 MPa, T = 120°C, exposition 10 min.	6.425	4.768	1.348	6.553	1.99	6.393	–0.50
2.	PhES-80 (3%) + PE (3%) + Sawdust (90%) + VinSi (OEt) (4%) P = 12 MPa T = 120C, Exp. 10 min	7.443	5.475	1.359	7.535	1.24	7.348	–1.28
3.	PhES-80 (5%) + Sawdust (95%), T = 100°C, P = 17 MPa, Exp. 10 min	4.757	3.532	1.347	4.935	Destroyed partially	–	–
4.	PhES-50 (3%) + Sawdust (92%) + PE (3%), T = 120°C, P = 12 MPa, Exp. 10 min.	5.267	3.797	1.387	6.864	Destroyed	–	–
5.	PhES-50 (3%) + Sawdust (92%) + PE (5%), T = 120°C, P = 12 MPa, Exp. 10 min.	5.469	3.885	1.408	7.220	Destroyed	–	–
6.	PhES-50 (3%) + Sawdust (92%)+ PE (5%), T = 110°C, P = 15 MPa, Exp. 10 min.	4.234	3.179	1.332	5.173	22.17	4.309	–1.77
7.	PhES-50 – 5% + PE (5%) + Sawdust (90%), T = 120°C, P = 10 MPa, Exp. 10 min.	5.096	3.885	1.312	5.260	3.22	5.099	0.058

TABLE 20.8 *(Continued)*

SL. No.	Composite	Mass (g)	Volume (cm^3)	Density (g/cm^3)	Mass in Water 3 h Exposition (g)	Water Absorption, 3 h Exposition (%)	Mass in Water 24 h Exposition (g)	Water Absorption, 24 h Exposition (%)
8.	PhES-50 – 5%) + PE (5%), Sawdust (90%), T = 120°C, P = 12 MPa, Exposition 10 min	4.770	3.709	1.286	4.893	2.570	4.721	–1.02
9.	PhES-50 (5%) + PE (5%) + Sawdust (90%), T = 120C, P = 13 MPa, Exp. 10 min	5.304	4.238	1.252	5.517	4.02	5.270	–0.64
10.	PhES-50 (5%) + Sawdust (95%), T = 120C, P = 15 MPa, Exp. 10 min	4.980	3.885	1.282	5.266	5.74	5.036	1.124
11.	PhES-50 (3%) + PE (3%) + Sawdust (90%), VinSi $(OEt)_3$ (4%), T = 120°C, P = 12 MPa, Exposition 10 min	5.098	3.797	1.343	5.240	14.20	5.027	–1.39

TABLE 20.9 Water Absorption of Composites on the Basis of Wood Sawdust

SL. No.	Composite (Mass%)	Mass (g)	Volume (cm^3)	Density (g/cm^3)	Mass in Water, 3 h Exposition (g)	Mass in Water, 24 h Exposition (g)	Water Absorption, 24 h Exposition (%)
1.	Sawdust 97% + LG 3%	4.47	3.18	1.41	–	–	Destroyed
2.	Sawdust 95% + LG 5%	4.72	3.52	1.34	–	–	Destroyed
3.	Sawdust 90% + LG 10%	3.76	2.81	1.31	–	–	Destroyed
4.	Sawdust 85% + LG 15%	3.78	2.81	1.35	4.13	4.45	17.72
5.	Sawdust 97% + PhES-50 – 3%	3.84	2.81	1.35	–	–	Destroyed
6.	Sawdust 95% + PhES-50 – 5%	3.83	2.81	1.36	5.21	5.49	43.34
7.	Sawdust 90% + PhES-50 – 10%	3.64	2.81	1.29	5.86	6.51	78.84
8.	Sawdust 85% + PhES-50 – 15%	3.46	2.46	1.40	5.98	–	Destroyed
9.	Sawdust 97% + PhES-80 – 3%	3.30	2.29	1.44	4.45	4.68	41.81
10.	Sawdust 95% + PhES-80 – 5%	3.83	2.81	1.36	5.75	6.06	58.22
11.	Sawdust 90% + PhES-80 – 10%	4.33	3.18	1.36	4.46	5.61	29.50
12.	Sawdust 85% + PhES-80 – 15%	4.24	3.18	1.33	4.24	4.47	50.00
13.	Sawdust 95% + PE 5%	3.48	2.64	1.32	3.53	3.65	4.89
14.	Sawdust 90% + PE 10%	3.99	2.99	1.33	4.37	4.70	17.8

TABLE 20.9 *(Continued)*

SL. No.	Composite (Mass%)	Mass (g)	Volume (cm^3)	Density (g/cm^3)	Mass in Water, 3 h Exposition (g)	Mass in Water, 24 h Exposition (g)	Water Absorption, 24 h Exposition (%)
15.	Sawdust 85% + PE 15%	3.23	2.64	1.22	3.40	3.53	9.30
16.	Sawdust 80% + PE 20%	3.68	2.64	1.33	3.53	3.60	2.86
17.	Sawdust 95% + Colophony 5%	3.82	2.64	1.45	3.86	3.97	3.93
18.	Sawdust 90% + Colophony 10%	3.68	2.64	1.39	3.74	3.82	3.80
19.	Sawdust 85% + Colophony 15%	3.64	2.64	1.37	3.83	4.50	23.50
20.	Sawdust 80% + Colophony 20%	3.44	2.46	1.4	3.44	3.48	8.14
21.	Sawdust 95% + 5% Wood glue	3.40	2.46	1.38	3.49	3.87	13.82
22.	Sawdust 90% + 10% Wood glue	3.84	2.64	1.45	3.95	4.13	7.55
23.	Sawdust 85% + 15% Wood glue	3.80	2.64	1.43	3.98	4.34	14.2
24.	Sawdust 80% + 20% Wood glue	3.23	2.29	1.41	3.27	3.47	7.43
25.	Sawdust 94% + PhES-50 – 3% + LG 3% Hardener	3.92	2.64	1.48	4.03	4.33	10.2
26.	Sawdust 90% + PhES-50 – 5% + LG 5% + Hardener	3.87	2.64	1.47	4.54	5.24	33.3

TABLE 20.9 *(Continued)*

SL. No.	Composite (Mass%)	Mass (g)	Volume (cm^3)	Density (g/cm^3)	Mass in Water, 3 h Exposition (g)	Mass in Water, 24 h Exposition (g)	Water Absorption, 24 h Exposition (%)
27.	Sawdust 80% + PhES-50 – 10% + LG 10%	3.68	2.64	1.39	3.80	4.00	8.1
28.	Sawdust 80% + PhES-50 – 15% + LG 5%	3.18	2.29	1.39	3.29	3.68	15.72
29.	Sawdust 80% + PhES-50 – 5% + LG 15%	3.41	2.29	1.39	3.49	3.65	7.04
30.	Sawdust 94% + PhES-80 – 3% + LG 3% + Hardener	3.87	2.81	1.38	4.02	4.99	28.94
31.	Sawdust 90% + PhES-80 – 5% + LG 5%	3.89	2.81	1.38	4.00	4.36	12.08
32.	Sawdust 91% + PhES-50 – 3% + LG 3% + PE – 3%	3.89	2.81	1.38	4.00	4.38	12.88
33.	Sawdust 85% + PhES-50 – 5% + LG 5% + PE – 5%	3.81	2.81	1.35	3.90	4.36	14.43
34.	Sawdust 91% + PhES-80 – 3% + LG 3% + PE – 3%	3.98	2.81	1.42	4.08	4.56	14.57
35.	Sawdust 85% + PhES-80 – 5% + LG 5% + PE – 5%	3.72	2.81	1.32	3.82	4.26	14.51
36.	Sawdust 94% + PhES-50 – 3% + PE – 3%	3.92	2.81	1.37	3.92	4.95	28.90
37.	Sawdust 90% + PhES-50 – 5% + PE – 5%	3.76	2.81	1.33	3.82	4.17	10.90

TABLE 20.9 *(Continued)*

SL. No.	Composite (Mass%)	Mass (g)	Volume (cm^3)	Density (g/cm^3)	Mass in Water, 3 h Exposition (g)	Mass in Water, 24 h Exposition (g)	Water Absorption, 24 h Exposition (%)
38.	Sawdust 85% + PhES-50 – 5% + PE – 10%	3.77	2.81	1.34	3.84	4.51	19.62
39.	Sawdust 94% + PhES-80 – 3% + PE – 3%	3.67	2.81	1.31	3.76	4.43	20.70
40.	Sawdust 90% + PhES-80 – 5% + PE – 5%	3.86	2.81	1.37	3.94	4.76	23.31
41.	Sawdust 85% + PhES-80 – 5% + PE – 10%	3.89	2.99	1.30	3.94	4.11	5.65
42.	Sawdust 85% + PhES-50 – 5% + PVA 10%	3.56	2.81	1.27	3.70	3.88	8.98
43.	Sawdust 88% + PhES-50 – 3% + LG 3% + PE – 3% + VinSi(OEt$_{)3}$ – 3%	3.84	2.81	1.37	3.93	4.32	12.5
44.	Sawdust 80% + PhES-50 – 5% + LG 5% + PE – 5% + VinSi(OEt$_{)3}$ – 5%	3.61	2.81	1.28	3.71	3.95	9.41
45.	Sawdust 88% + PhES-80 – 3% + LG 3% + PE – 3% + VinSi(OEt)$_3$ – 3%	3.90	2.81	1.39	4.02	4.36	11.79
46.	Sawdust 80% + PhES-80 – 5% + LG 5% + PE – 5% + VinSi(OEt)$_3$ – 5%	3.57	2.64	1.35	3.66	3.88	8.68
47.	Sawdust 90% + LG 5% + PE – 5%	3.81	2.81	1.36	4.09	4.34	13.91

TABLE 20.9 *(Continued)*

SL. No.	Composite (Mass%)	Mass (g)	Volume (cm^3)	Density (g/cm^3)	Mass in Water, 3 h Exposition (g)	Mass in Water, 24 h Exposition (g)	Water Absorption, 24 h Exposition (%)
48.	Sawdust 94% + LG 3% + PE – 3%	3.93	2.81	1.40	4.22	4.88	24.17
49.	Sawdust 85% + LG 5% + PE – 5% + Colophony 5%	3.53	2.64	1.34	3.57	3.70	4.81
50.	Sawdust 94% + LG 4% + PhES-80 – 2%	3.86	2.64	1.46	4.09	5.95	53.62
51.	Sawdust 94% + LG 2% + PhES-80 – 4%	3.88	2.81	1.38	4.05	5.14	32.47
52.	Sawdust 85% + LG 5% + PE 10%	3.70	2.64	1.40	3.76	4.50	21.62
53.	Sawdust 85% + LG 10% + PE 5%	3.59	2.64	1.35	3.68	4.21	17.27
54.	Sawdust 85% + LG 7.5% + PE 7.5%	3.70	2.64	1.40	3.77	4.46	20.54
55.	Sawdust 85% + PhES-50 – 7.5% + PE – 7.5%	3.78	2.61	1.35	4.24	6.20	64.02
56.	Sawdust 85% + PhES-50 – 10% + PE – 5%	3.55	2.64	1.34	4.30	5.92	66.76

of strengthening on the binder concentration has the maximum at definite concentrations of the choice binder and at increasing of it concentration the mechanical properties decrease. Here we see, that the mechanical properties of composite containing the binder higher than definite amount (for the given composite) are worse in comparison with same composites, but containing less concentration of same binder. This fact may be explained in terms of the creation of the associates (clusters) of the binder molecules at relatively high concentrations of lasts, which usually leads to weakening of the composite.

5. Technological factor is essentially reflected on the thermal stability of composites. For example, by method Vica, it is shown that thermal stability of the composite based on sawdust containing 5% PhES-80 and obtained at temperature 110°C and 15 MPa increases 5 times in comparison with the composite obtained at 10 MPa and at same temperature. Some good results are obtained for composites with polyethylene and colophony.
6. The value of the water absorption of the composites based on sawdust is in good corresponding with data of mechanical strengthening and thermal stability. Namely the higher is density of composites, the lower is their water absorption.

KEYWORDS

- **binders**
- **differential scanning microscopy**
- **electron beam specimens**
- **Fourier transition IR**
- **polymer composites**
- **thermogravimetric analysis**
- **water resistance**

REFERENCES

1. Sundarraj, A. A., & Ranganathan, Th. V., (2018). A review on cellulose and its utilization from agro-industrial waste. *Drug Invention Today, 10*(1), 89–94.
2. Puri, V. P., (1984). Effect of crystallinity and degree of polymerization of cellulose on enzymatic saccharification. *Biotechnol. Bioeng., 26*, 1219–1222.

3. Wyman, C. E., & Goodman, B. J., (1993). Biotechnology for production of fuels, chemicals and materials from biomass. *J. Appl. Biochem. Biotechnol., 39*, 39–59.
4. Van, W. J. P., (2001). Biotechnology and the utilization of biowaste as a resource for bioproduct development. *Trends Biotechnol., 19*, 172–177.
5. Okonko, I. O., Adeola, O. T., Aloysius, F. E., Damilola, A. O., & Adewale, O. A., (2007). Utilization of food wastes for sustainable development, electronic. *J. Environ. Agric. Food Chem., 8*, 263–286.
6. Kosseva, M. R., (2009). Chapter 3 processing of food wastes. *Adv. Food Nutr. Res., 58*, 57–136.
7. Kumar, S., (2010). *Conversion of Food Waste to Useful Chemicals/Products*. Bachelor of technology (chemical engineering). Thesis, National Institute of Technology, Rourkela, India.
8. David, G. B., & Ray, L. W., (2008). Cellulose. In: *Access Science* (p. 419). Cellulose: McGraw-Hill.
9. Ichazo, M. N., Albano, C., Gonzalez, J., Perera, R., & Candal, M. V., (2001). Polypropylene/wood flour composites: Treatments and properties. *Composite Structures, 54*(2, 3), 207–214. https://doi.org/10.1016/S0263-8223(01)00089-7.
10. Seung-Hwan, I., & Tsutomu, O., (2003). Mechanical and thermal flow properties of wood flour – biodegradable polymer composites. *Journal of Applied Polymer Science, 90*, 1900–1905. https://doi.org/10.1002/app.12864.
11. Torres, F. G., & Cubillas, M. I., (2005). Study of the interfacial properties of natural fibre reinforced polyethylene. *Polymer Testing, 24*(6), 694–698. http://dx.doi.org/10.1016/j.polymertesting.2005.05.004.
12. Arrakhiz, F. Z., Elachaby, M., Bouhfid, R., Vaudreuil, S., Essassi, M., & Qaiss, A., (2012). Mechanical and thermal properties of polypropylene reinforced with Alfa fiber under different chemical treatment. *Materials and Design, 35*, 318–322. http://dx.doi.org/10.1016/j.matdes.2011.09.023.
13. Rosa, S. M. L. R., Santos, E. F., Ferreira, C. A., & Nachtigall, S. M. B., (2009). Studies on the properties of rice-husk filled PP composites: Effect of maleated PP. *Materials Research, 12*, 333–338. http://dx.doi.org/10.1590/S1516-14392009000300014.
14. Poletto, M., Dettenborn, J., Pistor, V., Zeni, M., & Zattera, A. J., (2010). Materials produced from plant biomass. Part I: Evaluation of thermal stability and pyrolysis of wood. *Materials Research, 13*, 375–379. http://dx.doi.org/10.1590/S1516-14392010000300016.
15. Kim, H. S., Lee, B. H., Choi, S. W., Kim, S., & Kim, H. J., (2007). The effect of types of maleic anhydride grafted polypropylene (MAPP) on the interfacial adhesion properties of bio-flour filled polypropylene composites. *Composites Part A: Applied Science and Manufacturing, 38*(6), 1473–1482. https://doi.org/10.1016/j.compositesa.2007.01.004.
16. Sain, M., Suhara, P., Law, S., & Bouilloux, A., (2005). Interface modification and mechanical properties of natural fiber polyolefin composite products. *Journal of Reinforced Plastics and Composites, 24*, 121–130. https://doi.org/10.1177/0731684405041717.
17. Özmen, N., (2012). A study of the effect of acetylation on hemp fibres with vinyl acetate. *BioResources, 7*(3), 3800–3809.
18. Manikandan, N. K. C., Sabu, T., & Groeninckx, G., (2001). Thermal and dynamic mechanical analysis of polystyrene composites reinforced with short sisal fibres. *Composites Science and Technology, 61*, 2519–2529. https://doi.org/10.1016/S0266-3538(01)00170-1.

19. Bachtiar, D., Sapuan, S. M., Khalina, A., Zainudin, E. S., & Dahlan, K. Z. M., (2012). Flexural and impact properties of chemically treated sugar palm fiber reinforced high impact polystyrene composites. *Fibers and Polymers, 13*, 894–898. https://doi.org/10.1007/s12221-012-0894-1.
20. Grigoriev, A. P., & Fedotova, O. Ya., (1977). Laboratory workshop on plaltics technology, in two parts; Part 1. *Polymerized Plastics Textbook Manual for Chemical-Technological Universities* (2nd edn., p. 248). Rev. & add., Moscow, Higher school. In Russian.
21. Mineev, V. N., & Mineev, A. V., (1997). Viscosity of metals under shock-loading conditions, *J. Phys. IV France, 7*, C3–583–585. doi: 10.1051/jp4:19973100.
22. Liu C., Tanaka Y., & Fujimoto, Y., (2015). Viscosity transient phenomenon during drop impact testing and its simple dynamics model. *World J. Mech., 5*, 33–41. https://doi.org/10.4236/wjm.2015.53004.
23. Kliosov, A. A., (2007). *Wood-Plastic Composites* (p. 698). John Wiley & Sons, Hoboken, New Jersey.
24. Ellis, W. D., & O'Dell, J. L., (1999). Surface morphology of PECVD fluorocarbon thin films from hexafluoropropylene oxide, 1,1,2,2-tetrafluoroethane and difluoromethane. *J. Appl. Polym. Sci., 73*(12), 2493–2505.
25. Patil, Y. P., Gajre, B., Dusane, D., Chavan, S., & Mishra, S., (2000). Effect of maleic anhydride treatment on steam and water absorption of wood polymer composites prepared from wheat straw, cane bagasse, and teak wood sawdust using Novolac as matrix. *J. Appl. Polym. Sci., 77*(13), 2963–2967.
26. Brember, K. I., & Shneider, M. H., (1985). The significance of the glass transition of lignin in thermomechanical pulping. The significance of the glass transition of lignin in thermomechanical pulping. *Wood Sci. and Technology, 19*, 75–81.
27. Hausen, B. M., Kuhlwein, A., & Schulz, K. H., (1982). Colophony allergy. A contribution to the origin, chemistry, and uses of colophony and modified colophony products, 1. *Derm Beruf. Umwelt., 30*(4), 107–115.
28. Aneli, J., Buzaladze, G., Markarashvili, E., Tatrishvili, T., Kemkhadze, L., & Mukbaniani, O., (2015). Composites based on sawdust and some organic binders. *Abstracts of Communications of 4th International Caucasian Symposium on Polymers and Advanced Materials* (p. 10). Batumi. http://www.icsp4.tsu.ge/data/file_db/icsp/ICSP.pdf (accessed on 02 January 2022).
29. Mukbaniani, O., Aneli, J., Buzaladze, G., Markarashvili, E., & Tatrishvili, T., (2015). Wood polymer composites based on the renewable raw materials. *Abstracts of Presentation of the First SDSU – Georgia Stem Workshop on Nanotechnology and Environmental Sciences*. Tbilisi, Georgia. https://www.tsu.ge/data/file_db/faculty_zust_sabunebismetk/TOTAL-final-2.pdf (accessed on 02 January 2022).
30. Muraganatham, S., Anbalagan, G., & Ramamurthy, N., (2009). FTIR and SEM (EDS) comparative analysis of medical plants. Eclipta alba HassK and *Eclipta prostrate* Linn. *Rom J. Biophysics, 19*, 285–294.
31. Shirwaikar, A., Malini, S., & Kumari, S. C., (2003). Protective effect of *Pongamia pinnata* flowers against cisplatin and gentamicin induced nephrotoxicity in rats. *Indian J. Exp. Biol., 1*, 58–62.
32. Venkateshwaran, N., Peruma, A. E., & Arunsundaranayagam, D., (2013). Fiber surface treatment and its effect on mechanical and visco-elastic behavior of banana / epoxy composite. *Materials and Design, 47*, 151–159. http://dx.doi.org/10.1016/j.matdes.2012.12.001.

33. Kakou, C. A., Arrakhiz, F. Z., Trokourey, A., Bouhfid, R., Qaiss, A., & Rodrigue, D., (2014). Influence of coupling agent content on the properties of high-density polyethylene composites reinforced with oil palm fibers. *Materials and Design, 63*, 641–649. http://dx.doi.org/10.1016/j.matdes.2014.06.044.
34. Singha, A. S., & Raj, R. K., (2012). Natural fiber reinforced polystyrene composites: Effect of fiber loading, fiber dimensions and surface modification on mechanical properties. *Materials and Design, 41*, 289–297. http://dx.doi.org/10.1016/j.matdes.2012.05.001.
35. Asumani, O. M. L., Reid, R. G., & Paskaramoorthy, R., (2012). The effects of alkali silane treatment on the tensile and flexural properties of short fibre non-woven kenaf reinforced polypropylene composites. *Composites: Part A, 43*, 1431–1440. http://dx.doi.org/10.1016/j.compositesa.2012.04.007.
36. Merkel, K., Rydarowski, H., Kazimierczak, J., & Bloda, A., (2014). Processing and characterization of reinforced polyethylene composites made with lignocellulosic fibres isolated from waste plant biomass such as hemp. *Composites: Part B, 67*, 138–144. http://dx.doi.org/10.1016/j.compositesb.2014.06.007.
37. Nekkaa, S., Guessoum, M., Benamara, R., & Haddaoui, N., (2013). Influence of surface flour treatment on the thermal, structural and morphological properties of polypropylene / *Spartium junceum* flour composites. *Polymer-Plastics Technology and Engineering, 52*, 175–181. http://dx.doi.org/10.1080/03602559.2012.734363.
38. Xie, Y., Callum, A. S. H., Xiao, Z., Holger, M., & Carsten, M., (2010). Silane coupling agents used for natural fiber / polymer composites: A review. *Composites: Part A, 41*, 806–819. http://dx.doi.org/10.1016/j.compositesa.2010.03.005.
39. Abdelmouleh, M., Boufi, S., Belgacem, M. N., & Dufresne, A., (2007). Short natural fibre reinforced polyethylene and natural rubber composites: Effect of silane coupling agents and fibres loading. *Composites Science and Technology, 67*(7, 8), 1627–1639.

PART IV
Sustainable and Green Chemistry

CHAPTER 21

Structure of Bis(Lidocaine) Tetrachloridoferrate(III) Chloride

KOBA AMIRKHANASHVILI,[1] VLADIMER TSITSISHVILI,[1] ALEXANDRE N. SOBOLEV,[2] and NANI ZHORZHOLIANI[1]

[1]Petre Melikishvili Institute of Physical and Organic Chemistry, Tbilisi State University, Tbilisi, Georgia

[2]Center for Microscopy, Characterization and Analysis, University of Western Australia, Perth, Australia

ABSTRACT

The present chapter reports on the synthesis and structure of bis(2-(diethylamino)-N-(2,6-dimethylphenyl)acetamide) or bis(lidocaine) tetrachloridoferrate(III) chloride. The complex with the formula $(C_{14}H_{23}ON_2)_2[FeCl_4]{\cdot}Cl$ (or $(LidH)_2[FeCl_4]{\cdot}Cl$), crystallizes in the monoclinic space group $P2_1/c$ with $a = 11.0597(1)$, $b = 23.0083(2)$, $c = 14.6629(2)$ Å, $\beta = 109.378(2)°$, $V = 3519.82(8)$ Å^3, $Z = 4$, and $D_c = 1.328$ Mg/m^3. The coordination of the Fe^{3+} ion with four chlorine anions generates slightly distorted tetrahedral anion $[FeCl_4)]^-$, while two protonated cations $LidH^+$, as well as one chlorine anion Cl^- remain in an outer coordination sphere. The formation of a network of hydrogen bonds involves amine and amide nitrogen atoms, as well as carbonyl oxygen atoms of lidocaine cations: the amide nitrogen atom of one cation and the protonated nitrogen atom of the amino group of another cation form relatively weak hydrogen bonds with the "isolated" chlorine atom, while the nitrogen atoms of the amino groups also participate in the formation of strong intramolecular N–H···O hydrogen bonds. The anion

Advanced Polymer Structures: Chemistry for Engineering Applications. Omar Mukbaniani, DSc, Tamara Tatrishvili, PhD & Marc Jean M. Abadie, DSc (Eds.)

is associated with cations only by weak "non-classical" hydrogen bonds of the C–H···Cl type, but the relatively strong intermolecular N–H···O hydrogen bonds form a "bridge" connecting two protonated lidocaine molecules of neighboring $(LidH)_2[FeCl_4]{\cdot}Cl$ complexes with the formation of molecular dimers. "Bridges" are located in the center and in all eight corners of the unit cell, half of the ligands of the complex form dimers, the other half stacks with the metal group and chlorine anions. The dimers form a chain lying in the ***ac*** plane and on the diagonal passing through the origin at an angle of 54.69° (0.5*β*) to the ***a*** and ***c*** axes; the chains form layers lying in the ***ac*** plane with an interlayer distance of 11.5 Å (0.5*b*), and layers containing anions and cations not involved in the formation of dimers are arranged between the dimer layers.

21.1 INTRODUCTION

Lidocaine or lignocaine (2-(diethylamino)-N-(2,6-dimethylphenyl)acetamide, Lid) is the most common and important local anesthetic and antiarrhythmic drug [1]; its base ($C_{14}H_{22}ON_2$) is easily soluble in diethyl ether, but poorly soluble in water, and thus lidocaine is used in form of water-soluble lidocaine hydrochloride monohydrate $C_{14}H_{22}ON_2{\cdot}HCl{\cdot}H_2O$ (*see* Figure 21.1).

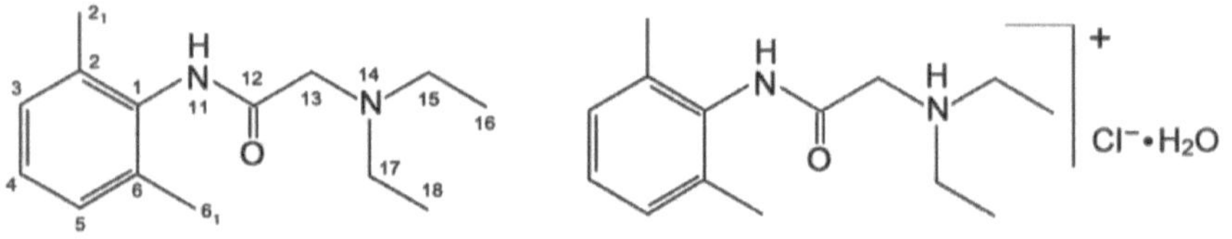

FIGURE 21.1 Lidocaine base (left) and hydrochloride monohydrate (right).

The molecular mechanism of action of local anesthetics on the nervous system is still unclear, but the ability to hydrogen bond donation is essential to their action, as was assumed more than half a century ago [2]. The crystal structure, hydrogen-bonding arrangement and conformation of the lidocaine molecule are significantly different for the free base, hydrochloride, and other salts, as well as for coordination compound with lidocaine as a ligand.

Thus, the structure of lidocaine is characterized by the presence of two independent molecules in the asymmetric unit, the intermolecular hydrogen bond is formed between the amido nitrogen and aceto oxygen atoms [3], while the predominant structure in monoclinic crystals of lidocaine hydrochloride monohydrate is fully hydrogen bonded, with adjacent lidocaine cations linked by water molecules into endless chains parallel to ***b*** axis; adjacent chains related by the screw axes are joined in pairs by chlorine ions, which bind $N^{+}H$ and H_2O groups in different chains [4].

A review of studies on lidocaine complexes is given in Ref. [5] and in our recent publications [6–8], and it was shown that the arrangement of hydrogen-bonding strongly depends on the nature of the complex even in cases where the lidocaine ligand is protonated at the amino nitrogen atom and exists in the $LidH^{+}$ cationic form.

The modern technique of X-ray diffraction analysis makes it possible to establish the presence of weak hydrogen bonds and obtain an accurate description of the supramolecular structure. The crystal structure of the complex described in 1990's [9] was re-refined recently [6], and it was shown that in bis(lidocaine) diaquatetrathiocyanatonickelate (II), $(LidH)_2[Ni(NCS)_4(H_2O)_2]$, coordination of the Ni^{2+} ion with thiocyanate ions and water molecules generates slightly distorted octahedral anion $Ni[(NCS_4)(H_2O)_2]^{2-}$ with *N*-bonded thiocyanate groups, while two protonated cations $LidH^{+}$ remain in an outer coordination sphere. The anion and cations are associated by hydrogen bonds formed by sulfur atoms with an amido nitrogen atoms; water molecules and an amino nitrogen atom are involved in the formation of hydrogen bonds with sulfur atoms of neighboring unit cells arranging alternating $Ni[(NCS)_4(H_2O)_2]^{2-}$ anions and $LidH^{+}$ cations into endless sheets lying in the ***ac*** plane.

The establishment of the structure of complexes with the participation of lidocaine began with tetrachlorozincate $(LidH)_2[ZnCl_4]$ [10], but it was noted that the structure was refined to an *R* value of 0.114, and the interatomic distances and angles are not highly accurate. Around 12 years later, the crystal structure of lignocaine hydrochloride – zinc chloride complex with strange Brutto-formula $ZnCl_4C_{28}N_4O_2H_{44}$ was reported [11], but the protonation of the amino nitrogen atom was not taken into account and led to conflicting conclusions. According to the latest data [7], this complex has the structure shown in Figure 21.2.

The anion and cations are associated by N–H···Cl hydrogen bonds, protonated amino nitrogen atoms participate in intramolecular and intermolecular hydrogen bonds with the oxygen atoms, combining molecules of the charge-transfer complex in pairs $2\{(LidH)_2[ZnCl_4]\}$. Each pair forms intermolecular N–H···Cl hydrogen bonds with four adjacent pairs, arranging them into endless sheets lying in the *bc* plane.

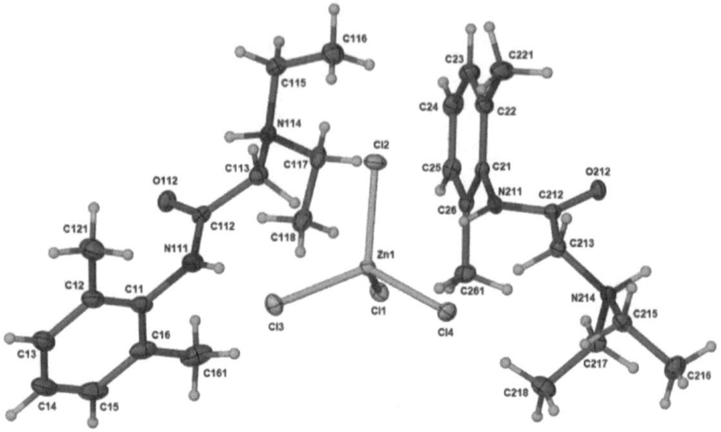

FIGURE 21.2 Molecular structure of $(LidH)_2[ZnCl_4]$.

The study of charge transfer complexes of lidocaine continues to this day due to their role in medical and other applications. The purpose of our work was to obtain and study new complexes of lidocaine; this contribution concerns a new iron complex of lidocaine, and special attention is paid to the formation of hydrogen bonds in this complex.

21.2 EXPERIMENTAL METHODS AND MATERIALS

21.2.1 SYNTHESIS

Iron(III) complex of lidocaine was prepared in a weakly acidic (pH=5–6) water-methanol solution with 1:2 molar ratio of iron(III) chloride ($FeCl_3$) and lidocaine hydrochloride monohydrate ($C_{14}H_{22}ON_2\cdot HCl\cdot H_2O$). The prepared mixture was filtered, placed on a magnetic stirrer with heating for a certain time, and then left at room temperature for slow evaporation. Brown prismatic crystals suitable for the X-ray measurements started to form after 21 days. The resulting crystals were washed with ether and dried in air; isolated yield 63%, melting point 156°C, the product is easily soluble in water, ethanol, acetone, and other organic solvents. Elemental analyzes were performed using a Labertherm CHN elemental analyzer and a Perkin-Elmer atomic adsorption spectrometer. Elemental analyzes data (wt.%): calculated for $C_{28}H_{46}Cl_5N_4O_2Fe$: C 47.78; H 6.59; N 7.96; Cl 25.19; Fe 7.93; found: C 47.56; H 6.30; N 7.94; Cl 24.87; Fe 7.95.

21.2.2 X-RAY DIFFRACTION ANALYSIS

The crystal structure of the complex was determined by single crystal X-ray diffraction (XRD) method using crystal with sizes 0.34 × 0.28 × 0.21 mm^3. XRD measurements were carried out with an Oxford Diffraction XCALIBUR E CCD diffractometer equipped with graphite-monochromated MoKα radiation ($\mu = 0.84$ mm^{-1}. $F_{000} = 1476$, $\lambda = 0.71073$ Å, $T = 100(2)$ K. The data collection, cell refinement and data reduction were carried out with the $CrysAlis^{PRO}$ package of Rigaku Oxford Diffraction (version 1.171.38.46, 2015); 2θ range for data collection was from 4.2 to 65.4°, 76404 reflections collected, 12220 independent ($R_{int} = 0.046$). The structure was solved by direct methods and refined against F^2 with full-matrix least-squares using the software complex SHELXL-2014 [12]; final $GOF = 1.00$, $R[F^2 > 2\sigma(F^2)] = 0.037$, $wR_2 = 0.096$, R indices based on 12220 reflections with $I > 2\sigma(I)$ (refinement on F^2), $|\Delta\rho|_{max} = 0.50$ eÅ^{-3}, 385 parameters, 0 restraints. CCDC 2109673 contains the supplementary crystallographic data for this contribution and can be obtained free of charge from the Cambridge Crystallographic Data Center *via* https://www.ccdc.cam.ac.uk/conts/retrieving.html or deposit@ccdc.cam.ac.uk.

21.3 RESULTS AND DISCUSSION

21.3.1 MOLECULAR STRUCTURE

21.3.1.1 COORDINATION OF THE IRON ATOM

According to the XRD crystallography, bis(lidocaine) tetrachloroferrate(III) chloride, $(LidH)_2FeCl_4Cl$, crystallizes in the monoclinic space group $P2_1/c$ with $a = 11.0597(1)$, $b = 23.0083(2)$, $c = 14.6629(2)$ Å, $\beta = 109.378(2)°$, $V = 3519.82(8)$ Å^3, $Z = 4$, and $D_c = 1.328$ Mg/m^3, and the complex has a molecular crystal structure, in which coordination of the Fe^{3+} ion with four chlorine anions generates slightly distorted tetrahedral anion $[FeCl_4)]^-$, while two protonated lidocaine cations $LidH^+$ 1 and 2, as well as one chlorine anion Cl^- remain in an outer coordination sphere (Figure 21.3).

Chlorine atoms in the $FeCl_4$ tetrahedron are located at distances from the central iron atom, slightly larger than 2.185 Å known for tetrachloroferrates(III), and much smaller than average 2.292 Å for tetrachloroferrates(II) [13]; the angles between the Cl–Fe–Cl bonds deviate from tetrahedral with a minimum of 107.594(17)° for angle Cl_3–Fe–Cl_4 and a maximum value of 111.261(17)° for angle Cl_2–Fe–Cl_4 (*see* Figure 21.4).

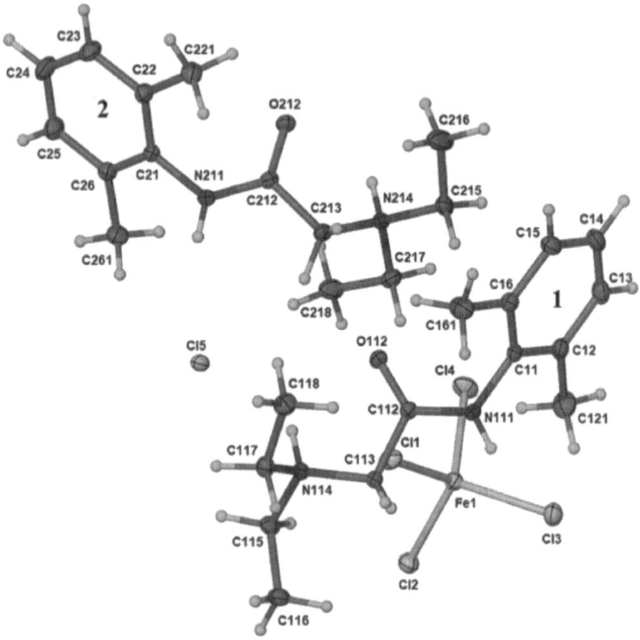

FIGURE 21.3 Molecular structure of the $(LidH)_2[FeCl_4]{\cdot}Cl$ complex with atom labeling according to CCDC 2109673.

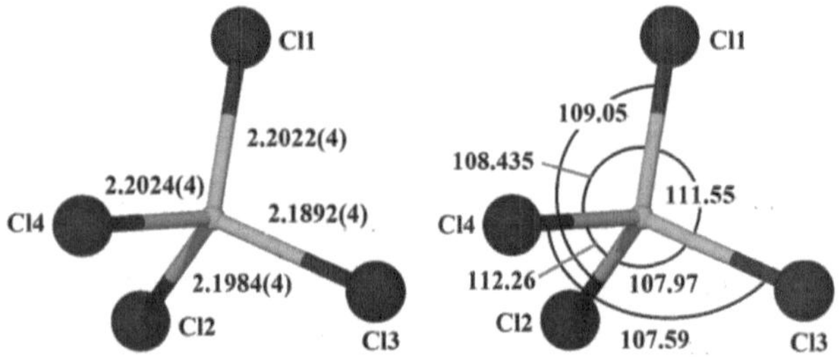

FIGURE 21.4 Interatomic distances (Å) and angles (°) in $FeCl_4$ tetrahedron.

Probably, the insignificance of distortions of the $FeCl_4$ tetrahedron is explained by the fact that chlorine atoms Cl_1–Cl_4 do not participate in the formation of "classical" intramolecular and intermolecular hydrogen bonds (*see* Table 21.1).

21.3.1.2 HYDROGEN BONDING

The SHELXL software generates four intramolecular (two N–H···O and two N–H···Cl), two intermolecular (N–H···Cl and N–H···O), and seven "non-classical" (five C–H···Cl and two C–H···O) hydrogen bonds listed in Table 21.1.

TABLE 21.1 Geometry of Intramolecular, Intermolecular, and "Non-Classical" Hydrogen Bonds

D–HA	D–H (Å)	H···A (Å)	D···A (Å)	D–HA (°)
N214–H214···O212	0.92(2)	2.176(19)	2.7092(16)	116.1(15)
N114–H114···O112	0.916(19)	2.237(18)	2.7129(15)	111.7(14)
N114–H114···Cl5	0.916(19)	2.417(19)	3.1855(12)	141.4(15)
N211–H211···Cl5	0.80(2)	2.49(2)	3.2607(13)	163.4(19)
N111–H111···Cl5^{i}	0.810(19)	2.41(2)	3.2182(13)	178.3(19)
N214–H214···O212^{iii}	0.92(2)	1.958(19)	2.8081(16)	153.1(17)
C213–H21*B*···Cl5	0.99	2.69	3.5576(15)	147
C115–H11*C*···Cl1	0.99	2.93	3.7980(16)	147
C217–H21*E*···Cl1	0.99	2.83	3.5751(15)	133
C217–H21*E*···O112	0.99	2.65	3.4124(19)	134
C113–H11*B*···Cl4^{ii}	0.99	2.87	3.8027(15)	156
C118–H11*L*···Cl4^{ii}	0.98	2.95	3.6733(17)	132
C218–H21*K*···O212^{iii}	0.98	2.57	3.2931(19)	131

Symmetry codes: (i) $x - ½, -y + ½, z - ½$; (ii) $x + ½, -y + ½, z - ½$; (iii) $-x + 1, -y + 1, -z + 1$.

The formation of a network of hydrogen bonds involves amine and amide nitrogen atoms, as well as carbonyl oxygen atoms of lidocaine cations. The amide nitrogen atom N211 of cation **2** and the protonated nitrogen atom N114 of the amino group of cation **1** form relatively weak (the distance D···A ~3.2 Å) hydrogen bonds with the "isolated" chlorine atom Cl_5, while the nitrogen atoms of the amino groups N114 and N214 also participate in the formation of strong (D···A ~2.7 Å) hydrogen bonds with oxygen atoms O112 and O212, respectively (*see* Figure 21.5).

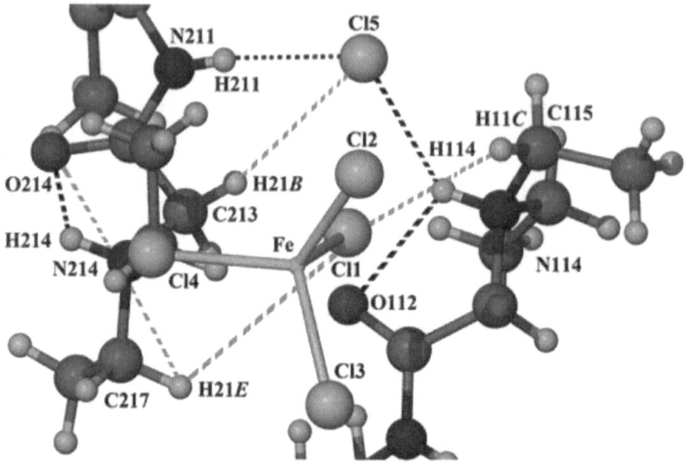

FIGURE 21.5 Intramolecular "classical" (black dotted lines) and "non-classical" (blue dotted line) hydrogen bonds generated by the SHELXL software.

Along with the "classical" hydrogen bonds of the N–H···Cl and N–H···O type, the SHELXL software generates four "non-classical" intramolecular hydrogen bonds with the participation of carbon atoms C213, C115, and C217 as donors. The chemical shift of the H21*B* proton of the methylene group located between the carbonyl and amino groups is δ 3.4, which indicates a significant deshielding and, therefore, a high probability of the formation of a hydrogen bond. The chemical shift of the methylene protons H11*C* and H21*E* of the diethylamine groups is δ 2.8, and the probability of hydrogen bond formation is much lower, although the distances between the donor and acceptor are comparable in all four cases.

21.3.1.3 CONFORMATION OF FLEXIBLE CHAIN

In the gaseous and liquid states, free rotation is possible along chemical bonds in the chain C1–C11–C12–C13–N14, but in the crystalline state the chain takes on a stable conformation. The formation of intramolecular N–H···O hydrogen bonds is due to the *synperiplanar* conformation of the carbonyl oxygen atom and the nitrogen atom of the amino group, as evidenced by the values of the O112–C112–C113–N114, and O212–C212–C213–N214 torsion angles (*see* Table 21.2).

TABLE 21.2 Torsion Angles in the Chain from Aromatic Rings to Diethylamino Group in the $LidH^+$ Cations

Atoms of Cation 1	Angle (°)	Atoms of Cation 2	Angle (°)
C16–C11–N111–C112	77.00(18)	C26–C21–N211–C212	112.97(15)
C11–N111–C112–O112	−3.8(2)	C21–N211–C212–O212	2.4(2)
C11–N111–C112–C113	174.53(12)	C21–N211–C212–C213	−178.91(12)
N111–C112–C113–N114	168.15(12)	N211–C212–C213–N214	152.20(12)
O112–C112–C113–N114	−13.26(18)	O212–C212–C213–N214	−29.05(18)

In general, the arrangement of atoms in the flexible chains of lidocaine cations **1** and **2** has the same character but differs in details. According to the values of the corresponding torsion angles, the amide group is twisted out from the plane of the aromatic ring by 77° in cation **1** and by ~113° in cation **2**; for both cations the aromatic ring and the oxygen atom adopt a *synperiplanar* (cis) conformation with respect to the N111–C112 and N211–C212 bonds, while the aromatic ring and diethylamino chain adopt the *antiperiplanar* (trans) conformation (*see* Figure 21.6).

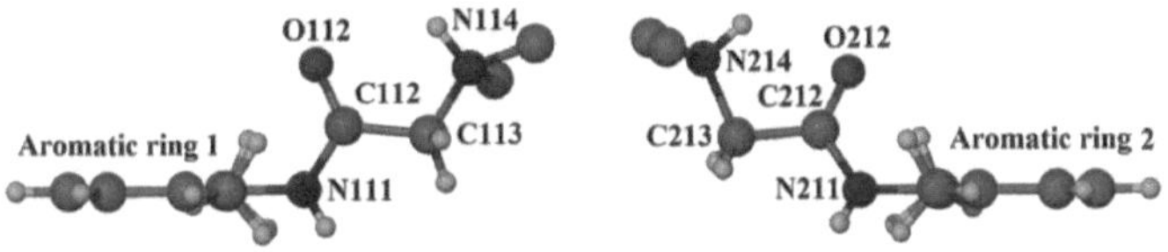

FIGURE 21.6 The arrangement of atoms in the chain between the aromatic ring and the amino group of lidocaine cations.

21.3.1.4 INTERMOLECULAR HYDROGEN BONDS

Nitrogen atoms are also involved in the formation of intermolecular hydrogen bonds – the amide nitrogen atom N111 of cation **1** forms a weak hydrogen bond with the "isolated" chlorine atom Cl_5^{i} of the neighboring complex, determined by the operation of symmetry (i) $x - ½, -y + ½, z - ½$ (*see* Figure 21.5), and the nitrogen atom N214 of the amino group of cation **2** forms a stronger hydrogen bond with the oxygen atom $O212^{iii}$ of the neighboring complex, determined by the operation of symmetry (iii) $-x + 1, -y + 1, -z + 1$ (*see* Figure 21.7).

Taking into account the "non-classical" hydrogen bonds generated by the SHELXL software (*see* Table 21.1), the oxygen atom $O212^{iii}$ as an acceptor participates in the formation of one more hydrogen bond of the C–H···O type. The distance between the donor and acceptor for the C218–H21*K*···O212^{iii} hydrogen bond meets the criteria for "nonclassical" bonds (<3.3 Å), but the C218 carbon atom is not bonded to an electronegative substituent, which is a prerequisite for the formation of a C–H···O hydrogen bond. The same applies to the C118–H11*L*···Cl4^{ii} hydrogen bond, only the C113–H11*B*···Cl4^{ii} hydrogen bond satisfies this criterion, but for it the distance between the donor and acceptor (>3.8 Å) is the largest of the "non-classical" hydrogen bonds generated by the SHELXL software for the $(LidH)_2FeCl_4{\cdot}Cl$ structure.

Intra- and intermolecular N–H···O hydrogen bonds, as well as "non-classical" intermolecular C–H···O and C–H···Cl hydrogen bonds are shown in Figure 21.7, and the scheme of "classical" and "non-classical" intermolecular hydrogen bonds formed in the $(LidH)_2[FeCl_4]Cl$ complex is shown in Figure 21.8.

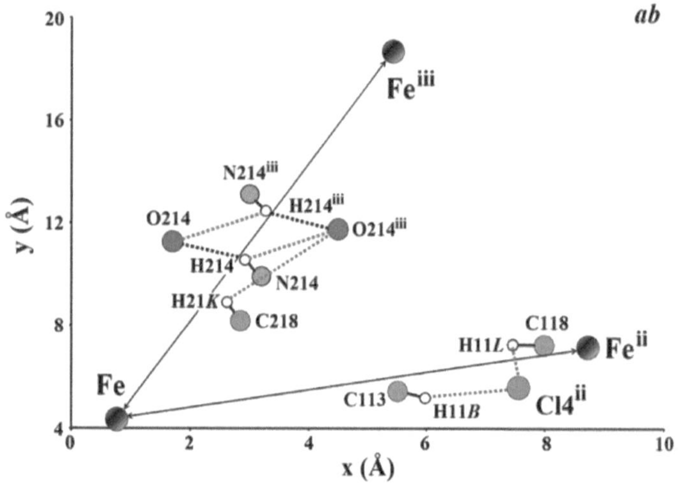

FIGURE 21.7 Intramolecular (black dotted lines) and intermolecular (red dotted lines) N–H···O hydrogen bonds, and "non-classical" intermolecular (blue dotted lines) C–H···O and C–H···Cl hydrogen bonds viewed along [001] (the two-pointed arrows show the lines connecting the "central" Fe atom with atoms Fe^{ii} and Fe^{iii}).

As follows from considerations of symmetry and is shown in Figures 21.7 and 21.8, in the neighboring complex of the iron atom Fe^{iii}, an intramolecular hydrogen bond N214^{iii}–H214^{iii}···O212^{iii} is formed, while the hydrogen atom H214^{iii} also enters the N214^{iii}–H214^{iii}···O212 intermolecular hydrogen bond. Taking into account the possibility of the formation of "non-classic" hydrogen bonds, the carbonyl oxygen atoms O212 and O212^{iii} enter into a trifurcated hydrogen bond.

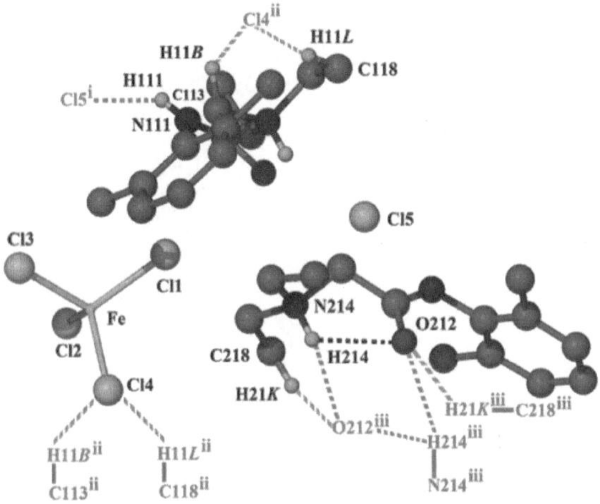

FIGURE 21.8 "Classical" (red dotted lines) and "non-classic" (blue dotted lines) intermolecular hydrogen bonds in the $(LidH)_2[FeCl_4]Cl$ complex.

21.3.2 SUPRAMOLECULAR STRUCTURE AND PACKING

The "bridge" of N214–H214···O212^{iii} and N214^{iii}–H214^{iii}···O212 hydrogen bonds shown in Figure 21.7 unites two protonated lidocaine molecules into a pair, as in the bis(lidocaine) tetrachloridozincate(II) [7], but with certain differences.

First, in zincate(II) the molecules of the complex are combined into pairs $2\{(LidH)_2[ZnCl_4]\}$ as a whole, including the organic and inorganic parts, while in ferrate(III) only the lidocaine cations **2** are combined forming molecular dimers 2LidH shown in Figure 21.9.

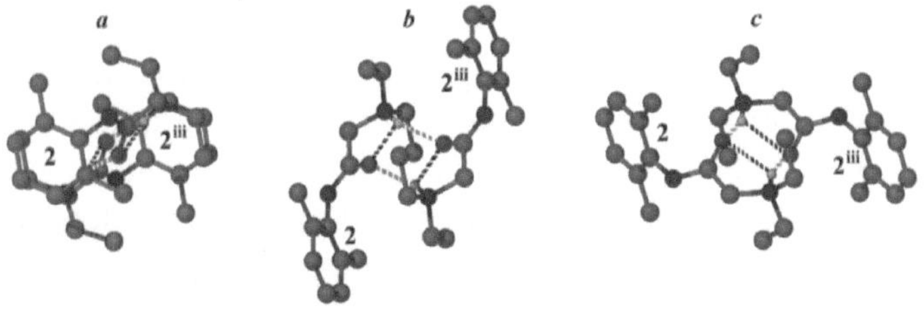

FIGURE 21.9 Molecular dimer $2LidH^+$ viewed along [100] (a); [010] (b); and [001] (c).

The aromatic rings of the lidocaine cations lie approximately in the ***bc*** plane, the amide groups in this plane form a sort of six-membered ring, and one of the ethyl groups is located along the opposite ring, and the dimer depicted along the ***a*** crystallographic axis looks like a molecule consisting of three condensed six-membered rings (*see* Figure 21.9(a)).

Translation of an asymmetric unit along the ***a*** crystallographic axis leads to the formation of a system of chains, looking like triple "channels" along the ***a*** axis; the outermost "channels" are "empty," while oxygen atoms are located in the middle "channel" (Figure 21.10(a)).

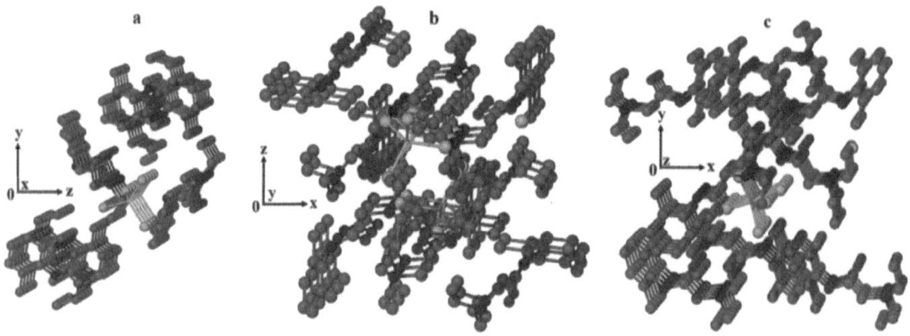

FIGURE 21.10 Translation of asymmetric unit along the (a), (b), and (c) crystallographic axes.

The inorganic part of the complex ($[FeCl]_4^-$ and Cl^- anions) is associated with type **1** cations by weak "non-classic" hydrogen bonds, translation of an asymmetric unit along the ***a*** and ***c*** axes leads to one row of anions (*see* Figures 21.10(a) and (c)), and translation along the ***b*** axis to two rows (*see* Figure 21.10(b)).

An important difference between bis(lidocaine) tetrachloroferrate(III) chloride and bis(lidocaine) tetrachlorozincate(II) is that in ferrate(III) the center of dimer coincides with the center of symmetry Fe–Fe^{iii} and inversion of the unit cell (½, ½, ½ in fractional coordinates (*see* Figures 21.11 and 21.12)).

The unit cell (Z = 4) contains four $[FeCl_4]^-$ anions, four Cl^- anions, and eight lidocaine cations $LidH^+$, of which six are located in the body of the unit cell as a whole, and two are in the forms of fragments in such a way that the "bridges" that combine lidocaine molecules into dimers are located in the center and in all eight corners of the unit cell (*see* Figures 21.11 and 21.12).

Figures 21.10 and 21.11 show the unit cell of the $(LidH)_2[FeCl_4]Cl$ complex and six "additional" lidocaine molecules, parts of which enter the cell body to "compensate" those parts of the eight lidocaine molecules that go beyond the unit cell facets (for example, for cation **1**, atoms H15, H16A, and H16C have fractional coordinates $z<0$, and for cation **2**, atoms C24, H23, and H24 have fractional coordinates $x>1$, etc.).

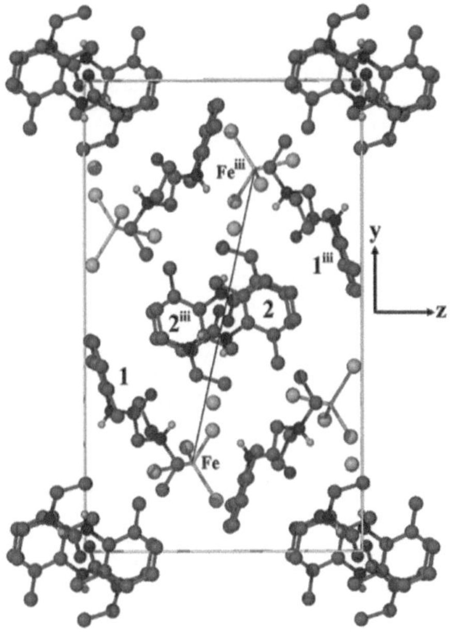

FIGURE 21.11 Unit cell of the $(LidH)_2[FeCl_4]Cl$ complex viewed along [100].

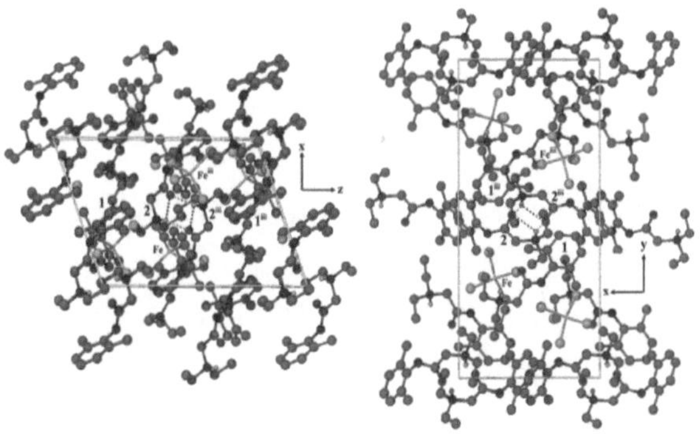

FIGURE 21.12 Unit cell of the $(LidH)_2[FeCl_4]Cl$ complex viewed along [010] and along [001].

Half of the ligands of the complex form dimers, the other half stacks with the metal group and chlorine anions, and this is clearly seen in Figure 21.13 showing the packing of the complex molecules along the ***a*** axis.

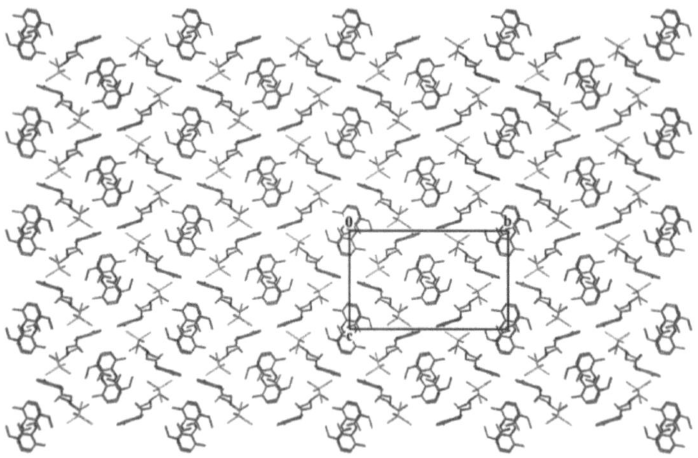

FIGURE 21.13 Packing of the $(LidH)_2[FeCl_4]Cl$ complex viewed along [100].

The dimers form a layer lying in the ***ac*** plane (*see* Figure 21.14), the distance between the layers is 0.5*b* = 11.5 Å, chains of dimers (triple "channels" along the **a** axis) lie on a diagonal passing through the origin, that is, at an angle of $0.5\beta = 54.69°$ to the ***a*** and ***c*** axes.

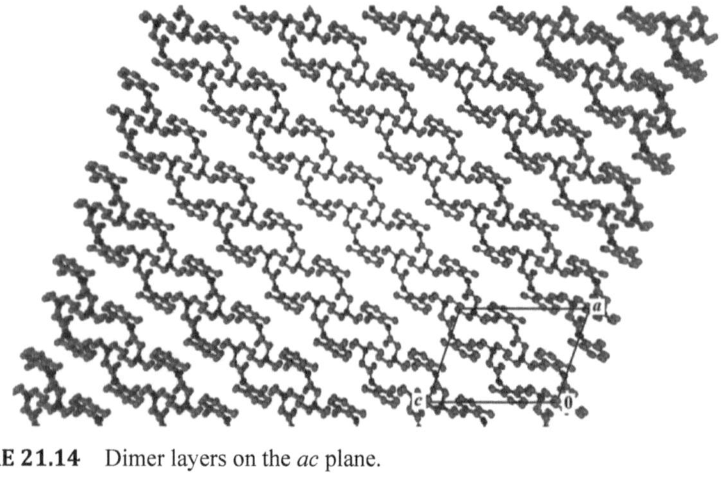

FIGURE 21.14 Dimer layers on the *ac* plane.

Between the dimer layers are arranged layers containing $[FeCl_4]^-$ and Cl^- anions, as well as type **1** cations. Along with the Coulomb attraction forces, the connection between $[FeCl_4]^-$ anions and lidocaine cations is provided by weak (the distance between the donor and acceptor ~3.3–3.8 Å) "non-classic" C–H···Cl hydrogen bonds, and the connection between the layers is provided by somewhat stronger (distance between donor and acceptor ~3.2 Å) hydrogen bonds N–H···Cl.

The layered nature of the crystal structure of the $(LidH)_2[FeCl_4]Cl$ complex is clearly seen in the figure depicting the packing of the complex in crystallographic planes perpendicular to the dimer and inorganic layers (*see* Figure 21.15).

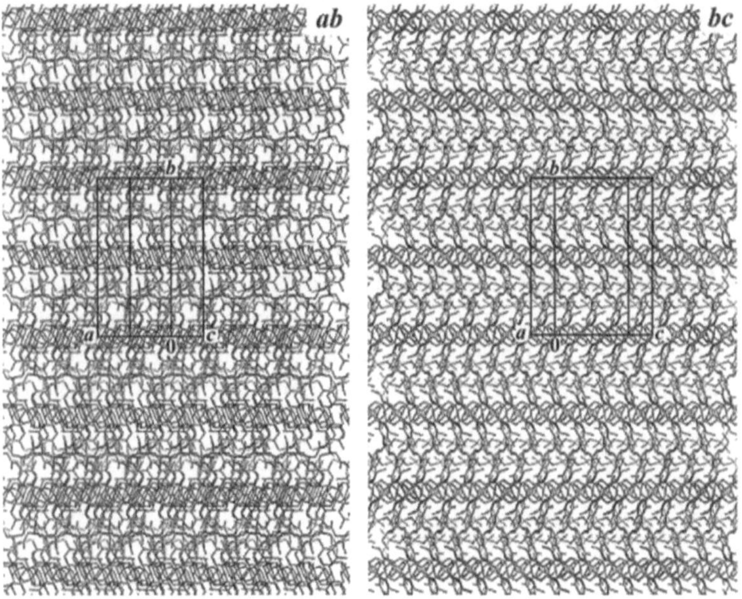

FIGURE 21.15 Packing of the $(LidH)_2[FeCl_4]Cl$ complex viewed on [110] and [011] planes.

21.4 CONCLUSION

This work reports the synthesis and structure refinement of lidocaine hydrochloride – iron(III) chloride complex. The synthesis of complex was carried out in water-methanol solution with 1:2 molar ratio of iron(III) chloride and lidocaine, resulting in prismatic brown crystals with chemical composition corresponding to brutto-formula $C_{28}H_{46}Cl_5N_4O_2Fe$.

The single-crystal XRD characterization shows that the complex crystallizes in the monoclinic space group $P2_1/c$ with $a = 11.0597(1)$, $b = 23.0083(2)$, $c = 14.6629(2)$ Å, $\beta = 109.378(2)°$, and $Z = 4$, consists of slightly distorted tetrahedral anion $[FeCl_4)]^-$, as well as of two protonated cations $LidH^+$ and one chlorine anion Cl^- remaining in an outer coordination sphere, and is named as bis(lidocaine) tetrachloridoferrate(III) chloride. The amide nitrogen atom N114 of cation 1 and the protonated nitrogen atom N211 of the amino group of cation 2 form relatively weak N–H···Cl hydrogen bonds with the "isolated" chlorine atom Cl5, thereby forming an intermolecular network of hydrogen bonds; the nitrogen atoms of the amino groups are also involved in strong intramolecular N–H···O hydrogen bonds with corresponding carbonyl oxygen atoms of lidocaine cations. In addition to the Coulomb attraction forces, the anion and cations are associated only by weak "non-classical" hydrogen bonds of the C–H···Cl type, but the relatively strong intermolecular N–H···O hydrogen bonds form a "bridge" that unites two protonated lidocaine molecules of neighboring $(LidH)_2[FeCl_4]Cl$ complexes, which leads to the formation of molecular dimers. "Bridges" are located in the center and in all eight corners of the unit cell, half of the protonated lidocaine molecules of the complex form dimers, the other half forms stacks with the metal group and chlorine anions. The dimers form a chain lying in the ac plane and on the diagonal passing through the origin at an angle of $0.5\beta = 54.69°$ to the a and c axes. The chains form a layer lying in the ac plane, the distance between the dimer layers is $0.5b = 11.5$ Å, and between the dimer layers there are layers of anions and cations that do not participate in the formation of dimers.

KEYWORDS

- **chlorine atoms**
- **crystal structure**
- **crystallography**
- **hydrogen bond**
- **hydrogen bonding**
- **lidocaine complex**
- **tetrahedral anion**
- **X-ray analysis**

REFERENCES

1. Jang, Y. J., Lee, J. H., Seo, T. B., & Oh, S. H., (2017). Lidocaine/multivalent ion complex as a potential strategy for prolonged local anesthesia. *European Journal of Pharmaceutics and Biopharmaceutics, 115*, 113–121. doi: https://doi.org/10.1016/j.ejpb.2017.02.007.
2. Sax, M., & Pletcher, J., (1969). Local anesthetics: Significance of hydrogen bonding in mechanism of action. *Science, 166*(3912), 1546–1548. doi: https://doi.org/10.1126/science.166.3912.1546.
3. Bambagiotti-Alberti, M., Bruni, B., Di Vaira, M., Giannellini, V., & Guerri, A., (2007). 2-(diethylamino)-*N*-(2,6-dimethylphenyl)acetamide, a low-temperature redetermination. *Acta Crystallographica, E63*, 768–770. doi: https://doi.org/10.1107/S1600536807001523.
4. Hanson, A. W., & Röhrl, M., (1972). The crystal structure of lidocaine hydrochloride monohydrate. *Acta Crystallographica, B28*, 3567–3571. doi: https://doi.org/10.1107/S0567740872008350.
5. Maulvi, F. A., Pillai, L. V., Patel, K. P., Desai, A. R., Shukla, M. R., Desai, D. T., Patel, H. P., et al., (2020). Lidocaine tripotassium phosphate complex laden microemulsion for prolonged local anaesthesia: *In vitro* and *in vivo* studies. *Colloids and Surfaces. B, Biointerfaces, 1*(185), 110632–110647. https://doi.org/10.1016/j.colsurfb.2019.110632.
6. Amirkhanashvili, K., Sobolev, A., Zhorzholiani, N., & Tsitsishvili, V., (2020). Re-refinement of cristal structure of bis(lidocaine) diaqvatetrathiocyanatonickelate(II). *Chemistry Journal of Moldova, 15*(1), 67–74. http://dx.doi.org/10.19261/cjm.2019.675.
7. Amirkhanashvili, K., Sobolev, A., Tsitsishvili, V., & Zhorzholiani, N., (2020). Molecular and crystal structure of bis(lidocaine) tetrachlorozincate(II). *Bull. Georgian Natl. Acad. Sci., 14*(2), 42–49. http://science.org.ge/bnas/t14-n2/07_Amirkhanashvili_Chemistry.pdf (accessed on 02 January 2022).
8. Amirkhanashvili, K., Sobolev, A., Tsitsishvili, V., & Zhorzholiani, N., (2021). Structure of bis(lidocaine) tetrachloridocuprate(II). *Bull. Georgian Natl. Acad. Sci., 15*(3), 34–40. http://science.org.ge/bnas/t15-n3/05_Amirkhanashvili_Physical%20Chemistry.pdf (accessed on 10 January 2023).
9. Indira, A., Sridhar, M. A., Shashindara, P. J., & Cameron, T. S., (1994). Crystal structure of lignocaine hydrochloride – nickel thiocyanate complex. *Zeitschrift für Kristallographie – Crystalline Materials, 209*(5), 440–442. **doi:** https://doi.org/10.1524/zkri.1994.209.5.440.
10. Główka, M. L., & Gałdecki, Z., (1981). Crystal structure of bis[2-(diethylammonium)-N-(2,6-dimethylphenyl)acetamide] tetrachlorozinc (lidocaine hydrochloride zinc chloride). *Polish Journal of Chemistry, 55*, 651–658.
11. Indira, A., Sridhar, M. A., Bellad, S. B., Babu, A. M., & Shashidhara, P. J., (1993). Crystal structure of lignocaine hydrochloride – zinc chloride complex. *Molecular Crystals and Liquid Crystals Science and Technology, A237*(1), 377–388. https://doi.org/10.1080/10587259308030150.
12. Sheldrick, G. M., (2015). Crystal structure refinement with SHELXL. *Acta Crystallographica, C71*(1), 3–8. http://dx.doi.org/10.1107/S2053229614024218.
13. Lauher, J. W., & Ibers, J. A., (1975). Structure of tetramethylammonium tetrachloroferrate(II), $[N(CH_3)_4]_2[FeCl_4]$. Comparison of iron(II) and iron(III) bond lengths in high-spin tetrahedral environments. *Inorganic Chemistry, 14*(2), 348–352. doi: 10.1021/ic50144a029.

CHAPTER 22

Synthesis and Hydrosilylation of 2-Methyl(Ethyl)-1-Allyl Pyrrole

O. B. ASKEROV, V. A. DZHAFAROV, D. R. NURULLAYEVA, R. V. ASADOV, and A. YA. KAGRAMANOVA

Institute of Polymer Materials, Azerbaijan National Academy of Sciences, Sumgait, Azerbaijan

ABSTRACT

The method of synthesis of organosilicon 2-methyl(ethyl)-1-propyl pyrrole by the interaction of trialkyl(aryl)hydride silanes with 2-methyl(ethyl)-1-allyl pyrrole in the presence of the catalyst (0.1 n of platinohydrochloric acid or rhodium acetylacetonatedicarbonyl) has been developed; there have been obtained the new organosilicon pyrroles, the structure of which has been established by a method of GLC, IR spectroscopy and elemental analysis. It has been shown that trialkyl(aryl)hydride silanes are added to unsaturated compounds on Farmers' rule.

The antimicrobial activity of the synthesized organosilicon pyrroles has been studied. It has been established in this case that from studied organosilicon pyrroles, 2-methyl(ethyl)-1-propyl pyrroles exhibit expressed antimicrobial activity against *E. coli* and *Pseudomonas aeruginosa* bacteria.

22.1 INTRODUCTION

It was known from domestic and foreign literature that the pyrrole and its derivatives occur in nature in the free state, and also as fragments in the

Advanced Polymer Structures: Chemistry for Engineering Applications. Omar Mukbaniani, DSc, Tamara Tatrishvili, PhD & Marc Jean M. Abadie, DSc (Eds.)

composition of complex natural compounds. In the organism, they are present during fulfillment of the most important physiological functions [1]. In this connection, the synthesis of functionally substituted 2-methyl (ethyl) allyl pyrroles, as well as their silicon derivatives and the study of their antimicrobial activity is an actual task possessing scientific and practical value.

In connection with above-stated ones and in continuation of the investigations in the field of synthesis of organosilicon 2-methyl(ethyl)-1-allyl pyrrole-containing compounds [1, 2], the study of the addition reaction of trialkyl(aryl)hydride silanes with 2-methyl(ethyl)-1-allyl pyrrole in the presence of 0.1 n of platinohydrochloric acid or rhodium acetylacetatedicarbonyl and, thereby, an elucidation of influence of the nature of substituents at the multiple bond of allyl radical on the yield and structure of the reaction products is of scientific interest.

We have developed the method of synthesis of pyrrole-containing organosilicon compounds by reaction of trialkyl(aryl)hydride silanes with 2-methyl(ethyl)-1-allyl pyrrole in the presence of the catalyst (0.1 n of platinohydrochloric acid or rhodium acetylacetatedicarbonyl) on the following scheme:

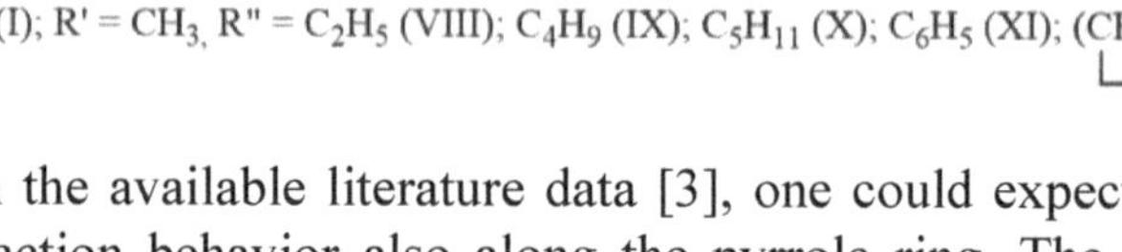

I, II (III-XII)

R = CH_3 (I); R' = CH_3, R'' = C_2H_5 (III); C_4H_9 (IV); C_5H_{11} (V); C_6H_5 (VI); $(CH_2)_4$ (VII)

R = C_2H_5 (I); R' = CH_3, R'' = C_2H_5 (VIII); C_4H_9 (IX); C_5H_{11} (X); C_6H_5 (XI); $(CH_2)_4$ (XII)

Based on the available literature data [3], one could expect the hydrosilylation reaction behavior also along the pyrrole ring. The presence of the pyrrole ring is confirmed by availability of the absorption band in the field of 3,440–3,400 cm^{-1}, belonging to the vibrations of the pyrrole ring in the spectrum. Consequently, triorganosilanes are added to the investigated unsaturated pyrrole compounds exclusively by C = C bond of the allyl radical, without affecting the pyrrole cycle [3].

The structure of the obtained adducts (III-XI) has been studied by modern spectral methods. The study of spectrum of adduct (III) showed that there are absorption bands in the field of 1,251–1,231 cm^{-1} referring to the valence vibrations of C–N bond, and a frequency at 1,620 cm^{-1} is referred to the valence vibrations of Si–$C_{alk.}$ bond. The deformation (1,358, 1,421, 1,458 cm^{-1}) and valence (2,873, 2,911, 2,951 cm^{-1}) vibrations are referred to C–H-bond in CH_2-group. The deformation vibrations (1,458, 1,421, and 1,350 cm^{-1}) are referred to vibration of C–N-bond, and the valence vibration at 2,873 cm^{-1} – to C–H-bond].

Thus, under the conditions accepted by us, the addition reaction of silicon hydrides with 2-methyl (ethyl)-1-allyl pyrrole, catalyzed by 0.1 n of platinohydrochloric acid or rhodium acetylacetatedicarbonyl proceeds on C = C bond of allyl radical with the formation of organosilicon compounds of pyrrole, and the triorganosilyl group is fixed at the peripheral carbon atom.

22.2 EXPERIMENTAL PART

About 13 g (0.1 mol) of 2-methyl-1-allylpyrrole, 25 ml of anhydrous benzene and catalyst are loaded into a three-neck flask equipped with a reflux condenser, a thermometer and a dropping funnel and heated for 24 hours. After distillation of the solvent and unreacted components, 8 g (72%) of the compound (III) was isolated from the residue by vacuum. B.p. – 100–101°C (3 mm merc.c.), n_D^{20} 1.4820, d_4^{20} 0.9480, MR_D 72.45 (found), 72.00 (calculated). $C_{13}H_{25}NSi$. Found %: C 70.0, H 11.30, N 6.30, Si 12.30. Calculated, %: C 69.80, H 11.0, N 6.21, Si 12.60.

Similarly, 8.6 g (78%) of the compound (VIII) has been obtained from 8 g of 2-methyl-1-allyl pyrrole, 10.25 g (0.1 mol) of methyldiethyl silane and 0.1 ml of the catalyst. B.p. – 108°C (3 mm merc.c.), n_D^{20} 1.4816, d_4^{20} 0.8993, MR_D 75.73 (found), 75.00 (calculated). $C_{14}H_{27}NSi$. Found, %: C 70.30, H 11.21, N 5.62, Si 11.40. Calculated, %: C 70.81, H 11.46, N 5.89, Si 11.825.

Similarly, there have been obtained the compounds (IV-VII) and (IX-XII), the properties of which are shown in Table 22.1.

TABLE 22.1 Physical-Chemical Constants of Organosilicon Pyrroles

Compounds	Yield (%)	B.p. °C (mm merc.c.)	n_D^{20}	d_4^{20}	MR_D	
					Found	Calculated
IV	83	99–100(3)	1.4850	0.8785	91.05	91.00
V	68	96–97(3)	1.4802	0.8697	100.42	100.00
VI	70	97(3)	1.6181	1.0731	103.07	103.00
VII	78	136–137(3)	1.4900	0.9471	92.73	92.00

TABLE 22.1 *(Continued)*

Compounds	Yield (%)	B.p. °C (mm merc.c.)	n_D^{20}	d_4^{20}	MR$_D$	
					Found	Calculated
IX	70	110(3)	1.4851	0.8950	94.30	94.00
X	72	97(3)	1.4821	0.8545	102.90	102.00
XI	74	140(3)	1.5667	1.088	100.13	100.03
XII	78	112(3)	1.4870	0.7933	89.89	89.00

22.2.1 DETERMINATION OF ANTIMICROBIAL ACTIVITY OF SYNTHESIZED PYRROLE-CONTAINING ORGANOSILICON COMPOUNDS

In this work, the antimicrobial activity of the obtained compounds was determined by a method of serial dilution. Gram-positive (Staphylococcus aureus), gram-negative (Escherichia coli and *Pseudomonas aeruginosa*) bacteria and yeast-like fungi (Candida) were used as test cultures. Meat-peptone agar for bacteria and Saburo medium for fungi were used as the nutrient medium. In control experiments, the alcohol, phenol, and nitrofungin were used. The microbial load in all tests was 5×10^6 microbial body/ml.

The incubation time for bacteria was 18–24 hours at 37C, and for fungi – 1–2 days at 28°C, after which the minimum suppressing concentration (MPK) of the tested compounds was determined (Table 22.2).

As can be seen from Table 22.2, the compound (IV, VI) at MPK = 500 mcg/ml causes the death of Staphylococcus bacteria within 60 min., and the compound (VI) at MPK = 250 mcg/ml within 50 min. Both compounds exhibit high antimicrobial activity against Gram-negative bacteria for 30 min., and against *Pseudomonas aeruginosa* for 40 min. The compound (V) at the same concentration causes the death of Escherichia coli bacteria within 30 min., and *Pseudomonas aeruginosa* – within 20 min.

Thus, the organosilicon 2-methyl (ethyl)pyrroles exhibit expressed antimicrobial activity against *E. coli* and *Pseudomonas aeruginosa* bacteria.

22.3 CONCLUSIONS

The addition reaction of trialkyl (aryl)silanes to 2-methyl(ethyl)-1-alkyl pyrroles in the presence of Spier catalyst (platinohydrochloric acid) or rhodium acetylacetatedicarbonyl. It has been established the reaction proceeds on C = C-bond of allyl radical. It has been also revealed that the yield of the addition reaction in use of rhodium dicarbonylacetylacetonate as the catalyst is higher than in use of the Spier catalyst.

KEYWORDS

- **antimicrobial activity**
- ***Escherichia coli***
- **IR spectroscopy**
- **organosilicon**
- **pyrroles**
- ***Staphylococcus aureus***

REFERENCES

1. Ioffe, B. V., (1968). CHC, № 6, 1089.
2. *Certificate of Authorship*. USSR 152444 // BI. 1989. № 7.
3. Yuryev, V. P., & Salimqareeva, I. M., (1982). *Reaction of Hydrosilylation of Olefins* (p. 224). Nauka, Moscow.
4. Kazitsyna, A. A., & Kupletskaya, N. B., (2012). *Use of UV, IR and NMR Spectroscopy in Organic Chemistry* (p. 262). M.: Vysshaya shkola.
5. Gordon, A., & Fors, R., (1976). *Sputnik Khimika* (p. 541). M.: Mir.

CHAPTER 23

Condensed Phosphates as Analogs of Inorganic Polymeric Compounds, Geopolymers: The Bilateral Materials Reciprocal to Organic and/or Inorganic Polymers

MARINA AVALIANI,[1] ELENA SHAPAKIDZE,[2] VAJA CHAGELISHVILI,[1] NANA BARNOVI,[1] KETEVAN CHIKOVANI,[1] MARIAM VIBLIANI,[1] GULNARA TODRADZE,[2] and NANA ESAKIA[3]

[1]*R. Agladze Institute of Inorganic Chemistry and Electrochemistry, Ivane Javakhishvili Tbilisi State University, Tbilisi, Georgia*

[2]*Al. Tvalchrelidze Caucasian Institute of Mineral Resources, Tbilisi, Georgia*

[3]*Faculty of Exact and Natural Sciences, Ivane Javakhishvili Tbilisi State University, Tbilisi, Georgia*

ABSTRACT

Double-condensed phosphates linked to monovalent and polyvalent metals represent an important part of phosphate's chemistry. Throughout last many experimental studies we have been capable to synthesize a diverse new group of condensed phosphates, in fact new inorganic polymers in the systems $M^{I}_2O–M^{III}_2O_3–P_2O_5–H_2O$ (where M^I represents monovalent metals including Ag, Li, K, Na, Rb, Cs, and M^{III} signifies gallium, indium, scandium, and aluminum). It was established the crystallization ranges of various

Advanced Polymer Structures: Chemistry for Engineering Applications. Omar Mukbaniani, DSc, Tamara Tatrishvili, PhD & Marc Jean M. Abadie, DSc (Eds.)

condensed phosphates and the dependency of formation of crystalline phases VS molar ratio P_2O_5:$M^I{}_2O$:$M^{III}{}_2O_3$ is revealed. The elemental composition of the condensed compounds was established. The point of view to the properties and similarities of double compounds containing phosphorus, Ga, In, Sc, Al, and monovalent metals, as analogs of inorganic polymers is marked. An interesting approach to the geopolymers, as bilateral of organic and/or inorganic polymers is pronounced and declared. As it is well-known, geopolymers are in fact inorganic polymers consisting of repeating chains, such as silicon oxide (-Si-O-Si-O-), silicoaluminate (-Si-O-Al-O-), ferro-silicoaluminate (-Fe-O-Si-O-Al-O-) or alumophosphate (-Al-OPO-) chains formed through the polymerization process. In other words, georpolimers are the bilateral advanced materials, reciprocals of organic and/or inorganic polymers. Primarily the influence of inherent characteristics of volcanic rocks on the creation of geopolymeric binders' structure was studied.

23.1 INTRODUCTION

The worsening environmental situation in the world has pushed scientists to find optimal ways for the development of new technologies that will ensure appropriate sustainable development [1–5].

From a general point of view main properties of inorganic polymers, belief, do not distinctly differ from those of organic ones, which emphasizes the cohesion of the essential regularities determining the polymeric nature of a substance. For inorganic polymers it is specific the formation of crystalline polymeric groups with a regular three-dimensional structure of macromolecules. Such bodies can be considered as one giant macromolecule, all atoms of which are linked by covalent bonds.

However, the existence of a rigid framework of chemical bonds gives to polymeric compounds a remarkable solidity, and it is not by chance that they are the ones who are first in the list of materials in the hardness scale. Inorganic polymers are an appreciated source of new heat-resistant materials. There is an important progress in the development of new polymeric materials [1–5]. The cause seems to be due to the progressive growth of various researches in this domain and the increasing applications of such materials [6–11].

These suppositions are also valid for the chemistry of phosphorus compounds. This domain namely the chemistry of inorganic compounds of phosphorous: of condensed phosphates, (so called inorganic polymers) has progressed intensively in the last decades. This is not surprising – because condensed compounds of phosphorus are utmost relevant, convenient, and suitable to promote development of inorganic polymers' chemistry; and even more important – they are reasonably recognized as the best fertilizers, detergents, and materials used in engineering, construction, and other areas, such as raw materials for creation of phosphates glasses, thermo-resistant constituents, nanomaterials, effective applying nourishments, cleaners, cement substances, ion-exchange ingredients and also catalytic agents.

It's impossible not to name the famous scientist A. Durif and his works [1, 2, 7]. Since 1966, many French scientists, namely Professor André Durif and colleagues' works in fact are dedicated to the structural chemistry of condensed phosphates. With his collaborators he has been able in a few years to create and study the large area of phosphate chemistry: the characterization of more than 600 new compounds, a coherent classification of condensed phosphates based on the geometry of their anions, the classes of many compounds have been clearly established and a coherent nomenclature has also been adopted [1, 7]. This exceptional accumulation of results obtained by A. Durif Varambon [1] and his colleagues has led to valuable technological spin-offs such as rare earth ultra-phosphates as example of stoichiometric laser crystals; the discovery of one of the best crystals for non-linear quadratic optics; the preparation of a new ammonium phosphate-tellurate which is the best competitor of triglycine-sulfate due to its pyroelectric properties and the diagrams which set up a solid basis for the advance of retardant fertilizers. A. Durif executed the preparation of the only biodegradable substituents of asbestos for which they possess all the mechanical and thermal qualities. A. Durif and M.T. Averbuch-Pouchot studied the large domain of condensed compounds, namely cyclohexa-, cycloocta-, cyclodeca-, and dodecaphosphates [1, 2, 7].

In Georgia our team of university scientists has been working for inorganic synthesis to develop an environment-friendly chemistry of condensed compounds – the technology for used chemicals less harmful in the form of decreased wastes and not producing damaging outputs) [10–18]. These polymeric materials that we have obtained have been recognized and recorded in academic works of scientists, even outside our country [1, 5, 7, 9, 19]. Over the years we were conducting fundamental researches on M^{I}_2O-$M^{III}_2O_3$-P_2O_5-H_2O poly-component systems for purpose to obtaining double, triple, polymeric, cyclic, and substituted phosphates, we have synthesized and investigated at least 85–90 new inorganic polymers between temperature range 120°C–650°C (where M^{I}: alkali metals and M^{III}: Al, Ga, In, and Sc) [4, 15–18, 20, 21].

The choice of the optimal temperature range, as well as the initial components and their preliminary ratio are the main factors influencing the final results of the experiments, which allowed to obtain new numerous types of double condensed phosphates/so-called inorganic polymers with various programmed and predetermined properties.

In the construction sector, new materials have emerged as an appropriate and suitable alternative, the production of which is characterized by a lower degree of negative impact on the environment. One of these construction materials are geopolymers, which are considered as additional replacement for Portland cement because they are characterized by properties similar to those of the above-mentioned material, exactly such as high physical-mechanical properties and durability/sturdiness. At the same time, the production of geopolymers is not associated with high carbon dioxide emissions. As it is well-known, geopolymers are in fact inorganic polymers consisting of repeating chains, such as silicon oxide (-Si-O-Si-O-), silicoaluminate (-Si-O-Al-O-), ferrosilicoaluminate (-Fe-O-Si-O-Al-O-) or alumophosphate (-Al-OPO-) chains formed through the polymerization process. In other words, georpolimers are the bilateral advanced materials, reciprocals of organic and/or inorganic polymers. Primarily the influence of inherent characteristics of volcanic rocks on the creation of geopolymeric binders' structure was studied. Analyzing numerous scientific publications, we concluded that some condensed phosphates can improve the binding properties of certain materials. Considering this fact our idea for further research in the aim of searching diverse forms of additive binders is focused on the study of the possibility to obtain new geopolymer binders based on slag, metallurgical discharge, and phosphoric acid, and by adding ahead of time synthesized by us double condensed phosphates [25]. Although, we are conscious: it would be very interesting to study in details some interactions with taking part of double condensed phosphates containing glassy or crystalline phases in curing of geopolymeric binders. This objective consists basically of two-stage studies; these experiments are on their way to implementation. At this stage the opportunity of activation of crystallized slag, with double condensed phosphates of mono- and polyvalent metals joint thermo-processing was examined, as well as influence of different modes of mechanical-activation and curing on the reaction-ability of components of geopolymeric binders. Of course, first of all, it was compulsory to synthesize the necessary phosphates that has already been done.

23.2 EXPERIMENTAL METHODS AND MATERIALS

23.2.1 OBJECTS OF STUDY AND USED EXPERIMENTAL METHOD OF SYNTHESIS FROM SOLUTION-MELTS OF PHOSPHORIC ACIDS (CALLED TANANAEV'S METHOD)

The main goals of our work were as follows:

- To obtain the new inorganic polymers – many double oligo-, poly-, cyclo, and/or ultra-phosphates of monovalent and polyvalent metals from solution-melts of phosphoric acids at various temperature range by targeted chemical synthesis.
- Analysis by various methods and comprehensive study of experimental records.
- Determination and evaluation of the several properties of synthesized compounds.
- Examination of the correlation with the achievements and improvements in the area of inorganic polymer's chemistry.
- Conducting preliminary research studies for search of various forms of additive binders focused on the possibility to obtain new geopolymer binders based on slag, metallurgical discharge by adding phosphoric acid and what is the most important – with the involvement of double condensed phosphates containing vitreous phase in curing of geopolymeric binders. Or in other words – by adding of a new phosphates that we have synthesized to investigate the opportunity of activation of the crystallized slag, as well as the examination of the influence of different modes of mechanical activation and hardening on the capability of components of geopolymeric binders.

For obtaining of condensed inorganic polymers, the method (so-called Acad. I. V. Tananaev's method) of synthesis of double phosphates from solution-melts of phosphoric acids was applied. Characterization of appropriate technique: ortho-phosphoric acid of percentage 85%, trivalent metal oxide and monovalent metal salt were mixed in a glassy carbon vessel of approximately volume 50 mL at different molar ratios and heated to temperatures between 120°C and 650°C. In our experimental studies, the duration of the synthesis process was very variable, up 2–3 days to 15–28 days, depending on the temperature of the experiments. Having placed in a glassy carbon crucible and having mixed gallium oxide, or scandium oxide, or indium oxide, ortho-phosphoric acid (85%), and silver nitrate and/or alkali metal carbonates in different molar ratios the synthesis is started which lasts from 2 to 28 days. The melting solution must be mixed 3–4 times per 24 H. The crystalline phases formed in the melts were removed by extraction of 20–30 g samples with 500–700 mL of distilled water and after washing with alcohol and then with acetone.

23.2.2 GRAVIMETRIC ANALYSIS

In the course of the investigation of poly-component systems in the various temperature range we have synthesized numerous inorganic polymeric compounds – double condensed phosphates. Gallium, indium, and scandium were determined by the hydroxyquinoline precipitation method. Silver was determined using argentometry by means of Volhard titration method [16–18].

23.2.3 METHOD OF PAPER CHROMATOGRAPHY

It is mentioned in the scientific literature that condensed phosphates are quite stable and can therefore be well-classified by paper chromatography. Considering this reality, we used this method to determine the degree of condensation of obtained oligo- and polymeric phosphates. For this purpose, the crystalline samples were decomposed with the cationic (H-form) at about 0°C. After decomposition, the resulting solution was neutralized by $NaHCO_3$ and chromatographed on FN-11 paper using the acidic solvent. Then the chromatograms were treated with ammonium molybdate solution and irradiated by UV light [15–18, 20, 21].

23.2.4 OTHER METHODS: THERMO-GRAVIMETRIC ANALYSIS (TGA), SCANNING MICRO-SPECTROSCOPIC ANALYSIS AND ROENTGEN PHASE'S ANALYSIS

Above-mentioned techniques together with the chemical analysis, IR spectroscopy, thermo-gravimetric analysis (TGA), X-ray diffraction measurement, structural analysis was used by us to study the process of formation and composition of many normal, basic, and/or acid of both simple and double di-, tri-, tetra-, octa-, and dodecaphosphates of polyvalent metals [13–18,

20–22]. For thermogravimetric analysis (TGA) it was used a Derivatograph Q1500-D with a heating rate of 10 degree/min., in air atmosphere and maximum temperature of 1,280°C (sometimes to 1,600°C). Scanning electronic microscope measurements were performed on a JEOL scanning electronic microscope JSM-6510LV (well-appointed by energy-disperse X–max N 20 micro-X-ray spectral analyzer produced by Oxford Instruments). SEM measurements were carried out by means of reflected (BES) as well as secondary (SEI) electrons at an accelerating voltage (at 20 kV). The employed distance was approximately 15 mm. Micrographs have been taken at the diverse enlargements. Micro-spectroscopic analysis was performed from the sampling point zones and its surface area. The structural organization of synthesized double condensed compounds is determined by X-ray diffraction method and IR spectroscopy. X-ray phase analysis was carried out on the DRON–3M diffractometer (Cu–Kα radiation) [16–18, 20, 21].

23.2.5 MATERIALS

The initial components were oxides of gallium, indium, scandium, and occasionally-Al, silver nitrate, sodium carbonate of monovalent (alkali) metals and ortho-phosphoric acid. However, mostly the synthesis temperature range is high enough, we generally present the studied system as oxides of appropriate elements, such as $M^{III}_2O_3$, M^I_2O, and P_2O_5, H_2O. The variation of the molar ratio of initial components $n_m = M^I/M^{III}$ was very large: from 1.0 to 12.0, but constant for each experiment – for each temperature range. The molar ratio of phosphorus to trivalent metal was always the same during all synthesis processes and was equal to $n_p = P/M^{III} = 15.0$; (Every now and then, more volume of phosphoric acid was taken during some experiments – up 16.0 to 18.0).

23.3 RESULTS AND DISCUSSION

23.3.1 DETAILED STUDY OF POLY-COMPONENT SYSTEMS CONTAINING SOME MONOVALENT CATIONS IN THE PRESENCE OF TRIVALENT METALS

Trivalent metal oxide, ortho-phosphoric acid with 85% and monovalent metal salt – silver nitrate and/or carbonate of alkali metals – were mixed in a glassy carbon container (crucible) at different molar ratios and heated to temperatures from 100–120°C to 620–650°C. The duration of the synthesis process was very variable, from 3–5 days to 15–28 days, depending on the temperature and the main goal of the experiments. The crystals formed in the solution-melts were removed by extraction of 20–30 g samples with 700–750 mL of distilled water, and after were rinsed with ethanol and ether. Table 23.1 shows the dependence of the composition of the samples synthesized in the M^I_2O-$M^{III}_2O_3$-P_2O_5-H_2O system on the synthesis temperature T and on the initial molar ratio of monovalent and trivalent metals n = Ag_2O/Sc_2O_3.

TABLE 23.1 Dependence of Composition from Temperature *T* and Molar Ratio *n* = Ag_2O/Sc_2O_3 = 1.5–2.5, 3.5–5.0, 6.0–7.5, and 8.0–10.0

T (°C)	n = 1.5–2.5	n = 3.5–5.0	n = 6.0–7.5	n = 8.0–10.0
130–150	$Sc(PO_3)_3$–C	$AgScHP_3O_{10}$	$AgSc(H_2P_2O_7)_2$	$AgSc(H_2P_2O_7)_2$
180–200	$Sc(PO_3)_3$–C	$AgScHP_3O_{10}$	$AgScHP_3O_{10}$	$AgScHP_3O_{10}$
220–240	$Sc(PO_3)_3$–C	$AgScP_4O_{12}$	$AgScP_4O_{12}$	$AgScP_4O_{12}$
310–335	$Sc(PO_3)_3$–C^I	$AgScP_4O_{12}$, $AgScHP_3O_{10}$	$AgScP_4O_{12}$	$AgScP_4O_{12}$
340–355	$Sc(PO_3)_3$–C^I	$AgScP_4O_{12}$	$AgScP_4O_{12}$ + $Ag_3Sc_3P_{12}O_{36}$	$AgScP_4O_{12}$
400–450	$Sc(PO_3)_3$–C^I	$AgScHP_3O_{10}$	$Ag_3Sc_3P_{12}O_{36}$	$Ag_2ScP_3O_{10}$
500–550	$Sc(PO_3)_3$–C^I	$Ag_2ScP_3O_{10}$	$Ag_3Sc_3P_{12}O_{36}$	$AgScP_2O_7$
600–630	$Sc(PO_3)_3$–C^I	$Sc(PO_3)_3$–C^I + mixt phases	Mixture of phases	$AgScP_2O_7$

Various acid and normal triphosphates, double acid diphosphates and some ultra-phosphates were synthesized by us in the systems Na^I_2O-$M^{III}_2O_3$-P_2O_5-H_2O (M^{III} = Ga, In, and Sc) at different temperature range. The main obtained compounds formulas are shown in Table 23.2: double acidic triphosphates $M^IM^{III}HP_3O_{10}$, (forms I, II, III), $M^IM^{III}HP_3O_{10}H_2O$, double normal triphosphates $M^I_2M^{III}P_3O_{10}$, double acidic diphosphates $M^I_2M^{III}H_3(P_2O_7)_2$, $M^IM^{III}(H_2P_2O_7)_2$, crystalline hydrate of double acidic diphosphate $M^IM^{III}(H_2P_2O_7)_2 2H_2O$, double diphosphate $M^IM^{III}P_2O_7$, long chain double polyphosphates with general formula $[M^IM^{III}(PO_3)_4]_x$, ultra-phosphate $M^I_3M^{III}P_8O_{23}$, cyclic octaphosphates $M^I_2M_2^{III}P_8O_{24}$, cyclic dode-caphosphates $M^I_3M^{III}_3P_{12}O_{36}$, diverse long chain polyphosphates $M^{III}(P0_3)_3$ form (A), forms (C) and (C^I) [10, 12–21].

TABLE 23.2 Main Classes of Synthesized Inorganic Polymers – Condensed Compounds of Phosphorus in the Systems Na^I_2O-$M^{III}_2O_3$-P_2O_5-H_2O

Various Double Acid Triphosphates	
$M^IM^{III}HP_3O_{10n}$	$M^IM^{III}P_3O_{10n}$
Double Acid Diphosphates $M^I_2M^{III}H_3(P_2O_7)_2$	
Double Normal Diphosphates $M^IM^{III}P_2O_7$	
Double Cyclotetraphosphates $M^IM^{III}P_4O_{12}$	
Double Cyclododecaphosphates $M_3^I\ M_3^{III}P_{12}O_{36}$	
Long Chain Polyphosphates $[M^IM^{III}(PO_3)_4]_x$ **Forms A, C, and C^I^**	
Ultra-phosphates $M^I_3M^{III}P_8O_{23}$	
Normal (Middle) Triphosphates $M^{III}H_2P_3O_{10}$	

As we have already noted numerous synthesized condensed compounds were wholly studied by chemical analysis, also are observed by X-ray structural techniques. Some of the data have been presented in our recent papers [13, 16, 17, 20, 21].

23.3.2 STUDY OF SYNTHESIZED DOUBLE CONDENSED PHOSPHATES OF GALLIUM AND SCANDIUM BY MICRO-ROENTGEN-SPECTROSCOPIC ANALYSIS

We've taken a lot of measurements using a scanning electronic microscope of the company JEOL equipped with a scanning electronic technic JSM-6510LV, which in turn was well-appointed by energy-disperse micro-roentgen spectral analyzer. Electronic micrographs were carried out by means of reflected (BES) and as well as secondary (SEI) electrons at an accelerating voltage (at 20 kV). The working distance was approximately 15 mm; micrographs has been taken at the diverse enlargements. Micro-spectroscopic analysis was performed from the sampling point zones and its surface area.

Some interesting results of mentioned crystals are shown on the Figures 23.1–23.4. Electron images have been taken at the diverse enlargements. In general very small crystals of diverse forms (prisms, dipyramides, trigonal trapezoids, trigonal scalenohedres, and/or tetragonal scalenohedrals, lamellar, and needle-like shape with a thickness about of 90–100 nm) were observed.

Figure 23.1 characterize for condensed compound-triphosphate, obtained on molar ratio n = Ag_20/Sc_2O_3 = 5 at the synthesis' temperature 160°C (at enlargement x2700).

Electron image (SEI) presented in Figure 23.1 illustrate condensed compound: double triphosphate of argentum-scandium, obtained on molar ratio n = Ag_20/Sc_2O_3 = 5 at the synthesis' temperature 160°C (enlargement x5500).

Electron images (BES) for condensed compound-double acid diphosphate, obtained on molar ratio n = Ag_20/Sc_2O_3 = 7.5 at the synthesis' temperature 160°C (enlargements: 550 and 1,300, respectively) are presented in the Figure 23.3 and 23.4.

FIGURE 23.1 Electron image (SEI) for condensed compound-triphosphate, obtained on molar ratio n = Ag_20/Sc_2O_3 = 5 at the synthesis' temperature 160°C (enlargement x2700).

FIGURE 23.2 Electron image (SEI) for condensed compound-triphosphate, obtained on molar ratio n = Ag_20/Sc_2O_3 = 5 at the synthesis' temperature 160°C (enlargement x5500).

FIGURE 23.3 Electron image (BES) for condensed compound-double acid diphosphate, obtained on molar ratio n = Ag_20/Sc_2O_3 = 7.5 at the synthesis' temperature 160°C (enlargement x550).

FIGURE 23.4 Electron image (BES) for condensed compound-double acid diphosphate, obtained on molar ratio n = Ag_20/Sc_2O_3 = 7.5 at the synthesis' temperature 160°C (enlargement x1300).

Interesting results are found for compounds obtained at the synthesis temperature 330–335°C. Figures 23.5 and 23.6 illustrate electron images (on enlargements 1,000 and 2,700 correspondingly) for silver scandium cyclotetraphosphate. This polymeric compound was obtained on molar ratio n = Ag_20/Sc_2O_3 = 7.5 and at a temperature of synthesis above 328–330°C.

Furthermore, if we analyze this micrograph and XRD data in detail, we can perceive the beginning of phosphate's formation with a higher condensation rate. By roentgen phase' analysis this other mix phase is just cyclododecaphosphate. In Figure 23.6, we can see the formation of crystalline structure similar approximately to trigonal scalenohedron.

FIGURE 23.5 Electron image (SEI) for condensed compound-double cyclotetraphosphate, obtained on molar ratio n = Ag_20/Sc_2O_3 = 7.5 at the synthesis' temperature 330–335C (enlargement x1000).

FIGURE 23.6 Electron image (SEI) for condensed compound-double cyclotetraphosphate, obtained on molar ratio n = Ag_2O/Sc_2O_3 = 7.5 at the synthesis' temperature 330–335°C (enlargement x2700).

Figures 23.7 and 23.8 presented show the formation of cyclic crystals: dodeca-phosphate of silver-scandium $Ag_3Sc_3P_{12}O_{36}$ and represent SEI micrographs for enlargements 1,000 and 2,700, respectively.

FIGURE 23.7 Electron (SEI) image of cyclo-dodecaphosphate $Ag_3Sc_3P_{12}O_{36}$ (enlargement 1000).

FIGURE 23.8 Electron (SEI) image of cyclo-dodecaphosphate $Ag_3Sc_3P_{12}O_{36}$ (enlargement 2700).

At the same temperature, the situation is slightly different for the system containing gallium. Very detailed XRD analysis of the inorganic polymers formed in the system $Ag_2O–Ga_2O_3–P_2O_5–H_2O$ permits to determine that the product obtained at temperature 330–335°C and for initial molar ratio n=Ag/Ga=7.5 contains two phases such as silver gallium cyclo-dodecaphosphate $Ag_3Ga_3P_{12}O_{36}$ (as primary phase) and silver-gallium cyclo-tetraphosphate $AgGaP_4O_{12}$. Figure 23.9 shows the electron image for silver-gallium cyclo-dodecaphosphate with a small mix phase – the minor quantity of silver-gallium tetra-phosphate [16, 21].

During our experiments it was revealed that optimum performance for the realization of big cyclic anions is the correlation of the big monovalent cations versus trivalent metals with a comparatively small ionic radius [15–18, 20, 21].

Data for one of the spectrums of the silver-gallium tetraphosphate $AgGaP_4O_{12}$ synthesized at molar ratio n=Ag/Ga = 5.0 is offered in Table 23.3.

FIGURE 23.9 Electron (SEI) image of cyclo-dodecaphosphate $Ag_3Ga_3P_{12}O_{36}$ (enlargement 1000).

Very noticeable and attention-grabbing SEM picture is presented in Figure 23.10. Around the great crystals similar approximately to ditetragonal and/or dihexagonal prisms very small crystals of lamellar forms about of 100 nm are observed.

FIGURE 23.10 Electron (SEI) image of cyclo-dodecaphosphate $Ag_3Ga_3P_{12}O_{36}$ (enlargement 2700).

TABLE 23.3 Data for Spectrum 22, Site 1, Molar Ratio n = Ag/Ga = 5.0, Synthesis at 335°C, Approximate Formula $AgGaP_4O_{12}$

Atom	Atomic Percentage
O	73.41
P	14.71
Ga	4.07
Ag	7.81

Some EDAX data and one of the spectrums of obtained by us cyclo-dodecaphosphate $Ag_3Sc_3P_{12}O_{36}$ can be seen in Figure 23.11.

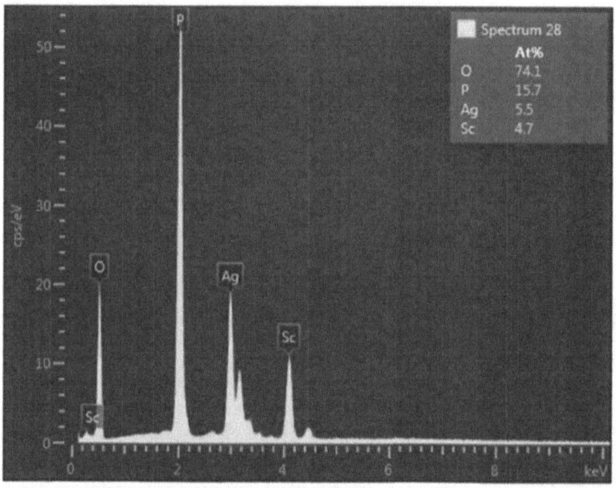

Result Type	Atomic %			
Spectrum Label	Spectrum 28	Spectrum 25	Spectrum 26	Spectrum 27
O	74.08	74.15	73.92	73.81
P	15.67	14.07	13.82	13.38
Sc	4.73	3.29	2.78	1.45
Ag	5.52	8.49	9.47	11.36
Total	100.00	100.00	100.00	100.00

FIGURE 23.11 The EDAX results: Spectrum and atomic percentage data vs. spectrum labels for crystals obtained at 330–335C in the system containing Ag and Sc for initial ratio $n = M^I/M^{III} = 7.5$.

This double compound – ciclo-dodecaphosphate of silver-scandium $Ag_3Sc_3P_{12}O_{36}$ was crystallized at n=Ag/Sc = 7.5 (synthesis at 330–335°C). Very small crystals around of 95–100 nm were observed.

The cyclic phosphates of gallium-silver and scandium-silver are isomorphs among themselves and are iso-structural with the sodium-gallium and sodium-indium cyclo-phosphates [1, 16, 21].

Assumed our experiences with the involvement of monovalent and trivalent metals, such as Ga, In, Sc, and partially – Al we can conclude that for obtaining condensed compounds with large cycles it is optimal the correlation of big monovalent cations with respect to trivalent metals having the ionic radius relatively less [15–18, 20, 21].

23.3.3 STUDIES OF GEOPOLYMERS AS AN INNOVATIVE AND BILATERAL MATERIALS

It is no secret that polymers, in the general sense of the term, are a very broad concept. Real polymers were discovered in the middle of the 19th century, but it is only in the 20th century that scientific research on the structure of polymers started. The term geopolymer, coined by the I. Davidowitz [22, 23], appeared at the end of the 20th century. This term designates a material called inorganic polymer obtained by activation of acid or alkaline components of an aluminosilicate or other raw materials. It can be not only aluminosilicate rocks, metakaolin, but also slags, fly ash, etc. [24–36]. In other words, geopolymers are an analog of organic and/or inorganic polymers. One could even say that it is a bilateral material. Instead of petroleum derivatives and various compounds, containing carbon chains, some mineral material is used, for example, constituted by silicon dioxide and aluminum oxide (or even industrial wastes of anthropomorphic origin) and also the mineral binder. Their polymerization can take place in air temperature or at relatively high temperatures, i.e., between 20°C and 120°C [22–25]. This geosynthesis makes it possible to obtain materials whose properties are partly similar to those of plastics (moldable, more or less extrudable/pressed materials), but above all, without the use of dangerous solvents, which is essential – they are non-flammable, even quite fireproof and do not emit gases, vapors or toxic gaseous components. Like natural stone, they are resistant to chemical influences as well as to erosion and allow for environmentally friendly production [37–46].

However, it should be noted that the behavior of geopolymers during their curing and the subsequent nature of their possible transformation (or non-transformation) during the process of their application are insufficiently studied.

We should not forget that in the modern world, the production of new ecological and economically attractive materials from local industrial wastes is of the major and even vital importance [22–26, 46–49]. Generally in the

21st century the fire-resistant and corrosion-resistant materials are of the utmost importance [44, 50].

The above-mentioned geopolymers are innovative materials and are progressively becoming an alternative – partial or complete – to the so-called "Portland" cement in concrete production. This production is currently considered as one of the main causes of the degradation of the environmental system and can destroy the model of Sustainable Development [23, 38–42].

In general, it can be said that the development of technologies in the field of magnetite-based geopolymers opens wide perspectives for the utilization of iron-rich industrial wastes, such as metallurgical slag, which will have a long-term effect and of course will also serve to improve the current rather worrying ecological situation. Currently, different types of types of rocks are generally used as additives. Based on the above, one of the aims of our work is to investigate the possibility of producing new geopolymer binders based on metallurgical slag and phosphoric acid.

Under laboratory conditions, we have obtained porous heat-insulating geopolymeric materials [25, 34–37, 40]. By selecting the ratio between slag and sand, it is possible to adjust the properties of the binder, such as porosity and mechanical strength. To specify in detail: thermally insulating porous geopolymeric materials were obtained by adding phosphoric acid to the composition of slag waste and sand [39–41].

In our experiments, the interaction of waste slag and phosphoric acid (of 85% concentration) was accompanied by intense heat release and rapid setting, which resulted from the ongoing reactions between magnetite and phosphoric acid, making molding impossible.

The size of the molded samples was 20 × 20 × 20 mm, they were demolded on the 3rd day after casting and heat treated at 80°C for 20 hours. After cooling, they were tested in the press. In order to avoid rapid water evaporation, the samples in the molds were wrapped with polyethylene for three days. The polyethylene was then removed, and the model specimens were heat treated as described above. In addition to durability and resistance, the materials were tested for thermal conductivity. We performed a series of experiments with different proportions of the initial components and with variations in the heat treatment regime [40–42].

From the result of these experiments, it can be concluded that the porous geopolymeric thermal insulation materials obtained by adding H_3PO_4 to the waste slag + sand and selecting their initial ratios it is possible regulate/control the binder's properties, such as the porosity and hardness. The development of magnetite-based geopolymer technology opens up a wide choice of possibilities for the use of iron-rich industrial wastes, such as metallurgical slag, which would also contribute to the improvement of the ecological situation.

23.3.4 GEOPOLYMERS' APPLICATIONS

Referring to the well-known scientists Tananaev [11]; and Wagh [27], chemically bounded phosphates are the real materials of the 21st century. Nevertheless, the development of their processing technology still lacks the necessary attention.

Advances in magnetite-based geopolymer's technology opens up wide prospects for the utilization of iron-rich industrial wastes, such as metallurgical slag, which will also contribute to the improvement of the current environmental situation.

Geopolymers – phosphated cements – are essentially two-component systems consisting of a reinforcing/hardening fluid and a solid phase. Their mixture results in a ceramic material with a three-dimensional lattice structure, which is strong, durable, and fire resistant. Since phosphate cements do not form an alkaline zone in which glass and basalt fibers are unstable, they create excellent preconditions for the production of textile-reinforced concrete based on phosphate cements. Mineralogical analysis of the dumped slag [39–41] showed that most of it contained silicate groups of the general formula $Am(X_2O_7)n$ (60–80 wt.%), quartz (up to 6 wt.%), $CaCO_3$ (10 wt.%), iron, manganese, and calcium sulfides (up to 5 wt.%), magnetite $FeO\text{-}Fe_2O_3$ (up to 5 wt.%), metal ions and aqueous compounds. The following silicates are also mentioned in the literature such as: Larnite, Merwinite, Monticellite, Ockermanite, etc.

The oil reserves are just as depleted as the coal reserves, and this is the basis for supplementing these conventional systems with polymers that contain inorganic components. Thus, there is no doubt that our future search for new types of polymers – in many combinations containing inorganic elements – is a major defy and even a challenge for modern science. Hence the great interest of researchers all over the world in the synthesis of new inorganic polymers and to their implementation on the geopolymers processing.

The marketable requests for geopolymers' applications: the production of fire-retardant and heat-resistant coatings and sealants, in medicine, in high temperature ceramics, for the encapsulation of new types of fibrous or toxic, and radioactive waste components, and as a component of cement and concrete [40–46]. By citing Dr. Terry Gourley – “Geopolymers or alumino-silicate inorganic polymers are a family of materials that only in the last few decades, have been given their rightful place in the spectrum of inorganic materials…and place them within the general family of inorganic polymers” [47].

23.4 CONCLUSIONS

We have conducted the fundamental researches of poly-component systems $M^{I}_{2}O$-$M^{III}_{2}O_{3}$-$P_{2}O_{5}$-$H_{2}O$ for the purpose to obtaining double, triple, polymeric, and substituted phosphates, we have synthesized and investigated at least 87 new inorganic polymers, between temperature range 130°C–650C (where M^{I}: alkali metals and M^{III}–AL, Ga, In, and Sc).

It can be declared and concluded that the choice of the optimal temperature range, as well as the initial components and their preliminary ratio are the main factors influencing the final results of the experiments, which allowed to obtain new various and different types of double condensed phosphates/inorganic polymers with some programmed and predetermined properties.

In order to study the influence of trivalent and monovalent cations on the formation of inorganic polymer’s anionic radicals and the level of condensation, we have studied multicomponent systems by various methods, namely chemical analysis, roentgen phase analysis, TGA, scanning electronic microscope measurements, etc. Overall reliance of essential structural composition and stability of double condensed phosphates from ion radius of M^{I} – have also been examined.

The study to obtain the wanted geopolimeric materials as bilateral compounds have been taken. Heat-insulating porous geopolymer materials were obtained under laboratory conditions. Studies of the possibilities of obtaining new geopolymer binders on the basis of industrial waste and phosphoric acid were conducted. When selecting the ratio of dump slag to sand, it is possible to regulate the properties of the binder, for example, porosity, and mechanical strength.

It was revealed that it’s possible to outline the prospects for the production of geopolymer binders based on local rocks as an alternative to PC.

Some condensed phosphates can improve the binding properties of certain materials. The first steps to obtain the desired materials have been taken. Considering this fact second steps for further study is searching for diverse forms of additive binders focused on the possibility of obtaining new geopolymer binders based on slag, metallurgical discharge, and phosphoric acid by adding double condensed phosphates already synthesized by us.

ACKNOWLEDGMENTS

Our deepest gratitude to Professor André Durif (France) for his major contribution in the chemistry of condensed phosphates and for his involvement in the classification of condensed compounds obtained by our team. We would also like to express our sincere recognition and appreciation for the review and systematization of the numerous inorganic polymers synthesized by us.

Our special thanks to Dr. Temuri Berikashvili (Georgian Technical University) for his significant advice.

KEYWORDS

- **activation**
- **condensed phosphates**
- **crystallized slag**
- **cyclo-phosphate**
- **geopolymer**
- **inorganic polymer**
- **multi-component systems**
- **reaction-ability**

REFERENCES

1. Durif, A., (2014). *Crystal Chemistry of Condensed Phosphates* (3rd edn., p. 415). Plenum Press.
2. Ribero, D., & Kriven, W. M., (2015). Synthesis of LiFePO4 powder by the organic–inorganic steric entrapment method. *J. Mater. Res., 30,* 2133–2143.
3. Averbuch-Pouchot, M. Th., & Durif, A., (1993). Préparation chimique et structure cristalline d'une nouvelle variété de monohydrogéno-monophosphate d'éthylène-diammonium: $[C_2N_2H_{10}]HPO_4$ (in french). *Acta Cryst. B: Structural Science, Crystal Engineering and Materials*, 45–93.

4. Avaliani, M. A., (2021). Investigation and thermal behavior of double condensed phosphates of gallium, scandium and silver. *Bull. of Intern. Nucl. Inform. Syst. INIS-IAEA, 52*, 31, 32.
5. Murashova, E. V., & Chudinova, N. N., (2001). Double condensed phosphates of cesium-indium. *J. Inorg. Mater., 37*(12), 1521–1524.
6. Mukbaniani, O., Londaridze, L., Markarashvili, E., Tatrishvili, T., & Aneli, J., (2021). Triethoxysilylated styrene as a new coupling agent in wood composites (symposium speech). *Proceedings of the 7th International Caucasian Symposium on Polymers and Advanced Materials – "ICSP&AM 7"* (p. 73).
7. Averbouch-Pouchot, M. Th., & Durif, A., (1996). *Topics in Phosphate Chemistry* (pp. 174–209). World Scientific.
8. Kvinikadze, N., Londaridze, L., Zedgenidze, A., Dzidziguri, D., & Mukbaniani, O., (2021). Wood polymer composites on the basis of new coupling agent. *Proceedings of the 7th International Caucasian Symposium on Polymers and Advanced Materials – "ICSP&AM 7"* (p. 60).
9. Strutinska, N. Y., Zatovsky, I., & Ogorodnyk, O. V., (2013). Slobodyannik, N., Rietveld refinement of $AgCa_{10}(PO_4)_7$ from x-ray powder data. *Acta Crystallogr. Sect. E Struct Rep Online, 69*, 23–25.
10. Avaliani, M. A., Tananaev, I. V., & Gvelesiani, M. K., (2003). Synthesis and investigation of double condensed phosphates of scandium and Akali metals; Abstracts for synthesis chemists, FIZ Chemie Berlin. *J. Phosphorus, Sulfur Silicon & Relat. Elem., 51.*
11. Grunze, I., Cudinova, N. N., Palkina, K. K., Avaliani, M. A., Guzeeva, L. S., & Maksimova, S., (2009). Structure and thermal rearrangements of binary cesium-gallium phosphates. *Energy Citations Database. J. Inorg. Mater., 23*(4), 539–544.
12. Avaliani, M. A., & Gvelesiani, M. K., (2006). Areas of crystallization of condensed scandium and cesium phosphates and regularities of their formation, *J. Proc. Georgian Acad. Sci. Chem. Ser., 32*, 52–58.
13. Avaliani, M. A., Todradze, G. Shapakidze, Ukleba, M., Chikovani, N., Vibliani, M., & Magradze, G., (2020). Optimization of the gravimetric method for determination of trivalent metals using oxiquinoline. *Proceed. Intern. Conf. on Analyt. Chem.* Modern Trends 2020, Kyiv, Ukraune. http://kcacmt.univ.kiev.ua/en (accessed on 02 January 2022).
14. Avaliani, M., Gvelesiani, M., Barnovi, N., Purtseladze, B., & Dzanashvili, D., (2016). New investigations of poly-component systems. *J. Proc. Georgian Acad. Sci. Chem. Ser., 42*, 308–311.
15. Avaliani, M., (2016). Main types of condensed phosphates synthesized in open systems from solution-melts of phosphoric acids. *Nano Studies, 1*, 135–138.
16. Avaliani, M., Chagelishvili, V., Shapakidze, E., Gvelesiani, M., Barnovi, N., Kveselava, V., & Esakia, N., (2019). Crystallization fields of condensed scandium-silver and gallium-silver phosphates. *Eur. Chem. Bull., 5*, 164–170.
17. Avaliani, M., Shapakidze, E., Barnovi, N., Dznashvili, D., Todradze, G., Kveselava, V., & Gongadze, N., (2019). Regio-controlled synthesis of double condensed oligo-, poly- and cyclo-phosphates, their characterization and possibilities of applications, due to their solid-state properties. *J. Nano Studies-Eur. Chem. Bull., II, 19*, 273–284.
18. Avaliani, M., Shapakidze, E., Barnovi, N., Gvelesiani, M., & Dzanashvili, D., (2017). About new inorganic polymers-double condensed phosphates of silver and trivalent metals. *J. Chem. Chem. Eng. (USA), 11*, 60–64.

19. Zanello, P., (2012). *Chains, Clusters, Inclusion Compounds, Paramagnetic Labels* (2nd edn., pp. 100–141). Elsevier.
20. Avaliani, M., Todradze, G., Shapakidze, E., & Kveselava, V., (2020). Synthesis of condensed phosphates of mono- and polyvalent metals, development of the optimal methods for the study of their properties and composition. *J. Nano Studies-Eur. Chem. Bull., 20,* 71–94.
21. Avaliani, M., Chagelishvili, V., Barnovi, N., Shapakidze, E., & Esakia, N., (2021). Condensed phosphates: New inorganic polymers with a variety of applications and improvement of their gravimetric determination methods. *Apple Acad. Press, Adv. Mater., Polymers and Composites; II; Chapt., 18,* 255–276.
22. Davidovits, J., (2015). *Geopolymers Chemistry and Applications* (4th edn., p. 644). Saint-Quentin, France: Geopolymer Institute.
23. Shapakidze, E., Nadirashvili, M., Maisuradze, V., Avaliani, M., Gejadze, I., & Petriashvili, T., (2021). Development of compositions of geopolymer binders based on rocks of Georgia. *Proceedings of 6th International Conference "Nanotechnology" (GTU-Nano20).* Tbilisi, Georgia.
24. Duxson, P., Fernandez-Jimenez, A., Provis, J. L., Lukey, G. C., Palomo, A., & Van, D. J. S., (2007). Geopolymer technology: The current state of the art. *J. Mater. Sci., 42,* 2917–2933.
25. Davidovits, J., (2017). Geopolymers: Ceramic-like inorganic polymers. *J. Ceram. Sci. Technol., 8,* 335–350.
26. Davidovits, J., (2002). 30 years of successes and failures in geopolymer applications. Market trends and potential breakthroughs. *Proceedings of Geopolymer Conference* (pp. 1–9). Saint-Quentin, France.
27. Wagh, A. S., (2005). Chemically bonded phosphate ceramics – a novel class of geopolymers. *Ceram Trans., 165,* 107–116.
28. Gualtieri, M. L., Romagnolia, M., & Gualtieri, A. F., (2015). Preparation of phosphoric acid-based geopolymer foams using limestone aspore forming agent – thermal properties by in situ XRPD and Rietveld refinements. *J. Eur. Ceram. Soc.,* 215.
29. Gualtieri, M., Romagnoli, Pollastri, S., & Gualtieri, A. F., (2015). Inorganic polymers from laterite using activation with phosphoric acid and alkaline sodium silicate solution: Mechanical and microstructural properties. *Cem. Concr. Res., 67,* 259–270.
30. Perera, D. S., Hanna, J. V., Davis, J., Blackford, M. G., Latella, B. A., Sasaki, Y., & Vance, E. R., (2008). Relative strengths of phosphoric acid-reacted and alkali-reacted metakaolin materials. *J. Mater. Sci., 43,* 6562–6566.
31. Liu, L. P. X. M., Cui, X., Qiu, S., Yu, J. L., & Zhang, L., (2010). Preparation of phosphoric acid-based porous geopolymers. *J. Appl Clay Sci., 50,* 600–603.
32. Shaqouri, F., Ismeik, M., & Esa, I. M., (2017). Alkali activation of natural clay using a $Ca(OH)_2/Na_2CO_3$ alkaline mixture. *Clay Minerals, 52,* 485–496.
33. Celerier, H., Jouin, J., Mathivet, V., Tessier-Doyen, N., & Rossignol, S., (2018). Composition and properties of phosphoric acid-based geopolymers. *J. Non. Cryst. Solids, 493,* 94–98.
34. Shapakidze, E., Nadirashvili, M., Maisuradze, V., Gejadze, I., Avaliani, M., & Petriashvili, T., (2019). Geopolymers – the future alternative to Portland cement. *Proceedings 6th International Caucasian Symposium on Polymers and Advanced Materials* (p. 219). Batumi, Georgia.

35. Avaliani, M., Gvelesiani, M., Barnovi, N., Purtseladze, B., & Dzanashvili, D., (2016). New investigations of poly-component systems. *J. Proc. Georgian Acad. Sci. Chem. Ser., 42,* 308–311.
36. Shapakidze, E., Nadirashvili, M., Maisuradze, V., Gejadze, I., Avaliani, M., & Todradze, G., (2019). Elaboration of the optimal mode for heat treatment of shales for obtaining metakaolin. *J. Eur. Chem. Bull., 8,* 31–33.
37. Shapakidze, E., Avaliani, M., Nadirashvili, M., Maisuradze, V., Gejadze, I., & Petriashvili, T., (2021). Geopolymers based on local rocks as a future alternative to Portland cement. *Materials Science, Composite Materials Engineering, Modeling and Technology* (p. 351–358). Apple Academic Press, USA, part III, Chapter 25.
38. Avaliani, M., Shapakidze, E., Chagelishvili, V., & Todradze, G., (2021). Investigating the influence of the trivalent and monovalent metals ionic radius on the structure of the synthesized double condensed compounds. *Proceedings of IX International Conference on Chemistry and Chemical Education-Sviridov Readings* (pp. 109, 110). Minsk.
39. Aneli, J., Zaikov, G., & Mukbaniani, O., (2011). Physical principles of the conductivity of electrical conducting polymer composites: A review. *Chem. & Chem. Technol., 5,* 75–87.
40. Shapakidze, E., Avaliani, M., Nadirashvili, M., Maisuradze, V., Gejadze, I., & Petriashvili, T., (2020). Obtaining of geopolymer binders based on thermally modified clay rocks of Georgia. *Nano Studies, 20,* 43–52.
41. Shapakidze, E., Avaliani, M., Nadirashvili, M., Maisuradze, V., Gejadze, I., & Petriashvili, T., (2020). Study of the possibility of obtaining new geopolymer binders on the basis of industrial waste and phosphoric acid. *Nano Studies, 20,* 53–64.
42. Avaliani, M., Shapakidze, E., Chagelishvili, V., Gvelesiani, M., Barnovi, N., Vibliani, M., & Chikovani, K., (2020). Properties of the synthesized inorganic polymeric phosphate materials and the possibilities for production of new environmentally friendly and economically viable supplies from local industrial waste. *Nano Studies-Eur. Chem. Bull., 20,* 63–70.
43. Gourley, J. T., & Johnson, G. B., (2005). Developments in precast geopolymer concrete. In: Davidovits, J., (ed.), *Proceedings of 4th International Geopolymer Conference, Saint-Quentin, France, Geopolymer Workshop* (pp. 139–143). Perth, Australia, Geopolymer Institute.
44. Gourley, J. T., & Johnson, G. B., (2019). The corrosion resistance of geopolymer concrete sewer pipe. *Concrete in Australia, 43,* 39–44.
45. Gourley, J. T., (2014). Geopolymers in Australia. *J. Austr. Ceram. Soc., 50*(1), 102–111.
46. Avaliani, M., (2019). Some innovative results-oriented scientific researches which lead to the development in the field of inorganic polymer's science (In: *Materials Science/ Composite Materials Engineering: Modeling and Technology* (pp. 61–72). Apple Academic Press. doi: 10.13140/RG.2.2.16514.32969.
47. Gourley, T., (2020). Geopolymer cement – environmental considerations. *Geopolymer Alliance, Research Gate, Project "Geopolymers,"* p. 15.
48. Kriven, W. M., (2021). Geopolymers and geopolymer-derived composites. In: *Reference Module in Materials Science and Materials Engineering.*
49. Katsiki, A., Hertel, T., Tysmans, T., Pontikes, Y., & Rahier, H., (2019). Metakaolinite phosphate cementitious matrix: Inorganic polymer obtained by acidic activation. *J. Materials, 12*(3), 442.

50. Brostow, W., Cañadas, I., Fałtynowicz, H., Levinskas, R., & Garcia, J. R., (2021). Fire resistance of materials. *Proceedings of the 7th International Caucasian Symposium on Polymers and Advanced Materials – "ICSP&AM 7,"* 14.

CHAPTER 24

Study of the Process of Countercurrent Extraction of Vegetable Oils via Mathematical Modeling

SIRADZE MANANA,[1] BERDZENISHVILI IRINE,[1] CHKHAIDZE EKATERINA,[1] and ANTIA GIORGI[2]

[1]*Department of Chemical and Biological Technologies, Georgian Technical University, Tbilisi, Georgia*

[2]*Department of Medical Chemistry, Tbilisi State Medical University, Tbilisi, Georgia*

ABSTRACT

The chapter considers and evaluates the mathematical model of the process of countercurrent extraction. It was evaluated that mathematical models of countercurrent extraction were carried out in connection with the study of the kinetics of the process and the development of methods for calculating the extraction of vegetable oils by solvents.

There were formulated recommendations on the appropriate use of models adequate to real conditions in various periods of the process of extracting vegetable oils.

24.1 INTRODUCTION

Extraction of oil by an extragent (solvent) is carried out by mass transfer inside a solid body-suitably prepared oilseed, and transport of oil from the

Advanced Polymer Structures: Chemistry for Engineering Applications. Omar Mukbaniani, DSc, Tamara Tatrishvili, PhD & Marc Jean M. Abadie, DSc (Eds.)

surface of particles into the depths of the solvent and, ultimately, entrainment by the extragent of the oil dissolved in it.

The transfer of material in a moving fluid is conditioned by two completely different mechanisms. Firstly, in the presence of a concentration difference in a liquid molecular diffusion occurs; secondly, particles of a substance dissolved in a liquid are carried away by the latter in the process of its movement and are transferred with it.

Since the extraction process is a mass transfer process, the transfer of a substance in it occurs through molecular diffusion, convective diffusion, a combination of both types of diffusion and other types of mass transfer.

As applied to the oil extraction process extraction of oil from one isolated particle, its transfer from inside the particle to its surface occurs as a result of molecular diffusion. In this case, the coefficient of internal diffusion is a summing characteristic of the structure of the extracted material [1].

Countercurrent extraction – multiple countercurrent contacting of raffinate and extract solutions of adjacent stages. Countercurrent extraction can be carried out in several mixer-settlers, or in a column apparatus. Countercurrent extraction provides good separation with high raffinate yield, while with multiple extraction, the yield of high quality raffinate is low.

24.2 EXPERIMENTAL METHODS AND MATERIALS

For studying the mechanism of countercurrent extraction, it is advisable to use the method of mathematical modeling. The mathematical description of the process was created on the basis of the following provisions: models which were obtained from the diffusion equations by using balance relations or the functional dependence of the concentration of the external solution on time in the boundary conditions [2, 3].

24.3 RESULTS AND DISCUSSION

During the extraction process, the extractable material and the extragent may be in a relatively immobile state (infusion), move relative to each other in co-current, countercurrent or cross-current. Since the transition of the extractable substances occurs continuously, the nature of the mutual motion of the phases will determine the difference in concentrations over time [4, 5]. At infusion and co-current:

$$\Delta C\tau = \Delta C_0 exp\left[-kF\left(\frac{1}{G_1}+\frac{1}{G_2}\right)^{\tau}\right] \quad (1)$$

At countercurrent:

$$dC\tau = dC_0 \ \exp-\left[kF\left(\frac{1}{G_2}-\frac{1}{G_1}\right)^{\tau}\right] \quad (2)$$

where; G_1 is the amount of miscella within the extractable material; G_2 is the amount of miscella of the liquid phase; and τ is the extraction time and the nature of the movement of the extracted material along the length or height of the extractor.

As follows from the analysis of dependencies, the nature of the mutual motion of the solid and liquid phases is of essence in maintaining the highest concentration difference throughout the entire process. The advantage of countercurrent is obvious.

Kinetic patterns characterize the speed of the process under given conditions and describe the relationship between the driving force of the process and the amount of transferred substance.

The rate of extraction of oil from the extracted material will depend on the rate of the process in each of the phases: in the extracted material and the extragent. The patterns of mass transfer in them differ significantly from each other.

Balance ratio of mathematical modeling of the extraction process mechanism can be expressed:

$$V_2 \frac{\partial C_2(\tau)}{\partial \tau} = -cS\int_0^1 \frac{\partial C_1(\tau)}{\partial \tau}dx \quad (3)$$

$$V_1[C_{10} - \bar{C}_1(\tau)] = V_2[C_2(\tau) - C_{20}] \quad (4)$$

$$-\frac{\partial \bar{C}_1(\tau)}{\partial \tau} = m\frac{\partial C_2(\tau)}{\partial \tau} \quad (5)$$

where; $c = dQ/C_1dV_1$; Q is the absolute content of substance (oil) in the solid phase; C_1, C_2 are the current micelle concentration inside and outside of particles; C_{10}, C_{20} are the initial concentration of the micelle inside and outside of particles; $\bar{C}_1$ – is the average current value of the concentration inside the particles (C_1 depends on τ and x, $\bar{C}_1$ – depends only on τ); V_1, V_2 are the internal and external volumes of the micelle; is the time; x is the current coordinate; l is the plate thickness (of petal); and S is the area of plate (of petal); $m = V_2/V_1$.

The purpose of the above relations is to determine included in the boundary conditions of the third kind. Relationship in Eqn. (3) differs from relationships in Eqns. (4) and (5) by the presence of the distribution coefficient of the substance between the solid and liquid phases, which relates the extraction of the substance from the solid phases with a change in the volume of the internal solution, which is quite regular, since with a change in the concentration of the solution, its specific volume changes. Under extraction conditions, the partition coefficient tends to increase, so that part of the internal solution, due to the unchanged porosity of the particles, passes outside. With a decrease in the specific volume of the internal solution, a certain amount of the external solution penetrates into the particles.

These features are not taken into account in some ratios where the volume is constant [4, 6]. Because of this, the balance equations do not include a part of the substance that needs to be refined.

By the way, relation in Eqn. (3) is written as the second equation after the diffusion equation (before the formulation of the boundary conditions), which determines the main (diffusion) transfer of substance.

While using the functional dependence [4] as a periodic function (exactly corresponding to the real state) with the duration of the process, which is equal to a quarter of the oscillation period, there is no need to take into account corrections for changes in the specific volume of the solution, since it is given by the functional dependence itself $C_2(\tau) = f(\tau)$ as far as in turn, between C_2 and V_2 and, therefore V_1, there is a functional dependence.

Using the experimental data [7] industrial belt extractor DC-70, industrial belt extractor DC-70, which is intended for the extraction of vegetable oils by the method of multiple step irrigation of the extracted material in countercurrent, we assessed the mathematical models of extraction (Figure 24.1). It is obvious that models in Eqns. (4) and (5) do not correspond to the experimental data.

Figure 24.1 shows the results of calculations obtained when considering the source material as a "medium" in which diffusion occurs. The initial concentration of the substance in the solid phase in this case was equal to the content of the substance (oil) in the material (oil content of the material). Calculations carried out taking into account the filling of the pores of the material with the extracting liquid and formation of a solution with a certain initial concentration inside the particles, gave the following residual contents of the substance at the end of the process: according to ratio in – Eqn. (3) – 0.7%; Eqn. (4) – 8.25%; Eqn. (5) – 9.1% and according to model estimates – 0.44% (the experimental value is – 0.48%).

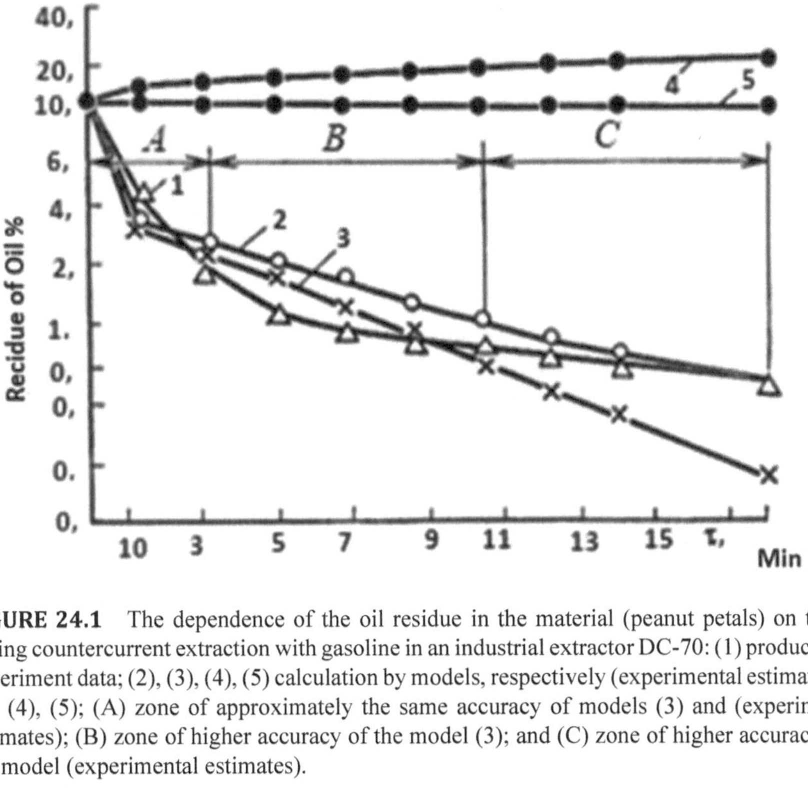

FIGURE 24.1 The dependence of the oil residue in the material (peanut petals) on time during countercurrent extraction with gasoline in an industrial extractor DC-70: (1) production experiment data; (2), (3), (4), (5) calculation by models, respectively (experimental estimates), (3), (4), (5); (A) zone of approximately the same accuracy of models (3) and (experiment estimates); (B) zone of higher accuracy of the model (3); and (C) zone of higher accuracy of the model (experimental estimates).

24.4 CONCLUSION

Finally we can conclude using of balance relations in the boundary conditions requires knowledge of such process parameters (phase ratio, solid phase porosity, etc.), which can be completely dispensed during calculating the remainder of the substance if a given function is used. But, on the other hand, the latter approach for calculating the productivity and some indicators of the operation of industrial extractors requires the use balance equations of the process.

KEYWORDS

- **balance ratios**
- **extractable material**
- **kinetic regularities**
- **mass transfer**
- **mathematical modeling**
- **molecular diffusion**
- **solid state**

REFERENCES

1. Beckman, I. N., (2016). *Mathematics of Diffusion: Textbook* (pp. 23–29). I. N. Beckman. – M.: Publishing house "OntoPrint."
2. Abiev, R. Sh., (2001). Modeling of the extraction process from a capillary-porous particle with a bidisperse structure. In: Abiev, R.Sh., & Ostrovsky, G. M., (eds.), *Theoretical Foundations of Chem. Technology* (Vol. 35, No. 3, pp. 270–275).
3. Shorstkiy, I. A., Meretukov, Z. A., Koshevoy, E. P., & Kosachev, V. S., (2015). Evaluation of the influence of the type of solvent and the preparation of oilseed material on the kinetic dependences of the extraction process. *New Technologies*, (2), 46–50.
4. Babenko, Yu. I., (2005). Mathematical model of extraction from a body with a bidisperse porous structure. In: Babenko, Yu. I., & Ivanov, E. V., (eds.), *Teor. Basics of Chem. Technol.* (Vol. 39, No. 6, pp. 644–650).
5. Meretukov, Z. A., Zaslavets, A. A., Koshevoy, E. P., & Kosachev, V. S., (2012). Methods for solving differential equations of hydrodynamics. *New Technologies*, (1), 36–41.
6. Babenko, Yu. I., (2007). Extraction of target components from a porous body into a moving liquid. In: Babenko, Yu. I., & Ivanov, E. V., (eds.), *Theoretical Foundations of Chem. Technology* (Vol. 41, No. 2, pp. 225–227).
7. Sergeyev A.G., (1974). Equipment of the extraction line DS-70 and its operation. *Guidelines for the Technology of Obtaining and Processing Vegetable Oils and Fats* (Extraction Method for the Production of Vegetable Oils). Under the general scientific editorship of Doctor of Technical Sciences, Leningrad, All-Union Research Institute of Fats, Book 2, (Vol. 1, pp. 282–295).

CHAPTER 25

Electrosynthesis of Nanomagnetite and Application for Purification Phenol Previously Contaminated Water

M. DONADZE and N. MAKHALDIANI

Faculty of Chemistry and Metallurgy, Georgian Technical University, Tbilisi, Georgia

ABSTRACT

Nanoiron oxide–magnetite (Fe_3O_4) is a prospective material for water purification and application in the biomedicine field. It enables us to purify water from bacteria and toxic, heavy metals, such as Hg, Pb, Cd, Tl, and so on. It is one of the best sorbents. Dyes, pesticides, and other organic pollutants can be removed by means of magnetic nanoparticles.

The aim of the study is the electrosynthesis of Fe_3O_4 nanomagnetite and purification of from phenol pre-contaminated water. The main component of the filter is magnetite nanoparticles stabilized with oleic acid, obtained by electrosynthesis in a two-layer bath. An aluminum arc was used as a rotating cathode and optimal electrolysis parameters were determined. A porous filter was obtained after impregnation of boehmite with magnetic nanoparticles and its subsequent burning at 450°C.

The optimal parameters of electrolysis are determined. The resulting nanomagnetite was characterized by X-ray analysis (XRD), Infrared Spectroscopy (FT-IR), Elementary Analysis and Scanning Microscopy (SEM-EDS). Particle size determined by dynamic light scattering (DLS Malvern). A filter based on nanomagnetite shows a significant effect in the process of purifying drinking water from phenol.

Advanced Polymer Structures: Chemistry for Engineering Applications. Omar Mukbaniani, DSc, Tamara Tatrishvili, PhD & Marc Jean M. Abadie, DSc (Eds.)

A porous γAl_2O_3 filter containing nanomagnetite can be used to purify water contaminated with phenol at the place of consumption.

25.1 INTRODUCTION

Nano-iron oxide – magnetite (Fe_3O_4) is a promising material in the water purification and biomedical fields. It can be used to clean water from bacteria and toxic, heavy metals such as Hg, Pb, Cd, Tl, and others. It is one of the best sorbents. Magnetic nanoparticles decorated with polyethylene can also eliminate traces of Cu^{2+} ions. With the help of magnetic nanoparticles, it is possible to get rid of dyes, pesticides, and other organic contaminants. It has both sorbtion and high oxidizing properties and, due to the combined presence of Fe^{2+} and Fe^{3+} ions, participates in Fenton-like reactions and used for mineralization heavy organic compounds such as phenol and its derivatives [1, 2].

There are many methods for obtaining magnetite (Fe_3O_4) nanoparticles. Nanomagnetite is mainly obtained by wet chemistry, chemical coagulation, thermochemistry, electrochemistry, and others [3–5].

Obtaining magnetite by electrosynthesis from both simple and complex electrolytes has been described in many articles. Electrolysis parameters – electrolyte concentration, current density, temperature, voltage, can change the particle size of the magnetite, magnetic parameters and other properties [6–10].

25.2 EXPERIMENTAL METHODS AND MATERIALS

25.2.1 PREPARATION OF NANOMAGNETITE

The magnetite sol is obtained by electrosynthesis in a two-layer bath on a rotating cathode. Nanosilver, nanozinc, nanocopper, and other metal's nanoparticles are obtained in the two-layer bath. The process of making nanosilver is described in articles and patents [11–16].

To obtain the nanomagnetite sol an iron (99.19% Fe, 0.75% Mn; 0.053% Cu) anode and an arc type aluminum cathode were used. The residence time of the cathode in the organic phase is $\tau r = 36$ s, and the current density – $i = 60\ A/dm^2$. In the experiment, as a solution, iron sulfate ($FeSO_4 \times 7H_2O$) and as an organic solution, hexane is used in which the surfactant substance is oleic acid.

The iron in the oleic acid is attached to the carboxyl group by a bidentate bond (Figure 25.1).

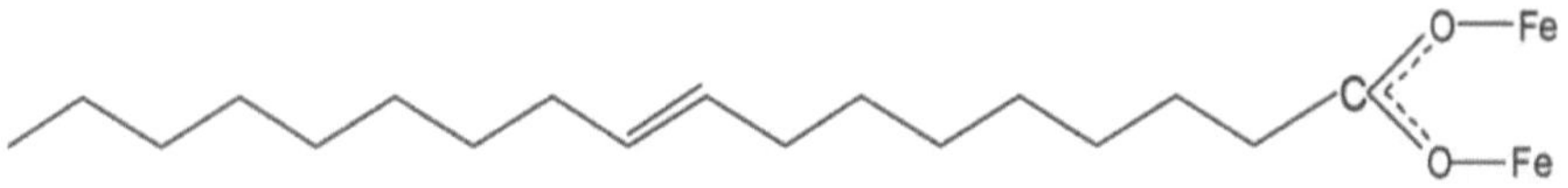

FIGURE 25.1 Oleic acid stabilized magnetite.

Processes in electrolysis:

i. **Anode:**
$Fe \rightarrow Fe^{2+} + 2e$; $4OH^- \rightarrow 2H_2O + O_2 + 4e$,
$Fe^{2+} + 2OH^- \rightarrow Fe(OH)_2$,
$3Fe(OH)_2 + 1/2\ O_2 \rightarrow Fe(OH)_2 + 2FeOOH + H_2O$.

ii. **Cathode:**
$2H_2O + 2e^- \rightarrow H_2 + 2OH^-$,
$3(FeOH)_3$ (Solid) + $H^+ + e \rightarrow Fe_3O_4$ (Solid) + $5H_2O$,
$6\ Fe + 8H_2O \rightarrow 2Fe_3O_4 + 8H_2$.

Initially, a brown solution forms in the organic layer, which indicates the formation of $Fe(OH)_2$. Oxygen released at the anode during electrolysis contributes to its oxidation and the production of trivalent-FeOOH. As a result of the reduction of hydrogen at the cathode, the solution is enriched with hydroxyl ions. Acidity also changes from neutral (pH = 7) to alkaline (pH = 9), which leads to active oxidation of ferrous iron and the production of ferric hydroxide. The oxidation of iron is facilitated by the active evolution of oxygen due to the partial passivation of the iron anode. Magnetite is obtained by solid-phase reduction and dehydration of ferric hydroxide in the cathode region.

To determine the optimal parameters of electro synthesis magnetite electrolyte of various concentrations $FeSO_4 \times 7H_2O$ g/l: 30; 20; 10; 5; (Fe^{2+}) is used; Different concentrations of superficially active (oleic acid) %: 0.8; 1; 1.5; 2 and different synthesis temperatures: 25; 30; 45°C. Nanoparticles in organosol are characterized by the method of dynamic dispersion. The residence time in the organic phase cathode $\tau_r = 36$ s and the current density i = 60 a/dm² did not change. As showed the measurements, at high electrolyte concentrations, the particle size increases (Figure 25.2).

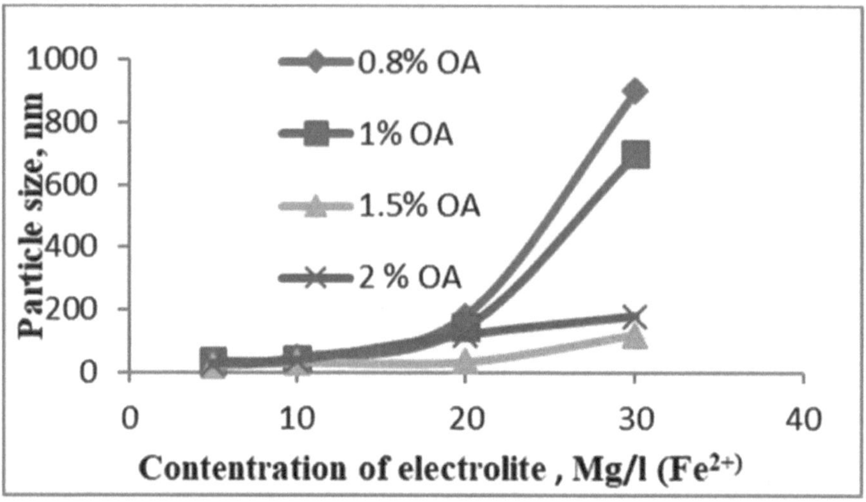

FIGURE 25.2 Particle size dependence of magnetite (Fe_3O_4) on electrolyte and oleic acid concentration, τ_r = 36 sec.

As the results show, the optimal electrolyte concentration is 5 g/l (Fe^{2+}) and 1.5% of oleic acid. As can be seen from the graphs (Figure 25.3), to obtained the 25–30 nm magnetite nanoparticles the optimal parameters of electrolysis are as follows: electrolysis temperature 45°C; Electrolyte concentration 5 g/l (Fe^{2+}) oleic acid concentration 1.5%; Current efficiency of nanomagnetite in organic phase is η=15% and full (organic phase and organic-water phase boundary) current efficiency is η=48%.

The magnetite nanoparticles obtained by means of electrolysis were washed with hot distilled water, ethanol, and again distilled water to remove excess oleic acid and dried at 120°C for 5 h. In order to determine and identify the magnetite, a study was performed by X-ray analysis (XRD), infrared spectroscopy (FT-IR) and elemental analysis (SEM-EDS).

X-ray diffraction analysis was used to study magnetite samples dried at 120°C and calcined for 2 h at 400°C (Figures 25.4 and 25.5). As can be seen from the X-ray diffraction analysis, the sample dried at 120°C has a protective layer of oleic acid, which prevents the fixation of the crystal structure, while the X-ray analysis of magnetite fired at 400°C shows that the main phase is magnetite only.

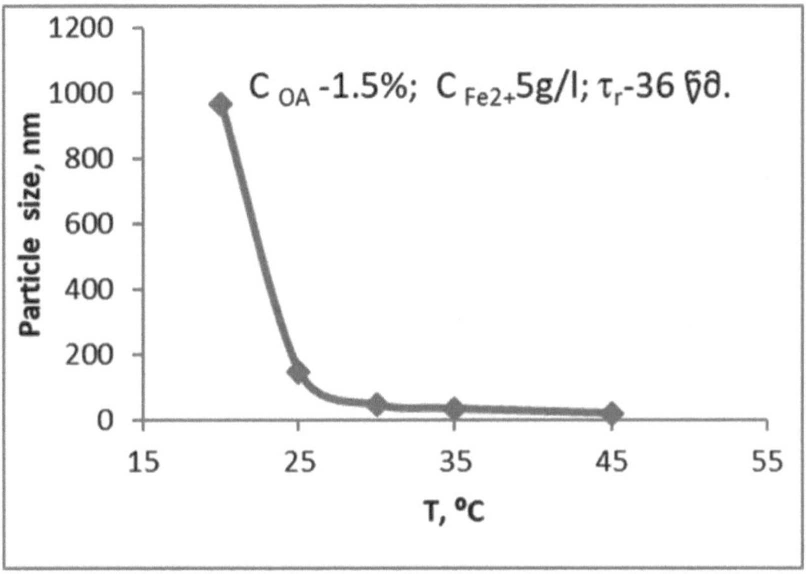

FIGURE 25.3 Dependence of the size of magnetite (Fe_3O_4) particles on the electrolysis temperature: C_{OA} – 1.5%; C_{Fe2+} – 5 g/l; and τ_r – 36 sec.

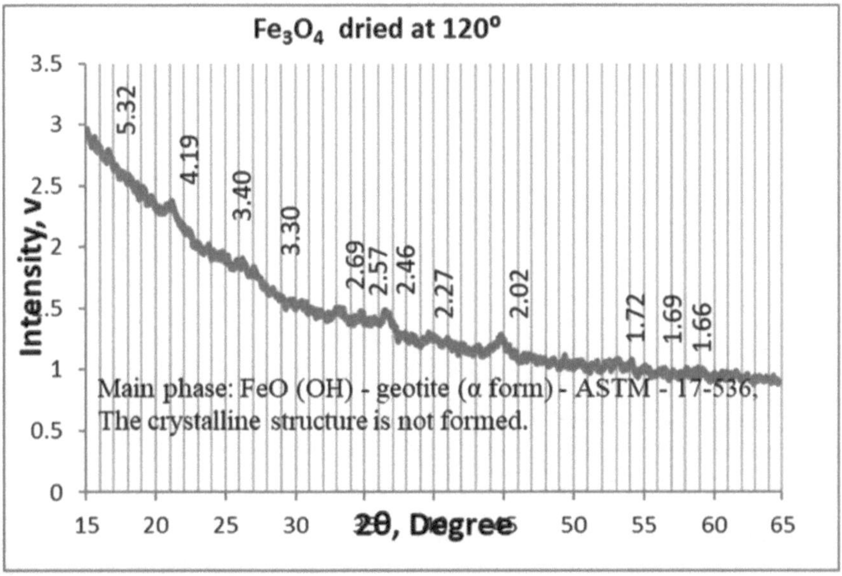

FIGURE 25.4 X-ray analysis of 120°C dried magnetite.

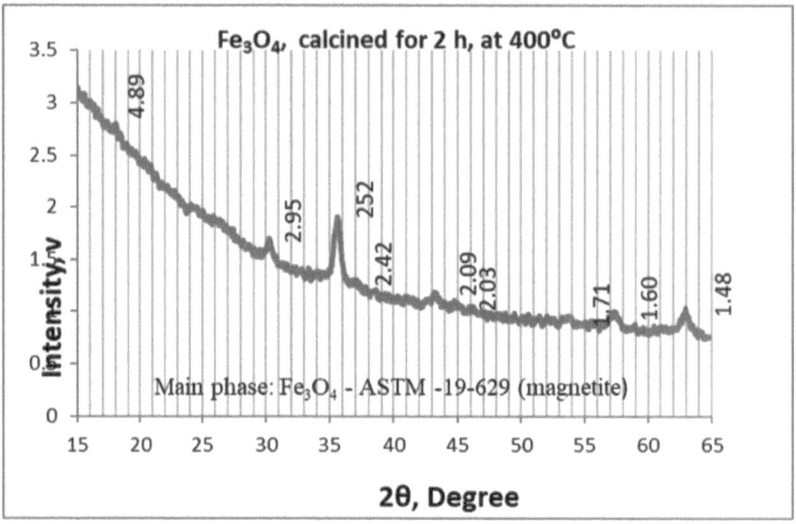

FIGURE 25.5 X-ray analysis of magnetite calcined at 400°C.

Electrochemically synthesized magnetite is characterized by infrared spectroscopy (FT-IR, Thermo Nicolet, Avatar 370, range: 400–4,000 cm^{-1}; Measurement accuracy: 0.5 cm^{-1}). The spectra were recorded in Vaseline. A finely dispersed powder of the sample, after mixing in Vaseline oil with an agate stick, was applied to a KB plate (Figure 25.6).

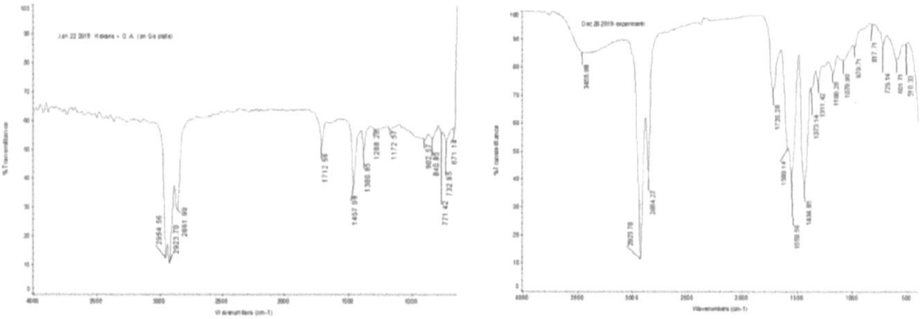

FIGURE 25.6 Fourier-infrared spectra: (a) Infrared spectrum of oleic acid dispersed in hexane; and (b) magnetite infrared spectrum Fe_3O_4 and 1.5% OA.

It is known that the frequency of fluctuations is related to C = O carboxylic groups of unsaturated carbonic acid in the range 1,700–1,725 cm^{-1}. There are two types of intermediates COOH-group oleic acid nanoparticles, and monolayer oleic acid is formed. In the case of high concentrations of oleic acid in the FT-IR spectrum, a peak of 1,710–1,725 cm^{-1}, which is typical for a free carboxyl group of COOH. In the spectrum Fe_3O_4, modified with oleic acid (high concentration of oleic acids), the peak of C = O, is 1,720 cm^{-1}, which corresponds to the bilayer model. The two peaks, 1,589 and 1,550 cm^{-1}, correspond to asymmetric and symmetric fluctuations of carboxylic (COO-) groups. The peaks 2,923 cm^{-1} and 2,854 cm^{-1} correspond to the asymmetric and symmetric fluctuations of the CH_2 group, while 601 cm^{-1} and 510 cm^{-1} correspond to the Fe-O bond. A recent fluctuation of 1079.99 cm^{-1} indicates the chemosorbtion of Fe_3O_4 oleic acid with the CO group [17].

Magnetic nanoparticles after washing and firing at 400°C were analyzed by scanning microscopy and elemental analysis (SEM-EDS) (Figure 25.7).

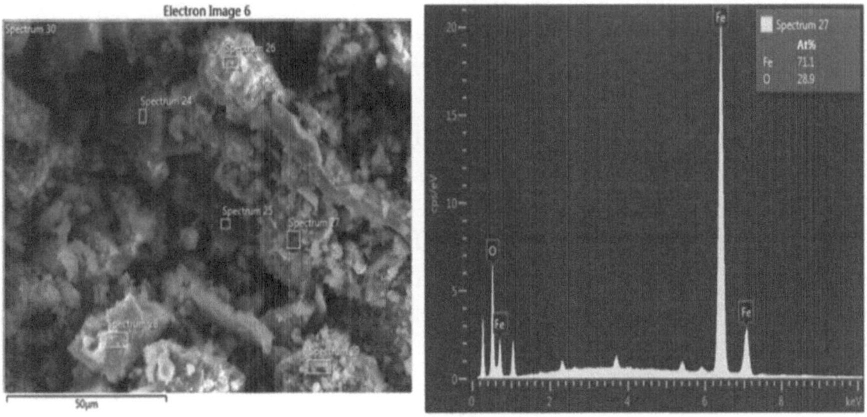

FIGURE 25.7 SEM–EDS of magnetite.

As elemental analysis shows, iron oxide nanoparticles correspond in composition to magnetite (Fe_3O_4). The iron content in magnetite is 71.79%, and the iron content in the obtained sample is 71.1%.

25.3 PURIFICATION OF WATER CONTAMINATED WITH PHENOL USING THE FENTON METHOD

Phenol and its compounds are widely used in the manufacture of antioxidants, biocides, disinfectants, pesticides, polymers, dyes, paper, pharmaceuticals, and other organic materials. Contaminants are often found in the industrial wastewater of the oil refining, petrochemical, and general chemical industries.

Over the past decade, the use of advanced oxidation processes (AOP) has been increasing due to the efficient removal of contaminants from water. It includes environmentally friendly chemical, photochemical, and electrochemical methods based on the action of the main oxidant(hydroxyl radical) (˙OH) formed in the reaction zone. The source of ˙OH is hydrogen peroxide. This is called "Green" – a reagent, the decomposition products of which are oxygen and water, is widely used for bleaching cellulose, paper, and fiber, as a disinfectant in electrical engineering, medicine, an oxidizing agent in organic synthesis, etc.

A mixture of Fe^{2+} and H_2O_2 (Fenton reagent) is used to remove persistent organic contaminants. The oxidizing ability of this method can be significantly improved by the addition of an artificial source of ultraviolet (UV) light. Photo-Fenton method (PFM), or by performing the process in the sunlight (solar photo-Fenton method: SPFM).

Advanced oxidation methods are used to treat groundwater and urban wastewater, to disinfect water, to neutralize volatile organic compounds (VOCs), and to remove odors.

In chemistry and biology ˙OH is an important free radical. It is formed directly at the place of consumption and reveals the properties of a non-selective oxidizer. After fluoride •OH is the second oxidizing agent with strength, with standard reducing potential E_0 (•OH/H_2O) = 2.8 V. It can oxidize most organic and organometallic pollutants to complete mineralization.

The Fenton process involves the formation of a hydroxyl radical by the classical reaction of Fenton:

$$Fe^{2+} + H_2O_2 \rightarrow Fe^{3+} + \bullet OH + OH^-,$$

In an acidic area:

$$Fe^{2+} + H_2O_2 + H^+ \rightarrow Fe^{3+} + H_2O + \bullet OH.$$

The method is especially effective for purifying contaminated water in the pH range of 2.8–3.0. It should be noted that even a small amount of Fe^{2+} catalyst is sufficient, because the current between Fe^{3+} and H_2O_2 is so-called. A "Fenton-like" reaction still produces Fe^{2+}:

$$Fe^{3+} + H_2O_2 \rightarrow Fe^{2+} + HO_2^{\bullet} + H^+$$

Magnetite has long been used to purify water from various contaminants. It removes heavy organic impurities until complete mineralization.

Electrosynthe-sized magnetite has been tested for the purification of phenol-contaminated water. The kinetics of purification of water contaminated with phenol by magnetite has been studied, quality of cleaning has been analyzed based on the determination of the chemical oxygen consumption.

25.3.1 USE OF NANOMAGNETITE TO PURIFY WITH PHENOL CONTAMINATED WATER

As Model water use with phenol 1 mg/l contaminated distilled water. A solution of hydrogen peroxide with a concentration of 5 g/l is placed in a biuret, and contaminated water in a beaker. In the first case, to the glass as a filtering material was added magnetite (1 g) dried at 120°C, in the second case, magnetite (1 g) treated at 400°C, and in the third case, the Al_2O_3 substrate impregnated magnetite (1 g). Hydrogen peroxide is added to the beaker with constant stirring, and the chemical oxygen demand (COD) is determined every 30 minutes.

The degree of conversion of phenol into a model solution is estimated by the change in the COD. The COD of distilled water, which was used to prepare the model solution, is taken into account.

25.3.1.1 COURSE OF ANALYSIS

About 100 ml of distilled water was placed in a 500 ml three-neck round flask, 10 ml of concentrated H_2SO_4 was added, followed by a solution of 1 ml of model solution and 20 ml of 0.025 N $KMnO_4$. The round flask is placed on the stove and boiled for half an hour. During boiling, the organic matter in the flask may evaporate, leading to analysis errors. Therefore, a reflux condenser is attached to the top of the flask, with cold water circulating in the outer shirt. After boiling for half an hour, 20 ml of 0.025 N oxalic acid was added to the flask after evaporation from the heater, and under constant stirring was heated hot with a solution of 0.025 N $KMnO_4$ until the solution turned pink. Chemical oxygen consumption is calculated by the following formula:

$$COD\ mg/L = \frac{(V_1 - V_0)\times K\times 0.01\times 8\times 1000}{V} = \frac{(V_1 - V_0)\times K\times 80}{V}$$

where; V_1 and V_0 are the volumes of 0.025 N $KMnO_4$ spent on titration of the study and background samples, respectively. *K* correction coefficient. 0.025 N for a solution of $KMnO_4$, *V is* the volume of the sample taken ml.

Based on the obtained results, the purified water quality is evaluated through the curve τ (time) – COD (chemical consumption of oxygen).

For the decomposition of phenol by the Fenton reaction, was used magnetite dried at 120°C and treated at 400°C. As can be seen from Figure 25.8, the efficiency of the calcined form exceeds that of the dried form, but phenol does not decompose completely. The dried form (at 120°C) undoubtedly retains a protective layer of oleic acid, which does not allow the magnetite to actively interact with phenol.

At the next stage of the study, to prevent the coarsening of magnetite particles is used, an inert, porous, and thermostable substrate of γ Al_2O_3.

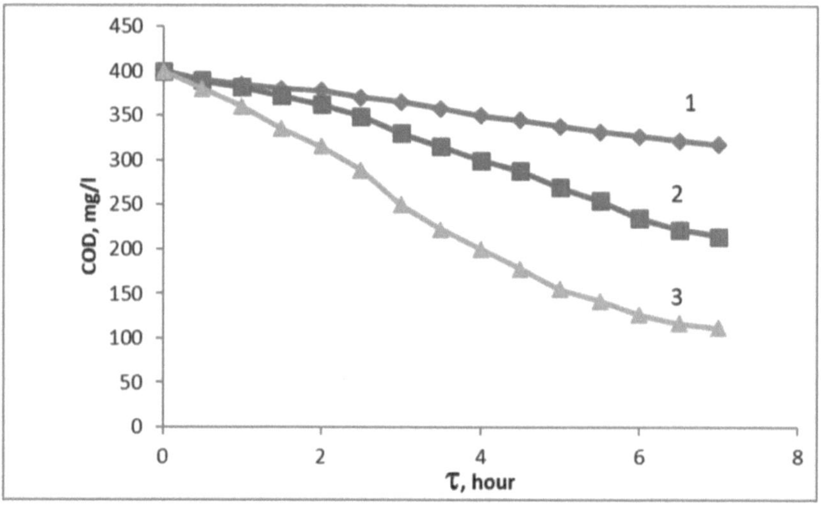

FIGURE 25.8 Kinetics of phenol degradation by magnetite: (1) Fe_3O_4 dried at 120°C; (2) Fe_3O_4 at 400°C; and (3) $Fe_3O_4-\gamma Al_2O_3$ filter material.

To prepare a suspension of boehmite and Al_2O_3, 20% of solid mass (10% γAl_2O_3 + 10% AlOOH-boehmite) is slowly added to distilled water and stirred for 24 hours. Of nanomagnetite sol with a total iron content of 1 g (determined by permanganatometry) is added to the prepared suspension. Stirring is continued for 24 hours at 55°C. The sol is evaporated, and a homogeneous suspension is obtained. The suspension is dried at 120°C to remove the aqueous phase and the powder is calcined at 400°C for 3 hours to obtain a filter material. The water contaminated with phenol and the filter material are placed in a beaker, to which hydrogen peroxide is added with constant stirring, and the COD is determined every 30 minutes.

Magnetite-Al_2O_3 composite filter material gives better results. As can be seen from the preliminary data, the chemical oxygen consumption at neutral acidity decreases sharply.

25.4 CONCLUSION

The main goal of the study was achieved: nanomagnetite (Fe_3O_4), stabilized with oleic acid in hexane, was obtained by electrolysis in a two-layer bath. The main parameters of electrosynthesis have been established, and a method for impregnating of Al_2O_3 with a layer of nanomagnetite has been developed. The filter made of nanomagnetite, used as a Fenton reagent for purifying water contaminated with phenols, actively oxidizes phenol and reduces the COD from 400 to 80 mg/l.

KEYWORDS

- **electrochemical synthesis**
- **Fenton method**
- **magnetite**
- **nano-iron oxide**
- **oleic acid**
- **phenol**

REFERENCES

1. Zhang, Sh., Zhao, X. I., Niu, H., Yali, Shi., Cai, Y., & Jiang, G., (2009). Superparamagnetic Fe_3O_4 nanoparticles as catalysts for the catalytic oxidation of phenolic and aniline compounds *Journal of Hazardous Materials, 16*, 560–566.
2. Hsueh, C. L., Huang, Y. H., Wang, C. C., & Chen, C. Y., (2005). Degradation of azo dyes using low iron concentration of Fenton and Fenton-like system. *J. Chemosphere, 58*, 1409–1414.
3. Silva, P. Da, S., Costa, De Moraes, D., & Samios, D., (2016). Iron oxide nanoparticles coated with polymer derived from epoxidized oleic acid and Cis-1,2-cyclohexanedicarboxylic anhydride: Synthesis and characterization. *J. Material Sci. Eng., 5*(3), 1–7.

4. Baharuddin, A. A., Ang, B. Ch., Hussein, N. A., Andriyana, A., & Wong, Y. H., (2018). Mechanisms of highly stabilized ex-situ oleic acid-modified iron oxide nanoparticles functionalized with 4-pentynoic acid. *J. Materials Chemistry and Physics, 203*, 212–222.
5. Zhang, L., He, R., & Gu, H. Ch., (2006). Oleic acid coating on the monodisperse magnetite nanoparticles. *Appl. Surf. Sci. 253,* 2611–2617.
6. Fajaroh, F., Setyawan, H., Widiyastuti, W., & Winardi, S., (2012). Synthesis of magnetite nanoparticles by surfactant-free electrochemical method in an aqueous system. *Advanced Powder Technology, 23*, 328–333.
7. Marques, R. F. C., Garcia, C., Lecante, P., Ribeiro, L., Noe, N. J. O., Silva V. S., Amaral, A., Millan, & Verelst, M. (2008). Electro-precipitation of Fe_3O_4 nanoparticles in ethanol, *J. Magn. Magn. Mater., 320*, 2311–2315.
8. Cabrera, L., Gutierrez, S., Menendes, N., Morales, M. P., & Herrasti, P., (2008). Magnetite nanoparticles: Electrochemical synthesis and characterization. *J. Electrochem. Acta, 53,* 3436–3441.
9. Franger, S., Berthet, P., & Berthon, J., (2004). Electrochemical synthesis of Fe_3O_4 nanoparticles in alkaline aqueous solutions containing complexing agents, *J. Solid State Electrochem., 8,* 218–223.
10. Franger, S., Berthet, P., & Dragos, O., (2007). Large influence of the synthesis conditions on the physico-chemical properties of nanostructured Fe_3O_4. *J. Nanopart. Res., 9*, 389–402.
11. Doandze, M., & Agladze, T., (2017). Strategy for nanohybridezed synthesis of MaMbOx system. In: *Chemical Engineering and Polymers Production of Functional and Flexible Materials.* Apple Academic Press, Canada, 13,17.
12. Agladze, T., Donadze, M., Gabrichidze, M., Toidze, P., Shengelia, J., Boshkov, N., & Tsvetkova, N., (2013). *Z. Phys. Chem., 227*, 1187–1198.
13. Donadze, M., Gabrichidze, M., & Agladze, T., (2016). Novel method of fabrication of hybrid metal(I)/metal(II) oxides nanoparticles. *Transaction of the IMF: The International Journal of Surface Engineering and Coatings, 94*, 16–23.
14. Khutsishvili, S., Toidze, P., Donadze, M., Gabrichidze, M., Agladze, T., & Makhaldiani, N., (2019). Structural and magnetic properties of silver oleic acid multifunctional nanohybrids. *Annals of Agrarian Science, 17,* 153–157.
15. Agladze, T., & Donadze, M., (2019). *Method of Obtaining Silver Monodisperse Nanoparticles.* GE P, 7022 B.
16. Agladze, T., Batsikadze, M., & Donadze, M., (2011). *Electrosynthesis of Silver Nanoparticles with Bactericidal Properties in a Two-Layer Bath.* GEP, 5254 B.
17. Yang, K., Peng, H. I., Wen, Y., & LiRe, N., (2010). Examination of characteristic FTIR spectrum of secondary layer in bilayer oleic acid-coated Fe_3O_4 nanoparticles. *Appl. Surf. Science, 256*, 3093–3097.

CHAPTER 26

Synthesis of Some New Azo Dyes on the Base of 6-Aminocoumarine

K. DZULIASHVILI and N. OCHKHIKIDZE

Agricultural University of Georgia, Tbilisi, Georgia

ABSTRACT

Diazotization and azo coupling reactions are versatile tools of fine organic synthesis. Azo dyes have very interesting physical, spectral, and chemical properties and are widely used in the different fields of science and technology. Moreover, some of them have very significant biological activity and are used in healthcare or medical diagnostics. Coumarin belongs to chromene-type dyes, but it may be used as an azo partner in the azo coupling reaction for the synthesis of azochromenes dyes.

In the current research, we have obtained four new dyes bearing two π-conjugated azo and chromene chromophores. The target compounds were synthesized by diazotizaton of 6-aminocarine (**1**) and with following azo coupling to 2-hydroxybenzoic acid (**3a**), naphthalen-2-ol (**3b**), 4-amino-5-hydroxynaphthalene-2,7-disulfonic acid (**3c**) and (E)-3-(4-hydroxyphenyl) acrylic acid (**3d**). Obtained azo dyes have been used for dyeing wool fiber and various technical and spectral properties have been studied.

26.1 INTRODUCTION

Diazotization and azo coupling reactions are versatile tools of fine organic synthesis. These consecutive reactions are commonly used for the generation

Advanced Polymer Structures: Chemistry for Engineering Applications. Omar Mukbaniani, DSc, Tamara Tatrishvili, PhD & Marc Jean M. Abadie, DSc (Eds.)

of azo chromophore and colored materials, having the ability to color other substances. Azo dyes are a very interesting class of chemical compounds that contain heterocyclic moieties and have drawn the attention of many researchers in recent years. for the reason that, they have very interesting physical, spectral, and chemical properties. Azo dyes are widely used in the food, pharmaceutical, cosmetic, textile, and leather industries. Azo dyes usually are in close contact to the human body and potentially may penetrate into an organism in some quantity and undergo metabolism process forming mutagenic and toxic primary aromatic amines. Therefore, it is very important to choose the non-toxic, eco- or biofriendly diazo and azo partners during azo compound construction. On the other hand, the conjunction strategy of different chromophores into one molecule structure for the aim of physical, chemical, and biological properties synergizes, is another modern technique in the design of the dyes.

Coumarin belongs to chromene type dyes, but it may be used as an azo partner in the azo coupling reaction for the synthesis of azochromenes dyes [1, 2]. While 6-aminocoumarin easily reacts to sodium nitrite in the presence of hydrochloric acid and gives corresponding diazonium salt, able to couple aromatic substances and form azo compounds. A series of sulfocoumarin-, coumarin-, and 4-sulfamoylphenyl-bearing indazole-3-carboxamide hybrids with the selective Inhibition properties of tumor-associated carbonic anhydrase isozymes IX and XII have been synthesized by S. Angapelly and co-workers [3]. Moreover, coumarin azo derivatives which may be used as a dyes [4], fluorescent probes [5], antimicrobial [6–9], gelling [10], antithrombotic [11] agents, Xa inhibitors [11], etc. [13].

26.2 RESULTS AND DISCUSSION

In the current research we have obtained four new dyes bearing two π-conjugated azo and chromene chromophores. The target compounds were synthesized in accordance with two sequential stages of diazotization-azo coupling, as shown in Scheme 26.1. 6-aminocarine (**1**) was chosen as the diazopartner, and 2-hydroxybenzoic acid (**3a**), naphthalen-2-ol (**3b**), 4-amino-5-hydroxynaphthalene-2,7-disulfonic acid (**3c**) and (E)-3-(4-hydroxyphenyl)acrylic acid (**3d**) for azo partnets.

Diazotization of **1** has been carried out in the diluted hydrochloric acid media by the action of sodium nitrite at 0–5C under rotation for a period of 60 min. The finishing of the process was checked by the positive test on starch-iodine paper (generation of blue color indicates the excess of nitrous acid). The excess of nitrous acid have been removed by the addition of solid urea until terminating gas evaluation from the reaction mass. Finally, diazonium salt in the form of water solution was filtered off quickly on the filter paper in the ice bath to avoid decomposition of **2**. Purified solution of **2** has been used immediately in the azo coupling reaction with preliminarily prepared and cooled to 0°C alkali solutions of azo partners **3a-d**.

The azo coupling reactions were carried out by adding diazopartner solution to the azo partner solution under vigorous stirring and careful monitoring of pH value. To convert phenolic and naphthol compounds into more reactive forms of phenolates and naphotates, it is imperative to maintain a slightly alkaline pH value of about 9–10. Thus, a constant pH was adjusted from time to time by adding a 10% NaOH solution throughout the azo coupling process. The final compounds have been precipitated by adding 10% hydrochloric acid solution to pH value of 7 and isolated by filtration. The solid remains have been washed out by cold water on the filter paper, transferred to Petri dish and dried at ambient temperature in the vacuum.

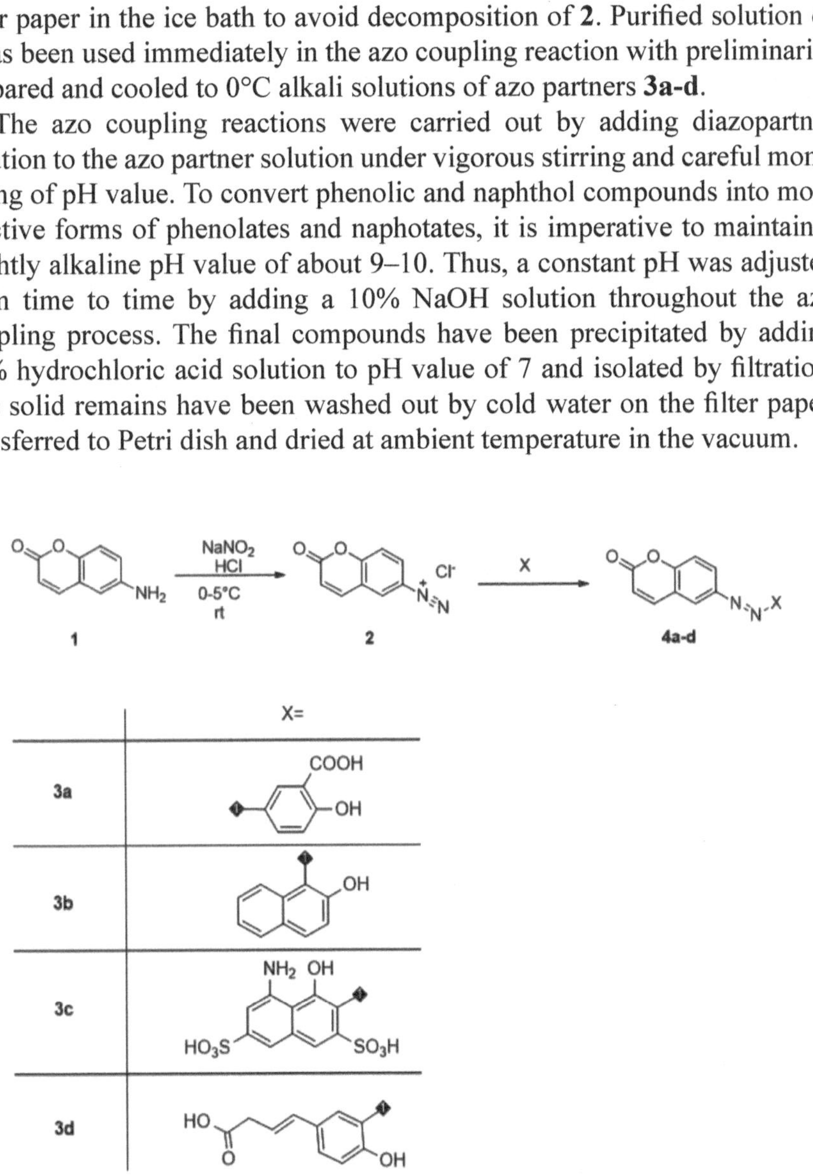

SCHEME 26.1 Synthesis of azo-coumarin dyes by diazotization-azo coupling reactions.

The isolated azo dyes have been used for dyeing without additional purifications except of **4d,** which was recrystallized in 1% HCl solution. The yields of final products were 40–55% (Table 26.1 and Figure 26.1).

Analysis of the UV-vis spectra shows that all obtained dyes have absorption in the visible range. **4c** absorbs on the 520 nm wavelength because of the longest π-conjugated system and bearing two strength electron donating groups (OH and NH_2). The absorption of **4d** is bathochromic ally shifted in comparison of **4a** absorption value which caused by participation of exocyclic double bond in the π-conjugated system.

TABLE 26.1 The Yields and Properties of 4a–d

SL. No.	Dye	Yield (%)	λ_{max}(ε), nm, solvent	Dye Uptake (%)	Lab-Coordinates
1.	4a	55	440 (1.50×10^5), water	80	L = 78, b = 1, b = 62
2.	4b	53	480 (1.486×10^5), ethanol	60	L = 50, b = 31, b = 56
3.	4c	54	520 (1.9412×10^5), water	75	L = 11, b = 27, b = 2
4.	4d	40	460 (1.7647×10^5) water	74	L = 78, b = 1, b = 70

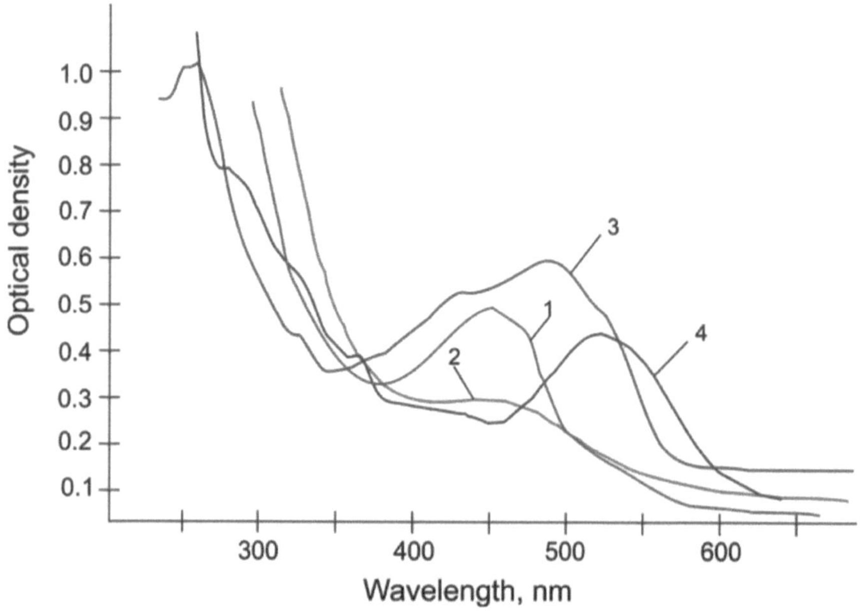

FIGURE 26.1 UV-Vis spectra: (1) 4a; (2) 4b; (3) 4c; and (4) 4d.

Coumarin-Azo dyes have been used for dying of wool fiber in accordance with the method described by Toussirot et al. [12]. The preliminarily washed up, dyed, and weight up fiber was put and soaked in pre-dyeing bath, containing potassium aluminum sulfate (0.3% weight of fiber). The temperature of the bath was increased up to 60°C and kept for 45 min. Then the dying bath was cooled to room temperature. Mordanted fiber was washed out with tap water to remove the excess of potassium aluminum sulfate.

For the dyeing bath, a M:L (material to liquor ratio) at 1:40 was used. The mordanted wrung out fiber was put in the dyeing bath and heated again up to 60°C for a period of 45 min with manual periodical stirring. The fiber in the bath was allowed to cool down, and then rinsed with tap water and left for drying at room temperature.

Dye uptake properties of all dyes **4a-d** have been determined. For this aim, the absorbance of the dyeing bath have been measured before and after dyeing. The dyebath was cooled to room temperature prior to measures. The percentage of dye exhaustion (DE) was calculated according to the given formula:

$$DE = [(A_0 - A_1)/A_0] \times 100$$

where; A_0 and A_1 are absorbances of the dye bath, before and after dyeing, respectively.

The color strength and color depth of the dyed samples were determined in CIELab coordinates. L* corresponds to the brightness (100% white, 0 – black, a* is red-green balance (+a* = red, –a* = green) and b* is the yellow-blue balance (+b* = yellow, –b* = blue). For the aim of color coordinate measurement, the high-resolution standardized screen has been used with graphical software. The dyed samples have been positioned on graphical square of the horizontally located screen surface. The color of the graphical square was changing programmatically until achieving corresponding color and the Lab-coordinate values have been recorded (*see* Table 26.1).

The dyed samples have been studied light fastness and stability against wet treatment. For the light-fastness test 4 cm^2 dyed fiber sample was places of the white surface, irradiated by UV light for 2 hours, and re-measured the color according to the above-described method. Finally, the fastness has been calculated using the following relationship:

$$\Delta E = \sqrt{(\Delta L)^2 + (\Delta a)^2 + (\Delta b)^2}$$

where; $\Delta L = L^*_{after} - L^*_{before}$; $\Delta a = a^*_{after} - a^*_{before}$; $\Delta b = b^*_{after} - b^*_{before}$

The fastness against wet treatment has been performed and resistance against extraction by Soap, HCl, NaOH, and ethanol have been carried out. The color changes have been calculated as described above and the results are listed in Table 26.2.

TABLE 26.2 Light Fastness and Stability Against Wet Treatment of 4a–d

SL. No.	Dye	Light Fastness (%)	Soap, 5%	HCl, 5%	NaOH, 5%	Organic Solvents (EtOH)
1.	4a	90	94	92	86	94
2.	4b	85	93	92	88	94
3.	4c	80	91	90	81	96
4.	4d	93	95	93	87	94

26.3 CONCLUSION

We are able to suggest that 6-aminocoumarin may be used as a diazo partner in the azo coupling reaction for obtaining dyes, bearing both coumarin and azo chromophores and characterized good spectral and technical properties.

KEYWORDS

- **2-hydroxybenzoic acid**
- **azo dyes**
- **diazotization and azo coupling**
- **dye exhaustion**
- **hydrochloric acid**
- **indazole-3-carboxamide hybrids**

REFERENCES

1. Nikpassand, M., Fekri, L. Z., Changiz, N., & Imani, F., (2014). Synthesis of new 3-cyanocoumarins with C-6 azo function using ultrasound and grinding techniques in the presence of nano Fe_3O_4. *Letters in Organic Chemistry, 11*(1), 29–34.
2. Rasheed, O. K., & Quayle, P., (2018). Azo dyes: New palladium- and copper-catalyzed coupling reactions on an old template. *Synthesis, 50*(13), 2608–2616.

3. Angapelly, S., Sri Ramya, P. V., Angeli, A., Supuran, C. T., & Arifuddin, M., (2017). Sulfocoumarin-, coumarin-, 4-sulfamoylphenyl-bearing indazole-3-carboxamide hybrids: Synthesis and selective inhibition of tumor-associated carbonic anhydrase isozymes IX and XII. *ChemMedChem, 12*(19), 1578–1584.
4. Amjad, R., Munawar, M. A., Khan, S. R., & Naeem, M., (2009). Synthesis and spectral studies of some novel coumarin based disperse azo dyes. *Pakistan Journal of Scientific and Industrial Research, 52*(3), 117–121.
5. Guha, S., Lohar, S., Hauli, I., Mukhopadhyay, S. K., & Das, D., (2011). Vanillin-coumarin hybrid molecule as an efficient fluorescent probe for trace level determination of Hg(II) and its application in cell imaging. *Talanta, 85*(3), 1658–1664.
6. Choudhari, B. P., & Mulwad, V. V., (2005). Synthesis and antimicrobial screening of N-[coumarin-6-ylamino]thiazolidinone and spiro indolo-thiazolidinone derivatives. *Indian Journal of Chemistry, Section B: Organic Chemistry Including Medicinal Chemistry, 44B*(5), 1074–1078.
7. Mulwad, V. V., & Mayekar, S. A., (2007). Synthesis and antimicrobial screening of 5-(4,7-dimethyl-2-oxo-2H-benzopyran-6-ylazo)-2-methyl-6-morpholin-4-yl-2,3-dihydro-3Hpyrimidin-4-one and 5-(4,7-dimethyl-2-oxo-2H-benzopyran-6-ylazo)-2-methyl-6-piperidin-1-yl-2,3-dihydro-3H-pyrimidin-4-one. *Indian Journal of Chemistry, Section B: Organic Chemistry Including Medicinal Chemistry, 46B*(11), 1873–1878.
8. Choudhari, B. P., & Mulwad, V. V., (2006). Synthesis and antimicrobial screening of 3H,11H-9-methyl-3-oxopyrano[2,3-f]cinnolino[3,4-c]pyrazole and its derivatives. *Indian Journal of Chemistry, Section B: Organic Chemistry Including Medicinal Chemistry, 45B*(1), 309–313.
9. Mulwad, V. V., Mhamunkar, Y., & Langi, B., (2010). Synthesis and biological activity of heterocycles synthesized from diazo compounds. *Indian Journal of Heterocyclic Chemistry, 19*(4), 349–352.
10. Okamoto, H., & Morita, Y., (2012). *Alkoxyphenylazocoumarin-Type Gelling Agents, Organic Substance Gels Formed Therewith, and Their Preparation.* Japan Patent JP2012017384, 26 01.
11. Amin, K. M., Abdel, G. N. M., Abdel, R. D. E., & El Ashry, M. K. M., (2014). New series of 6-substituted coumarin derivatives as effective factor Xa inhibitors: Synthesis, in vivo antithrombotic evaluation and molecular docking. *Bioorganic Chemistry, 52*, 31–43.
12. Toussirot, M., Nowik, W., Hnawia, E., Lebouvier, N., Hay, A. E., De La Sayette, A., Dijoux-Franca, M. G., Cardon, D., & Nour. M., (2014). Dyeing properties, coloring compounds and antioxidant activity of Hubera nitidissima (Dunal) Chaowasku (Annonaceae). *Dyes and Pigments, 102*, 278–284.
13. Datta, P., Sardar, D., Saha, R., Mondal, T. K., & Sinha, C., (2013). Structure, photophysics, electrochemistry and DFT calculations of [RuH(CO)(PPh3)2(coumarinyl-azo-imidazole)]. *Polyhedron, 53*, 193–201.

CHAPTER 27

The Determination of the Complex Formation Process of Lead(II) with Macromolecular Organic Substances-Fulvic Acids at pH = 9

TAMAR MAKHARADZE

Ivane Javakhishvili Tbilisi State University, R. Agladze Institute of Inorganic Chemistry and Electrochemistry, Tbilisi, Georgia

ABSTRACT

The complex formation process between Pb(II) and fulvic acids (FA) was studied by the solubility method at pH = 9.0. The suspension of $Pb(OH)_2$ was used as a solid phase. FA were isolated from Paravani lake by the adsorption-chromatographic method. During the calculation of molar concentrations of FA, was taking into consideration the value of molar mass of FA at pH=9 (Mw = 7,610).

In this chapter, is shown that, during the complex formation process, every 0.24 part of an associate of FA, inculcates into lead's inner coordination sphere as an integral ligand, so it may assume, that the average molecular weight (MW) of the associate of FA which takes part in complex formation process equals to 1826.4. This part of the associate of FA was conventionally called an "active associate." The average MW of the "active associate" was used for the determination of the concentration of free ligand and average stability constant (1:1), which equals to $\beta = 1.26\times10^7$, $\lg \beta = 7.10$.

Advanced Polymer Structures: Chemistry for Engineering Applications. Omar Mukbaniani, DSc, Tamara Tatrishvili, PhD & Marc Jean M. Abadie, DSc (Eds.)

27.1 INTRODUCTION

Fulvic acids (FA) are the major organic matter of natural waters [1–6]. FA form stable complexes with heavy metals and radionuclides [1, 3, 7–17] and stipulate migration forms thereof in natural waters and soils [1, 3, 18–23]. Therefore, FA may effect on the transport, accumulation, bioavailability, and toxicity of metals in the environment.

The carboxyl and phenol groups of FA take an active part in complex formation and sorption processes, proceeding in natural waters, bottom sediments and soils [1, 3, 24]. The participation of carboxyl or phenol hydroxide groups depends not only on the value of pH (which determines the dissociation degree of carboxyl and phenol groups), also on the nature of metals [24].

In spite of research, experimental data on stability constants (β) of complex compounds of FA with lead (as in the case of other metals) are heterogeneous, and they differ in several lines from each other [25–36]. This condition is mainly stipulated by the ignoring the average molecular weight (Mw) of the associates of FA, which value in its turn depends on the value of pH and finally causes the wrong results. Therefore, it's difficult to investigate complex formation processes, taking place in natural waters, identify migration forms of Pb(II) and evaluate and assess chemical-ecological condition of natural waters. The objective of the work was to investigate the complex formation process between the pure samples of FA, isolated from natural water and Pb and to calculate the average stability constant of lead fulvate complex. Complex formation process was studied at pH=9.0 by the solubility method. While using this method, in case of choosing correctly the solid phase, practically is impossible polynuclear hydrolysis and is easy for the determination the exact composition and stability constants of complex. The suspension of $Pb(OH)_2$ was used as a solid phase.

27.2 EXPERIMENTAL METHODS AND MATERIALS

For obtaining pure samples of FA, after filtration through membrane filters (0.45 μm pore size), the water of Paravani lake was concentrated by the frozen method. The concentrated water samples were acidified with 6 M HCl to pH 2 and was put for 2 hours on a water bath at 60°C for coagulation of humin acids. Then the solution was centrifuged for 10 min at 8,000 rpm (Centrifuge T-23). For the isolation of FA from centrifugate was used the adsorption-chromatographic method. Charcoal was used as a sorbent.

Desorption of amino acids and carbohydrates was performed by means of 0.1 M HCl. For desorption of polyphenols was used 90% acetone water solution. The elution of the fraction of FA was performed with 0.1 N NaOH solution [1, 9, 37]. Obtained alkalic solution of FA, for the purification was passed through a cation-exchanger (KU-2-8). For determination the concentration of FA in obtained solution was used gravimetric method, the part of the solution was dried under vacuum until the constant weight was obtained. Elemental composition of standard samples of FA isolated from natural waters and average meaning of dissociation constants respectively equals to C – 53.75%, H – 4.29%; O – 40.48%, N – 0.68%, S – 0.50%; P – 0.01%; ashing 0.35; $pK_{H,COOH} = 4.37$, $pK_{H,Ph-OH} = 10.4$. Then, model solutions of FA were prepared. The solution of fulvate complexes was obtained by the solubility method. About 0.1 ml suspension of lead hydroxide and increasing quantity of standard solution of FA were placed in 15 ml capacity fluoroplastic cylinders. The concentration of FA in model solutions changes from 1.11×10^{-5} to 4.44×10^{-5} mol/L (Mw(FA) = 7,610). The initial concentration of Pb(II) was 3.00×10^{-6} mol/L before adding ligand, pH=9.0, ionic strength μ=0.01 (KNO_3), v = 10 ml. Then, it was stirred in a mechanical mixer for 100 hours, until the balance was achieved and then suspension was filtered through the membrane filters (0.45 μm pore size). In filtrates, the concentration of lead was measured by atomic absorption spectrophotometer (Agilent 280Z AA).

27.3 RESULTS AND DISCUSSION

The data show, that in line with the increasing of concentration of FA in solution, the concentration of Pb(II) increases for several times as well due to formation of fulvate complex (Table 27.1). During the calculation of molar concentrations of FA, the fact that they (FA) form associates in water solutions was taking into consideration. While investigation of water solutions of FA by the gel chromatographic method was established, that in the interval pH 4–11, there is a line dependence between the average molecular weight FA (Mw) and the value of pH, which is expressed in the following way: Mw = 1350pH-4540 [1]. At pH = 9, Mw(FA) = 7,610.

In the diluted water solutions, the forms of lead ($\beta_{PbOH}{}^{+} = 1.84\times10^{6}$, $\beta_{Pb(OH)2}{}^{0} = 3.17\times10^{12}$ [38]) were calculated through the following Eqns. (1)–(3):

$$Pb^{2+}\% = 100/(1 + \beta_{1,1}\,[OH^-] + \beta_{1,2}\,[OH^-]^2) \quad (1)$$

$$PbOH^{+}\% = \beta_{1,1}[OH^{-}]100/(1 + \beta_{1,1}[OH^{-}] + \beta_{1,2}[OH^{-}] \quad (2)$$

$$Pb(OH)_2^{0}\% = \beta_{1,2}[OH^{-}]^2\, 100 \,/\, (1+ \ \beta_{1,1}[OH^{-}] + \beta_{1,2}[OH^{-}]^2 \quad (3)$$

As the results show, lead dihydroxy complex $Pb(OH)_2^{0}$ is the dominant form at pH = 9.0.

The reaction between lead dihydroxy complex and the anions of FA may be written in the following way:

$$Pb(OH)_2^{0} + m\ FA \leftrightarrows Pb(OH)_2\ (FA)m \quad (4)$$

$$\beta = [Pb(OH)_2\ FAm]/\{[Pb(OH)_2^{0}]\ [FA]^{m}\} \quad (5)$$

In solution, the concentration of lead dihydroxy fulvic complex equals to the difference between final $[Pb(II)_{total}]$ and initial $[Pb(II)_{free}]$ concentrations of lead received after formation of complex:

$$[Pb(OH)_2\ FAm] = [Pb(II)_{total} - Pb(II)_{free}] \text{ and } [Pb(OH)_2^{0}] = [Pb(II)_{free}] \quad (6)$$

Put these values in Eqn. (5):

$$\beta = [Pb(II)_{total} - Pb(II)_{free}]/\{[Pb(II)_{free}][FA]^{m}\} \quad (7)$$

From Eqn. (7):

$$\beta[Pb(II)_{free}] = [Pb(II)_{total} - Pb(II)_{free}]/[FA]^{m} \quad (8)$$

At the fixed pH, the left part of the Eqn. (8) is a permanent value, and we mark it as K.'

$$K' = [Pb(II)_{total} - Pb(II)_{free}]/[FA]^{m} \quad (9)$$

The logarithm of this Eqn. (9) is:

$$lgK' = lg[Pb(II)_{total} - Pb(II)_{free}] - mlg[FA] \quad (10)$$

The numeral value (m) of the stoichiometric coefficient or the number of ligands in the inner coordination sphere of complex equals to tangent of the tilt angle of straight line built in coordinates.

$$lg[Pb(II)_{total} - Pb(II)_{free}] - lg[FA_{total}] \tag{11}$$

To calculate the exact value of tangents tilt angle of a straight line, for this purpose was used the least square method:

$$m = tg\alpha = (n\Sigma x_i y_i - \Sigma x_i \Sigma y_i)/(n\Sigma x_i^2 - (\Sigma x_i)^2) \tag{12}$$

where; $x_i = lg\ [FA_{total}]$ and

$$y_i = lg\ [Pb(II)_{total} - Pb(II)_{free}] \tag{13}$$

TABLE 27.1 The Dependence of the Solubility of Lead Hydroxide on the Concentration of FA and the Necessary Data for the Determination the Composition of Lead Dihydroxy Fulvate Complexes pH = 9.0; Mw(FA) = 7,610; [Pb(II)free] = 3.00×10^{-6} mol/L

Mol/L			$[Pb_{total}]:[FA_{total}]$	lg $[FA_{total}]$	lg $[PbFA_m]$
$[FA_{total}]$	$[Pb_{total}]$	$[PbFA_m]$			
1.11×10^{-5}	4.77×10^{-5}	4.47×10^{-5}	1:0.23	–4,9547	–4,3497
1.66×10^{-5}	6.86×10^{-5}	6.56×10^{-5}	1:0.24	–4,7799	–4,1831
2.22×10^{-5}	9.34×10^{-5}	9.04×10^{-5}	1:0.24	–4,6536	–4,0438
2.77×10^{-5}	11.45×10^{-5}	11.15×10^{-5}	1:0.24	–4,5575	–3,9527
3.33×10^{-5}	13.20×10^{-5}	12.90×10^{-5}	1:0.25	–4,4775	–3,8894
3.88×10^{-5}	15.66×10^{-5}	15.36×10^{-5}	1:0.25	–4,4112	–3,81,36
4.44×10^{-5}	18.36×10^{-5}	18.06×10^{-5}	1:0.24	–4,3526	–3,7433

After the calculation, was obtained the numeral value of m (Mw(FA) = 7,610) which equal to 0.98 (Tables 27.1 and 27.2), most likely can be said that the complex obtained at pH=9 in system $Pb(OH)_2$(solid)-Pb(II) (solution)-FA-H_2O is the lead dihydroxy fulvate, with the structure 1:1.

TABLE 27.2 The Calculation of the Composition of Lead Fulvate Dihydroxy Complex by the Least Square Method, Mw(FA)=7,610; pH=9.0 . Xi =lg $[FA_{total}]$, Yi= lg$[PbFA_m]$

X_i	Y_i	X_iY_i	X_i^2
–4,9547	–4,3497	21,5514	24,5490
–4,7799	–4,1831	19,9948	22,8474
–4,6536	–4,0438	18,8182	21,6560
–4,5575	–3,9527	18,0144	20,7708
–4,4775	–3,8894	17,4115	20,0480
–4,4112	–3,8136	16,8225	19,4587
–4,3526	–3,7433	16,2931	18,9451

$\sum X_i = -32,187$; $(\sum X_i)^2 = 1036,0029$; $\sum Y_i = -27,9756$; $\sum X_i^2 = 148,275$; $\sum X_iY$=128,9059; and m = 0.98.

So, the complex formation reaction (pH = 9.0) may be written:

$$Pb(OH)_2^0 + FA = Pb(OH)_2FA \tag{14}$$

Therefore, the formula for the stability constant can be expressed in the following way:

$$\beta = [Pb(OH)_2FA]/\{[Pb(II)_{free}]\,[FA]\} = [Pb(II)_{total} - Pb(II)_{free}]/\{[Pb(II)_{free}]\,[FA]\} \tag{15}$$

The complexation of FA to metal ions cannot be described in strict frames because of the ill-defined nature of FA in contrast with the complexation of single ligands. So, in order to succeed the calculation of stability constants of fulvate complexes, it is necessary to make some assumptions.

In balanced solutions, correlation $[Pb(II)_{total}]:[FA_{total}]$ on average equals to 1:0.24 (Table 27.1). This means, that during the complex formation process, the associate of FA, which Mw at pH=9 equals to 7,610, divides and every 0.24 parts of this associate inculcates into lead's inner coordination sphere, as an integral ligand. So it may assume, that Mw of the associate of FA which takes part in complex formation process equals to 1826.4. This part of the associate of FA was conventionally called an "active associate" [9, 39]. The meaning of Mw of the "active associate" of FA (Mw = 1826.4) was used for determination the concentration of free ligand ($[FA_{free}]$) and average stability constant. It should be noted, that in case of using the average molecular weight of the associate (7,610), it will be impossible to calculate the concentration of free ligand. Without it, it's impossible to calculate the stability constant of lead fulvate complex.

For the calculation of average stability constant of lead dihydroxy fulvate at pH=9.0 was used Leden function F(L) [40].

$$\text{Function } F(L) = F(FA) = [Pb(OH)_2FA]/\{[Pb(II)_{free}][FA_{free}]\} =$$

$$= ([Pb(II)_{total}] - [Pb(II)_{free}])/([Pb(II)_{free}][FA_{free}]) = \beta_1 + \beta_2[FA_{free}] \tag{16}$$

where $[FA_{free}] = [FA_{total}] - [Pb(OH)_2FA] = [FA_{total}] - \{[Pb(II)_{total} - Pb(II)_{free}]\}$ (17)

In solubility method the concentration of $[Pb(II)_{free}]$, equals to the initial concentration of metal before adding of ligand in solution. It should be emphasized, using average molecular weights of "active associates" in given systems are both real and forced, because as it is shown from data (Table 27.1), during using the solubility method, in any point the molar concentration of fulvate complex exceeds the molar concentration of total associate, which average molecular weight at pH=9 is big enough. That's why it could not be calculated the free ligand. Without it, is impossible to determine the stability constant of fulvate complex.

When $[FA_{free}]$ aspires to zero, β could be found by the graphical method. The section which is cut on the ordinate by the straight line built in coordinates F(FA)— $[FA_{free}]$ equals to the stability constant. The value of β was calculated by the square method:

$$\beta = (\Sigma yi - a\Sigma xi)/n \tag{18}$$

where $a = (n\Sigma xiyi - \Sigma xi\Sigma yi)/[(n\Sigma xi^2 - (\Sigma xi)^2)]$ (19)

$$xi = [FA_{free}] \text{ and } yi = F(FA) \tag{20}$$

The necessary data, for calculation the average stability constants of lead dihydroxy fulvate complex are given in Tables 27.3 and 27.4. $\beta(Pb(OH)_2FA) = 1.26\times10^7$, lg β=7.10.

TABLE 27.3 The Necessary Data of the Calculation of Conditional Stability Constants of Lead Fulvate Complex by the Leden Method, pH = 9.0; Mw(FA) = 1826.4, $[Pb(II)_{free}] = 3.00\times10^{-6}$ mol/L; $F(FA) = [Pb(OH)_2FA]/([Pb(II)_{free}][FA_{free}])$; $[FA_{free}] = [FA_{total}] - [Pb(OH)_2FA]$; $[pb(OH)_2FA] = [Pb(II)_{total}] - [Pb(II)_{free}]$

Mol/l				**F (FA)**
$[FA]_{total}$	$[Pb(II)_{total}]$	[PbFA]	$[FA_{free}]$	
4.63×10^{-5}	4.77×10^{-5}	4.47×10^{-5}	0.16×10^{-5}	9.31×10^{6}
6.94×10^{-5}	6.86×10^{-5}	6.56×10^{-5}	0.38×10^{-5}	5.75×10^{6}
9.26×10^{-5}	9.34×10^{-5}	9.04×10^{-5}	0.22×10^{-5}	13.70×10^{6}
11.57×10^{-5}	11.45×10^{-5}	11.15×10^{-5}	0.42×10^{-5}	8.85×10^{6}
13.89×10^{-5}	13.20×10^{-5}	12.90×10^{-5}	0.99×10^{-5}	4.34×10^{6}
16.20×10^{-5}	15.66×10^{-5}	15.36×10^{-5}	0.84×10^{-5}	6.09×10^{6}
18.52×10^{-5}	18.36×10^{-5}	18.06×10^{-5}	0.46×10^{-5}	13.08×10^{6}

TABLE 27.4 The Calculation of Conditional Stability Constant of Lead Fulvate Complex by the Least Square Method*

X_i	Y_i	X_iY_i	X_i^2
0.16×10^{-5}	9.31×10^{6}	14.90	0.0256×10^{-10}
0.38×10^{-5}	5.75×10^{6}	21.85	0.1444×10^{-10}
0.22×10^{-5}	13.70×10^{6}	30.14	0.0484×10^{-10}
0.42×10^{-5}	8.85×10^{6}	37.17	0.1764×10^{-10}
0.99×10^{-5}	4.34×10^{6}	42.97	0.9801×10^{-10}
0.84×10^{-5}	6.09×10^{6}	51.15	0.7056×10^{-10}
0.46×10^{-5}	13.08×10^{6}	60.16	0.2116×10^{-10}

*Mw (FA) = 1826.4; pH = 9.0; X_i = [FA_{free}]; and Y_i = F(FA).
$\sum X_i = 3.47\times10^{-5}$; $(\sum X_i)^2 = 12.0409\times10^{-10}$; $\sum Y_i = 61.12\times10^{6}$; $\sum Xi^2 = 2.2921\times10^{-10}$; $\sum X_iY_i$ = 258.34; and $\beta = 1.26\times10^{7}$.

27.4 CONCLUSION

Calculations have approved that in diluted water solutions lead dihydroxy complex $Pb(OH)_2^0$ is the dominant form at pH = 9.0. It was shown that, during complex formation process, an associate of FA, which Mw at pH=9 equals to 7,610 divides and every 0.24 parts of this associate inculcates into lead's inner coordination sphere, as an integral ligand. This part of the associate of FA was conventionally called an "active associate." The meaning of Mw of the "active associate" of FA (Mw = 1826.4) was used for determination the concentration of free ligand ([FA_{free}]) and average stability constant. It is shown that using average molecular weights of "active associates" in given systems are both real and forced.

It was established, that in the $Pb(OH)_2$(solid) —Pb(II)(solution) —FA—H_2O system at pH=9.0, dominates the lead dihydroxy fulvate complex with the structure 1:1, which average stability constant $\beta=1.26\times10^7$, lg β=7.10.

ACKNOWLEDGMENTS

"This research was supported by Shota Rustaveli National Science Foundation of Georgia (SRNSFG) [grant number №YS-21-3728]"

KEYWORDS

- **active associates**
- **fulvic acids**
- **gravimetric method**
- **heterogeneous**
- **lead fulvate**
- **molecular weight**
- **stability constants**

REFERENCES

1. Varshal, G. M., (1994). *Migration Forms of Fulvic Acids and Metals in Natural Waters*. Dissertation. Vernad sky Institute of Geochemistry and Analytical Chemistry of Russian Academy of Sciences.
2. Dulaquas, G., Waeles, M., Gerringa, L. A., Midag, R., Rijkenberg, M., & Riso, R. G., (2018). The biogeochemistry of electroactive humic substances and its connection to iron chemistry in the north east Atlantic and the Western Mediterranean Sea. *J. Geophys. Res: Oceans, 123,* 5481–5499.
3. Osadchyy, V., Nabyvanets, B., Linnik, P., Osadcha, N., & Nabyvanets, Y., (2016). *Processes Determining Surface Water Chemistry*. Springer International Publishing Switzerland.
4. Makharadze, G. A., Supatashvili, G. D., & Varshal, G. M., (1989). Humic acids in surface waters of Georgia. *Hydrochemical Materials, 106*, 22–30.
5. Makharadze, G., Goliadze, N., Khaiauri, A., Makharadze, T., & Supatashvili, G., (2016). Fulvic and humin acids in surface waters of Georgia. *High-performas Polymers for Engineering-Based Composites* (pp. 167–179). Apple Academic Press, Waretown, NJ USA.
6. Pisarek, I., & Glowacki, M., (2015). Quality of groundwater and aquatic humic substances from main reservoire of ground water no. 333. *J. Ecol. Eng., 16*, 46–53.
7. Bertoli, A. C., Garcia, J. S., Trevisan, M. G., Ramalho, T. C., Matheus, P., & Freitas, M. P., (2016). Interactions fulvate-metal (Zn^{2+}, Cu^{2+} and Fe^{2+}):theoretical investigation of thermodynamic, structural and spectroscopic properties. *Biometals, 29*, 275–285.
8. Xu, H., Xu, D. C., & Wang, Y., (2017). Natural indices for the chemical hardness/softness of metal cations and ligands. *ACS Omega, 2*, 7185–7193.
9. Makharadze, T., & Makharadze, G., (2020). Investigation of complex formation process of copper with macromolecular organic substances, isolated from natural water. *Organic Chemistry Plus, 1,* 1–5.
10. Maccarthy, P., & O'Cinneide, S., (2006). Fulvic acids: Interactions with metal ions. *European Journal of Soil Science, 25,* 429–437.
11. De Oliveira, V. D., Fernandes, A. N., & Szpoganicz, B., (2018). Complexations of divalent metallic ions with fulvic acids. *Eclética Química, 43*, 54–58.

12. Boguta, P., & Sokolowska, Z., (2020). Zinc binding to fulvic acids: Assessing the impact of pH, metal concentrations and chemical properties of fulvic acids on the mechanism and stability of formed soluble complexes. *Molekules, 25*, 1297–1321.
13. Zhu, B., & Ryan, D. K., (2016). Characterizing the interaction between uranyl ion and fulvic acid using regional integration analysis (RIA) and fluorescence quenching. *J. Environ. Radioact., 153,* 97–103.
14. Wang, J., Lü, C., He, J., & Zhao, B., (2016). Binding characteristics of Pb^{2+} to natural fulvic acids extracted from the sediments in ake Wuliangsuhai, inner Mongolia plateau, P.R. China. *Environmental Earth Sciences, 75*, 768–779.
15. Schnitzer, M., & Skinner, S. I. M., (1967). Stability constants of Pb, Ni, Mn, Co, Ca and Mg fulvic acid complexes. *Soil Science, 103,* 247–252.
16. Schnitzer, M., & Hansen, E. H., (1970). An evaluation of methods for the determination of stability constants of metal-fulvic acid complexes. *Soil Science, 109,* 333–340.
17. Dinu, M., & Shkinev, V. M., (2020). Complexation of metal ions with organic substances of humus nature: Methods of study and structural features of ligands, and distribution of elements between species. *Geochemistry International, 58*, 200–211.
18. Moiseenko, T. I., Dinu, M. I., Gashkina, N. A., & Kremlevaa, T. A., (2013). Occurrence forms of metals in natural waters depending on water chemistry. *Water Resour., 40*, 407–416.
19. Adusei-Gyamfi, J., Ouddane, B., Rietveld, L., Cornard, J., & Criquet, J., (2019). Natural organic matter-cations complexation and its impact on water treatment: A critical review. *Water Research, 160*, 130–147.
20. Dinh, Q. T., Li, Z., Tran, T. A., Wang, D., & Liang, D., (2017). Role of organic acids on the bioavailability of selenium in soil: A review. *Chemosphere, 184,* 618–635.
21. Makharadze, G. A., Supatashvili, G. D., & Varshal, G. M., (1998). The research of the forms of copper in surface waters. *Hydrochemical Materials, 103,* 3–16.
22. Whitby, H., Planquette, H., Cassar, N., Bucciarelli, E., Osburn, C. L., Janssen, D. J., Jay, T., et al., (2020). A call for refining the role of humic-like substances in the oceanic iron cycle. *Scientific Reports, 10*, 6144–6156.
23. Linnik, P., Zhezheva, V., Linnik, R., & Ivanchenko, Ya., (2013). Influence of the component composition of organic matter on relationship between dissolved forms of metals in the surface waters. *Hydrobiological Journal, 49*, 91–108.
24. Rey Castro, C., Mongin, S., Huidobro, C., David, C., Salvador, J., Garces, J., Galceran, J., Mas, F., & Puy, J., (2009). Effective affinity distribution for the binding of metal ions to a generic fulvic acid in natural waters. *Environmental Science and Technology, 43*, 7184–7191.
25. Orsetti, S., Marco-Brown, J. L., Andrade, E. M., & Molina, F. V., (2013). Pb(II) binding to humic substances: An equilibrium and spectroscopic study. *Environ. Sci. Technol., 47,* 8325–8333.
26. Saar, R. A., & Weber, J. H., (1980). Lead(II)-fulvic acid complexes. Conditional stability constants, solubility, and implications for lead(II) mobility. *Environ. Sci. Technol., 14*, 877–880.
27. Sahu, S., & Banerjee, D. K., (1996). Complexation of copper (II), cadmium (II) and Lead(II) with humic and fulvic acids of Yamuna River sediments. In: *Chemistry for the Protection of the Environment* (pp. 375–388). NY, USA.

28. Turner, D. R., Varney, M. S., Whitfield, M., Mantoura, R. F. C., & Riley, J. P., (1986). Electrochemical studies of copper and lead complexation by fulvic acid. *Geochim. Cosmochim. Acta, 50*, 289–297.
29. Chakraborty, P., & Chakraborty, Ch. L., (2008). Competition from Cu(II), Zn(II) and Cd(II) in Pb(II) binding to Suwannee river fulvic acid. *Water Air Soil Pollut., 195*, 63–71.
30. Pinheiro, J. P., Mota, A. M., & Benedetti, M. F., (1999). Lead and calcium binding to fulvic acids: Salt effect and competition. *Environmental Science & Technology, 33*, 3398–3404.
31. Christl, I., Metzger, A., Heidmann, I., & Kretzschmar, R., (2005). Effect of humic and fulvic acid concentrations and ionic strength on copper and lead binding. *Environ Sci Technol., 39*, 5319–5326.
32. Xiong, J., Koopal, L. K., Tan, W. F., Fang, L. C., Wang, M. X., Zhao, W., Liu, F., et al., (2013). Lead binding to soil fulvic and humic acids: NICA-Donnan modeling and XAFS spectroscopy. *Environmental Science & Technology, 47*, 11634–11642.
33. Gondar, D., López, R., Fiol, S., Antelo, J. M., & Arce, F., (2006). Cadmium, lead, and copper binding to humic acid and fulvic acid extracted from an ombrotrophic peat bog. *Geoderma, 135*, 196–203.
34. Orsetti, S., Marco-Brown, J. L., Andrade, E. M., & Molina, F. V., (2013). Pb(II) binding to humic substances: An equilibrium and spectroscopic study. *Environ. Sci. Technol., 47*, 8325–8333.
35. Quan, G., & Yan, J., (2010). Binding constants of lead by humic and fulvic acids studied by anodic stripping square wave voltammetry. *Russian J. of Electrochemistry, 46*, 90–94.
36. Gurjia, Zh., Supatashvili, G. D., Makharadze, G. A., Varshal, G. M., & Chitiashvili, Z. D., (1989). Complex formation of lead ions with fulvic acids, derived from natural waters. *Bull. Georgian. Natl. Acad. Sci., 136*, 65–68.
37. Revia, R., & Makharadze, G., (1999). Cloud-point preconcentration of fulvic and humic acids. *Talanta, 48*, 409–413.
38. Gurjia, Zh., Supatashvili, G. D., Varshal, G. M., Makharadze, G. A., & Chitiashvili, Z. D., (1992). Monohydrolysis of lead ions in dilute solutions. *Proceedings Georgian Natl. Acad. Sci., 18*, 22–26.
39. Makharadze, G., Supatashvili, G., & Makharadze, T., (2018). New version of calculation of stability constant of metal-fulvate complexes on the example of zinc fulvate. *International Jl. of Environmental Science and Technology, 15*, 2165–2168.
40. Beck, M. T., & Nagypal, I., (1990). *Chemistry of Complex Equilibria*. Chichester, Horwood, New York.

CHAPTER 28

Investigation of Complex Formation Process of Nickel(II) with Fulvic Acids at pH = 5 by the Gel Filtration Method

TAMAR MAKHARADZE, LIA NADAREISHVIL, GIORGI MAKHARADZE, and NAZI GOLIADZE

Ivane Javakhishvili Tbilisi State University, Tbilisi, Georgia

ABSTRACT

Fulvic acids (FA) take an active part in complex formation processes and stipulate migration forms of heavy metals in natural waters. In spite of research, experimental data on stability constants of complex compounds of FA with heavy metals (among them nickel) are heterogeneous, and they differ in several lines from each other. One of the reasons of such a condition is ignoring an average molecular weight of the associates of FA, which finally causes the wrong results. The complex formation process of nickel with FA was studied by the gel filtration method (sephadex G-25) at pH=5.0. FA were separated from the water of the Paravani lake by the adsorption-chromatographic method. The charcoal was used as a sorbent. The average molecular weight of the FA at pH = 5.0 (Mw = 2,210) was used for determination the composition of nickel fulvate complex (1:1), the concentration of free ligand and average stability constant (β). $\beta = 1.47 \times 10^4$; $\lg \beta = 4.17$.

28.1 INTRODUCTION

Macromolecular organic substances-fulvic acids (FA) are natural organic complexing agents found in the environment, which originate from

Advanced Polymer Structures: Chemistry for Engineering Applications. Omar Mukbaniani, DSc, Tamara Tatrishvili, PhD & Marc Jean M. Abadie, DSc (Eds.)

chemically and microbially modified plant and animal matter. They are the major organic matter of natural waters. FA concentrations are ranging from less than 1 mg/L to more than 100 mg/L [1–6]. The carboxyl and phenol groups of FA take an active part in complex formation and sorption processes, proceeding in natural waters, bottom sediments and soils [1, 7–10]. $pK_{H,COOH}$ changes from 4.1 to 4.9, and $pK_{H,Ph-OH}$ changes from 9.3 to 10.6 [1, 4, 5, 9, 11]. In complex formation process, the participation of carboxyl or phenol groups of fulvic acids (FA) depends on different factors, first of all on the value of pH and the nature of metal.

FA form stable complexes with heavy metals and radionuclides [1, 10, 12–22] and stipulate migration forms thereof in natural waters and soils [1, 3, 23–28].

In spite of research, experimental data on stability constants (β) of complex compounds of FA with nickel (II) are heterogeneous and they differ in several lines from each other [29–36]. This condition is mainly stipulated by the ignoring the average molecular weight (Mw) of the associates of FA, which value in its turn depends on the value of pH and finally causes the wrong results. Therefore, it's difficult to investigate complex formation processes, taking place in natural waters, identify migration forms of nickel and evaluate and assess chemical-ecological condition of natural waters.

The objective of the work was to investigate the complex formation process between the pure samples of FA, isolated from natural water and nickel and to calculate the average stability constant of nickel fulvate complex. The complex formation process between FA and nickel were studied by the gel filtration method (sephadex G-25) at pH=5.0.

28.2 EXPERIMENTAL METHODS AND MATERIALS

For preparing pure samples of fulvic acids from the water of Paravani Lake was used the adsorption-chromatographic method. Charcoal was used as a sorbent. Desorption of amino acids and carbohydrates was performed by means of 0.1 M HCl. For desorption of polyphenols was used 90% acetone water solution. The elution of the fraction of FA was performed with 0.1 N NaOH solution [1, 37]. Obtained alkalic solution of FA, for the purification was passed through a cation-exchanger (KU-2-8). For determination the concentration of FA in obtained solution was used gravimetric method, the part of the solution was dried under vacuum until the constant weight was obtained. Then, model solutions of FA were prepared.

The parameters of sephadex G-25: the mass of dry gel – 17 g, the height of swelled layer of gel – 42 cm, the inner diameter of column – 1.6 cm. For the calibration of sephadex G-25 was used blue dextran, polyethylene glycols with molecular weights 300, 600, 1,000, and glucose. The titer of standard substances – 1 mg/ml, transmission speed 3 ml/min, apply volume of solution – 2 ml.

In fluoroplastic cylinders (volume 15 ml) the increasing quantity of standard solutions of FA were added to the solutions of nickel of the same concentrations, which were prepared from the standard solutions, intended for atomic absorption method. The constant ionic strength was made by adding 0.1 M 1 ml potassium nitrate (KNO_3). The final volume of solutions was 10 ml, T(Ni)=10 mkg/ml. Concentration of hydrogen ions was regulated by 0.01 M potassium hydroxide and 0.01 M nitric acid, in model solutions at pH = 5.

The aliquots (2 ml) of different solutions with the same pH were placed in the top part of the column. The elution process was done by 0.1 M KNO_3, that has the same pH as the aliquots of solution.

We determined the quantity of metals connected with FA in fractions, which releasing volume fits substances with molecular weight 300≤ Mw>5,000. The concentration of nickel was measured by atomic absorption spectrophotometer (Perkin Elmer 200).

28.3 RESULTS AND DISCUSSION

FA form associates in water solutions. It was established, that in the interval pH 4-11, there is a line dependence between Mw and the value of pH, which is expressed in the following way: Mw = 1350 pH – 4540 (1) [1]. Molar concentrations of FA at pH = 5 (Mw = 2,210) could be calculated.

As it is shown from the results (Table 28.1), the containing of nickel in high weight molecular fraction (300 ≤ Mw > 5,000) increases with the increasing of the concentration of FA, which could only be explained by the formation of fulvate complexes.

If it is not taken into account charges of ions, the reaction of formation of nickel fulvate complexes, could be written in the following way:

$$Ni(II)_{free} + mFA_{total} = NiFA_{m} \tag{2}$$

$$\beta = [NiFA_{m}]/([Ni(II)_{free}]\,[FA_{total}]^{m}) \tag{3}$$

In homogeneous systems, where $[Ni(II)_{free}]$ is not constant value, the numeral value of stoichiometric coefficient (m) equals to tangent of tilt angle of straight built-in coordinates $lg([NiFAm]/[Ni(II)_{free}])$ and lg[FA]. During the investigation by the gel chromatographic method, [NiFAm] equals to the concentration of metal in high weight molecular fraction ($300 \leq Mw > 5{,}000$) and $[Ni(II)_{free}] = [Ni(II)_{total}] - [NiFAm]$ [4]. To calculate the exact value of tangents tilt angle of a straight line, for this purpose was used the least square method:

$$m = tg\alpha = (n\Sigma x_i y_i - \Sigma x_i \Sigma y_i)/(n\Sigma x_i^2 - (\Sigma x_i)^2) \quad (5)$$

where; $x_i = lg\ [FA_{total}]$ and $y_i = lg([NiFAm]/[Ni(II)_{free}])$

After the calculation, was obtained the numeral value of m which equal to 0.98 (Tables 28.1 and 28.2).

TABLE 28.1 The Necessary Data for the Identification of the Composition of Nickel Fulvate Complexes, pH = 5.0; Mw(FA) = 2210; $[Ni(II)_{total}] = 17.17\times10^{-5}$

mol/L			$[NiFA_m]:[Ni(II)_{free}]$	lg $[FA_{total}]$	$lg([NiFA_m]:[Ni(II)_{free}])$
$[FA_{total}]$	$[NiFA_m]$	$[Ni(II)_{free}]$			
3.82×10^{-5}	2.39×10^{-5}	14.28×10^{-5}	0.17	–4.4179	–0.7695
5.73×10^{-5}	3.73×10^{-5}	13.44×10^{-5}	0.28	–4.2418	–0.5528
7.64×10^{-5}	4.91×10^{-5}	12.26×10^{-5}	0.40	–4.1169	–0.3979
9.55×10^{-5}	5.17×10^{-5}	12.00×10^{-5}	0.43	–4.0199	–0.3665
11.46×10^{-5}	6.16×10^{-5}	11.01×10^{-5}	0.56	-3.9408	–0.2518
13.37×10^{-5}	6.42×10^{-5}	10.75×10^{-5}	0.60	-3.8738	–0.2218
15.28×10^{-5}	6.89×10^{-5}	10.28×10^{-5}	0.67	-3.8159	–0.1739

TABLE 28.2 The Calculation of the Composition of Nickel Fulvate Complex by the Least Square Method, Mw(FA) = 2210; pH = 5.0; $X_i = lg\ [FA_{total}]$, $Y_i = lg([NiFA_m]:[Ni(II)_{free}])$

X_i	Y_i	X_iY_i	X_i^2
–4.4179	–0.7695	3.3995	19.5178
–4.2418	–0.5528	2.3449	17.9928
–4.1169	–0.3979	1.6381	16.9488
–4.0199	–0.3665	1.4733	16.1596
-3.9408	–0.2518	0.9923	15.5299
-3.8738	–0.2218	0.8592	15.0063
-3.8159	–0.1739	0.6636	14.5611

$\Sigma X_i = -28.427$; $(\Sigma X_i)^2 = 808.0943$; $\Sigma Y_i = -2.7342$; $\Sigma X_i^2 = 115.7163$; $\Sigma X_iY_i = 11.3709$; m = 0.98.

So, in Ni (II)(solution) —FA—H_2O system at pH=5.0, dominates nickel fulvate complex, with the structure 1:1. So the complex formation reaction (pH = 5.0) may be written:

$$\text{Ni(II)} + \text{FA} = \text{NiFA} \quad (6)$$

$$\beta = [\text{NiFA}]/([\text{Ni(II)}_{\text{free}}]\,[\text{FA}_{\text{total}}]) \quad (7)$$

For the calculation of average stability constant of nickel fulvate at pH=5.0 was used Leden function F(L) [38]. The necessary data, for calculation of the average stability constants of nickel fulvate complex are given in Tables 28.3 and 28.4.

TABLE 28.3 The Necessary Data for the Calculation of Average Stability Constants of Nickel Fulvate Complex by the Leden Method, pH = 5.0; $[\text{Ni(II)}_{\text{total}}] = 17.17\times10^{-5}$M; Mw(FA) = 2210; F(FA) = $[\text{NiFA}]/([\text{Ni}_{\text{free}}][\text{FA}_{\text{free}}])$; $[\text{FA}_{\text{free}}] = [\text{FA}_{\text{total}}] - [\text{NiFA}]$; $[\text{Ni(II)}_{\text{free}}] = [\text{Ni(II)}_{\text{total}}] - [\text{NinFA}]$

Mol/L				F(FA)
$[\text{FA}_{\text{totsal}}]$	[NiFA]	$[\text{Ni(II)}_{\text{free}}]$	$[\text{FA}_{\text{free}}]$	
3.82×10^{-5}	2.39×10^{-5}	14.28×10^{-5}	1.43×10^{-5}	1.17×10^{4}
5.73×10^{-5}	3.73×10^{-5}	13.44×10^{-5}	2.00×10^{-5}	1.39×10^{4}
7.64×10^{-5}	4.91×10^{-5}	12.26×10^{-5}	2.73×10^{-5}	1.47×10^{4}
9.55×10^{-5}	5.17×10^{-5}	12.00×10^{-5}	4.38×10^{-5}	0.98×10^{4}
11.46×10^{-5}	6.16×10^{-5}	11.01×10^{-5}	5.3×10^{-5}	1.05×10^{4}
13.37×10^{-5}	6.42×10^{-5}	10.75×10^{-5}	6.95×10^{-5}	0.86×10^{4}
15.28×10^{-5}	6.89×10^{-5}	10.28×10^{-5}	8.39×10^{-5}	0.80×10^{4}

$\beta = 1.47\times10^{4}$; lg $\beta = 4.17$.

TABLE 28.4 The Calculation of Average Stability Constant of Nickel Fulvate Complex by the Least Square Method, Mw(FA) = 2210; pH = 5.0; $X_i = [\text{FA}_{\text{free}}]$; Y_i = F(FA)

X_i	Y_i	X_iY_i	X_i^2
1.43×10^{-5}	1.17×10^{4}	0.1673	2.2449×10^{-10}
2.00×10^{-5}	1.39×10^{4}	0.2780	4.0000×10^{-10}
2.73×10^{-5}	1.47×10^{4}	0.4013	7.4529×10^{-10}
4.38×10^{-5}	0.98×10^{4}	0.4292	19.1844×10^{-10}
5.3×10^{-5}	1.05×10^{4}	0.5565	28.0900×10^{-10}
6.95×10^{-5}	0.86×10^{4}	0.5977	48.3025×10^{-10}
8.39×10^{-5}	0.80×10^{4}	0.6712	70.3921×10^{-10}

$\sum X_i = 31.18\times10^{-5}$; $(\sum X_i)^2 = 972.1924\times10^{-10}$; $\sum Y_i = 7.72\times10^{4}$; $\sum X_i^2 = 179.6670\times10^{-10}$; $\sum X_iY_i = 3.1012$; $\beta = 1.47\times10^{4}$.

The calculation of characteristics of Leden's function F(L) = F(FA) = [NiFA]/([Ni(II)$_{free}$] × [FA$_{free}$]) = $\beta_1 + \beta_2$[FA$_{free}$] [8] is easy during the investigation of fulvate complexes by the gel chromatographic method. The concentration of metal connected to fulvic acids [NiFA] quantitatively equals to the quantity of metals (mol/l) determined in high weight molecular fraction (300 ≤ Mw > 5,000), [FA$_{free}$] = [FA$_{total}$] – [NiFA], and [Ni(II)$_{free}$] = [Ni(II)$_{total}$] – [NiFA], where [FA$_{total}$] – the total amount of FA in solution, [FA$_{free}$] – the concentration of free ligand, and [Ni(II)$_{total}$] – the total amount (mol/l) of metal in the sample, which is constant value according to the condition of the experiment.

When [FA$_{free}$] aspires to zero, β could be found by the graphical method. The section which is cut on the ordinate by the straight-line built-in coordinates F(FA)—[FA$_{free}$] equals to the stability constant. The value of β was calculated by the square method.

$$\beta = (\Sigma y_i - a\Sigma x_i)/n \tag{9}$$

where;

$$a = (n\Sigma x_i y_i - \Sigma x_i \Sigma y_i)/(n\Sigma x_i^2 - (\Sigma x_i)^2) \tag{10}$$

$$x_i = [FA_{free}] \text{ and } y_i = F(FA)$$

$$\beta = 1.47\times10^4;\ \lg \beta = 4.17$$

28.4 CONCLUSION

It was established, that in the Ni(II) (solution)—FA–H_2O system at pH = 5.0, dominates the nickel fulvate complex with the structure 1:1, which β = 1.47×10^4; lg β = 4.17.

ACKNOWLEDGMENTS

The work was done by supporting the World Federation of Scientists and the World Laboratory.

KEYWORDS

- **frozen method**
- **fulvic acids**
- **gel filtration**
- **macromolecular organic substances**
- **nickel fulvate**
- **stability constants**

REFERENCES

1. Varshal, G. M., (1994). *Migration forms of Fulvic Acids and Metals in Natural Waters.* Dissertation. Vernadsky Institute of Geochemistry and Analytical Chemistry of Russian Academy of Sciences.
2. Dulaquas, G., Waeles, M., Gerringa, L. A., Midag, R., Rijkenberg, M., & Riso, R. G., (2018). The biogeochemistry of electroactive humic substances and its connection to iron chemistry in the north east Atlantic and the Western Mediterranean Sea. *J. Geophys. Res., 123,* 5481–5499.
3. Osadchyy, V., Nabyvanets, B., Linnik, P., Osadcha, N., & Nabyvanets, Y., (2016). *Processes Determining Surface Water Chemistry.* Springer International Publishing Switzerland.
4. Makharadze, G. A., Supatashvili, G. D., & Varshal, G. M., (1989). Humic acids in surface waters of Georgia. *Hydrochemical Materials, 106,* 22–30.
5. Makharadze, G., Goliadze, N., Khaiauri, A., Makharadze, T., & Supatashvili, G., (2016). Fulvic and humin acids in surface waters of Georgia. *High-performas Polymers for Engineering-based Composites* (pp. 167–179). Apple Academic Press, Waretown, NJ USA.
6. Pisarek, I., & Glowacki, M., (2015). Quality of groundwater and aquatic humic substances from main reservoire of ground water No. 333. *J. Ecol. Eng., 16,* 46–53.
7. Buffle, J., Deladoey, P., Greter, F. L., & Haerdi, W., (1980). Study of the complex formation of copper(II) by humic and fulvic substances. *Analitica Chimica Acta, 116,* 255–274.
8. Joris, W. J., Van, S., Dan, B. K., & Jon, P. G., (2010). Acid-base and copper-binding properties of three organic matter fractions isolated from a forest floor soil solution. *Geochimica et Cosmochimica Acta, 74,* 1391–1406.
9. Rey-Castro, C., Mongin, S., Huidobro, C., David, C., Salvador, J., Garces, J., Galceran, J., et al., (2009). Effective affinity distribution for the binding of metal ions to a generic fulvic acid in natural waters. *Environmental Science and Technology, 43,* 7184–7191.
10. Boguta, P., & Sokolowska, Z., (2020). Zinc binding to fulvic acids: Assessing the impact of pH, metal concentrations and chemical properties of fulvic acids on the mechanism and stability of formed soluble complexes. *Molekules, 25,* 1297–1321.

11. Lenoir, T., & Manceau, A., (2010). Number of independent parameters in the potentiometric titration of humic substances. *Langmuir, 26*, 3998–4003.
12. Bertoli, A. C., Garcia, J. S., Trevisan, M. G., Ramalho, T. C., Matheus, P., & Freitas, M. P., (2016). Interactions fulvate-metal (Zn^{2+}, Cu^{2+} and Fe^{2+}): Theoretical investigation of thermodynamic, structural and spectroscopic properties. *Biometals, 29*, 275–285.
13. Xu, H., Xu, D. C., & Wang, Y., (2017). Natural indices for the chemical hardness/softness of metal cations and ligands. *ACS Omega, 2*, 7185–7193.
14. Makharadze, T., & Makharadze, G., (2021). Measurement of complex formation process of nickel (II) with freshwater fulvic acids using the solubility method. *Fine Chemical Engineering, 2*, 54–61.
15. Makharadze, T., & Makharadze, G., (2020). Investigation of complex formation process of copper with macromolecular organic substances, isolated from natural water. *Organic Chemistry Plus, 1*, 1–5.
16. Makharadze, G., & Makharadze, T., (2014). Method of calculation of stability constants of fulvic complexes on the example of copper. *J. of Chemistry and Chemical Engineering, 8*, 108–111.
17. Maccarthy, P., O'Cinneide, S., (2006). Fulvic acids: Interactions with metal ions. *European Journal of Soil Science, 25*, 429–437.
18. *De Oliveira, V. D., Fernandes, A. N., & Szpoganicz, B., (*2018*).* Complexations of divalent metallic ions with fulvic acids. *Eclética Química, 43*, 54–58.
19. Saldana-Robles, A., Saldana-Robles, N., Saldana-Robles, A. L., Damian-Ascencio, C., Rangel-Hernandez, V. H., & Cuerraa-SSanchez, R., (2017). Arsenic removal from aqueous solutions and the impact of humic and fulvic acids. *J. of Cleaner Production, 159*, 425–431.
20. Zhu, B., & Ryan, D. K., (2016). Characterizing the interaction between uranyl ion and fulvic acid using regional integration analysis (RIA) and fluorescence quenching. *J. Environ Radioact., 153*, 97–103.
21. Town, R. M., Van, L. H. P., & Buffle, J., (2012). Chemodynamics of soft nanoparticulate complexes: Cu(II) and Ni(II) complexes with fa and aquatic humic acids. *Environmental Science and Technology, 46,* 10487–10498.
22. Wang, J., Lü, C., He, J., & Zhao, B., (2016). Binding characteristics of Pb^{2+} to natural fulvic acids extracted from the sediments in Ake Wuliangsuhai, inner Mongolia plateau, P.R. China. *Environmental Earth Sciences, 75*, 768–779.
23. Moiseenko, T. I., Dinu, M. I., Gashkina, N. A., & Kremlevaa, T. A., (2013). Occurrence forms of metals in natural waters depending on water chemistry. *Water Resour., 40*, 407–416.
24. Adusei-Gyamfi, J., Ouddane, B., Rietveld, L., Cornard, J., & Criquet, J., (2019). Natural organic matter-cations complexation and its impact on water treatment: A critical review. *Water Research, 160*, 130–147.
25. Dinh, Q. T., Li, Z., Tran, T. A., Wang, D., & Liang, D., (2017). Role of organic acids on the bioavailability of selenium in soil: A review. *Chemosphere, 184,* 618–635.
26. Makharadze, G. A., Supatashvili, G. D., & Varshal, G. M., (1988). The research of the forms of copper in surface waters. *Hydrochemical Materials, 103,* 3–16.
27. Whitby, H., Planquette, H., Cassar, N., Bucciarelli, E., Osburn, C. L., Janssen, D. J., Jay, T., et al., (2020). A call for refining the role of humic-like substances in the oceanic iron cycle. *Scientific Reports, 10*, 6144–6156.

28. Mostofa, K. M. G., Wu, F., Liu, C. Q., Vione, D., Yoshioka, T., Sakugawa, H., & Tanue, E., (2011). Photochemical, microbial and metal complexation behavior of fluorescent dissolved organic matter in the aquatic environments. *Geochemical Journal, 45*, 235–254.
29. Cheam, V., & Gamble, D. S., (1974). Metal-fulvic acid chelation equilibrium in aqueous $NaNO_3$ solution Hg(II), Cd(II) and Ni(II) fulvate complexes. *Can. J. Soil Science, 54,* 413–417.
30. Goncharova, T. O., Kolosov, I. V., & Kaplin, B. T., (1978). Hydrolysis and complexation of nickel ions in solutions of FA. *Hydrochemical Materials, 71*, 64–72.
31. Kostić, I., Anđelković, T., Anđelković, D., Nikolić, R., Bojić, A., Cvetković, T., & Nikolić, G., (2016). Interaction of cobalt(II), nickel(II) and zinc(II) with humic-like ligands studied by ESI-MS and ion-exchange method. *J. Serb. Chem. Soc., 81*, 255–270.
32. Mantoura, R. F. C., Dixon, A., & Rilly, J. P., (1978). The speciation of trace metals with humic compounds in natural waters. *Thalassia Jugoslavica, 14,* 127–145.
33. Makharadze, G., Goliadze, N., Makharadze, T., & Supatashvili, G., (2014). The determination of average stability constant of nickel-FA complex at pH = 8.0 by the solubility method. *J. of Chemistry and Chemical Engineering, 8*, 344–348.
34. Schnitzer, M., & Skinner, S. I. M., (1967). Stability constants of Pb, Ni, Mn, Co, Ca and Mg fulvic acid complexes. *Soil Science, 103,* 247–252.
35. Schnitzer, M., & Hansen, E. H., (1970). An evaluation of methods for the determination of stability constants of metal-fulvic acid complexes. *Soil Science, 109,* 333–340.
36. Schnitzer, M., & Kerndorff, H., (1981). Reactions of fulvic acid with metal ions. *Water Air Soil Pollut., 15*, 97–108.
37. Revia, R., & Makharadze, G., (1999). Cloud-point preconcentration of fulvic and humic acids. *Talanta, 48*, 409–413.
38. Beck, M. T., & Nagypal, I., (1990). *Chemistry of Complex Equilibria.* Chichester, Horwood, New York.

CHAPTER 29

Study of Water-In-Oil Emulsions on the Basis of Sodium Cholate and Tetraethylene Glycol Dodecyl Ether: Stability Estimation

RUSUDAN LAZASHVILI, NINO LOMINADZE, NINO TAKAISHVILI, MANANA KEKENADZE, GEORGE BEZARASHVILI, and MARINA RUKHADZE

Faculty of Exact and Natural Sciences, Ivane Javakhishvili Tbilisi State University, Tbilisi, Georgia

ABSTRACT

Water-in-oil or reverse emulsions were prepared based on the biological surfactant sodium cholate and nonionic surfactant tetraethylene glycol dodecyl ether. The decomposition kinetics of an emulsion stabilized with sodium cholate is described with satisfactory accuracy by a first-order equation, and the decomposition kinetics of an emulsion stabilized with tetraethylene glycol dodecyl ether obey a second-order kinetic equation. Modification of water droplets of emulsion with sodium fluoride increases the stability of reversed emulsions. A suspension of barium sulfate in the water droplets of water/hexane/sodium cholate emulsion was obtained. It has been established that the emulsion is stable for 20 minutes and its decomposition proceeds according to the second-order kinetic equation.

Advanced Polymer Structures: Chemistry for Engineering Applications. Omar Mukbaniani, DSc, Tamara Tatrishvili, PhD & Marc Jean M. Abadie, DSc (Eds.)

29.1 INTRODUCTION

Emulsions are microheterogeneous systems in which one liquid is dispersed as fine droplets (>100 nm) into another immiscible liquid. Natural emulsions include most products of plant and animal origin. Milk is an important raw material for many products. The natural emulsion is egg yolk. Emulsions in the food industry include margarine, mayonnaise, various sauces. In the pharmaceutical industry, many preparations are used in the form of emulsions, while preparations for internal use are prepared mainly as oil/water emulsions, and for external use in the form of water/oil emulsions. Cosmetics, pesticides, dyes, lubricants, etc. Are usually used in the form of emulsions. The main function of emulsions is to transport water-insoluble substances in a stable and therefore well-dispersed form, although they have other specific functions [1]. Therefore, the study of emulsions as potential carriers of vitamins, drugs, dietary supplements, carotenoids, curcumin, and other bioactive substances is an urgent issue of scientific research.

The purpose of this work was: to obtain reversed or water-in-oil emulsions and determine the optimal conditions for their preparation; to study the dynamics of the decomposition of the resulting reversed emulsion systems and evaluate their stability; to plot kinetic curves that reflect the decrease in optical density with time; determination of kinetic parameters that quantitatively characterize the stability of water-in-oil emulsions; preparation of lyophobic colloids in the water droplets of reversed emulsions and evaluating the expediency of application of turbidimetric method for their determination.

29.2 EXPERIMENTAL METHODS AND MATERIALS

29.2.1 REAGENTS AND EQUIPMENT

The intensity of light scattering by water-in-oil emulsions was measured on a KFK-2 photocolorimeter (Optical-mechanical factory, Zagorsk, Russia). Wavelength 400 nm.

Reagents used: biologically active substance sodium salt of cholic acid, empirical formula $C_{24}H_{39}O_5Na$ (HLB = 18); Tetraethylene glycol dodecyl ether Brij-30, empirical formula $CH_3(CH_2)_{11}(OCH_2CH_2)_4OH$ (HLB=9.7); Hexane and cyclohexane were applied as the oil phase, butanol, and isoamyl alcohol were used as co-surfactants.

To study the properties of emulsions, aqueous solutions of NaF, $CaCl_2$, KI, Na_2HPO_4, but aqueous solutions $BaCl_2$ and Na_2SO_4 salts were used for turbidimetric determination.

29.2.2 CALCULATIONS

The results were processed according to the following formulae. The reproducibility of the results was evaluated by known statistical methods [2]. An average of the obtained results was calculated by the number (n) of experiments performed:

$$\bar{X} = \frac{\Sigma X_i}{n} \tag{1}$$

where; X_i is i^{th} result of trial.

The standard deviation was found using the formula:

$$SD = \left[(1/n-1)\Sigma(\text{Xi X})^2\right]^{\frac{1}{2}} \tag{2}$$

The value of the confidence interval was determined using the following formula:

$$\bar{X} \pm t_\alpha \bullet \frac{\text{SD}}{\sqrt{n}} \tag{3}$$

where; t_α is the student's coefficient for the significance level α.

In the kinetic processing of the results, the equations of the first and second order are used:

$$D = D_0 e^{-k(t-\tau)} \text{ I order} \tag{4}$$

$$\frac{1}{D} - \frac{1}{D_0} = k'(t-\tau) \text{ II order} \tag{5}$$

where; D and D_0 are the optical densities at time 't' and at the beginning of measurements, respectively; 'k' is the conversion rate constant; and 'k'' is the proportional value of the conversion rate constant.

The emulsion rate constant and half-life were calculated using known formulas [3].

29.3 RESULTS AND DISCUSSION

29.3.1 WATER-IN-HEXANE EMULSIONS STABILIZED WITH SODIUM CHOLATE: STABILITY ESTIMATION

The aim of this subsection was to obtain water-in-oil reverse emulsions using the anionic biologically active substance sodium cholate, to select the optimal amounts of sodium cholate and water to obtain a stable emulsion, and to study the effect of the co-surfactant butanol additive.

The effect of sodium cholate concentration on optical density has been studied. Due to this, an aqueous solution of sodium cholate was prepared, different amounts of which were introduced in hexane. The experiment showed that an increase of the concentration of sodium cholate in hexane at a constant content of water leads to an increase of optical density which indicates an increase of the number of emulsion droplets (Figure 29.1).

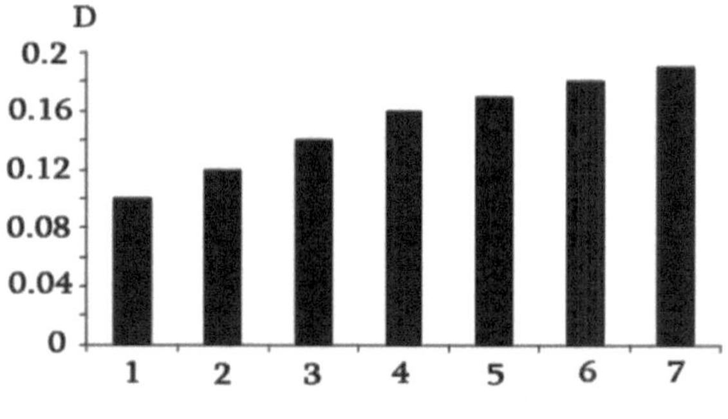

FIGURE 29.1 Diagram of optical density for water/hexane emulsions stabilized with different concentration of sodium cholate (numbers 1, 2, 3, 4, 5, 6, 7 correspond to concentrations of sodium cholate 1.0, 1.5, 2.0, 6.0, 7.0, 8.0, and 10.0%).

The effect of electrolytes introduced into the aqueous phase has been studied. As the experiment showed, an emulsion prepared on a sodium hydrophosphate buffer reveals less optical density than one prepared on water, and an emulsion prepared on water modified with sodium fluoride occupies an intermediate position by optical density between these two systems. Optical density is also reduced by the addition of calcium chloride. The decrease in optical density upon the introduction of electrolytes at a constant content of water is stipulated by reducing in the number of emulsion droplets due to droplet expansion (Figure 29.2).

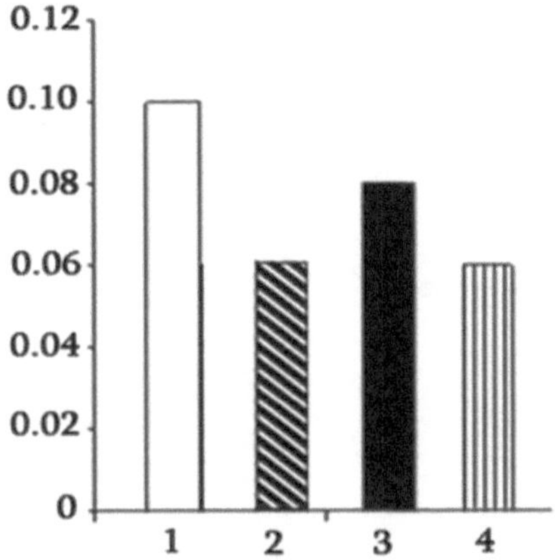

FIGURE 29.2 Diagram of optical density for water/hexane emulsions stabilized with sodium cholate and modified with different electrolytes: (1) water; (2) 0.025 M sodium hydrogen phosphate; (3) 0.025 M sodium fluoride; and (4) 0.025 M calcium chloride.

The influence of the co-surfactant on the optical density of the resulting emulsions was studied. Different amounts of organic modifier butanol were added to hexane: 10%, 20%, 30%, and 40%. Optical density of emulsions prepared in hexane modified with small additions of butanol (1%–2%) was also studied (Figure 29.3). The optical density value obtained by extrapolation of the obtained graph coincides with the optical density of the emulsion without butanol with considerable accuracy.

As can be seen from the figure, the optical density of the emulsion linearly reduces with increasing the amount of organic modifier up to 30%, probably due to a decrease in the number of droplets. Therefore, butanol can be used in combination with water-in-oil emulsions when its concentration does not exceed 10%.

It is known that ions effect on the structure of water in different ways. Kosmotropes (sulfate, phosphate, chloride) increase water order and they are called structure-making, while chaotropes (bromide, thiocyanate, perchlorate) do the opposite and they are called structure-breaking. Inorganic ions alter properties of micelles such as the aggregation number. In addition chaotropic anions For example, fluoride ions help strengthen the structure of water, while iodide ions help break it down [4]. The aqueous solutions of NaF and KI salts were added to emulsions stabilized with sodium cholate. As shown in Figure 29.4, the optical density decreases dramatically in the presence of KI compared to pure water, while optical density decreases slightly in the presence of NaF. However, emulsions modified with fluoride ions are more stable than emulsions prepared with pure water. Stability of emulsions was assessed by measuring the optical density over a certain period of time (Figure 29.5).

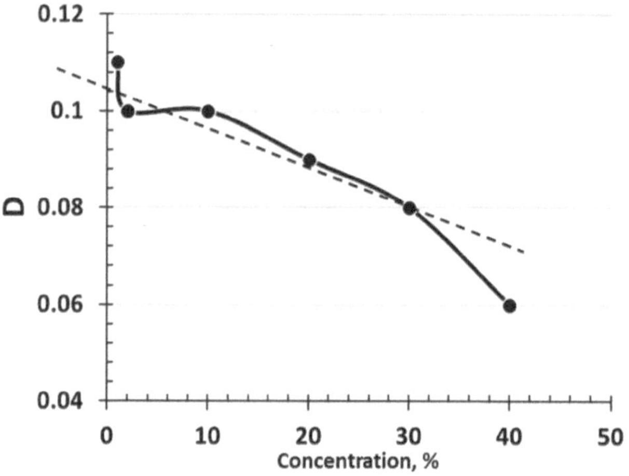

FIGURE 29.3 Optical density (D) – butanol concentration curve of a water/hexane emulsion stabilized with sodium cholate.

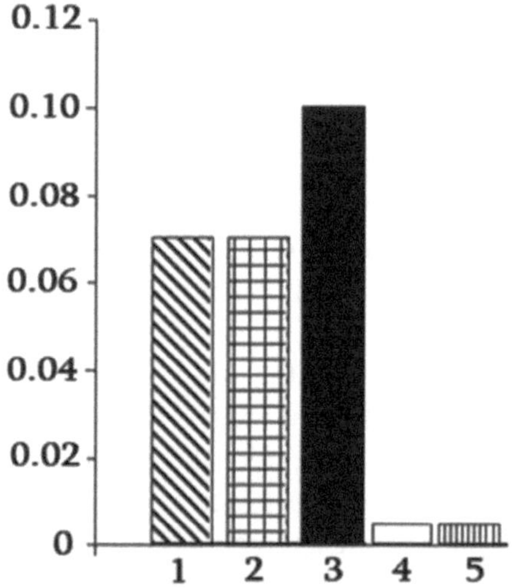

FIGURE 29.4 Diagram of optical density for water/hexane emulsions stabilized with sodium cholate and modified with sodium fluoride and potassium iodide: (1) 0.05 M NaF/hexane/sodium cholate; (2) 0.05 M NaF/cyclohexane/sodium cholate; (3) water/hexane/sodium cholate; (4) 0.05 M KI/hexane/sodium cholate; and (5) 0.05 M KI/cyclohexane/sodium cholate.

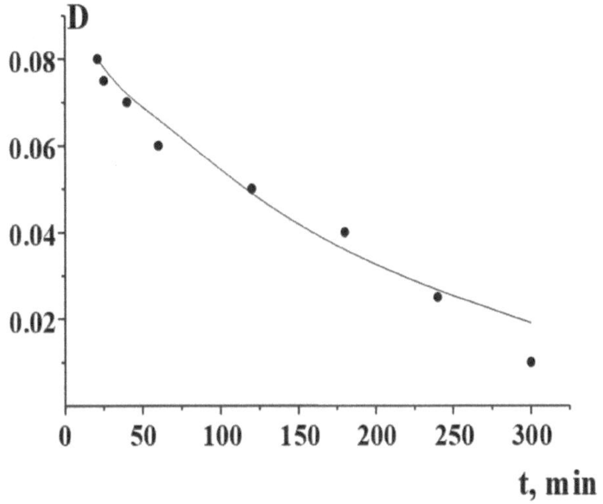

FIGURE 29.5 Optical density (D) – time curve in the water/hexane emulsion stabilized with sodium cholate.

Analysis of the data presented in Figures 29.6 and 29.8 showed that the half-life of emulsions stabilized with sodium cholate is:

$T_{1/2}$ (water/sodium cholate/oil) = 139 min.
$T_{1/2}$ (Sodium fluoride/water/sodium cholate/oil) = 136 min.

Both systems are very similar in terms of half-life and decomposition kinetics. In both cases, the decomposition presumably proceeds according to the first order equation (Figures 29.6 and 29.8), which indicates that large emulsion droplets break up into small droplets due to some tension inside the droplets. The tension is due to the presence of numerous inclusions of water/oil droplets corresponding to the size of the microemulsion inside the water droplets. It is well known that emulsions always contain microemulsions as a subsystem when micelle forming surfactants are used as emulsion stabilizers. Under such conditions, microdroplets of water phases coexist with swollen micelles [5] (Figures 29.6–29.8).

There is a difference between the steady state of water/sodium cholate/hexane and sodium fluoride/water/sodium cholate/hexane emulsions, after which the optical density decreases, i.e., decomposition of emulsion droplets proceeds. This period is 5 minutes longer in the case of an emulsion modified with sodium fluoride than the plateau of optical density of an emulsion prepared with pure water. The increased stability may be due to a decrease in the tension inside the water droplets of the emulsion under the influence of fluoride ions.

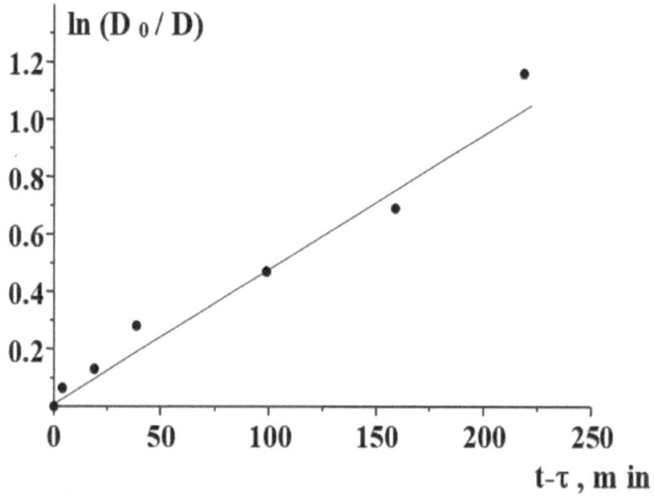

FIGURE 29.6 For the determination of the decomposition rate constant of a water/hexane emulsion stabilized with sodium cholate; $\tau = 14$ min.

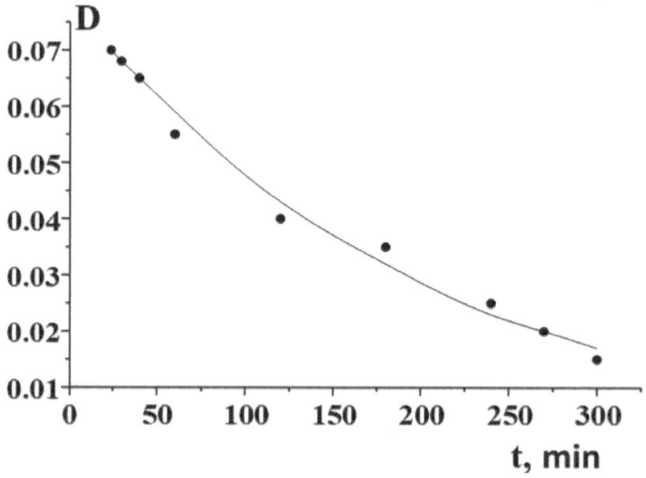

FIGURE 29.7 Optical density (D) – time curve in the water/hexane emulsion stabilized with sodium cholate and modified with sodium fluoride.

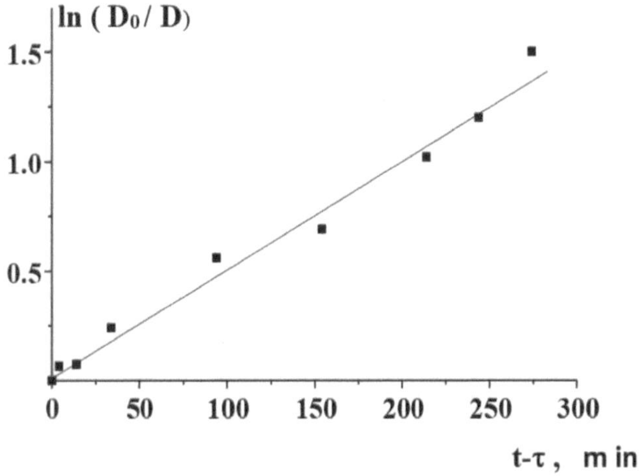

FIGURE 29.8 For the determination of the decomposition rate constant of a water/hexane emulsion stabilized with sodium cholate and modified with sodium fluoride; $\tau = 19$ min.

29.3.2 WATER-IN-OIL EMULSIONS MODIFIED WITH TETRAETHYLENEGLYCOL MONODODECYL ETHER: STUDY OF STABILITY

The aim of this subsection was to obtain water-in-oil reverse emulsions using the nonionic surfactant Brij-30, to determine the optimal amounts of water and Brij-30, to evaluate the stability of the obtained emulsions, to study the effect of isoamyl alcohol as co-surfactant additive.

The influence of the amount of water on the optical density of a reverse emulsion or an indirect emulsion in the condition of fixed values of Brij-30 and hexane (or cyclohexane) was studied (Figure 29.9). It can be seen in Figure 29.9 that optical density slightly increases with increasing the amount of water within 1–5%, confirming the well-known fact that an increase of water concentration only effects on the size of the droplets, that is, it increases their size, but does not affect their number.

A slight decrease in optical density occurs when the inverse emulsion is modified with isoamyl alcohol (Figure 29.10). If we compare the optical density of emulsions with and without isoamyl alcohol, we will see that the optical density decreases when modified with alcohol, which can be explained by a slight decrease in the number of water droplets. It is known in the literature that co-surfactants transform a conventional emulsion into a microemulsion state [6]. Therefore, the concentration of isoamyl alcohol was no longer increased to prevent the emulsion from transferring into a microemulsion, i.e., adding more than 4.5% isoamyl alcohol would significantly weaken the stability of the emulsion.

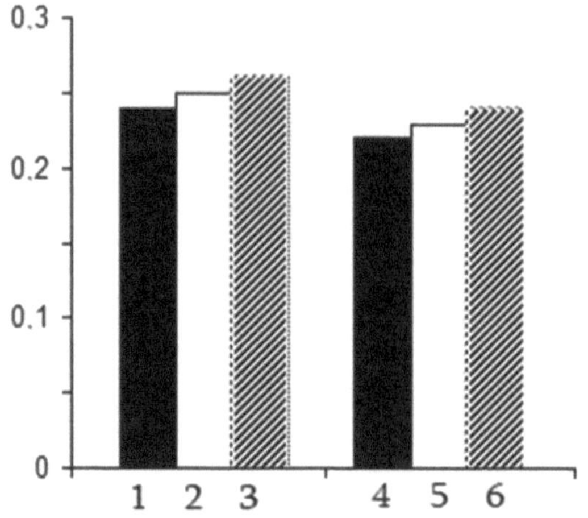

FIGURE 29.9 Diagram of optical density for water/oil emulsions stabilized with Brij-30 and modified with different amount of water: (1) 1%; (2) 3%; (3) 5% (hexane as oil phase); (4) 1%; (5) 3%; and (6) 5% (cyclohexane as oil phase).

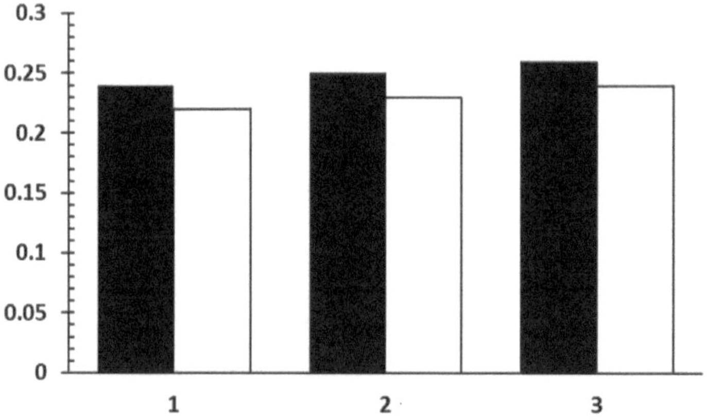

FIGURE 29.10 Diagram of optical density for water/oil emulsions stabilized with Brij-30 and modified with different amount of isoamyl alcohol: (1) ■ (1%Brij-30-hexane-water), □ (1%Brij-30- hexane-water – 1.5% isoamyl alcohol; (2) ■ (2%Brij-30- hexane- water), □ (2%Brij-30-hexane-water – 3% isoamyl alcohol; and (3) ■ (3%Brij-30-hexane-water), □ (3%Brij-30-hexane-water-4.5 isoamyl alcohol.

As in the case of sodium cholate, we also studied the effect of adding sodium fluoride on optical density of emulsion modified by Brij-30 (Figure 29.11). The increase of optical density is probably associated not with an increase of the number of emulsion droplets, but with an increase of the density of water droplets. The same results were obtained for the isoamyl alcohol/cyclohexane/Brij-30 emulsion (Figure 29.11).

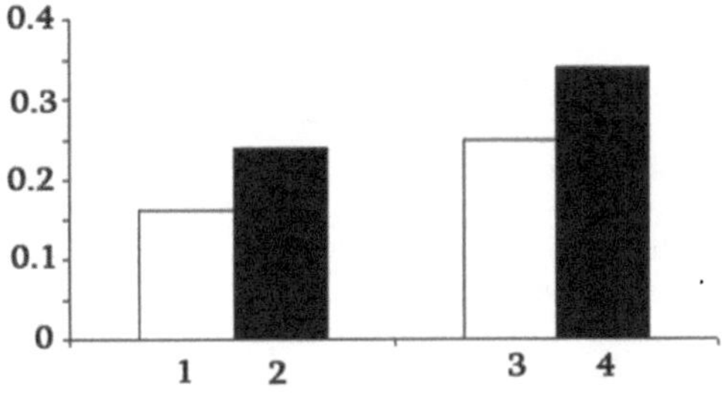

FIGURE 29.11 Diagram of optical density for water/oil emulsions stabilized with Brij-30 and modified with sodium fluoride: (1) isoamyl alcohol/cyclohexane/Brij-30/water; (2) isoamyl alcohol/cyclohexane/Brij-30/0.025 M NaF; (3) hexane/Brij-30/water; and (4) hexane/Brij-30/0.025 M NaF

As for the stability of the emulsion obtained on the basis of Brij-30, the process of its decomposition is well described by the second-order kinetic equation. The emulsion is stable for 15 min, then it decomposes with $T_{1/2} = 45$ min (Figures 29.12 and 29.13). The decomposition kinetics of the system modified with sodium fluoride also obeys the second order kinetic equation, although the emulsion is stable for 25 min. $T_{1/2} = 77$ min (Figures 29.14 and 29.15).

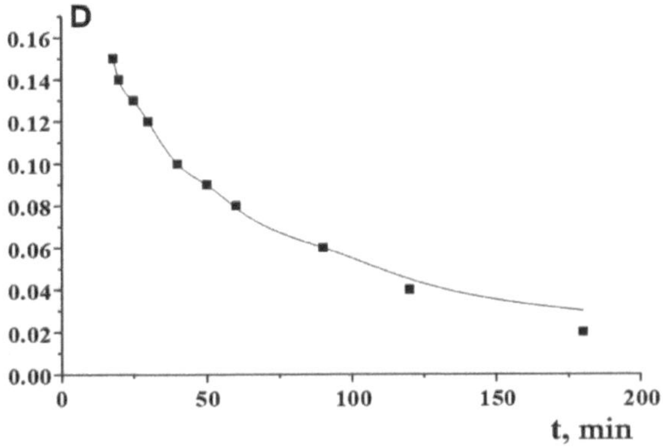

FIGURE 29.12 Optical density (D) – time curve in the water/hexane emulsion stabilized with Brij-30.

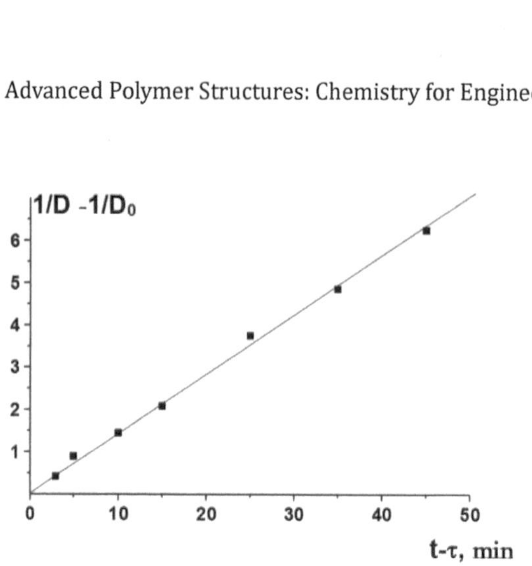

FIGURE 29.13 To determine the decomposition rate constant of a water/hexane emulsion stabilized with Brij-30; τ = 15 min.

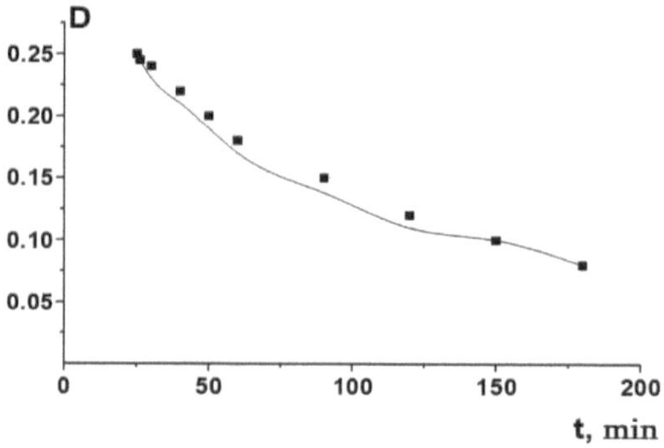

FIGURE 29.14 Optical density (D) – time curve in the water/hexane emulsions stabilized with Brij-30 and modified with 0.025 M sodium fluoride.

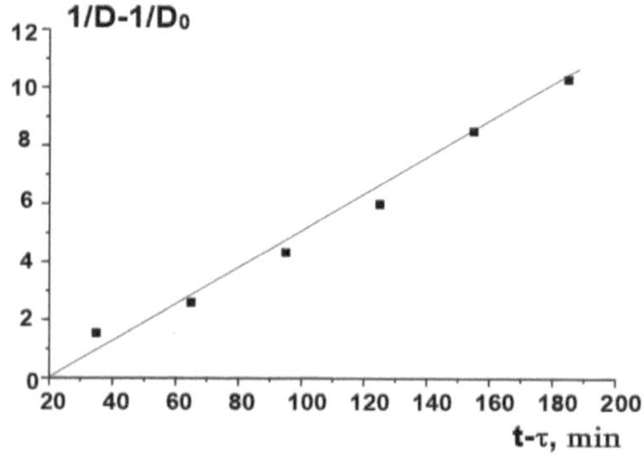

FIGURE 29.15 To determine the decomposition rate constant of a water/hexane emulsion stabilized with Brij-30 and modified with 0.025 M sodium fluoride; τ = 25 min.

The decomposition of an emulsion by a second-order equation, suggests that the basis of the emulsion decomposition is based on the collision of water droplets with each other and their subsequent merging or coalescence. In the case of sodium fluoride additive, an increase in the stability of the emulsion (the half-life increases by 1.7 times) indicates that, due to an increase of the density of water droplets, the collision of emulsion droplets does not always end in coalescence and the probability of elastic collisions is higher.

29.3.3 PREPARATION OF A SUSPENSION OF BARIUM SULFATE IN REVERSE EMULSIONS: A STUDY OF ITS DECOMPOSITION KINETICS

The purpose of the presented subsection was: to obtain a suspension of barium sulfate in the water droplets of reverse emulsions, i.e., an introducing of aqueous solutions of sodium sulfate and barium chloride into water droplets of reverse emulsions prepared on the basis of sodium cholate and hexane in order to embed barium sulfate particles, to study the influence of the concentration of components on the stability of emulsions, to develop a turbidimetric method for the determination of sulfate ions in reverse emulsions.

G The optimal concentration of barium chloride was determined, which should be added to the emulsion with a sulfate ions content of 10–80 μg/ml so that the optical density remains constant with a further increase in the concentration of $BaCl_2$. Experiments have shown that at a concentration of Na_2SO_4 in the range of 10–80 μg/ml, a 130 μg/ml of $BaCl_2$ solution is added to the sample. A calibration curve of sodium sulfate was built within indicated concentration range of sulfate ions, i.e., from 10 up to 80 μg/ml.

The appropriate amount of sodium cholate/hexane/butanol mixture was divided into two parts and the aqueous solutions of Na_2SO_4 and $BaCl_2$ were introduced in each part of emulsion, respectively. Salt-modified emulsions were mixed with each other; the optical density of the $BaSO_4$ suspension increased with increasing sulfate ions concentration. At the same time, the calibration curve has a linear character within the concentration of sulfate ions 10–80 μg/ml. Metrological and analytical parameters of the developed method are given in Table 29.1.

TABLE 29.1 Metrological Characteristics of Turbidimetric Method of Determination of the Sulfate Ions

Standard Deviation (SD, μg/ml)	Relative Standard Deviation (RSD, %)	Confidence Interval, $\overline{X} \pm \delta$ $\alpha = 0.05$; n = 7	Concentration of Sulfate Ions (μg/ml)	Linearity Interval (μg/ml)	Calibration Graph Equation
2.44	10.4	23.0±2.4	20.0	10–80	y = 0.5 + 0.43 x

The concentration of sulfate ions was determined in tap water according to the developed method. A sample of 20 μl was taken. As is known from the literature, the concentration of sulfate ions in drinking water is approximately≈20 μg/ml. Our value is equal to 18 μg/ml. The result was compared with the data available in the literature [7, 8].

The stability of the $BaSO_4$ suspension embedded into the water droplets of reverse emulsions was studied, since the weak point of turbidimetric analysis is the unstable nature of the obtained dispersions. It turned out that the suspension obtained in reverse emulsions is stable for 20 min (Figures 29.16 and 29.17).

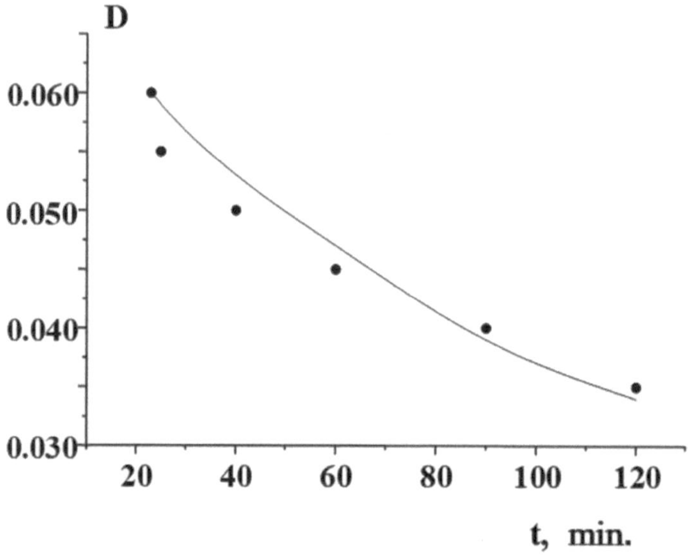

FIGURE 29.16 Optical density (D) – time curve in the water/hexane emulsions stabilized with sodium cholate and modified with barium sulfate.

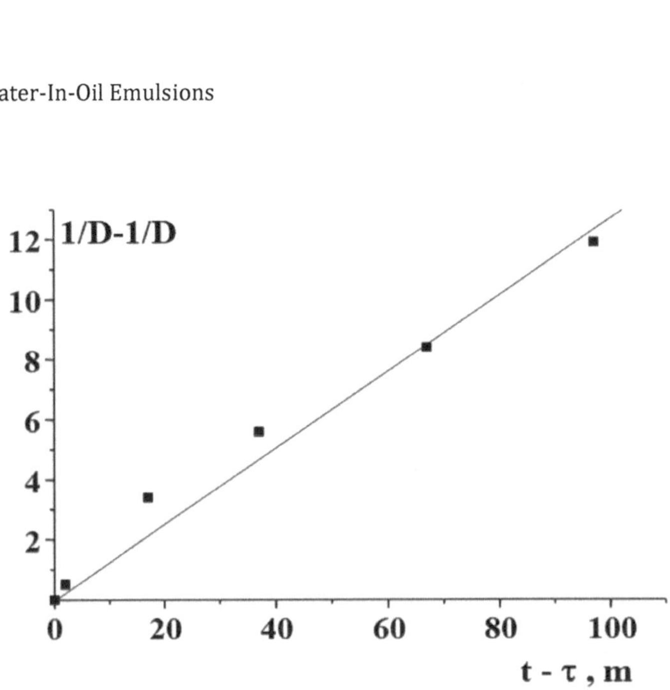

FIGURE 29.17 To determine the decomposition rate constant of a water/hexane emulsion stabilized with sodium cholate and modified with barium sulfate; $\tau = 20$ min.

It can be seen from the kinetic curves that the $BaSO_4$ particles introduced into the water droplets of the water/oil emulsion change the kinetics of the emulsion decomposition. If the decomposition of the emulsion obtained on the basis of sodium cholate is described by the first order equation (half-life $T_{1/2} = 139$ min), then the decomposition proceeds according to the second-order equation (half-life 130 min) due to embedded $BaSO_4$ particles. This suggests that the decomposition of sodium cholate/water/hexane reverse emulsions is mainly stipulated by the creation of some tension in the emulsion water droplets themselves, which is caused by droplets corresponding to the size of the water/oil microemulsion embedded in the water droplets of emulsions. The change in the order of decomposition after the imbedding of $BaSO_4$ particles into this emulsion system may indicate the collision of droplets with each other and, consequently, coalescence.

29.4 CONCLUSION

1. Reversed water/hexane emulsions based on a biological surfactant with a concentration of 1–2% have been obtained. Up to 10% butanol can be used as a co-surfactant in these emulsions.

2. The stability of the emulsion stabilized with sodium cholate was studied. The half-life of the emulsion is 139 minutes, and the decomposition kinetics is described by a first-order equation with satisfactory accuracy.
3. Reversed water-in-oil emulsions based on nonionic tetraethylene glycol dodecyl ether have been prepared. Changing the water content within 1–5% has a slight effect on the optical density of the emulsion. The optimum content of isoamyl alcohol as a co-surfactant is 4.5%.
4. The decomposition of an emulsion stabilized with tetraethylene glycol dodecyl ether obeys well a second-order kinetic equation. The emulsion is stable for 25 minutes.
5. Modification of water droplets with sodium fluoride showed that the stability of the emulsion was increased by 5 minutes in the case of sodium cholate and by 7 minutes in the presence of tetraethylene glycol dodecyl ether.
6. A suspension of barium sulfate in water droplets of a reversed emulsion was obtained. The emulsion stabilized with sodium cholate with embedded particles of barium sulfate decomposes according to the second order equation. The suspension is stable for 20 minutes.

KEYWORDS

- **barium sulfate suspension**
- **decomposition kinetics**
- **fluoride ions**
- **microheterogeneous systems**
- **pharmaceutical industry**
- **sodium cholate**
- **turbidimetric method**
- **water in oil emulsions**

REFERENCES

1. Walstra, P., (2005). Emulsions, Chapter 8. In: *Fundamentals of Interface and Colloid Science* (Vol. 5, pp. 8.1–8.94).
2. Bezarashvili, G., (1989). *Experiment Design* (pp. 49–51). Tbilisi, TSU.

3. Soustelle, M., (2011). *An Introduction to Chemical Kinetics* (p. 448). London, Wiley.
4. Hribar, B., Southall, N. T., Vlachy, V., & Dill, K. A., (2002). How ions affect the structure of water. *J. Am. Chem. Soc., 124*(41), 12302–12311.
5. Alexander, I., B., Turusbekovna, A. A., Demidova, M. G., Popovetskiy, P. S., Plyusnin, P. E., & Bulavchenko, O. A., (2018). Synthesis and concentration of organosols of silver nanopartcles stabilized by AOT: Emulsion vs Microemulsion. *Langmuir, 34*(8).
6. Sujatha, B., Himabindu, E., Battu, S., & Abbulu, K., (2020). Microemulsions: A review. *Int. J Pharm. Sci., 10*(1). https://ipharmsciencia.edwiserinternational.com/home.php (accessed on 02 January 2022).
7. Takaishvili, N., & Supatashvili, G., (2003). *Influence of Organic Solvents on the Degree of Dispersion and Optical Density of Barium Sulphate Suspension* (pp. 134–136). Georgian Engineering News, #1, (in Russian).
8. Takaishvili, N., (2003). Influence of surface-active substance on degree of dispersion and optical density of Barium sulphate suspension. *Georgia Chemical Journal, 3*(1), 10–12. In Russian.

CHAPTER 30

Synthesis and Study of Tetra-Substituted Arsonium Tetraiodcuprates(I) and Argentates(I)

M. CHIKOVANI,[1] M. RUSIA,[2] and KR. GIORGADZE[2]

[1]*Kutaisi Akaki Tsereteli State University, Department of Chemistry, Kutaisi, Georgia*

[2]*Ivane Javakhishvili Tbilisi State University, Department of Chemistry, Tbilisi, Georgia*

ABSTRACT

Based on numerous experimental data, it was found in earlier works that copper(I) and silver(I) form pseudo-halide complexes of anionic nature. Namely, it was shown that tetrasubstituted arsonium salts precipitate copper(I) and silver(I) in the form of $[R_4As][MX_2]$ salts, where M = Cu or Ag; X = SCN or CN and R is an organic radical and varies within a very wide range. Such composition of the synthesized substances does not contradict the literature data. As it is known that in such compounds, copper and silver also "lose" ns-valence electrons thus releasing ns-orbital. This orbital and one orbital of the np-sublevel becomes the basis for "settling" the pseudo halogen groups and at this time they undergo sp-hybridization. Thus, we can conclude that the $[MX_2]^-$-anion should have a linear structure [1], although the mentioned complexes have not yet been investigated in this respect.

Advanced Polymer Structures: Chemistry for Engineering Applications. Omar Mukbaniani, DSc, Tamara Tatrishvili, PhD & Marc Jean M. Abadie, DSc (Eds.)

30.1 INTRODUCTION

we made sure that the coordination number of complexing agents in the composition of these compounds is equal to 2 and set the goal of studying the preparation of copper(I) and silver(I) complex compounds with tetra-substituted arsonium salts. It is known that Cu^{2+}-ion forms monovalent copper iodide with iodine ion [2]. We knew that tetraalkyl (aryl) arsenium, as well as mixed-radical salts of arsonium, precipitate copper(I) and silver(I) in the form of a singly charged cationic-anionic complex. However, on the basis of numerous experimental data, it has been established that the products obtained in terms of chemical composition, are not diiodocuprates(I) and arsonium argentates(I). A detailed analysis carried out for arsenic, iodine, copper, and silver showed that the products of the interaction were tetraiodecuprates(I) and argentates(I). However, it was observed that the atomic ratios between arsenic, metal, and iodine were as follows: As:M:I = 1:1:4. If we take into account this ratio, the composition of the search complex could be $[R_4As][MI_4]$. The oxidation state (oxidation number) of copper and silver in this compound should have been +3, which means that the metal is oxidized in this environment. This was considered impossible, since none of the reagents exhibits oxidative capacity in other reactions. This prompted us to find a different composition structure for the compounds under study. Mass spectroscopic study convinced us that the objects under study contained potassium. Quantitative analysis of the latter by flame photometry brought complete clarity to our research activities. It was found that instead of the expected diiodocuprates(I) and argentates(I) of tetraalkyl (aryl) arsonium, tetraiodocuprates(I) and argentates(I) of dipotassium-tetrasubstituted arsonium appear. The formation of the final products is well explained by a combination of the following sequential reactions:

$$MI + 3KI \rightarrow K_3\,[M\,I_4]$$

$$K_3[MI_4] + [Ar_2As(R)R]I \rightarrow K_2[Ar_2Ar(R)R][MI_4] + KI$$

or in total:

$$2KI + MI + [Ar_2As(R)R]I \rightarrow K_2[Ar_2Ar(R)R][MI_4]$$

30.2 EXPERIMENTAL METHODS AND MATERIALS

30.2.1 MATERIALS

Tetra substituted arsonium salt, cupric(I) iodide, silver(I) iodide, potassium iodide, ethanol (96%), diethylether, phosphorus pentoxide.

30.2.2 SYNTHESIS OF DIPOTASSIUM TRIPHENYLMETHYLARSONIUM TETRAIODOCUPRATE(I)

A mixture of 1.06 g of copper(I) iodide and 9.24 g of potassium iodide is dissolved in 50 ml of distilled water. About 2.5 g of triphenylmethylsaronium iodide dissolved in 40 ml of ethyl ether is added to the resulting solution. A fine crystalline substance formed immediately and was left in solution for 3 hours until a perfect crystalline form was formed. The precipitate was filtered off, thoroughly washed with distilled water, then with ethanol and dried in a vacuum desiccator with phosphorus pentoxide until constant mass was obtained. We obtained 3.5 g of dipotassium triphenylmethylarsonium tetraiodocuprate(I), melting point of which is t_{melt}=116–117°, found,%: As 7.84; Cu 6.62; I 52.76. Other tetraiodocuprates and argentates of dipotassium tetra-substituted arsonium were prepared in a similar manner. Some physical-chemical characteristics of the reaction products are shown in Table 30.1.

30.2.3 CHARACTERIZATION

30.2.3.1 IR-SPECTROSCOPY

The structure of the tetraiodocuprates and argentates of synthesized tetra-substituted arsenium was determined by IR spectroscopy. The IR spectra of the final products are almost identical to the reagents – tetrasubstituted arsonium iodides. This, in turn, indicates that arsenic is also part of the cation in the test substance. The absorption band of the Cu—I bond is not observed in the spectrum. It is beyond the sensitivity of the device. The spectra of synthesized substances are similar to each other. An analysis of the spectra of cuprates(I) of dipotassium-triphenylmethylsaronium tetraiodine(I), and of dipotassium-triethylamidoarsonium is given. An intense absorption band of the spectrum in the range of 3,600–3,000 cm^{-1} is not observed for any of

TABLE 30.1 Yield, Molar Conductivity, Melting Point, and Elemental Analysis Data of Four Substituted Arsonium Tetraiodcuprates (I) and Argentates (I)

SL. No.	Ar	R	R^1	$t_{melt.}$ (°C)	Molar Electroconductivity in Dimethylformamide at 25°C (ohm^{-1} sm^2 mol^{-1})	Was Found (%)		
						As	M	I
1.	C_6H_5	C_6H_5	CH_3	116–117	130.1	7.84	6.62	52.76
2.	M-$CH_3C_6H_4$	M-$CH_3C_6H_4$	CH_3	125–126	136.9	7.93	6.34	50.53
3.	C_6H_5	C_6H_5	CH_2-CH=CH_2	173–273	123.9	7.98	6.45	51.37
4.	M-$CH_3C_6H_4$	M-$CH_3C_6H_4$	CH_2-CH=CH_2	108–109	115.3	7.52	6.24	49.42
5.	C_6H_5	izo-C_3H_7	CH_3	100–101	119.6	8.23	6.92	54.86
6.	C_6H_5	N-C_4H_9	CH_3	106–107	125.4	7.97	6.79	53.93
7.	C_6H_5	C_6H_5	CH_3	160–161	111.9	7.28	10.96	50.58
8.	M-$CH_3C_6H_4$	M-$CH_3C_6H_4$	CH_3	154–155	106.2	7.06	10.68	48.64
9.	C_6H_5	C_6H_5	CH_2-CH=CH_2	148–149	108.1	7.35	10.75	48.99
10.	M-$CH_3C_6H_4$	M-$CH_3C_6H_4$	CH_2-CH=CH_2	138–138.5	115.6	7.04	10.26	47.36
11.	C_6H_5	izo-C_3H_7	CH_3	140–141	117.0	7.98	11.44	52.28
12.	C_6H_5	N-C_4H_9	CH_3	118–119	124.3	7.97	11.37	51.45

TABLE 30.1 *(Continued)*

SL. No.	Gross Formula	Calculated (%)			Yield $K_2[Ar_2AsRR'][MI_4]$		
		As	M	I	g	mol	%
1.	$C_{19}H_{18}AsK_2CuI_4$	7.73	6.54	5.34	3.52	0.0036	65.0
2.	$C_{22}H_{24}AsK_2CuI_4$	7.61	6.27	50.17	4.05	0.0040	65.3
3.	$C_{21}H_{20}AsK_2CuI_4$	7.53	6.37	50.98	5.72	0.0057	68.0
4.	$C_{24}H_{26}AsK_2CuI_4$	7.22	6.12	48.92	4.72	0.0048	67.6
5.	$C_{16}H_{20}AsK_2CuI_4$	8.01	6.78	54.24	2.84	0.0030	62.8
6.	$C_{17}H_{22}AsK_2CuI_4$	7.89	6.68	53.45	3.93	0.0041	59.0
7.	$C_{19}H_{18}AsK_2AgI_4$	7.39	10.64	50.05	6.92	0.0068	61.1
8.	$C_{22}H_{24}AsK_2AgI_4$	7.10	10.22	48.06	4.85	0.0046	64.2
9.	$C_{21}H_{20}AsK_2AgI_4$	7.20	10.38	48.80	4.12	0.0040	62.5
10.	$C_{24}H_{26}AsK_2AgI_4$	6.93	9.97	46.91	2.53	0.0023	60.2
11.	$C_{16}H_{20}AsK_2AgI_4$	7.65	11.01	51.78	5.84	0.0060	61.6
12.	$C_{17}H_{22}AsK_2AgI_4$	7.54	10.85	51.06	3.63	0.0036	62.5

the compounds. This means that the test substances do not contain water, i.e., are not hydrates, although these compounds were formed in aqueous solutions. In the spectra of amide products, the absorption band of the C=O bond is observed in the region of 1,750 $cm^{-1,}$ and the N—H bond of the amid-group, in the region of 1,620–1,590 cm^{-1}. As already mentioned, the IR spectra of this type of compounds essentially correspond to the IR spectra of the initial iodides, which confirms the cation-anionic structure of the synthesized substances. In the spectrum of the studied substances, an absorption band in the range of 620–625 cm^{-1} is observed, which is due to the As—C bonding. Such absorption is typical for quaternized arsenic and is similar to that of initial arsenic iodides. Thus, it can be concluded that in the synthesized complex compounds, as well as in the initial iodides, arsenic is in a quaternary-substituted state and is part of the cation. The absorption bands in the regions of 750, 1,600, 3,600–3,000 cm^{-1} indicate the presence of a benzene nucleus and, in general, the content of aromatic groups. From the IR spectra of dicalium-allyltriphenylarsonium and dicalium-isopropylarsonium tetraiodarsenates(I) it appears that the presence of absorption bands in the 860, 880, and 1,000 cm^{-1} regions indicates the existence of a vinyl group =As—CH_2—CH=CH_2. The frequency of the valence vibrations of the C–H bond is observed by the absorption band near 8,000 cm^{-1}.

30.2.3.2 MOLAR CONDUCTIVITY

The ionic structure of the studied substances was confirmed by the study of the molar electrical conductivity. Since the test substances are not soluble in water, this conductivity was investigated in dimethylformamide solutions. The results are presented in Table 30.1. As is known [3] molar conductivity of 1: 2 type electrolytes in dimethylformamide varies in the range of about 77–104 ohm^{-1} sm^2 mol^{-1}. The electrical conductivity of the three-dimensional electrolyte does not exceed 104. In our case, this value ranges from 108 to 137 ohm^{-1} sm^2 mol^{-1}, which indicates that the objects under study contain more number of ions. It should also be noted that the molar conductivity of the investigated substances is much lower than the conductivity of 4-ionic compounds in dimethylformamide. The authors explain this case by the partial coordination of anions by metal ions. It is possible that in our case $[CuI_4]^{-3}$-ions are also partially surrounded by K^+ and $[R_4As]^+$-ions, which leads to a decrease in conductivity. Taking into account the fact that potassium salts are much more soluble than the corresponding arsonium salts, we can conclude that tetraiodocuprate(I) ions should probably be surrounded mainly by tetra-substituted arsonium cations. The molar conductivity of dipotassium-tetra-substituted arsonium arsenates is much lower than the electrical conductivity of the corresponding cuprates(I), which, in our opinion, should be explained by the lower electrolytic dissociation of tetraiodoargentate(I) products. However, the latter are also four-ionic and undergo dissociation according to the following scheme:

$$K_2\,[Ar_3AsR][AgI_4] \rightarrow 2K^+ + [Ar_3AsR]^+ + [AgI_4]^{3-}$$

Thus, by chemical analysis, infrared spectroscopy and the study of electrical conductivity it was established that the synthesized compounds are four-ionic complexes, and their composition can be expressed by the formula $K_2[Ar_3AsR][MI_4]$, where M= Cuor Ag.

30.2.3.3 THERMOGRAVIMETRIC ANALYSIS (TGA)

The thermal behavior of tetraiodocuprates and argentates of dipotassium tetra-substituted arsonium of synthesized complexes at high temperatures was also studied. It turned out that their thermograms are identical.

30.2.3.4 X-RAY DIFFRACTION

To confirm the individuality of the synthetic coordination compounds synthesized by us, an X-ray diffraction study was carried out. The results show that dipotassium-substituted tetraiodocuprates and arsonium argentates of the synthesized complexes are crystalline substances and are characterized by an individual combination of relative intensity and interplanar distance.

30.3 CONCLUSION

Our studies have shown that under the action of four-substituted arsonium iodides copper(I) and silver(I) iodides precipitate as corresponding tetraoidolprates(I) and argentates(I) from aqueous potassium iodide solution.

KEYWORDS

- **cupric(I) iodide**
- **ionic structure**
- **IR spectroscopy**
- **IR-spectrum thermographic analysis.**
- **molar conductivity**
- **tetrasubstituted arsonium salt**

REFERENCES

1. Akhmetov, N. C., (1975). *Neorhanicheskaya Ximia* (pp. 65, 66). M. Vishaiz shkola.
2. Brauer, G., (1985). *Guide to Inorganic Synthesis* (Vol. 4, pp. 1065, 1066). Ximii. M.: Mir.
3. Peuronel, G., Malavasi, W., & Pignedoli, A., (1982). Copper(I), silver(I) and mercury (II) halide complexes of the 3,5-diamino-1,2,4-dithazolium halides (thioret, hidrohalides). *Spectrochim. Acta, 10*, 1069–1072.

CHAPTER 31

Formation of Trimethine Cyanine of the Dipyrrolobenzoquinoxaline Series Under the Conditions of the Vilsmeier Reaction

SH. A. SAMSONIYA, M. V. TRAPAIDZE, and N. N. NIKOLEISHVILI

Department of Exact and Natural Sciences, Ivane Javakhishvili Tbilisi State University. Tbilisi, Georgia

ABSTRACT

We studied the formation of a bifunctional analog of the "Fischer base" – 1,4,5,8-tetrahydro-1,1,8,8-tetramethyl-2,7-dimethylidenedipyrolo[1,2,3-d,e:3,2,1-i,j]benzo[g]quinoxaline under Wilsmeyer reaction conditions at different ratios of substrate and Wilsmeyer complex (WC). At 60°C and a molar ratio of the reagents of 1:5, an abnormal product, trimethyl cyanine, was isolated. The di formyl derivative, 2,7-di(formylmethylladen)-1,4,5,8-tetrahydro-1,1,8,8-tetramethyldipyrrolo[1,2,3-d,e:3,2,1-i,j]benzo[g]quinoxaline, was obtained in 67% yield at 35°C with a large excess of CW (1:45).

31.1 INTRODUCTION

"Fischer bases" (2-methyleneindolines) are easily formed on the methylene group by the Wilsmeyer complex (WC) from DMFA and $POCl_3$. The so-called Fischer aldehyde obtained is an important intermediate for the preparation of cyanine and hemicyanine dyes [1]. Earlier, we synthesized isomeric bis-analogs of "Fisher's base" from the corresponding dipyrrolonaphthalenes [2, 3]. During the recent research we studied the formylation

Advanced Polymer Structures: Chemistry for Engineering Applications. Omar Mukbaniani, DSc, Tamara Tatrishvili, PhD & Marc Jean M. Abadie, DSc (Eds.)

reaction of the obtained bifunctional base of dipyrrolo-benzoquinoxaline series **1** at different ratios of the substrate and the WC.

During formylation of 1,4,5,8-tetrahydro-1,1,8,8-tetramethyl-2,7-dimethylidene-dipyrrolo[1,2,3-d,e:¬3,2,1-i,j]benzo[g]quinoxaline (1) at molar ratio of base and Wilsmaier complex – 1: 5, after 30 minutes of stirring at 40°C, along with the formylation of the expected product formylation in the reaction medium chromatographed still observed the presence of the starting base **1**. A further increase in temperature at 60°C causes a rapid color change in reaction medium. The initially yellowish-red solution acquires an intense blue coloring. As a result of the reaction blue crystals were yielded, which, on the basis of spectral data (UV, PMR-1H, and Mass spectra), were assigned the structure of the anomalous reaction product, trimethine cyanine **2**, by analogy with the compound **2a**, which is formed by the interaction of Fischer's indoline base with Fischer's aldehyde [4].

On this basis, we assume that the formation of compound **2** is due to the interaction of the bisdimethylaminovinyl derivative of structure **A**, the intermediate of bisaldehyde **3**, formed at the beginning of the reaction with the unreacted base **1** under reaction conditions according to Scheme 31.1. The intense blue coloring of the reaction medium can be caused by the formylation of dye of structure **B**, which transforms into trimethylincyanine **2** under the action of alkali. We do not exclude also partial one-reactor formation of base **1** followed by dimerization with formation of cyanine dye (Scheme 31.1).

Formylation of Fischer base **1** was also performed according to Helmut-Fritz [5] at 35°C with a large excess of CW (1:45) for 2 h. In this case, a bis-analog of Fischer aldehyde, 2,7-di(formylmethylidene)-1,4,5,8-tetrahydro-1,1,8,8-tetramethyldipyrrolo[1,2,3-d,e:3,2,1-i,j]benzo[g]quinoxaline (**3**), was isolated in 67% yield (Scheme 31.2) [6].

In the UV spectrum of trimethine cyanine **2** there is a bathochromic shift of absorption maximum in comparison with the spectrum of dialdehyde **3** and two intensive absorption maximum appear in the long-wave region of the spectrum at 539 and 613 nm, which is typical for cyanine compounds [4].

In the ^{1}H NMR spectrum of the putative cyanine compound **2**, the doublet signals at the weakest field (10.04 ppm) and at 5 ppm with identical SSIC are ascribed to the protons of aldehyde and methylidene groups of the =CH-CHO system; the signal of protons of the exocyclic =CH_2 group appears as a broadened singlet while the signals of protons of N-CH_2-CH_2– groups appear as triplets and are shifted toward the weakest field as compared with analogous signals of the starter compound. The spectrum also contains multiplet signals of aromatic protons and singlet signals of methyl groups in the corresponding areas. The signals of the protons of the cyanine =CH-CH=CH– group appear as one triplet and two doublets with the SSIC of 11–14 Hz which corresponds to the trans configuration of these protons similarly to the similar compound obtained from 2-methyleneindoline [4].

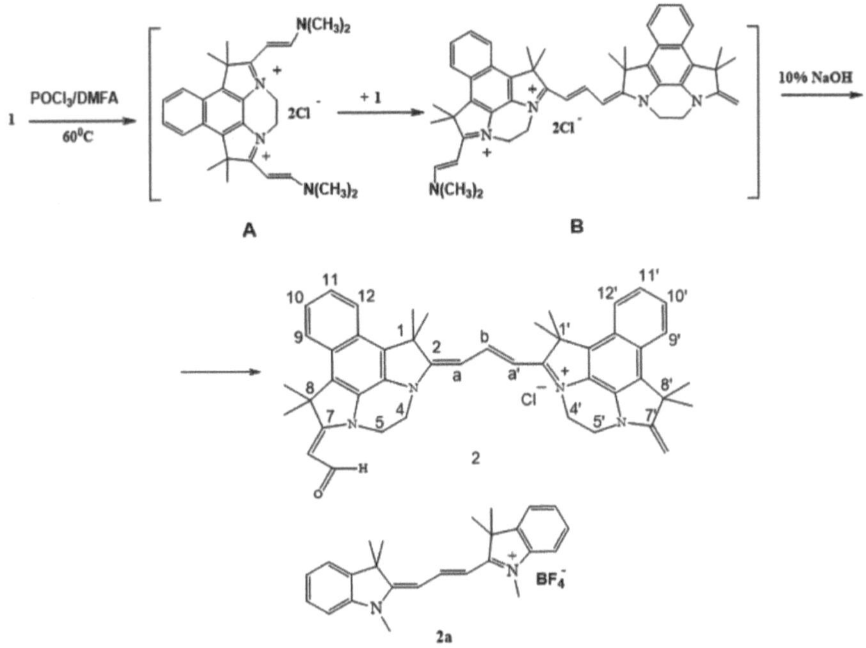

SCHEME 31.1 Proposed scheme for the formation of compound 2 (trimethine cyanine).

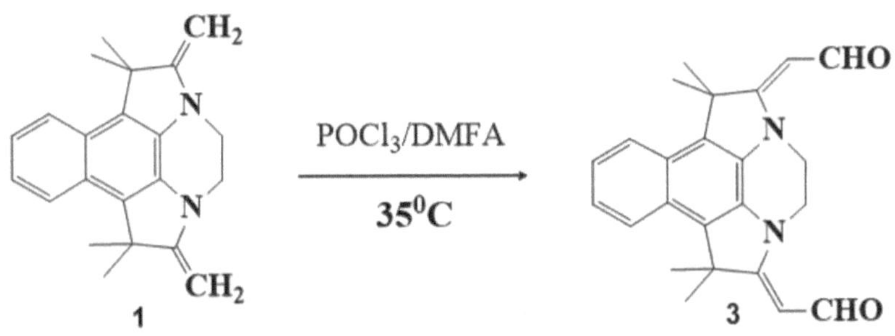

SCHEME 31.2 Formation of bissanalogue of Fischer's aldehyde.

We assigned the asymmetric triplet signal at 5.46 ppm consisting of 5 signals in the range of 5.478–5.446 ppm. in which the doublet with an SSIC of 8.8 Hz (-CH=exocyclic) is clearly distinguished, to the cyanine β-proton. Determination of the SSIC of the triplet is difficult.

The high-resolution mass spectrum of compound 2 does not capture the peak of the molecular ion. Maximum in the spectrum is a peak with the mass $[M\text{-}317]^+ = 354$ (100%) which can form at the decomposition of the cyanine **5** cation in the CH_α-CH_β bond and separation of the fragment **b** with the mass 317 (9.6%) – 1,1,2,8,8-pentamethyl-7-methylene-dipyrrolo[1,2,3-d,e:3,2,1-i,j]benzo[g]quinoxalinium cation. The stability of the remaining fragment with mass 354 (100%) is probably due to formylation of structure a which is fixed in the spectrum by maximum peak.

The given data from UV, ^1H NMR and mass spectra of compound 2 do not contradict our proposed structure:

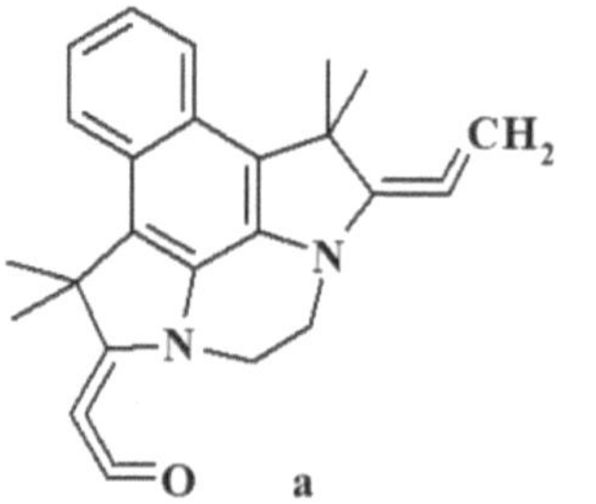

a

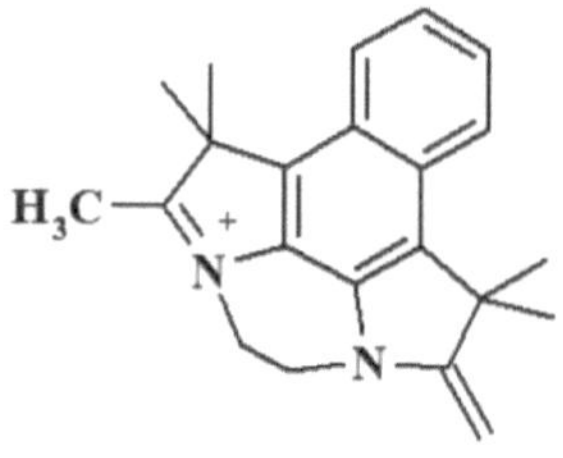

b

Thus, by the formylation of the Fischer base in the dipyrrolobenzoquinosaline series under the conditions of the Wilsmeier reaction at 60°C, we managed to isolate the anomalous reaction product – trimethine cyanine 2, the production of which can be explained by the condensation of the first the only formed dialdehyde intermediate 3-bis-dimethylaminovinyl derivative of structure A – with the initial base. We also do not exclude partial one-pot formylation followed by dimerization with the formation of a cyanine dye.

31.2 EXPERIMENTAL PART

The progress of the reactions and the purity of the compounds were monitored by TLC on "Silufol UV-254" plates. The IR spectra were recorded on a Thermo Nikolet Avatar 370 FTIR spectrometer in Vaseline oil. The UV spectra were obtained on a Varian, Carry 100, UV-vis instrument. ^1H and ^{13}C NMR spectra were recorded on a Bruker AM-400 spectrometer (400 and 100

MHz, respectively) in DMSO-d_6, internal TMS standard. Mass spectra were obtained on a MAT 95 v. Finnegen (USA).

31.2.1 SYNTHESIS OF TRIMETHYL CYANINE 5

During Preparation 1.5 mol CW (0.14 ml $POCI_3$ + 0.5 ml DMFA), cool to – 5°C and add to it a solution of 0.1 g (0.32 mmol) of Fischer base **1** in 4 ml DMFA. Stir for 30 min at 40°C. After chromatographic control it is found that the starter base **3** remains in the reaction medium. The heating is continued and at 60°C the reaction mixture changes color rapidly. Initial a yellowish-red solution becomes an intense blue color. The reaction mixture is heated at 60°C for 2 hours. After that the reaction mixture is transferred to ice water, 10% NaOH solution is added to pH 14. A blue precipitate precipitates. The residue is left to stand overnight. The precipitate is filtered off, washed with water to neutral reaction and dried. Blue crystals are obtained. The yield is 0.081 g (42%). T_d > 300°C (EtOH). R_f 0.56 (ethanol-ammonia, 10:1). R_f 0.56 (ethanol-ammonia, 10:1). IR spectrum, ν, cm^{-1}: 1623 (CHO), 1565, 1532 (>C=C<), 1304, 1157, 1049, 964, 601, 570. UV spectrum (EtOH), λ_{max}, nm (lg): 267 (0.92); 315 (0.7); 378 (0.72); 539.5 (0.21); 613 (0.27). ^1H NMR spectrum, (DMSO-d_6), δ, ppm. (J, Hz): 1.48(3H, c,C(CH_3)$_2$, at C-1'); 1.58(6H, c,C(CH_3)$_2$, at C-8'); 1.65 (3H, c,C(CH_3)$_2$, at C-1');1.93(12H, c,C(CH_3)$_2$, at C-1 and C-8); 2.09(2H,w.c, N'-CH_2); 2.33(4H,t, $J_{=CH2CH2}$ = 2.0, N- CH_2); 2.67 (2H,t, $J_{=CH2CH2}$= 1.6, N'-CH_2); 4.04 (2H,w.c, = CH_2 exocycle); 5.46(1H, t, β-H); 5.47 (1H, d, $J_{=CH-CHO}$ = 8.8, =CH exocycl.); 5.73 (1H, d, $J_{\alpha\beta}$ =11.2, α-H); 6.35 (1H, d, J α'β =14.4, α' -H); 7. 24-7.27(2H, 10-H,11-H); 7.36-7.39 (2H,10'-H,11'-H); 7.98-8.00(2H,9-H,12-H); 8.07–8.09 (2H, 9'-H, 12'-H); 10.04 (1H, d, $J_{=CH-CHO}$ = 8.8, CHO).

KEYWORDS

- **2-methyleneindolines**
- **diformyl derivative**
- **dipyrrolo-benzoquinoxaline**
- **dipyrrolo-naphthalenes**
- **Fischer base**
- **Wilsmeyer complex**

REFERENCES

1. (1990). 2-methyleneindoline compounds. *Synthesis and Properties [in Russian]* (Vol. 14, p. 127). Itogi Nauk Tekhn., Ser. Org. Khem.
2. Samsoniya, Sh. A., Trapaidze, M. V., Nikoleishvili, N. N., Japaridze, K. G., Maisuradze, J. P., & Kazmaier, U., (2010). *Chemistry of Heterocyclic Compounds, 46*(8), 1020–1022.
3. Samsoniya, Sh. A., Trapaidze, M. V., Nikoleishvili, N. N., Japaridze, K. G., Maisuradze, J. P., & Kazmaier, U., (2011). *Chemistry of Heterocyclic Compounds, 47*(9), 1098–1104.
4. Khan, M. N., Fleury, J. P., Baumlin, P., & Hubschwerlen, C., (1985). *Tetrahedron, 41*, 5341–5345.
5. Fritz, H., (1959). *Chem. Ber., 92*, 1809.
6. Samsoniya, Sh. A., Trapaidze, M. V., & Nikoleishvili, N. N., (2013). Bisindoles 43. Dipyrrolo quinoxalines 2. Synthesis of bis-analogs of Fischer's aldehyde under the conditions of the Wilsmeier reaction. *International Journal – Chemistry of Heterocyclic Compounds, 4*, 580–584.

CHAPTER 32

Anionic-Form Zeolite Material and Possibility of Its Use in Agriculture

GIORGI TSINTSKALADZE, TINATIN SHARASHENIDZE, TEIMURAZ KORDZAKHIA, MARINE ZAUTASHVILI, MANANA BURDJANADZE, and LUBA EPRIKASHVILI

Petre Melikishvili Institute of Physical and Organic Chemistry, Ivane Javakhishvili Tbilisi State, Tbilisi, Georgia

ABSTRACT

A new anionic-form zeolite nanomaterial was obtained in which a new anion-substituting phase was formed after the introduction of ammonium dihydrophosphate and potassium nitrate (KNO_3) into the zeolite structure. The structure and properties of the obtained nanomaterial were studied by chemical, IR-spectroscopy, X-ray diffractometric, and electron microscopy analysis methods. It was shown that after the introduction of salts into the zeolite structure, its structure practically does not change and undergoes only insignificant deformation. The obtained zeolite anionic nanomaterials can be used as long-term zeolite fertilizers of complex composition in open and closed ground and nurseries.

32.1 INTRODUCTION

Analysis of global trends in agriculture reveals main trends in increasing economic efficiency and strengthening environmental protection measures. This is evidenced by the constantly growing interest in technologies for the balanced consumption of mineral fertilizers with minimal losses and the least harm to the environment [1, 2].

Advanced Polymer Structures: Chemistry for Engineering Applications. Omar Mukbaniani, DSc, Tamara Tatrishvili, PhD & Marc Jean M. Abadie, DSc (Eds.)

Plant mineral nutrition issues and its regulation by the application of fertilizers have been studied for over 100 years. Nevertheless, the problem of mineral nutrition of plants is still far from being solved for two main reasons. The first of them is that plants need a complex of nutrients at the same time both macro- (N, P, K, Ca, Mg, and S) and microelements (Fe, Cu, Zn, Mn, Co, Mo, B, etc.) [3]. There is a close interdependence between nutrients in the process of their injection to plants, i.e., when changing the content of one element in the soil, the supply of plants of other elements changes as well as their nutritional regime. The second reason is the large heterogeneity of environmental factors (composition and properties of soils, humidity, temperature, etc.), which largely determine the amount of nutrients required to meet the needs of plants. All these factors are very insufficiently considered in practice and in experimental work [4].

In addition, the effect of mineral fertilizer application is low because after application to the soil a significant portion of nitrogen fertilizer is either washed away or evaporate, while most phosphorus fertilizers are easily absorbed by the soil and converted to the hardly soluble forms of calcium-, magnesium, iron- or aluminum phosphates. So, the utilization efficiency of this type fertilizers is very low and does not exceed 15–25%. In addition, the accumulation of hardly soluble phosphates in the soil leads to deterioration of its structure and ecological pollution [5–8].

Various natural minerals such as diatomite, perlite, marl, limestone, zeolite tuffs [9–11] and others are widely used to improve soil structure and thereby increase soil fertility.

Natural zeolites – crystalline hydrated aluminosilicates – are nanoporous materials with a pore size in the range of 1.5–10 nm [11, 12] and are of interest for nanotechnology [13]. They belong to the group of cation exchangers. Their enrichment with potassium, calcium, ammonium, and other cations is not difficult. This process is complicated for the anions present in the respective compounds in the form of anionic acid salts.

The main and important feature of zeolite structure is the system of intracrystalline pores and channels, where ion exchange, occlusion, and release of small molecules can easily take place, resulting in a great ability of zeolite adsorption and desorption. The three-dimensional anionic framework of zeolites consists of SiO_4^- and AlO_4^- tetrahedrons, which are connected to each other by solid Si–O–Si and Si–O–Al bridges (Figures 32.1 and 32.2) [14, 15]. Because the channels and bridges in the zeolite structure are arranged in a sequential and orderly manner, a zeolite porous regulated structure is formed, which creates a well-known molecular-sieve effect of the zeolite [11, 12].

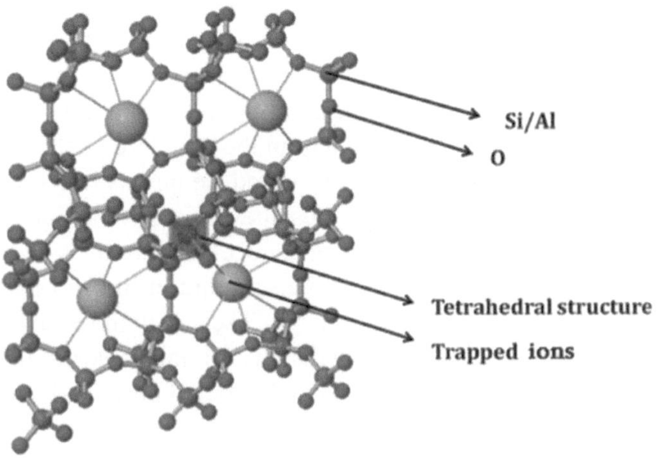

FIGURE 32.1 The tetrahedral framework of clinoptilolite zeolite [16].

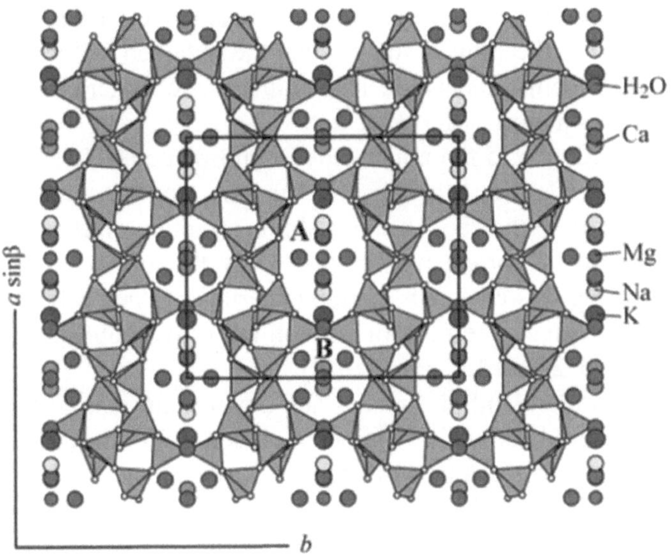

FIGURE 32.2 Position of cations in the clinoptilolite structure [17].

Source: Reprinted from [17]. The International Zeolite Association.

The zeolite structure is ionic and can be changed by adsorption of molecules. In this case, the intracrystalline space is filled with polar or ionic complexes, which reduce the free energy of the system. In the crystalline system of zeolites, it is possible to split and replace polar or ionic complexes, which in turn increases the efficiency of acid hydrogen centers in the zeolite frame. Therefore, the activity of zeolites may be due not only to the presence of acid centers in them, but also to reduction of free energy of the whole system. This is possible by the additional polarization or ionization of occluded molecules and their placement in the structure [13].

We used this property of zeolites to introduce anions into their structure. When obtaining anionic forms of zeolites, they must have a high resemblance to ionic compounds. They should be easily occluded in this system and have to be placed in the intracrystalline zeolite structure in such a way as to avoid excessive collision of similarly charged ions. The role of the zeolite structure in this process is primarily due to the fact that the zeolite is also a solid electrolyte and not just simply symmetrically arranged, ordered pores and channels.

32.2 EXPERIMENTAL METHODS AND MATERIALS

The nanotechnological method for obtaining anionic substituents is proposed. Method is based on the introduction of the respective desired salt into the zeolite structure in the form of ions so that the obtained material would not lose its zeolite structure and would acquire anion-exchanger properties based on new phase formation [16, 17].

Certain amounts of ammonium dihydrophosphate ($NH_4H_2PO_4$), KNO_3, ions and cations (Fe, Ca, Mn, Mg, Zn, Cu, Mo, Co, Sn) mixed with them were introduced into the zeolite-clinoptilolite structure on the nano-level at melting point of the corresponding salts. Only the amount of ammonium dihydrophosphate was changed. Zeolite materials of three different compositions were obtained. The ratio of ammonium dihydrophosphate ions and zeolite was 1:1, 1:2, and 1:3, respectively.

Fixation of nitrate and phosphate- and ammonium ions in the obtained zeolite nanomaterial, as well as cations included in this structure, observation of zeolite structure changes was carried out by methods of chemical and morphological analysis, IR-spectroscopy, and X-ray diffractometric analysis, as well as electronic scanning. X-ray diffractometric analysis was carried out on a Dron-4 device (Russia); infrared spectrometric analysis was carried out on an Agilent Cary 630 FTIR Spectrometer in the range of 350–5,000 cm^{-1} (USA); chemical analysis was carried out on a Spectroscout XEP-04 (Germany); Electron-microscopic scanning-morphological analysis was performed on a JEOL JSM-6510LV electron microscope equipped with an X-Max^N energy dispersive micro-X-ray spectral analyzer from the English company Oxford Instruments.

32.3 RESULTS AND DISCUSSION

32.3.1 ZEOLITE NANOMODIFICATION

32.3.1.1 FIXATION OF NITRATE AND PHOSPHATE- AND AMMONIUM IONS IN THE OBTAINED ZEOLITE NANOMATERIAL BY METHOD OF CHEMICAL AND MORPHOLOGICAL ANALYSIS

The results of the analysis are presented in Table 32.1.

TABLE 32.1 Chemical Analysis of Zeolite Nanomaterial Obtained

Ions Contained in Zeolite Nanomaterials	Chemical Composition of Zeolite Nanomaterials		
	1:1	1:2	1:3
SiO_2	5.95%	7.55%	6.08%
Al_2O_3	2.27%	2.38%	1.90%
P_2O_5	34.17%	33.80%	38.20%
K_2O	10.80%	13.50%	11.10%
$NH_4^+ + NO_3^-$	2.46%	2.46%	2.46%
Fe_2O_3	3.175%	1.08%	0.90%
CaO	1.40%	1.50%	1.20%
MnO	0.043%	0.04%	0.03%
Na_2O	0.135%	0.13%	0.13%
MgO	0.016%	0.016%	0.016%
Zn	59.50 ppm	52.60 ppm	43.50 ppm
Cu	46.40 ppm	30.10 ppm	26.50 ppm
Mo	13.90 ppm	14.70 ppm	13.30 ppm
Co	5.40 ppm	3.50 ppm	3.00 ppm
Sn	1.30 ppm	1.20 ppm	1.09 ppm

As seen from the table, a certain amount of cations and anions entered in the zeolite structure. The process is less dependent on the amount of ammonium dihydrophosphate ions, their increase is not proportional to the amount of ions entering the zeolite structure, because only a certain amount of ions can be introduced into the zeolite volume.

The results of the analysis are presented in Figure 32.1. The experiment showed that the penetration of the amorphous salt mass into the clinoptilolite structure as a result of melting causes some deformation of its structure.

32.3.1.2 THE RESULTS OF BOTH X-RAY DIFFRACTOMETRIC ANALYSIS AND IR-SPECTROSCOPY

As can be seen from the chemical analysis, the zeolite structure is deformed, this is evidenced by the results of both X-ray diffractometric analysis and IR-spectroscopy. It seems that desorption of water from zeolite channels as well as migration of cations occurred during melting. These places were occupied by ammonium dihydrophosphate, which caused a partial change in zeolite structure, because the salts included in the zeolite structure (ammonium dihydrophosphate, sodium nitrate) have a high degree of symmetry. The X-ray diffractogram mainly shows peaks of these salts. The intensities of zeolite-clinoptilolite peaks at 9.5; 22.5 and 31.7 – 2Θ decrease accordingly for ratios 1:1 and 1:2 and are almost not observed at ratio 1:3 (Figure 32.3).

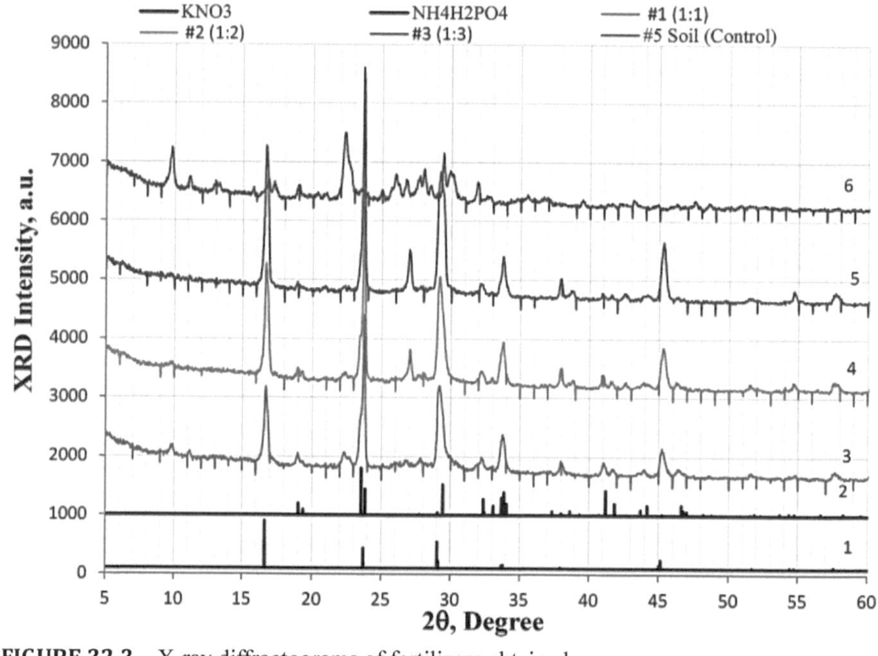

FIGURE 32.3 X-ray diffractograms of fertilizers obtained.

Source: Reprinted from [17]. The International Zeolite Association.

The IR-spectra of the samples clearly show the corresponding bands of NH_4^+, NO_3^-, and PO_4^{3-}-ions entering in the zeolite structure at 1,402, 1,387, 1,292, 925, 825, 547 cm^{-1} [18, 19]. The structure of the zeolite practically does not change after nanomodification. This is indicated by the presence of bands both deformation (449 cm^{-1}) and valence (1,117 cm^{-1}) [20] vibration of the Si–O–Si(Al) bonds (characteristic of zeolite) in the infrared spectrum of the zeolite. A slight shift in the frequencies of these bands, caused by the deformation of the zeolite structure is observed in the spectrum. This result agrees with the data of X-ray structural analysis. To study the mechanism of removal of anions from the zeolite structure, the obtained zeolite nano-materials were washed with water. The study of the zeolite structure by IR spectroscopy was carried out at a certain stage. Removal of anions from zeolite structure by washing occurs gradually; at the end the zeolite structure reverts to its initial form which is indicated by maximum decrease of intensities of bands at 1,402, 1,387, 1,292, 925, 825, 547 cm^{-1}, which correspond to the NH_4^-, NO_3^-, and PO_4^{3-}-ions (Figures 32.4–32.8).

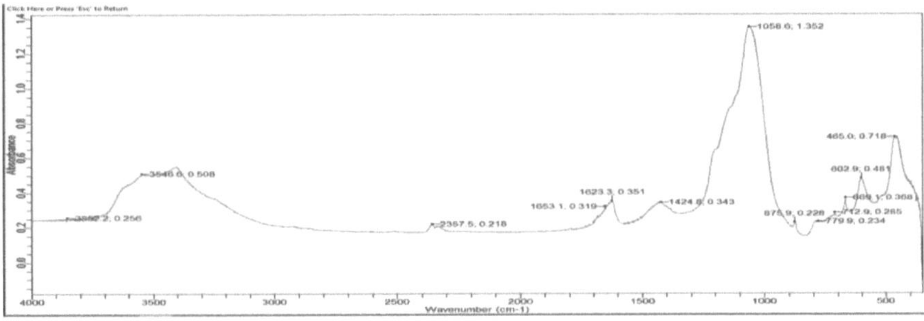

FIGURE 32.4 IR-spectrum of the natural zeolite-clinoptilolite initial form.

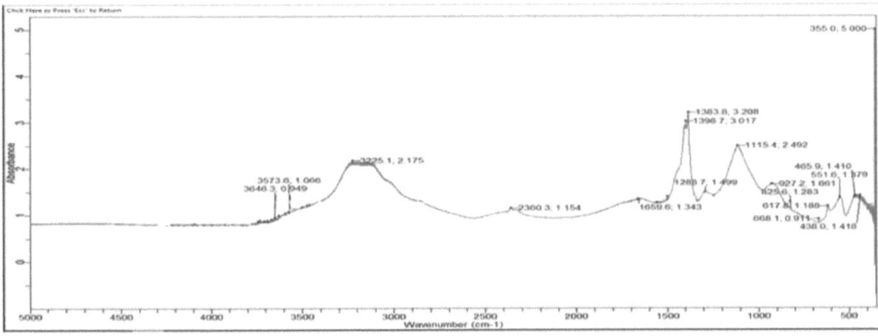

FIGURE 32.5 IR-spectra of anionic zeolite material obtained (for Ratio 1:1).

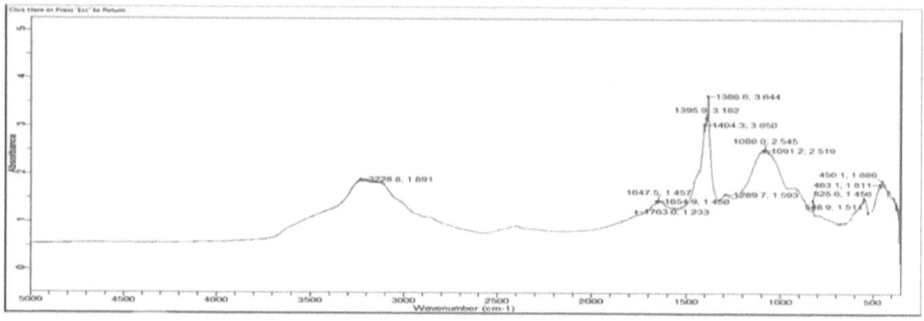

FIGURE 32.6 IR-spectra of anionic zeolite material obtained (for Ratio 1:2).

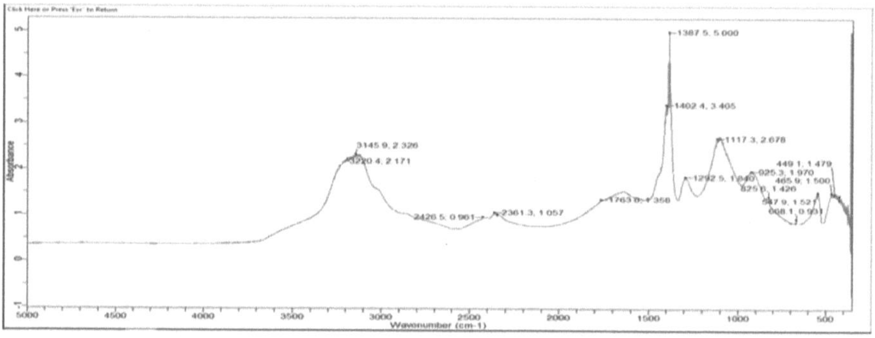

FIGURE 32.7 IR-spectra of anionic zeolite material obtained (for Ratio 1:3).

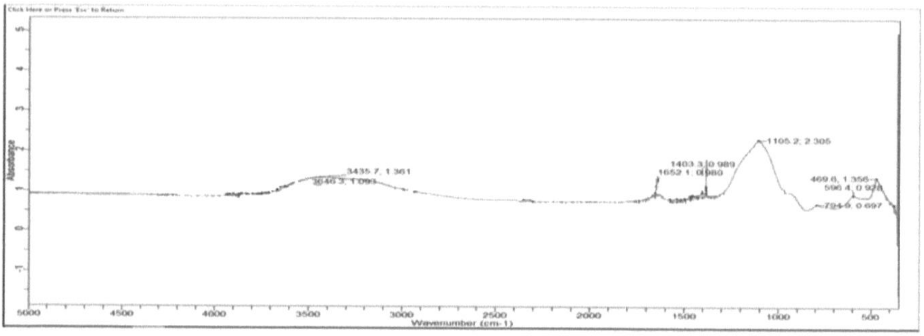

FIGURE 32.8 IR-spectra of the obtained anionic zeolite material after washing with distilled water during the certain time.

Thermal analysis of the obtained zeolite nanomaterials and water vapor adsorption study showed that in the first case, the amount of water lost in the original form was 11.9%, and for the obtained nanomaterials – 6.3%. In the second case, water vapor adsorption was reduced from 3.12 mmol/g to

2.44 mmol/g. Both processes are due to the presence of PO_4^{3-}-anions in the zeolite structure, which inhibits water adsorption in the zeolite because the ions are occupied by a large part of the zeolite channels. It is noteworthy that in comparison with the original zeolite, the thermal stability of the obtained nanomaterials was significantly increased [20].

32.3.1.3 THE RESULTS OF ELECTRON MICROSCOPIC SCANNING METHOD

Electron microscopic scanning method was used to study the surface morphology of the original and modified zeolite nanomaterials. The results obtained by observation under an electron microscopy method show that clinoptilolite zeolite crystals are plates of thinned shape (Figure 32.9(a)) [21].

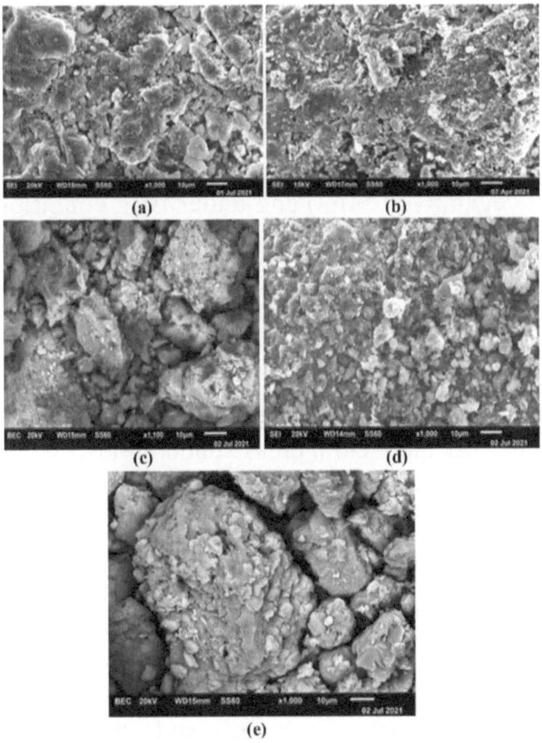

FIGURE 32.9 Microphotographs of initial and modified zeolite nanomaterials examined by electron microscopy: (a) zeolite-clinoptilolite – initial form; (b) anionic zeolite material (ratio 1:1); (c) anionic zeolite material (ratio 1:3); (d) washed anionic zeolite material (ratio 1:1); and (e) washed anionic zeolite material (ratio 1:3).

After modification of clinoptilolite with ammonium dihydrophosphate and KNO_3, its layered structure is not ordered in comparison with the original samples (Figure 32.9(b) and (c)).

After washing of modified samples with distilled water within a certain time, the main amount of PO_4^{3-} and NO_3^- ions were removed from the zeolite structure (Figure 32.9(d) and (e)).

Under an electron microscope, the structure of the washed clinoptilolite (after washing the modified samples from the zeolite structure) is also different from that of the original zeolite. As can be seen, after washing, the zeolite structure is deformed, which is be clearly seen on X-ray diffractograms and IR-spectra.

The study by means of electron scanning microscopy showed that cations and anions were non-uniformly distributed in the space of the zeolite framework.

32.3.2 NANOMODIFIED ZEOLITE AS A FERTILIZER

In order to study the effect of the obtained zeolite fertilizer on crop species, we choose winter wheat. The experiment was carried out in laboratory conditions (in a vegetative vessel) and under field conditions (in the open ground). Zeolite nanomaterials – fertilizers in which the ratio of ammonium dihydrophosphate and zeolite was 1:1, 1:2, and 1:3, and the amount of ammonium nitrate was the same in all three cases – were chosen for testing.

After harvesting wheat, it was found that under both growing and field conditions the best results were obtained with zeolite fertilizer obtained at a ratio of 1:1. In this case the yield increase under field-conditions was 36.6% compared to the variant without fertilizer while under vegetative conditions the yield increase was 26%.

The increase in the quality indicators of wheat should be especially noted. We obtained the highest quality wheat when using the fertilizer obtained at a 1:1 ratio. It has a higher content of protein, as well as fats and raw gluten.

After harvesting, samples of zeolite fertilizer processed during the season were taken from the vegetation container and chemically analyzed. The results are presented in Table 32.2.

TABLE 32.2 Chemical Analysis Results of Composition of Zeolite Nanomaterial (Fertilizer) After Their Use

Ions Contained in Zeolite Nanomaterials	Chemical Composition of Zeolite Nanomaterials		
	1:1	1:2	1:3
SiO_2	35.24%	36.16%	25.04%
Al_2O_3	8.30%	9.48%	6.73%
P_2O_5	18.97%	13.9%	35.4%
K_2O	5.5%	5.5%	8.6%
$NH_4^+ + NO_3^-$	–	–	–
Fe_2O_3	3.1%	4.7%	7.8%
CaO	9.6%	6.8%	3.5%
MnO	0.1%	0.2%	0.13%
Na_2O	0.13%	0.13%	0.13%
MgO	1.1%	1.9%	0.2%

The analysis showed that the remaining amount of potassium and phosphorus in the applied fertilizer is quite large, which once again proves the ability of prolonged action of the obtained zeolite fertilizer. The advantageous characteristic of prolonged-acting fertilizers is the safe release of nutrients even in unstable natural conditions. The release of nutrients occurs in accordance with the physiological needs of crops [1, 22–26].

32.4 CONCLUSION

The obtained zeolite nanomaterial can be successfully used in agriculture as a prolonged action complex composition fertilizer. Due to the molecular sieve properties of zeolite, phosphoric, and nitric acid anions injected in the zeolite structure, located both in the internal channels of zeolite and in mesopores, are gradually transferred to the soil. This is evidenced by the results of chemical analysis. We can say that the obtained nanomaterial is a capsule and phosphate and nitrate ions, as well as some types of cations, injected in it are gradually transferred to the plant. A good illustration of this process is an experiment in which the obtained nanomaterial was washed with certain portions of water. The structure of this material was studied by Fourier infrared spectroscopy, which clearly showed a decrease in the intensity of the bands characteristic of PO_4^{3-} and NO_3^- ions. At a certain stage of this experiment, the anions were completely removed from the zeolite structure, which was confirmed after taking infrared spectra of the sample (*see* Figures 32.4–32.8).

Thus, after fertilization, contact of PO_4^{3-}-anions with the ions present in the soil is minimized; accordingly, the formation of phosphorus ions in the form of insoluble or hardly soluble salts is minimal, and the efficiency of fertilizer application is raised. Besides, nitric acid anions are less likely washed down into wastewater or groundwater and cause less soil contamination.

ACKNOWLEDGMENT

The research was financially supported by the Shota Rustaveli National Science Foundation within the framework of the project AR-18-601.

KEYWORDS

- **electron microscopy analysis**
- **IR spectroscopy**
- **nanomodified**
- **natural zeolite**
- **X-ray diffractometry**
- **zeolite-anionic form**

REFERENCES

1. Abregova, F. M., & Bzhetseva, A. B., (2020). *"Smart" (Prolonged Action) Fertilizers are a Topical Area of Research in the Field of Agriculture.* FSBSI "Adyghe Scientific Research Institute of Agriculture" [online]. Website http://adygniish.ru/index.php/13-novosti-niiskh/316-umnye-prolongirovannogo-dejstviya-udobreniya-aktualnoe-napravlenie-issledovanij-v-sfere-selskogo-khozyajstva (accessed on 02 January 2022).
2. Ebrahim, R., Farzad, N., Taimoor, J., Saadi, S., & Jaime, A. T. Da. S., (2021). Potassium-enriched clinoptilolite zeolite mitigates the adverse impacts of salinity stress in perennial ryegrass (*Lolium perenne* L.) by increasing silicon absorption and improving the K/Na ratio. *Journal of Environmental Management, 285*, 1–11. https://doi.org/10.1016/j.jenvman.2021.112142.
3. White, P. J., & Brown, P. H., (2010). Plant nutrition for sustainable development and global health. *Annals of Botany, 105*(7), 1073–1080. https://doi.org/10.1093/aob/mcq085.

4. Egorov, V., (2021). *The Physiological Role of the Most Important Elements Found in the Soil in the Life of Plants*. [Online]. Website www.agro-archive.ru/pochvovedenie/ (accessed on 02 January 2022).
5. Zavalin, A. A., Blagoveshchenskaya, G. G., Chernova, L. S., & Shmyreva, N. Ya., (2012). Management of nitrogen nutrition of plants in the soil. *J. Agrochemical Bulletin, 4*, 38–40. In Russian.
6. Volynkin, V. I., & Volynkin, O. V., (2013). The effect of the fertilizer composition and nitrogen doses in systematic use in crop rotation and monoculture of wheat. *J. Fertility, 2*, 20–22. In Russian.
7. Kudashkin, M. I., (2011). Influence of nitrogen and micro-fertilizers on the productivity of winter wheat of different sowing dates in crop rotations of agricultural landscapes in the south of the non-black earth region. *J. Agrochemistry, 5*, 26–34. In Russian.
8. Tsitsishvili, G. V., Andronikashvili, T. G., Nesterov, M. B., & Labutin, V. G., (1997). Natural zeolites in crop production. Chapter 4. *Use Natural Zeolites in Plant Growing* (pp. 112–143). Tbilisi: "Metsniereba."
9. Soltys, L. M., Mironyuk, I. F., Tatarchuk, T. R., & Tsinurchyn, V. I., (2020). Zeolite-based composites as slow release fertilizers (review). *Physics and Chemistry of Solid State, 21*(1), 89–104. https://DOI.ORG/10.15330/pcss.21.1.89-104.
10. Lateef, A., Nazir, R., Jamil, N., Alam, Sh., Shah, R., Naeem, K. M., & Saleem, M., (2016). Synthesis and characterization of zeolite-based nano–composite: An environment friendly slow release fertilizer. *Microporous and Mesoporous Materials, 232*, 174–183. doi: 10.1016/j/micromeso.2016.06.020.
11. Sangeetha, C., & Baskar, P., (2016). Zeolite and its potential uses in agriculture: A critical review. *Agricultural Reviews, 37*(2), 101–108. doi: 10.18805/ar.v0iof.9627.
12. Tsitsishvili, G. V., Tsintskaladze, G. P., Tsitsishvili, V. G., Chipashvili, D. S., & Tsintskaladze, Z. P., (2006). New form of phosphorous containing clinoptilolite. *Azerbaijan Chemical Journal, 3*, 100–102.
13. Breck, D., (1974). *Ion Exchange Reactions on Zeolites* (pp. 544–603). Chapter 7; Zeolite molecular sieves. A Wiley-Interscience publication John Wiley & Sons, New York. London. Sydney. Toronto.
14. Sidamonidze, Sh., (2011). Nanochemistry. Chapter 5: *Synthesis of Nanoparticles* (pp. 52–72). Tbilisi: The Saint Andrew the First-Called Georgian University of the Patriarchate of Georgia (in Georgian).
15. Sudisha, J., Murali, K. P., Krishnan, V., Nagaraja, G., Muhammad, M., Shashikant, S. U., & Muthusamy, G., (2021). Phytotoxicological effects of engineered nanoparticles: An emerging nanotoxicology. *Science of The Total Environment, 801*, 149809. https://doi.org/10.1016/j.scitotenv.2021.149809.
16. International Zeolite Association, (2020). Available online: http://www.iza-online.org (accessed on 02 January 2022).
17. http://www.iza-online.org/natural/Datasheets/Clinoptilolite/clinoptilolite.htm (accessed on 02 January 2022).
18. Rabo, J. A., (1976). *Zeolite Chemistry and Catalysis* (Vol. 1, p. 504). Union Carbide Corp., Tarrytown Technical Center, American Chemical Society.
19. Flanigen, E. M., Khatami, H., & Szymanski, H. A., (1974). Infrared structural studies of zeolites frameworks; Chapter 16. In: Flanigen, E. M., & Sand, L. B., (eds.), *Molecular Sieve Zeolites: Advances in Chemistry* (Vol. 101, pp. 201–229). American Chemical Society, Washington DC.

20. Papko, L. F., & Kravchuk, A. P., (2013). *Physicochemical Methods for the Study of Inorganic Substances and Materials* (pp. 48–56). Workshop. Part 4: Infrared spectroscopy. Belarusian State Technological University, Minsk, (in Russian).
21. Maha, M. H., Khalil, M. K., & Salem, M. M., (2014). Characterization of raw zeolite and surfactant-modified zeolite and their use in removal of selected organic pollutants from water. *International Journal Chemical Science, 12*(3), 815–844. www.sadgurupublications.com (accessed on 02 January 2022).
22. Eprikashvili, L., Zautashvili, M., Kordzakhia, T., Pirtshkalava, N., Dzagania, M., Rubashvili, I., & Tsitsishvili, V., (2016). Intensification of bioproductivity of agricultural cultures by adding natural zeolites and brown coals into the soil. *Annals of Agrarian Science, 14*, 67–71.
23. Tsintskaladze, G., Eprikashvili, L., Zautashvili, M., Sharashenidze, T., Burdjanadze, M., Burkiashvili, N., & Gabunia, V., (2013). New biotechnological material based on natural zeolite acts as a fertilizer of a prolonged action and prospects of its application. *Proceedings, Symposium: II International Scientific-Practical Conference "Bioeconomy and Sustainable Development of Agriculture* (pp. 559–562). Tbilisi, Georgia.
24. Tsintskaladze, G., Tsitsishvili, V., Sharashenidze, T., Zautashvili, M., Iluridze, G., & Tsintskaladze, P., (2016). Possibilities of agricultural application of zeolite nanomaterials enriched by nitrate ions. *Proceedings of the Georgian National Academy of Sciences, Chemical Series* (Vol. 42, No. 1, pp. 78–80). Georgia.
25. Tsintskaladze, G. P., Eprikashvili, L. G., Urushadze, T. F., Kordzakhia, T. N., Sharashenidze, T. V., Zautashvili, M. G., & Burjanadze, M. N., (2016). Nanomodified natural zeolite as a fertilizer of prolog acting. *Annals of Agrarian Science, 14*, 163–168.
26. Tsintskaladze, G., Eprikashvili, L., Mumladze, N., Gabunia, V., Sharashenidze, T., Zautashvili, M., Kordzakhia, T., & Shatakishvili, T., (2017). Nitrogenous zeolite nanomaterial and the possibility of its application in agriculture. *Annals of Agrarian Science, 15*, 365–369.

CHAPTER 33

Bactericidal Metal-Containing Adsorbents Prepared from Georgian Natural Zeolites

VLADIMER TSITSISHVILI, NANULI DOLABERIDZE, NATO MIRDZVELI, MANANA NIJARADZE, ZURAB AMIRIDZE, and BELA KHUTSISHVILI

Petre Melikishvili Institute of Physical and Organic Chemistry, Tbilisi State University, Tbilisi, Georgia

ABSTRACT

The coronavirus pandemic has increased interest in antibacterial agents containing bioactive metals, for which zeolites are promising carriers. On the other hand, zeolite adsorbents and ion exchangers containing bioactive metals and endowed with bactericidal properties are promising for water treatment and other environmental and medical applications. Natural zeolites characterized by comparatively high ion-exchange capacity (phillipsite, analcime, heulandite) can be used to produce such materials. Silver-, copper-, and zinc-containing microporous materials have been prepared on their basis using ion-exchange reactions between zeolite microcrystals and a salt of a corresponding transition metal in the solid phase followed by washing with distilled water. Synthesized in such way adsorbent-ion-exchangers are characterized by chemical composition based on the X-ray energy dispersion spectra, powder X-ray diffraction patterns, and Fourier transform infra-red spectra. Obtained materials remain the zeolite crystal structure and contain 130–230 mg/g of silver, 65–72 mg/g of copper, and 58–86 mg/g of zinc, as compared with modified samples of synthetic zeolite A containing up to 290

Advanced Polymer Structures: Chemistry for Engineering Applications. Omar Mukbaniani, DSc, Tamara Tatrishvili, PhD & Marc Jean M. Abadie, DSc (Eds.)

mg/g of silver, 75 mg/g of copper, and 100 mg/g of zinc. Prepared metal-containing materials show bacteriostatic activity against gram-negative bacterium *Escherichia coli*, gram-positive bacteria *Staphylococcus aureus* and *Bacillus subtilis*, fungal pathogenic yeast *Candida albicans*, and a fungus *Aspergillus niger*, and natural zeolites enriched with biometals exhibit a synergistic effect – their mixtures have a higher bacteriostatic activity. It has been established that the bacteriostatic activity of metal-containing zeolites is determined not only and not so much by the ions of bioactive metals released into the liquid medium, but an important role in inhibiting the growth of microorganisms plays a type of zeolite matrix. Despite the relatively low ion-exchange capacity, heulandite turned out to be a fairly effective matrix for bioactive metals.

33.1 INTRODUCTION

The coronavirus pandemic has led to a sharp increase in both the consumption of disinfectants and interest in new antibacterial and antiviral substances. Research in this direction is carried out in many countries, along with liquid disinfectants, solid substances are being developed. In particular, germicidal powder materials are used in the production of polymeric materials and paper as fillers to protect the surface of an article from microbial contamination. Among advanced materials, zeolites (aluminosilicates of the general formula $M_x[Al_xSi_yO_{2(x+y)}]{\cdot}mH_2O$), in which ions of alkali (Na^+, K^+, etc.), or alkaline-earth (½Ca^{2+}, ½Mg^{2+}, etc.) metals M are partially replaced by ions of a bioactive metal (Ag^+, Cu^{2+}, Zn^{2+}, etc.) are recognized as promising [1]. At the same time, zeolites have been widely used for adsorption of ions and organics in the tap water and wastewater treatment [2, 3], and today it is recognized that natural zeolites outperform conventional granular materials in water purification [4], but it is necessary to provide the filtering materials with bacteriostatic properties in order to prevent the growth of microorganisms on their surface.

Studies that started at the beginning of the 21st century and continue to this day [5–26] have shown that natural and synthetic zeolites enriched with bioactive metals exhibit antimicrobial activity against a wide range of microorganisms, while silver-containing zeolites (AgZ) are the most active. It is believed that the porous structure of cation-exchanged and bioactive metal-containing zeolites enables metal cation to move freely from the lattice to the environment, and this seems to be responsible for their activity toward microorganisms [9], but it has recently been established that in some cases the antibacterial activity could be attributed to the zeolite matrix itself [14, 20, 23–26]. Moreover, the activity of metal-containing zeolites against a number of viruses, including human coronavirus, was established [10], and work in this direction continues [21, 22].

This chapter summarizes the earlier results [23–26] on the structure, chemical composition, and bacteriostatic activity of natural zeolites phillipsite, analcime, and heulandite enriched in silver, copper, and zinc, as well as synthetic zeolite of LTA type. In addition, new data on the bacteriostatic activity of mixtures of zeolite materials with respect to such microorganisms as gram-negative bacterium *Escherichia coli*, gram-positive bacteria *Staphylococcus aureus* and *Bacillus subtilis*, fungal pathogenic yeast *Candida albicans*, and a fungus *Aspergillus niger*.

33.2 EXPERIMENTAL METHODS AND MATERIALS

33.2.1 NATURAL AND SYNTHETIC ZEOLITES

Phillipsite-containing tuff rock from Shukhuti (Western Georgia), analcime from Chachubeti and heulandite-clinoptilolite from the Rkoni plot of Tedzami deposit (Eastern Georgia) were used as starting materials, their properties and preliminary treatment is described in Refs. [24, 25, 27], respectively. Crushed (planetary micro-mill Pulverisette 7, Fritsch Laboratory Instruments, Idar-Oberstein, Germany) and sieved (250 BBS mesh) zeolite-containing rocks were washed with a diluted solution of HCl (0.025 N) to remove clay impurities, and named as P_{SH} (phillipsite from Shukhuti), A_{CH} (analcime from Chachubeti), H_{RT} (heulandite from Rkoni-Tedzami). Hydrothermal synthesis of the LTA zeolite from analcime is described in Ref. [28]; in the experiments, crystallites of its sodium form NaA with sizes from 3 to 5 μm have been used. Ion exchange was carried out as follows: powder of zeolite and the corresponding salt were mixed in a molar ratio zeolite aluminum: biocide metal of 1:3–6 and thoroughly grinded in an agate mortar for 10 minutes for $AgNO_3$-containing mixture, and for 15 minutes for $CuCl_2$ or $ZnCl_2$-containing mixtures. The solid mixture was then transferred to a filter and washed with distilled water until the complete disappearance of nitrate or chlorine anions, after which the modified samples were first dried in air and then at 100–105°C in a thermostat; samples with a maximum content of corresponding bioactive metal are labeled as AgZ, CuZ (copper-containing zeolite), and ZnZ (zinc-containing zeolite).

33.2.2 CHARACTERIZATION

Chemical composition of raw material and prepared samples was determined by elemental analyzes carried out using X-ray energy dispersive (XRED) spectral analysis (Oxford Instruments X-Max 20 analyzer). The crystalline phase was identified by powder X-ray diffraction (XRD) patterns obtained from a modernized Dron-4 X-ray diffractometer (USSR) employing the Cu-K_α line (λ = 0.154056 nm). The samples were scanned in the 2Θ range of 5° to 50° with a 0.02° step at a scanning speed of 1 degree/min. Fourier transform infrared (FTIR) spectra were collected by a 10.4.2 FTIR spectrometer (Perkin-Elmer, UK) over the range of 400–4,000 cm^{-1} with a resolution of 2 cm^{-1} using the KBr pellet technique for sample preparation. Sample morphology was characterized by scanning electron microscope (SEM) images (JSM6510LV, Jeol, Japan); changes in the pore system and in the specific surface area were fixed using the Brunauer-Emmett-Teller (BET) and Barrett-Joyner-Halenda (BJH) methods measuring nitrogen adsorption-desorption on sorption analyzer (ASAP 2020 Plus, Micromeritics, USA).

Bacteriostatic properties of zeolite samples were determined by the disk diffusion (Kirby-Bauer) method in standard conditions using the cultures of gram-negative bacterium *Escherichia coli* (strain ATTC 8739), Gram-positive bacteria *Staphylococcus aureus* (ATTC 6538) and *Bacillus subtilis* (ATTC 6633), fungal pathogenic yeast *Candida albicans* (ATTC 10231) and a fungus *Aspergillus niger* (ATTC 16404 – *A. brasiliensis*) placed (10^9 CFU/cm^3) on Mueller–Hinton agar (3 mm deep) poured into 100 mm Petri dishes. Before testing the antibacterial activity, all dry zeolite products were sterilized at 70°C for 2 hours in a dry sterilizer. No microbial contamination of the prepared samples was found. About 0.2 g of zeolite in the form of pellets with 8 mm in diameter was placed into the plates. The plates contaminated with *E. coli*, *St. aureus* and *B. subtilis* were incubated at 37°C over 5% CO_2 medium and, finally, the width of inhibition zone of each sample in the plates was measured at the end of the first day. The plates contaminated with *Candida albicans* and *Aspergillus niger* were incubated at 25°C during 3–4 days. All experiments were done in triplicate, the values obtained were averaged to give the final data with standard deviations.

33.3 RESULTS AND DISCUSSION

33.3.1 CHEMICAL COMPOSITION AND STRUCTURE

33.3.1.1 CHEMICAL COMPOSITION

Chemical composition of native zeolites and their modified forms with a maximum silver, copper or zinc content are listed in Table 33.1 in terms of the empirical formulas $(M_aNa_bK_cCa_dMg_eMe_f)[Al_xSi_yO_z]{\cdot}nH_2O$, where z=48 for zeolites A, z=96 for analcimes, z=32 for phillipsites, and z=72 for heulandite, Me^+ ion corresponds to the impurity metals, M = Ag^+, ½Cu^{2+}, or ½Zn^{2+}, and deviations are given in parentheses. In addition, XRED spectra of heulandites show the presence of titanium atoms in amount of one atom per 240±20 aluminum and silicon atoms, and since this value remains unchanged for ion-exchange samples, there is reason to believe that titanium atoms are part of the zeolite crystal lattice.

TABLE 33.1 Chemical Composition of Native and Modified Zeolites

Sample	Empirical Formula	Si/Al	Biometal Content (mg/g)
A	$Na_{11.25(25)}(K,½Ca,½Mg)_{0.7(1)}(Al_{11.95(25)}Si_{12.3(3)}O_{48}){\cdot}18.0H_2O$	1.03(5)	–
AgA	$Ag_{10.3(4)}Na_{1.2(5)}(K,½Ca,½Mg)_{0.20(5)}(Al_{11.7(3)}Si_{12.3(3)}O_{48}){\cdot}18.6H_2O$	1.05(6)	382(23)
CuA	$Cu_{5.0(3)}Na_{1.35(25)}(K,½Ca,½Mg)_{0.35(7)}(Al_{11.7(3)}Si_{12.3(3)}O_{48}){\cdot}19.4H_2O$	1.05(6)	149(9)
ZnA	$Zn_{5.05(25)}Na_{1.2(3)}(K,½Ca,½Mg)_{0.20(5)}(Al_{11.5(3)}Si_{12.2(3)}O_{48}){\cdot}19.7H_2O$	1.06(6)	154(9)
A_{CH}	$(Na_{9.25}K_{2.25}Ca_{1.1}Mg_{1.0})Me_{0.50}[Al_{16.2}Si_{32.0}O_{96}]{\cdot}18.4H_2O$	1.98(15)	–
AgA_{CH}	$Ag_{9.5}(Na_{1.9}K_{1.7}Ca_{0.5}Mg_{0.5})Me_{0.10}[Al_{15.2}Si_{32.0}O_{96}]{\cdot}18.9H_2O$	2.10(20)	234(14)
CuA_{CH}	$Cu_{3.7}(Na_{3.2}K_{1.3}Ca_{0.75}Mg_{0.75})Me_{0.20}[Al_{15.0}Si_{32.0}O_{96}]{\cdot}19.5H_2O$	2.13(15)	65(4)
ZnA_{CH}	$Zn_{3.8}(Na_{3.7}K_{1.4}Ca_{0.75}Mg_{0.75})Me_{0.20}[Al_{15.1}Si_{32.0}O_{96}]{\cdot}19.2H_2O$	2.13(15)	69(4)
P_{SH}	$(Na_{1.30}K_{2.0}Ca_{0.30}Mg_{0.25})Me_{0.10}[Al_{4.50}Si_{11.80}O_{32}]{\cdot}11.4H_2O$	2.62(16)	–
AgP_{SH}	$Ag_{3.42}(Na_{0.01}K_{0.08}Ca_{0.25}Mg_{0.20})Me_{0.03}[Al_{4.44}Si_{11.20}O_{32}]{\cdot}14.2H_2O$	2.52(15)	230(15)
CuP_{SH}	$Cu_{1.80}(Na_{0.22}K_{0.17}Ca_{0.125}Mg_{0.12})Me_{0.075}[Al_{4.50}Si_{11.70}O_{32}]{\cdot}14.3H_2O$	2.60(16)	66(4)
ZnP_{SH}	$Zn_{1.84}(Na_{0.01}K_{0.35}Ca_{0.075}Mg_{0.125})Me_{0.025}[Al_{4.47}Si_{11.70}O_{32}]{\cdot}15.6H_2O$	2.64(16)	86(5)

TABLE 33.1 *(Continued)*

Sample	Empirical Formula	Si/Al	Biometal Content (mg/g)
H_{RT}	$(Na_{1.96}K_{0.47}Ca_{1.49}Mg_{1.17})Me_{0.08}[Al_{7.8}Si_{28.2}O_{72}]\cdot 24.0H_2O$	3.6(2)	–
AgH_{RT}	$Ag_{3.6}(Na_{0.85}K_{0.45}Ca_{0.8}Mg_{0.7})[Al_{7.8}Si_{28}O_{72}]\cdot 24.3H_2O$	3.6(2)	134(8)
CuH_{RT}	$Cu_{2.85}(Na_{0.2}K_{0.35}Ca_{0.15}Mg_{0.1})[Al_{6.85}Si_{28}O_{72}]\cdot 24.5H_2O$	4.10(25)	72(5)
ZnH_{RT}	$Zn_{2.3}(Na_{0.5}K_{0.35}Ca_{0.35}Mg_{0.5})[Al_{7.2}Si_{28}O_{72}]\cdot 25.2H_2O$	3.9(2)	58(4)

The degree of substitution in terms of biometal content is highest for synthetic zeolite and lowest for heulandite. Compared to the maximum possible metal content, which follows from the previously published [7] data on ion exchange isotherms on clinoptilolite from Gördez, Turkey (194, 77, and 65 mg/g for Ag, Cu, and Zn, respectively), the silver and copper content achieved in zeolite from Rkoni should be recognized as high. The hydrated ions of copper and zinc in solution are known to have the same regular configuration $M(H_2O)_6^{2+}$, size and diffusion coefficients [29], but under "solid" ion exchange conditions, the incorporation of zinc ions into the structure of heulandite-clinoptilolite encountered difficulties.

Nevertheless, the achieved metal content in H_{RT} (1.24, 1.14, and 0.89 mmol/g for Ag, Cu, and Zn, respectively) is significantly higher than that reported for the clinoptilolite-containing tuff from the Semnan deposit, Iran (0.24, 0.28, and 0.24 mmol/g of Ag, Cu, and Zn, respectively [20]); the silver content in AgHCR is one and a half times higher than in natural clinoptilolite (0.84 mmol/g [16]) used for the removal of bacteria *Escherichia coli* and heavy metals from aqueous solutions; the content of copper and zinc is at least twice as high as in natural clinoptilolite from the sedimentary deposit Zlatokop, Serbia (0.41 and 0.225 mmol/g of Cu and Zn, respectively [14]), used for disinfection of secondary effluent water.

According to the elemental analysis data, when silver, copper, and zinc ions are introduced into the crystal lattice of zeolites, the ratio Si/Al changes insignificantly, and the number of water molecules in the crystal lattice slightly increases.

The greatest change in the silicate modulus Si/Al and leaching of aluminum atoms (up to 10%) occurs when copper Cu^{2+} ions are introduced into the heulandite structure; when silver Ag^+ ions are introduced, the modulus does not change. Ion exchange mainly takes place with the participation of sodium Na^+ ions, relatively heavy and large potassium ions (sodium and potassium ion radii 0.098 and 0.138 nm, respectively) do not participate in ion exchange reactions.

The introduced ions of bioactive metals are evenly distributed over the surface of the zeolite, as evidenced by XRED mappings: at a small magnification (Figure 33.1), it can be seen that transition metal ions repeat the locations of aluminum atoms; with a higher magnification (Figure 33.2), the pattern of the arrangement of metal ions is absolutely uniform (details are given in Ref. [30]).

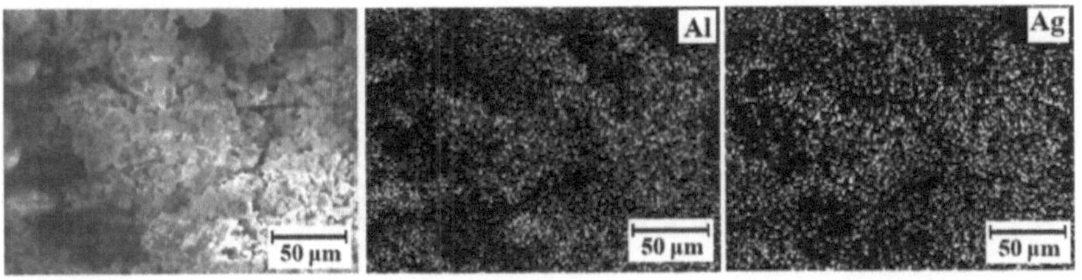

FIGURE 33.1 SEM image of AgH_{RT} and XRED mappings of aluminum (Al) and silver (Ag).

FIGURE 33.2 SEM image of CuH_{RT} and XRED mappings of aluminum (Al) and silver (Ag).

33.3.1.2 STRUCTURE

According to the XRD data, the crystal structure of native A, A_{CH}, P_{SH}, and H_{RT} zeolites corresponds to the LTA, ANA, PHI, and HEU types, respectively. Ion exchange reactions do not change the crystal structure of the zeolite, this is confirmed by the powder X-ray diffraction patterns of the modified samples: XRD pattern of the LTA structure [28] remains unchanged; characteristic peaks of the ANA, PHI, and HEU structures remain in XRD patterns, only their intensities change; for the silver-containing heulandite, a slight broadening of the peaks and a decrease in their intensity are observed

(Figure 33.3), associated with an increase in the degree of dispersion of the sample, recorded on SEM images [26, 27].

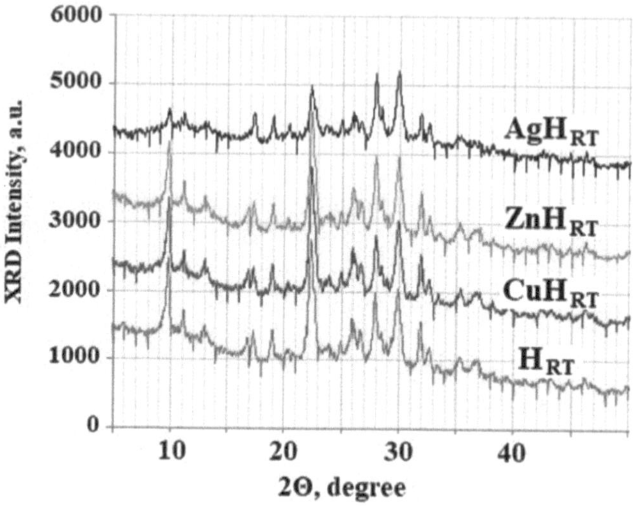

FIGURE 33.3 Powder X-ray diffraction patterns of the Rkoni zeolite (H_{RT}) and its modified forms enriched with silver (AgH_{RT}), copper (CuH_{RT}) and zinc (ZnH_{RT}).

With the exception of the Si/Al sensitive ν_{asym} vibration band for heulandites, no notable changes were observed in the IR spectra of the modified samples as compared with the vibration bands of starting zeolitic material, only the intensity of the broad band at 3,200–3,700 cm^{-1} corresponding to the asymmetric stretching of OH group is increased due to the larger number of water molecules in the samples containing silver, copper, and zinc.

33.3.2 BACTERICIDAL ACTIVITY

All original zeolites not containing bioactive metals (A, A_{CH}, P_{SH}, H_{RT}) were inactive (Tables 33.2–33.4), enriched synthetic zeolites equally inhibit the growth of gram-negative and gram-positive bacteria, but the copper form CuA is not active against yeast *Candida albicans* and fungus *Aspergillus niger*, while the zinc form ZnA is not inferior in activity to silver, and in relation to *E. coli* even shows a slightly higher activity (Table 33.2).

Previously, researchers from Serbia and Croatia [20] showed that the activity of synthetic type A zeolites with respect to Escherichia coli decreases in the series AgA>CuA>>>ZnA, but this discrepancy can be attributed to the use of commercial zeolite A in Ref. [20] and low degree of enrichment, about 0.25 mmol/g.

TABLE 33.2 Diameter (mm) of the Zone of Inhibition of the Growth of Microorganisms by Native and Modified Zeolites A

Microorganism	Zeolite			
	NaA	AgA	CuA	ZnA
Escherichia coli	0	23	17	26
Staphylococcus aureus	0	25	25	23
Bacillus subtilis	0	26	23	25
Candida albicans	0	20	0	24
Aspergillus niger	0	24	0	24

Of the enriched forms of analcime, only the silver form AgA_{CH} slows down the growth of all microorganisms used in the study, the copper form CuA_{CH} is not active at all, and the zinc form ZnA_{CH} is active only against *Bacillus subtilis*. However, a mixture of silver and zinc forms (Ag + Zn) A_{CH} in a 1:1 molar ratio shows a synergistic effect – increased activity against all microorganisms, a mixture of silver and copper forms (Ag + Cu)A_{CH} in the same 1:1 molar ratio shows a rather high activity, mixture of zinc and copper forms (Zn + Cu)A_{CH} is active against Gram-positive bacteria (Table 33.3).

TABLE 33.3 Diameter (mm) of the Zone of Inhibition of the Growth of Microorganisms by Native and Modified Analcimes

Microorganism	Zeolite						
	A_{CH}	AgA_{CH}	CuA_{CH}	ZnA_{CH}	(Ag + Cu) A_{CH}	(Ag + Zn) A_{CH}	(Zn + Cu) A_{CH}
Escherichia coli	0	21	0	0	25	28	0
Staphylococcus aureus	0	25	0	0	24	27	13
Bacillus subtilis	0	23	0	20	18	29	21
Candida albicans	0	20	0	0	16	24	0
Aspergillus niger	0	25	0	0	17	35	0

TABLE 33.4 Diameter (mm) of the Zone of Inhibition of the Growth of Microorganisms by Native and Modified Phillipsites

Microorganism	Zeolite						
	P_{SH}	AgP_{SH}	CuP_{SH}	ZnP_{SH}	(Ag + Cu) P_{SH}	(Ag + Zn) P_{SH}	(Zn + Cu) P_{SH}
Escherichia coli	0	20	11	15	30	20	0
Staphylococcus aureus	0	19	20	0	22	16	20
Bacillus subtilis	0	16	23	22	20	19	19
Candida albicans	0	29	0	18	14	26	24
Aspergillus niger	0	27	0	28	0	29	7

Synergism was also established for enriched forms of phillipsite – the greatest activity against *Escherichia coli* and staphylococcus is shown by a mixture (1:1) of silver and copper forms (Ag + Cu)H_{RT}, against an *Aspergillus niger* – by a mixture (1:1) of silver and zinc forms (Ag + Zn)H_{RT} (Table 33.4).

To the greatest extent, the synergistic effect is manifested in enriched forms of heulandite – despite the relatively low activity of copper and zinc forms, their mixtures are characterized by the highest activity against all microorganisms, as shown in Figures 33.4–33.6.

Despite the fact that the zinc form ZnH_{RT} is inactive to *E. coli*, even a small addition of the silver AgH_{RT} or copper CuH_{RT} form dramatically increases the activity; the highest bacteriostatic activity is observed for the ¼ZnH_{RT} + ¾CuH_{RT} mixture (Figure 33.4).

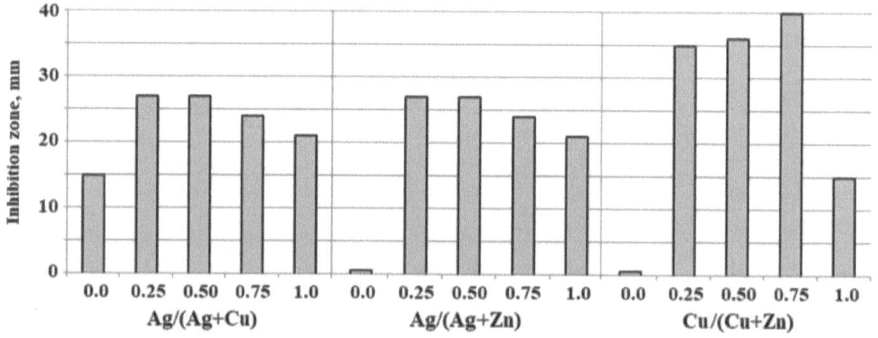

FIGURE 33.4 Inhibition zones of *E. coli* growth by AgH_{RT}, CuH_{RT}, ZnH_{RT}, and their mixtures in different molar ratios.

The synergistic effect is also manifested in relation to the both gram-positive bacteria (Figure 33.5), the mixtures ¼AgH_{RT} + ¾CuH_{RT}, ½AgH_{RT} + ½ZnH_{RT}, and ¼CuH_{RT} + ¾ZnH_{RT} have the greatest activity against staphylococcus, and mixtures of copper and zinc forms are most effective against hay bacillus *B. subtilis*.

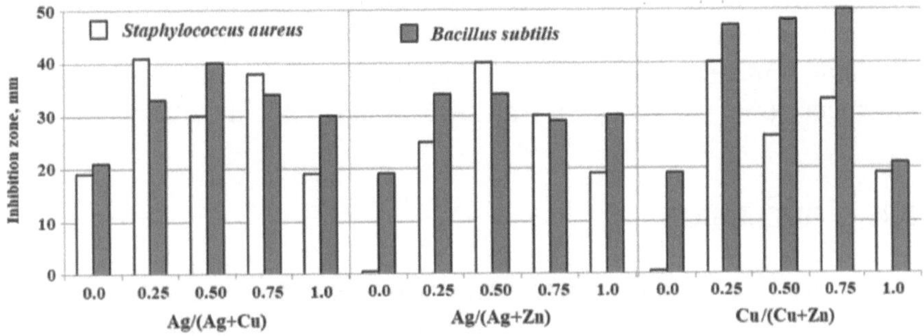

FIGURE 33.5 Inhibition zones of gram-positive bacteria growth by AgH_{RT}, CuH_{RT}, ZnH_{RT}, and their mixtures in different molar ratios.

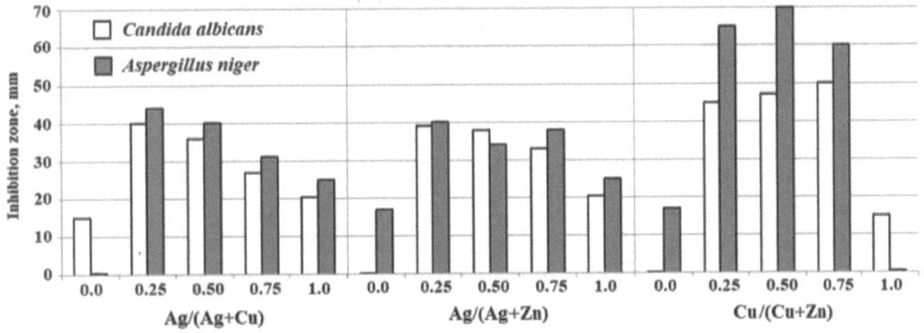

FIGURE 33.6 Inhibition zones of fungi growth by AgH_{RT}, CuH_{RT}, ZnH_{RT}, and their mixtures in different molar ratios.

In inhibiting the growth of fungi (Figure 33.6), mixtures with the AgHCR show low activity, the most effective are mixtures of CuHCR and ZnHCR, which is very important from a practical point of view, to replace expensive silver (800 $/kg) with cheaper copper (10 $/kg) and zinc (3 $/kg).

As follows from the data obtained, a mixture with a high copper content, Cu/Zn = 3, is the most effective against *E. coli* and hay bacillus, while the zone of inhibition is two times higher than that of the silver-containing form. On the contrary, against staphylococcus and *C. albicans*, the most effective is a mixture with a high zinc content, Zn/Cu = 3, and the maximum inhibition of the growth of *A. niger* is caused by a mixture with equal amounts of copper and zinc.

A similar synergistic effect was noted in the study of synthetic zeolite fibers [31]. It has been shown that the bacterial inhibition for nanofibers containing a three metal (Ag + Cu + Zn) nano-zeolite Y is higher than for containing bimetals (Ag + Cu, Ag + Zn, or Cu + Xn), while for containing monometals (Ag, Cu, or Zn), it was lower compared to the others.

33.4 CONCLUSION

As a result of the conducted research, it was established that solid-state ion-exchange reactions between Georgian natural zeolites phillipsite, analcime, and heulandite, as well as synthetic zeolite NaA, on the one hand, and salt of corresponding transition metal followed by washing results in zeolite materials with a high content of silver, copper, and zinc. The maximum amounts of introduced metals are achieved for synthetic zeolite (silver more than 380 mg/g, copper up to 150 mg/g, and zinc more than 150 mg/g), natural zeolites incorporate a smaller amount of bioactive metals (silver about 230 mg/g for phillipsite and analcime, 130 mg/g for heulandite, and copper 65–72 mg/g, zinc 58–86 mg/g). Thus, the incorporation of hydrated silver, copper, and zinc ions is determined by the size and accessibility of micropores and channels of the used zeolite.

According to the results of the Kirby-Bauer test on Gram negative bacterium *Escherichia coli*, Gram positive bacteria *Staphylococcus aureus* and *Bacillus subtilis*, fungal pathogenic yeast *Candida albicans*, and a fungus *Aspergilus niger*, all metal-containing samples of zeolite A, heulandite and phillipsite show bacteriostatic activity; in analcime, only the silver form is active, but the mixtures of all metal-containing natural zeolites act sinergistically showing higher activity against all microorganisms. The data obtained indicate the important role of the zeolite type, contrary to the opinion that the bacteriostatic activity of metal-containing zeolites is determined exclusively by bioactive metal ions released into the liquid medium. Despite the lower aluminum content and lower ion-exchange capacity than that of analcime and phillipsite, heulandite turned out to be a fairly effective matrix for bioactive metals.

KEYWORDS

- **analcime**
- **bacteriostatic activity**
- **Brunauer-Emmett-teller**
- **Fourier transform infrared**
- **heulandite**
- **phillipsite**
- **scanning electron microscope**

REFERENCES

1. Díez-Pascual, A. M., (2018). Antibacterial activity of nanomaterials. *Nanomaterials, 8*, 359–365.
2. Delkash, M., Bakhshayesh, B. E., & Kazemian, H., (2015). Using zeolitic adsorbents to cleanup special wastewater streams: A review. *Micropor. Mesopor. Mat., 214*, 224–241.
3. Nakhli, S. A. A., Delkash, M., Bakhshayesh, B. E., & Kazemian, H., (2017). Application of zeolites for sustainable agriculture: A review on water and nutrient retention. *Water, Air, & Soil Pollution, 228*(12), 464–476.
4. Wołowiec, M., Komorowska-Kaufman, M., Pruss, A., Rzepa, G., & Bajda, T., (2019). Removal of heavy metals and metalloids from water using drinking water treatment residuals as adsorbents: A review. *Minerals, 9*, 487–505.
5. Kawahara, K., Tsuruda, K., Morishita, M., & Uchida, M., (2000). Antibacterial effect of silver-zeolite on oral bacteria under anaerobic conditions. *Dental Mat. J., 16*, 452–455.
6. Rivera-Garza, M., Olguin, M. T., Garcia-Sosa, I., Alcantara, D., & Rodriguez-Fuentes, G., (2000). Silver supported on natural Mexican zeolite as an antibacterial material. *Micropor. Mesopor. Mat., 39*, 431–444.
7. Top, A., & Ülkü, S., (2004). Silver, zinc, and copper exchange in Na-clinoptilolite and resulting effect on antibacterial activity. *Applied Clay Science, 27*, 13. doi: 10.1016/j.clay.2003.12.002.
8. De La Rosa-Gomez, I., Olguin, M. T., & Alcantara, D., (2008). Bactericides of coliform microorganisms from wastewater using silver-clinoptilolite rich tuffs. *Applied Clay Science, 40*, 45–53.
9. Kwakye-Awuah, B., Williams, C., Kenward, M. A., & Radecka, I., (2008). Antimicrobial action and efficiency of silver-loaded zeolite X. *J. Appl. Microbiology, 104*(5), 1516. doi: 10.1111/j.1365-2672.2007.03673.x.
10. Bright, K. R., Sicairos-Ruelas, E. E., Gundy, P. M., & Gerba, C. P., (2009). Assessment of the antiviral properties of zeolites containing metal ions. *Food & Env. Virology, 1*(1), 37–41.
11. Ferreira, L., Fonseca, A. M., Botelho, G., Almeida-Aguiar, C., & Neves, I. C., (2012). Antimicrobial activity of faujasite zeolites doped with silver. *Micropor. Mesopor. Mat., 160*, 126–132.

12. Guerra, R., Lima, E., Viniegra, M., Guzman, A., & Lara, V., (2012). Growth of *Escherichia coli* and *Salmonella typhi* inhibited by fractal silver nanoparticles supported on zeolite. *Micropor. Mesopor. Mat., 147*, 267–273.
13. Hrenovic, J., Milenkovic, J., Ivankovic, T., & Rajic, N., (2012). Antibacterial activity of heavy metal-loaded natural zeolite. *J. Hazard. Mat., 201, 202*, 260–264.
14. Hrenovic, J., Milenkovic, J., Goic-Barisic, I., & Rajic, N., (2013). Antibacterial activity of modified natural zeolite against clinical isolates of *Acinetobacter baumannii. Micropor. Mesopor. Mat., 169*, 148–152.
15. Jiraroj, D., Tungasmita, S., & Tungasmita, D. N., (2014). Silver ions and silver nanoparticles in zeolite A composites for antibacterial activity. *Powder Technology, 264*, 418–422.
16. Akhigbe, L., Ouki, S., Saroj, D., & Min, L. X., (2014). Silver-modified clinoptilolite for the removal of Escherichia coli and heavy metals from aqueous solutions. *Env. Sci. Poll. Res., 21*, 10940–10948.
17. Demirci, S., Ustaoğlu, Z., Yılmazer, G. A., Sahin, F., & Baç, N., (2014). Antimicrobial properties of zeolite-X and zeolite-A ion-exchanged with silver, copper, and zinc against a broad range of microorganisms. *Appl. Biochem. Biotechn., 172*(3), 1652–1662.
18. Rossainz-Castro, L. G., De La Rosa-Gomez, I., Olguín, M. T., & Alcantara-Díaz, D., (2016). Comparison between silver- and copper-modified zeolite rich tuffs as microbicidal agents for *Escherichia coli* and *Candida albicans. J. Env. Manag., 183*, 763–770.
19. Jędrzejczyk, R. J., Turnau, K., Jodłowski, P. J., Chlebda, D. K., Łojewski, T., & Łojewska, J., (2017). Antimicrobial properties of silver cations substituted to faujasite mineral. *Nanomaterials (Basel), 7*(9), 240–251.
20. Milenkovic, J., Hrenovic, J., Matijasevic, D., Niksic, M., & Rajic, N., (2017). Bactericidal activity of Cu-, Zn-, and Ag-containing zeolites toward *Escherichia coli* isolates. *Env. Sci. Poll. Res., 24*, 20273–20281.
21. Assem, A. M., & Amin, M., (2019). Studies on antiviral activity of zeolite against foot and mouth disease and ephemeral fever viruses. *ARC J. Animal & Veterinary Sci., 5*(1), 31–34.
22. Balasubramaniam, B., Prateek, Ranjan, S., Saraf, M., Kar, P., Singh, S. P., Thakur, V. K., et al., (2021). Antibacterial and antiviral functional materials: Chemistry and biological activity toward tackling COVID-19-like pandemics. *ACS Pharmacol. Transl. Sci., 4*(1), 8–54.
23. Dolaberidze, N. M., Tsitsishvili, V. G., Khutsishvili, B. T., Mirdzveli, N. A., Nijaradze, M. O., Amiridze, Z. S., & Burjanadze, M. N., (2018). Silver- and zinc-containing bactericidal phillipsites. *New Materials, Compounds and Applications, 2*(3), 247–260.
24. Tsitsishvili, V. G., Dolaberidze, N. M., Nijaradze, M. O., Mirdzveli, N. A., & Amiridze, Z. S., (2019). Bactericidal adsorbents obtained by ion exchange modification of natural phillipsite. *Chem. Phys. Techn. Surf., 10*(4), 327–333.
25. Tsitsishvili, V., Dolaberidze, N., Mirdzveli, N., Nijaradze, M., & Amiridze, Z., (2020). Properties of bactericidal adsorbents prepared from Georgian natural analcime and phillipsite. *Bull. Georg. Natl. Acad. Sci., 14*(4), 25–33.
26. Tsitsishvili, V., Dolaberidze, N., Mirdzveli, N., Nijaradze, M., & Amiridze, Z., (2021). Properties of Georgian natural heulandite-clinoptilolite and its silver, copper, and zinc-containing forms. *Bull. Georg. Natl. Acad. Sci., 15*(2), 60–67.

27. Tsitsishvili, V., Dolaberidze, N., Nijaradze, M., Mirdzveli, N., Amiridze, Z., Khutsishvili, B., Virsaladze, K., & Kapanadze, T., (2021). Properties of Georgian heulandite-clinoptilolite and its application for production of bactericidal adsorbents. *Inter. Conf., 59*, 633–642.
28. Tsitsishvili, V., Dolaberidze, N., Mirdzveli, N., Nijaradze, M., Amiridze, Z., Gabunia, V., & Tsintskaladze, G., (2019). Hydrothermal transformation of natural analcime and phillipsite. *Bull. Georg. Natl. Acad. Sci., 13*(1), 66–73.
29. Persson, I., (2010). Hydrated metal ions in aqueous solution: How regular are their structures? *Pure Appl. Chem., 82*(10), 1901–1917.
30. Tsitsishvili, V., Dolaberidze, N., Mirdzveli, N., Nijaradze, M., & Amiridze, Z., (2021). Preparation of bactericidal fillers from Georgian heulandite-clinoptilolite and their application for paper production. I. Bactericidal fillers. *InterConf., 67*, 340–358.
31. Zendehdel, R., Goli, F., & Hajibabaei, M., (2020). Comparing the microbial inhibition of nanofibres with multi-metal ion exchanged nano-zeolite Y in air sampling. *J. Appl. Microbiol., 128*(1), 202–208.

CHAPTER 34

Synthesis of N-Substituted Dichlormaleinimides and Their Testing as Potential Non-Metallic Oxidation Catalysts

YAGUB NAGIYEV[1,2] and ELDAR ZEYNALOV[1]

[1]*Azerbaijan National Academy of Sciences, Institute of Catalysis and Inorganic Chemistry (named after academician M. Nagiyev), Baku, Azerbaijan*

[2]*Azerbaijan University, Baku, Azerbaijan*

ABSTRACT

The development of highly active, reusable, environmentally friendly heterogeneous catalysts for industrially important oxidation reactions is one of the most important challenges of the contemporary technical chemistry. In this regard, the use of non-metallic, easily, and cheaply available compounds as a catalyst for aerobic liquid-phase oxidation of hydrocarbons at low temperatures appears as the topical issue. The imide catalysts used in this investigation were obtained as a result of interaction of the amines with dichloromaleinic anhydride, their structure was proved by advanced analytical means. To access their catalytic activity the imides were involved in the environment of cumene model oxidation. Several parameters of the reaction were studied, including temperature, catalyst amount, and reaction time. The imides have demonstrated the distinguishing catalytic activity stemmed from the imides molecular structure An interesting fact is that no

Advanced Polymer Structures: Chemistry for Engineering Applications. Omar Mukbaniani, DSc, Tamara Tatrishvili, PhD & Marc Jean M. Abadie, DSc (Eds.)

cumolhydroperoxide detected in the reaction products. Moreover, it has been established the substantial weakening the reaction rates over the reaction time. Reason of these results are discussed. Thus, the imides can be enlisted as active non-metallic oxidation catalysts and can be applied as patterns of the specific selectivity.

34.1 INTRODUCTION

At present, the search for catalysts for the synthesis of industrially important organic compounds obtained by selective oxidation of hydrocarbons of various structures is of great importance. In recent decades, the oxidation of hydrocarbons with molecular oxygen and hydrogen peroxide on the basis of nanoscale, nanostructured, and nanocomposite materials is characterized by the presence of new generation catalysts and catalytic systems [1–4]. The activity of catalysts is due to the fact that its surface is both regular and defective. It is clear that by reacting with the microstructure and degree of defect of the surface of the catalyst, its reactivity and catalytic properties can be fundamentally changed. Materials with a high degree of defect include nano-sized metal powders whose properties differ from those of cod-dispersed metal powders as well as compact metals. The processes of oxidation of kumol with molecular oxygen in the presence of a catalyst consisting of nanoscale metal powders obtained by electric explosion and mechanical-chemical processing of wires were studied in Refs. [5–8], it was shown that oxidation processes using these catalytic systems are quite high at 30–60°C and atmospheric pressure – can be carried out quickly (without a catalyst these processes occur at 120°C and high pressure). This proves once again the high activity of nanoscale metal powders in the oxidation of cumene. However, the high activity of nanometallic powders is also a disadvantage. In this oxidation process, they are easily aggregated and lose their activity over time. To solve this problem in catalysis, the method of impregnation of nanoscale particles on various carriers is used. Physicochemical properties play a decisive role in the choice of carriers [9]. Given the requirements for the active phase of the catalyst carriers, various dispersed inorganic materials (e.g., metal oxides) may be used as the matrix in the fixation. Silica 4-oxide (SiO_2) and aluminum oxide (γ-Al_2O_3) are widely used as oxide carriers [10].

The chapter [11] determined the high catalytic effect of nano-titanium dioxide anatase (PC-500), studied the formal kinetics of the process of oxidation of kumol with oxygen in the liquid phase and low temperatures

(60–100°C), IR-analysis of the obtained products. Nano-titandioxide as a highly effective catalyst is recommended for industrial aerobic oxidation of alkylaromatic hydrocarbons. Most of the oxidation processes applied on an industrial scale usually proceed according to the self-oxidation regime in the presence of salts of transition metals (Cu, Co, Fe, Mn, Ni, Cr, Ti, etc.) [12–15]. The latter compounds allow the hydroperoxides to decompose non-radically with the formation of oxidation chains or non-oxidized products. Many difficulties for these reactions depend on the type of target products. For example, if hydroperoxide is to be obtained, this is a relatively easy task, but the oxidation process must be performed in a more difficult way to synthesize oxygenated organic compounds containing carbonyl, alcohol, carboxyl, or ether groups. It should be noted that the increase in temperature in such processes does not give the desired results, and therefore the further development of the scientific direction in the use of metal complexes and biomimetic catalysts is considered [16–18].

Nanocatalysts should be noted specifically because they have higher activity, productivity, and selectivity at temperatures close to the environment [19, 20]. Specific surface area of carbon nanomaterials, high porous structure, stability in acid-base environment and high temperature, ability to be easily modified, relatively low cost, etc. Such features have allowed them to be widely used in catalysis. In terms of the environment, it has been shown that processed and degraded nanocarbon materials can be an alternative to conventional metal oxide catalysts for more sustainable petrochemical production processes [21, 22]. Existing industrial processes are often carried out using homogeneous catalysts that cannot be reused. Clearly, from this point of view, it is very important to develop highly active, reusable heterogeneous catalysts for oxidation reactions [23]. It should be noted that there is a lot of research in this area, and they were still ongoing. Therefore, the aerobic oxidation process of hydrocarbons in the liquid phase is a very rich field for more research areas. Recently, metal-containing nano-carbon catalysts have been used successfully for several important chemical processes, such as the oxidative dehydrogenation of aromatic hydrocarbons [24], the decomposition of methane [3], and cyclobination, epoxidation, cross-adaptation, photocatalytic, and electrochemical reactions [25]. In these processes, metal particles (Fe, Ni, Co, Pd, Pt, Ag, Au, etc.) and their compounds are deposited on the surface or in the internal cavities of carbon nanotubes (KNB) [26]. Fe-containing KNBs, along with other oxidation processes [27], have catalytic activity in processes such as hydrogenation [13] and ammonia decomposition [29]. It has been shown that the catalytic activity of KNBs in the oxidative dehydrogenation reaction of butene-1 to butadiene is much higher and more stable than that of activated carbon and iron oxide [8]. Metal alloys can be separated at very high temperatures, which is a major drawback for strong metal catalysts [33, 34]. One of the proposed ways to overcome this problem is to use a new generation of compounds characterized by the unique properties of carbon nanotubes and non-metal.

The research work carried out in our laboratory studies the effects of multi-walled carbon nanotubes (CBN) and N-substituted dichloromaleinimides on the chain processes of aerobic and peroxide oxidation of hydrocarbons [35–40].

Thus, works on the processes of oxidation of cumene in the liquid phase with the participation of various catalysts have been analyzed, and in recent years imides have been proposed as new effective non-metallic catalysts in aerobic oxidation of organic compounds. It has been determined that the most studied, researched, and applied catalyst among imides is N-hydroxyphthalimide. An effective method of oxidation of cumene in the liquid phase with the participation of phthalimide catalysts was developed, the effect of the main technological parameters on the selectivity and rate of oxidation was studied [41]. The use of these catalysts lowers the average oxidation temperature by 10–15°C and increases the rate of hydroperoxide accumulation by up to 7% [42]. At the same time, such information about other hopes is almost non-existent. For this reason, it is expedient to study new types of imides in oxidation processes. The presented work presents the results of the process of oxidation of cumene with atmospheric oxygen at different temperatures (80°C, 100°C, and 130°C) in the catalytic presence of N-substituted dichloromaleinimides synthesized in our laboratory.

34.2 EXPERIMENTAL METHODS AND MATERIALS

34.2.1 SYNTHESIS PROCEDURES

At first, catalysts were obtained in the study. N-substituted imides of dichloromaleic acid were synthesized by the interaction of amines and dichloromalein anhydride when heated in glacial acetic acid. The reaction proceeds in two stages, in the initial stage amidobutyric A is formed, as a result of increasing the temperature of the reaction, the cycle is closed by the separation of water, and finally the corresponding imide is obtained [43–46] (Scheme 34.1).

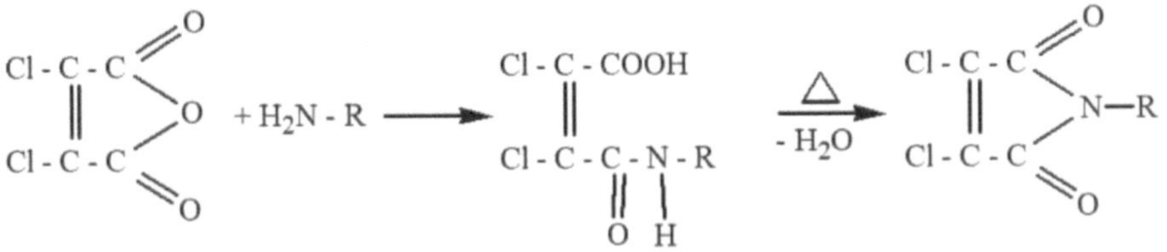

1 2a - b A 3a - b

SCHEME 34.1 Schematic presentation of reaction between primary amine and dichloromaleinic anhydride.
R = (a), $NHCOCH_3C_6H_4$; (б), 2,5-$Cl_2C_6H_3$; (v), p-OHC_6H_4.

The composition and structure of the obtained compounds were determined by spectral and analytical methods [28].

In the next stage of the study, the catalytic activity of the catalyst samples obtained was studied in the mode 1 oxidation reaction of cumene ("Alpha Aesar" purity rate – 99%). The oxidation process was carried out in a laboratory, in a gasometric device, using a standard amount of 10 cm^3 of cumene. The experiments were performed at three temperatures (80, 100, 130°C) and no additional initiator was used to initiate the oxidation process under these conditions. The reaction mixture was introduced into a container with a volume of 30–40 cm^3 and placed in a gasometric device system by means of a grinder, as shown in Figure 34.1. The rate of the reaction was measured volumetrically based on the amount of oxygen absorbed by the air. The experiment begins by inserting a vessel containing the reaction mixture into a preheated thermostat, at which point the vessel is shaken at a rate of about 3–7 cycles per second for better diffusion of the oxygen in the system air into the reaction mixture. The rate of the oxidation reaction was measured by the rise in the level of the liquid in the burette with the absorption of oxygen by the air.

34.3 RESULTS AND DISCUSSION

Thus, the catalytic activity of N-substituted imides synthesized in the model oxidation reaction of cumene was studied. It is clear from the results obtained that N-substituted imides have an active catalytic effect on the aerobic oxidation reaction of cumene in the liquid phase (Figures 34.2–34.4) with the production of a number of valuable organic products (Tables 34.1–34.4). Catalysts break down hydroperoxide by a free radical mechanism. That is, under the influence of the catalyst, additional active centers are formed in the system, and an increase in speed is observed at the beginning of the reaction. Oxidation products formed during the increase of the reaction time significantly affect the rate of the reaction. It is important to note that the higher activity of the catalyst in the oxidation process was observed at 80, 100, and 130°C.

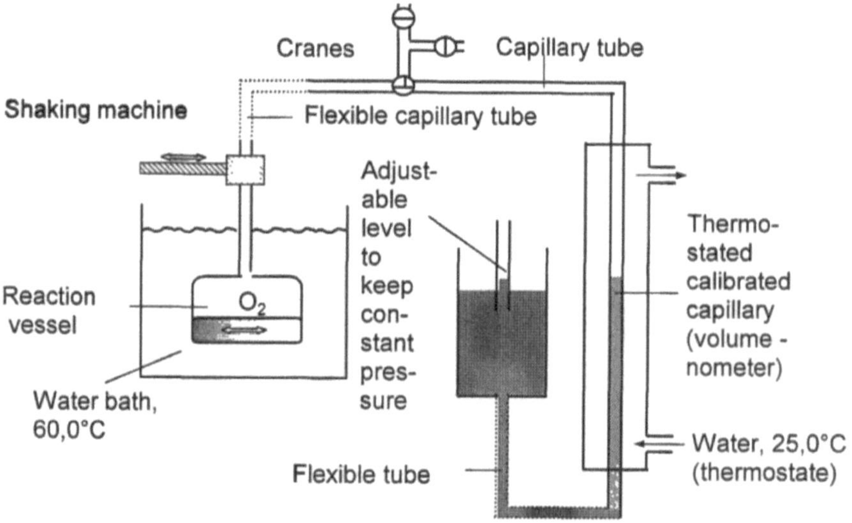

FIGURE 34.1 Schematic description of the device used to measure oxygen consumption at constant pressure (gasometric device).

In the process of oxidation, the N-(p-hydroxyphenyl) dichloromaleimide radical (N-p-OPDCMIR) is formed from the molecule of N-p-HPDCHMI [N-(p-Hydroxophenyl) dichlorormaleimide)] upon the interaction of an oxidizing substance with a peroxide radical, which is capable of selectively removing an atom hydrogen from the CH – bond of alkylarenes with the formation of the corresponding alkylarene radicals. This C-centered alkyl-arene radical, in the presence of atmospheric oxygen, is converted into a peroxide radical, which in turn interacts with N-p-GPDCHMI to form N-p-OPDCMIR and hydroperoxide. This oxidation route leads to the appearance of the corresponding peroxides:

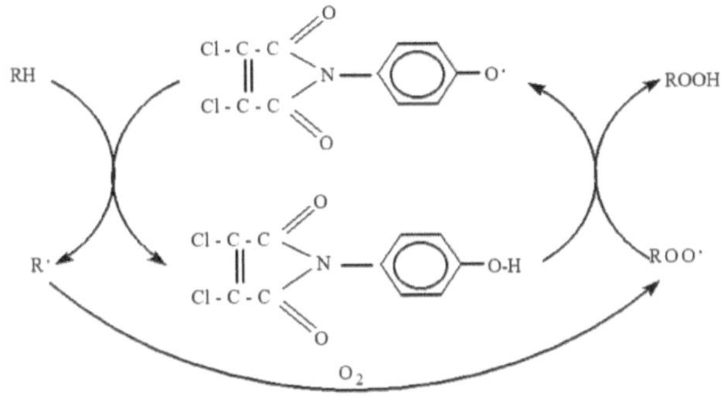

Typically, such processes are carried out using initiators or cocatalysts that cause the formation of free radicals by the decomposition of the resulting peroxides or hydroperoxides. However, from our experimental data, it can be seen that in the absence of initiators or cocatalysts, N-p-HPDCMI – is capable of activating the oxidation process itself; therefore, it is important to establish its role at this stage. Oxygen takes part in the chain initiation stage. In the absence of an initiator, N-p-GPDCMIR can be formed from N-p-GPDCMIR by the reaction [28]:

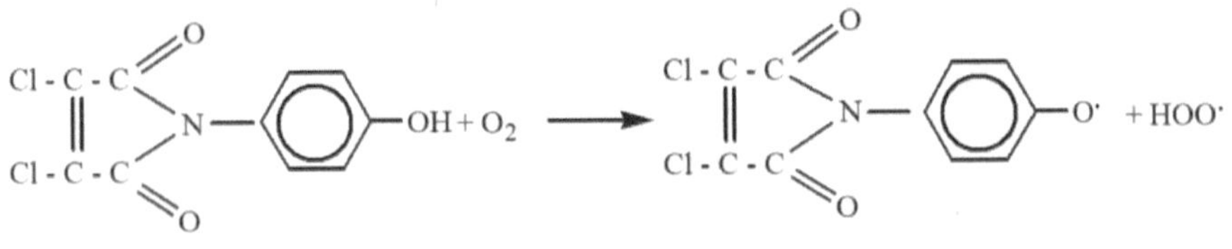

With the subsequent reduction of the original molecule by the reaction:

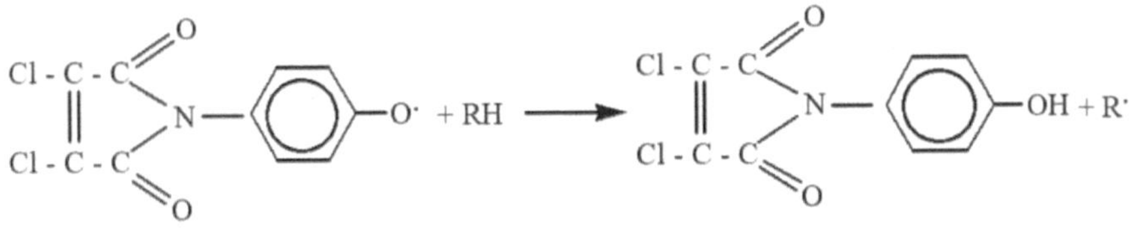

where; RH is hydrocarbon (cumene).

During the oxidation of cumene in the presence of imides, oxidative dehydrogenation reactions occur intensively, which, along with the production of oxygen-containing compounds – acetophenone, phenol, and a number of subsequent products, cause either the formation of unsaturated or alkyl-containing compounds. Considering the scale of phenol and acetone production, solving the problem of intensifying the oxidation of isopropyl benzene using N-substituted dichloromaleimide and its analogs as active components can significantly improve the economic performance of the known process.

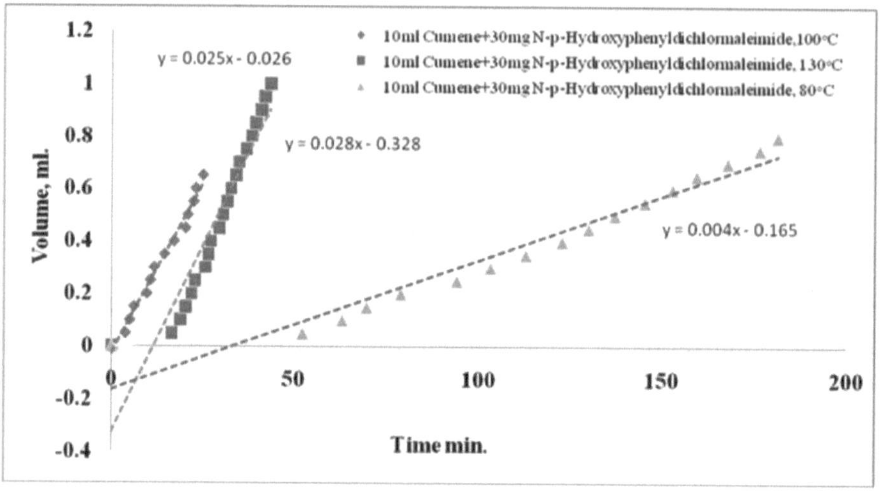

FIGURE 34.2 Kinetic dependences of oxygen uptake in the liquid phase aerobic oxidation reaction of kumol at different temperatures in the presence of N-p-hydroxyphenyl dichloromaleinimide catalyst.

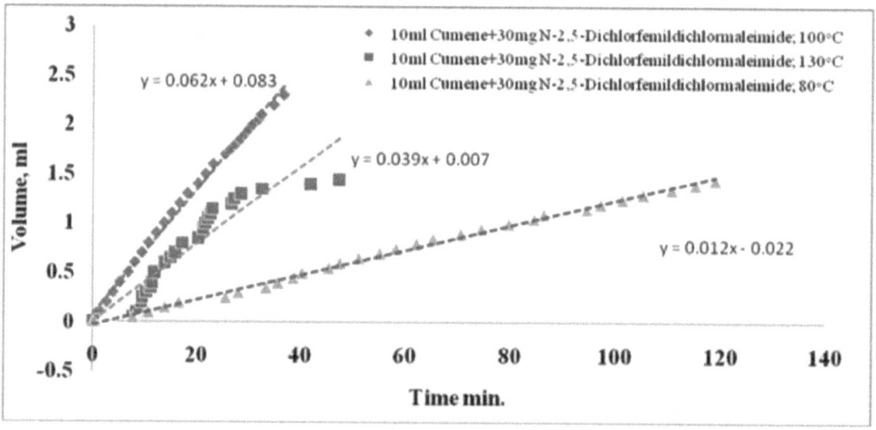

FIGURE 34.3 Kinetic dependences of oxygen absorption in the liquid phase aerobic oxidation reaction of kumol in the presence of N-2.5-chichlorophenyl chloromaleinimide catalyst at different temperatures.

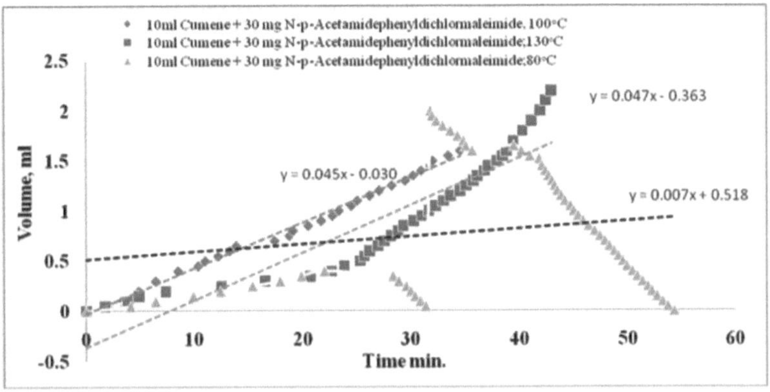

FIGURE 34.4 Kinetic dependences of oxygen absorption in the aerobic oxidation reaction of kumol in the liquid phase in the presence of N-p-acetamide phenyldichloromaleinimide catalyst at different temperatures.

After oxidation, the reaction mixture was analyzed on a mass spectrometer "Agilent Technologies, 7820A GS system ." Mass chromatography was used to determine the oxidation products in the oxide obtained after the reaction. The results obtained are shown in Tables 34.1–34.4 (Data are only available for products with more than 1% content).

TABLE 34.1 The Main Oxidation Products of Cumene Analyzed by Mass Chromatography (a) p-OHC_6H_4 [Sample A_1 30 mg (100°C); $A_2$30 mg (130°C)]

SL. No.	Products (A_1) 100C (в)	Quantity (%)	Products (A_2) 130C	Quantity (%)
1.	Cumene	26.177	+	40.671
2.	Acetophenone	16.941	+	20.071
3.	Phenol	2.08	+	0.093
4.	2-phenylpropene	4.918	–	–
5.	1.2.3.4-Tetramethylbenzene	1.159	–	–
6.	2.4-Dimethylbenzoic acid	3.478	–	–
7.	Dicumylperoxide	13.943	–	–
8.	1.1.2.2-Tetramethyl-1.2-diphenylethane	5.149	–	–
9.	1-Methylethyllbenzene	–	+	3.191
10.	1,1'-(1,1,2,2-tetramethyl-1,2-etandiole benzene	–	+	5.686
11.	Heksacosane	–	+	1.172
12.	Heptacosane	–	+	1.137
13.	1,1-dimethylpropylbenzene	–	+	11.063

TABLE 34.2 The Main Oxidation Products of Cumene Analyzed by Mass Chromatography (b), 2,5-$Cl_2C_6H_3$ [Sample A_1 30 mg (100°C); $A_2$30 mg (130°C)]

SL. No.	Products (A_1)30 mg 100°C (в)	Quantity (%)	Products (A_2) 30 mg 130C	Quantity (%)
1.	Cumene	10.586	+	55.823
2.	Acetophenone	2.499	+	6.220
3.	Phenol	4.166	+	5.528
4.	2-phenylpropene	3.911	–	–
5.	Phenyl-o-(alpha-alpha) dimethylbenzyl	1.482	–	–
6.	2.4-Diphenyl-4-methyl-1-pentene	3.906	+	0.402
7.	Phenol-4-(1-methyl-1-phenylethy)	2.396	+	0.375
8.	*1.2-Benzoldycarboxylic acid*	14.37	+	0.690
9.	Tretbutylbenzene	–	+	1.261
10.	α,α-dimethylbenzenemetanole	–	+	1.950
11.	1,1'-(1,1,2,2-tetramethyl-1,2-etandiole benzene	–	+	8.331

TABLE 34.3 The Main Oxidation Products of Cumene Analyzed by Mass Chromatography (v), ($NHCOCH_3C_6H_4$) [Sample $A_1$30 mg (100°C); $A_2$30 mg (130°C)]

SL. No.	Products (A_1)30 mg 100°C (в)	Quantity (%)	Products (A_2)30 mg 130C	Quantity (%)
1.	Cumene	21.538	+	17.083
2.	Acetophenone	15.206	+	4.154
3.	Phenol	3.324	+	1.376
4.	2-phenyl-2-propanol	3.752	–	–
5.	Tretbutylbenzene	29.875	+	20.049
6.	4-Dimethylbenzoic acid	1.732	–	–
7.	1.1.2.2-Tetramethyl-1.2-diphenylmethane	3.249	–	–
8.	Tetradecane	–	+	4.753
9.	Pentadecane	–	+	8.949
10.	Nonadecane	–	+	11.224
11.	2-methyipropilsiklopentane	–	+	2.542

TABLE 34.4 Basic Oxidation Products of Cumene by Mass Chromatography Without Catalyst (100°C); Reaction Time 5 Hours

SL. No.	Products (A_1) 30 mg 100°C (B)	Quantity (%)	Products (A_2) 30 mg 130C	Quantity (%)
1.	Cumene	43.846	–	–
2.	Acetophenone	11.561	–	–
3.	Phenol	0.17	–	–
4.	2.4-Dimethylbenzoic acid	1.045	–	–
5.	2.4.4.6-Tetramethyl-6-phenyl-1-heptene	2.9085	–	–
6.	1-Methylene propyl benzene	0.6045	–	–

As can be seen from the tables, all imides exhibit catalytic activity, resulting in higher conversion of hydrocarbons compared to the non-catalyzed oxidation process.

34.4 CONCLUSION

1. The details of the synthesis of N-substitute imides to be effectively used as selective catalysts for aerobic oxidation of alkylaromatic hydrocarbons have been determined.
2. N-substituted dichlormaleimides have a strong catalytic effect on the aerobic liquid-phase oxidation of cumene at moderate temperatures, resulting in production of a number of valuable organic products.
3. Due to full depletion of hydroperoxides, they are not detected as reaction product of cumene catalytic oxidation. The radical mechanism was proposed to explain the decomposing activity of N-substituted imides.
4. N-substituted dichloromaleimides are recommended as effective non-metallic catalysts for the aerobic oxidation of alkylaromatic hydrocarbons.

KEYWORDS

- **catalyst**
- **cumene**
- **cumolhydroperoxide**
- **kinetic parameters**
- **N-dichloromaleinimide**
- **oxidation**

REFERENCES

1. Maravin, G. B., Avdeev, M. V., & Bagriy, E. I., (2000). Oxidative functionalization of saturated hydrocarbons on metal complex catalysts of the porphyrin series (review). *Petrochemistry, 1*, 3–21.
2. Giannini, D. M., Shulpina, L. S., Shuhardt, U., & Shulpin, G. B., (2005). Oxidative functionalization of saturated hydrocarbons on metal complex catalists of the porphyrin series (review). *Petrochemistry, 45*, 447–452.
3. Bukhtiyarov, V. I., & Slinko, M. G., (2001). Metallic nanosystems in catalysis. *USP. Chem., 70*, 167–181.
4. Sergeev, G. B., (2001). Nanochemistry of metals. *USP. Chemistry, 70*, 915–933.
5. Skorokhodova, T. S., Kobotaeva, N. S., & Sirotkina, E. E., (2005). Copper in a model reaction of isopropyl benzene oxidation. *Journal. App. Chem., 78*, 767–771.
6. Sirotkina, E. E., Kodotaeva, N. S., Svarovskaya, N. V., Skorohodova, T. S., & Sedoi, V. S., (2000). In: 32nd *Int. Ann. Conf. of ICI* (pp. 97–105). Karlsruhe, Germany.
7. Feduschak, T. A., Sizova, N. V., & Velichkina, L. M., (2009). Oxidation of isopropyl benzene. *Journal of Physics Chemistry, 83*, 1531–1538.
8. Kobotaeva, N. S., Skorokhodova, T. S., Mikubaeva, E. V., & Sirotkina, E. E., (2008). Oxidation of isopropylbenzene with molecular oxygen using metal nanopowders. *Izb. Volume. Polytechnic Un-Ta, 313*, 55–59.
9. Bukhtiyarov, V. I., (2007). *Science and Technology: Materials of Scientific* (p. 264). Sessions of the Presidium of the SB RAS. Novosibirsk.
10. Ermakov, Yu. I., Zakharov, V. A., & Kuznetsov, B. N., (1980). *Fixed Complexes on Oxide Carriers in Catalysis* (pp. 17–45). Novosibirsk: Nauka.
11. Zeynalov, E. B., Naghiyev, Ya. M., Aliyeva, S. A., & Meherremova, M. Ya., (2017). *Liquid-Phase Aerobic Oxidation of Cumene Catalyzed by Nano-Qrade Titanium Dioxide* (Vol. 1, pp. 32–36). Reports of ANAS Academic news of the Academ.
12. Suresh, A. K., Sharma, M. M., & Sridhar, T., (2000). Engineering aspects of industrial liquid-phase air oxidation of hydrocarbons. *Ind. Eng. Chem., 39*, 3958–3997.
13. Ishii, Y., Sakaguchi, S., & Iwahama, T., (2001). Innovation of hydrocarbon oxidation with molecular oxygen and related reactions. *Adv. Synth. Catal., 343*, 393–427.
14. Sheldon, R. A., & Dakka, J., (1994). Heterogeneous catalytic oxidations in the manufacture of fine chemicals. *Catalysis Today, 19*, 215–245.
15. Litvintsev, I., (2004). Oxidation processes in industrial organic chemistry. *Soros Educational Journ., 8*, 24–31.
16. Matienko, L. I., Mosolova, L. A., & Zaikov, G. E., (2009). Selective catalytic oxidation of hydrocarbons. New prospects. *Russ. Chem. Reviews, 78*, 211–230.
17. Karakhanov, E. A., Maksimov, A. L., & Ivanova, E. A., (2007). Supramolecular catalytic systems in biomimetic oxidation. *Russian Chemical Bulletin, 56*, 621–630.
18. Shulpin, G. B., (2002). Metal-catalyzed hydrocarbon oxygenations in solutions: The dramatic role of additives: A review. *J. Molecul. Catalysis, 189*, 39–66.
19. Garrido-Ramírez, E. G., Theng, B. K. G., & Mora, M. L., (2010). Clays and oxide minerals as catalysts and nanocatalysts in Fenton-like reactions—A review. *Applied Clay Scienc., 47*, 182–192.
20. Akia, M., Yazdani, F., Motaee, E., Han, D., et al., (2014). A review on conversion of biomass to biofuel by nanocatalysts. *Biofuel Research Journal, 1*, 16–25.

21. Serp, P., & Castillejos, E., (2010). Catalysis in carbon nanotubes. *Chem. Cat. Chem., 2*, 41–47.
22. Su, D. S., Zhang, J., Frank, B., Thomas, A., Wang, X., & Paraknowitsch, J., (2010). Schlogl RMetal-free heterogeneous catalysis for sustainable chemistry. *Chem. Sus. Chem., 3*, 169–180.
23. Machado, B. F., & Serp, P., (2012). Graphene-based materials for catalysis. *Catal. Sci. Technol., 2*, 54–75.
24. Zhang, J., Su, D. S., Blume, R., Schlogl, R., Wang, R., Yang, X. G., & Gajovic, A., (2010). Surface chemistry and catalytic reactivity of a nanodiamond in the steam-free dehydrogenation of ethylbenzene. *Angew. Chem., 49*, 8640–8644.
25. Huang, L. P., Santiso, E. E., Nardelli, M. B., & Gubbins, K. E., (2008). Catalytic role of carbons in methane decomposition for CO- and CO_2-free hydrogen generation. *J. Chem. Phys., 128*, 214702–214708.
26. Miners, S. A., Rance, G. A., & Khlobystov, A. N., (2016). Chemical reactions confined within carbon nanotubes. *Chemical Society Reviews, 45*, 4727–4746.
27. Yang, X. X., Yu, H., Peng, F., & Wang, H. J., (2012). Confined iron nanowires enhance the catalytic activity of carbon nanotubes in the aerobic oxidation of cyclohexane. *Chem. Sus. Chem., 5*, 1213–1217.
28. Naghiyev, Ya. M., (2021). N-substituted imides of dichloromaleic acid as a catalyst for liquid-phase aerobic oxidation of isopropylbenzene. *Pross. of Petrochem. and Oil Refining, 22*, 90–100.
29. Zhang, J., Müller, J. O., Zheng, W. Q., Wang, D., Su, D. S., & Schlogl, R., (2008). Individual Fe-Co alloy nanoparticles on carbon nanotubes: *Structural and Catalytic Properties. Nano Lett., 8*, 2738–2743.
30. Liu, X., Su, D. S., & Schlogl, R., (2008). Oxidative dehydrogenation of 1-butene to butadiene over carbon nanotube catalysts. *Carbon, 46*, 547–549.
31. Rocha, R. P., Sousa, J. P., Silv, A. M., Pereira, M. F., & Figueiredo, J. L., (2011). Catalytic activity and stability of multiwalled carbon nanotubes in catalytic wet air oxidation of oxalic acid: The role of the basic nature induced by the surface chemistry. *Appl. Catal. B: Environ., 104*, 330–336.
32. Frank, B., Blume, R., Rinaldi, A., Trunschke, A., & Schlogl, R., (2011). Oxygen insertion catalysis by sp^2 carbon. *Angew. Chem. Int. Ed., 50*, 10226–10230.
33. Carmo, M., Paganin, V. A., Rosolen, J. M., & Gonzalez, E. R., (2005). Alternative supports for the preparation of catalysts for low-temperature fuel cells: The use of carbon nanotubes. *Journal of Power Sources, 142*, 169–176.
34. Serp, P., Corrias, M., & Kalck, P., (2003). Carbon nanotubes and nanofibers in catalysis. *Applied Catalysis A, General, 253*, 337–358.
35. Abdurahmanova, N. A., Zeynalov, E. B., Huseynov, A. B., Ahmadova, L. I., Nadiri, M. I., & Nagiyev, Y. M., (2018). Influence of multilayer carbon nanotubes obtained from cyclohexane by thermocatalytic pyrolysis on oxidation processes of alkyl aromatic hydrocarbons. *Scientific Works of AzTU, 3*, 78–85.
36. Naghiyev, Ya. M., Zeynalov, E. B., & Meherremova, M. Y., (2016). Investigation of catalytic properties of nano-particulated titanium dioxide in oxidation of isopropylbenzene. *Azerb. Chem. Journal, 1*, 97–101.
37. Naghiyev, Ya. M., Zeynalov, E. B., Zeynalov, S. B., & Budagova, R. N., (2016). Liguid-phase aerobic oxidation of cumene in the presence of N-phenylimide-endo-2,3-dichlorobicyclo [2.2.1]-hepta-5-en-2,3-dicarboxylicasid. *XX Mendeleev Congress on*

General and Applied Chemistry: Fundamental Problems of Chemical Science Abstract Book (Vol. 1, p. 470). Ekaterinburg.
38. Naghiyev, Ya. M., Nadiri, M. I., Akhmedova, L. I., Esedxade, G. Sh., & Zeynalov, E. B., (2020). Aerobic oxidation of cumene catalyzed with N-(2,5-dichlorofenyl) dichloromaleimide. *İnternational Conference on Actual Problems of Chemical Engineering APCE* (pp. 101–105). Baku (Azerbaiyan).
39. Zeynalov, E. B., Naghiyev, Ya. M., Nadiri, M. I., Ahmedova, L. I., & Abdurahmanova, N. A., (2019). Nano-catalysis in decaline oxidation. In: *6th International Caucasian Symposium on Polymers and Advanced Materials* (p. 117). Batumi, Georgiya.
40. Abdurahmanova, N. A., (2020). Study of catalytic activity of the modified multi-walled carbon nanotubes in the aerobic oxidation of cumene. *Pross. of Petrochem. and Oil Refining, 21*, 268–275.
41. Akhrem, A. A., Metelitza, D. I., Bielski, S. M., Kiselev, P. A., Skurko, M. E., & Usanov, S. A., (1977). *Croat. Chem. Acta, 49*, 223–233.
42. Kurganova, E. A., (2013). Liquid-phase oxidation of alkyl and cyclohexyl aromatic hydrocarbons to hydroperoxides in the presence of phthalimide catalysts. *Step Doctor of Chemical Sciences*, 17.
43. Nagiyev, Ya. M., (2006). Optimal conditions for the reactions of obtaining mono- and bicyclic compounds based on aminoisobutyronitrile. *Azerb. Chem. Journ., 3*, 38.
44. Nagiyev, Ya. M., (2015). Sinthesis of N-substituted imides of 2,3-dichlorobicyclo [2.2.1]hept-5-ene-2.3-dikarboxylic asid. *Russ. J. of Org. Chem., 51*, 1202–1205.
45. Nagiyev, Ya. M., (2012). Synthesis of N-carboxyphenylimides of 2,3-dichlorobicyclo [2.2.1]-hept-5-ene-2,3-dicarboxylic acid. *Russ. J. of Org. Chem., 48*, 301–303.
46. Nagiyev, Ya. M., (2012). Synthesis of N-substituted imides of 2,3-dichlorobicyclo [2.2.2]-oct-5-ene-2,3-dicarboxylic acid. *Russ. J. of Org. Chem., 48*, 467–469.

PART V
Constitutional Systems for Medicine

CHAPTER 35

The Role of Anthelmintic Compounds in Zoological Medicine

R. GIGAURI,[1] L. KHVICHIA,[2] and L. ARABULI[3]

[1]*TSU – R. Agladze Institute of Inorganic Chemistry and Electrochemistry, Tbilisi, Georgia*

[2]*GTU – Institute of Hydrogeology and Engineering Geology, Tbilisi , Georgia*

[3]*University of Georgia, Tbilisi, Georgia*

ABSTRACT

The Arsenic compounds, such as As_2O_3, As_2S_2, As_2S_3, $(RO)_3As$, $ZnHAsO_4$, $SnHAsO_4$, $MnHAsO_4$ are widely used as nanomaterials. One of the study directions of our research is to create and manufacture the compounds of nano-helminths. The parasitic diseases of plants and animals are very perfidious and trespass the producer. Nowadays, the modern and efficient anti-helminth Nano-compounds are scrutinized – Stannum (II) hydro arsenic with copper sulfate as an adjunct; Zinc hydro arsenic with albendazole. The active substance, Stannum (II) hydro arsenic as the Nano-compound, is obtained by the synthesis of pharmacopeial 7–10% arsenic acid and Stannum (II) chloride. The second active compound-zinc hydro arsenic, as the nano-substance, is produced by the combination of pharmacopeial 11–12% arsenic acid and zinc oxide.

Advanced Polymer Structures: Chemistry for Engineering Applications. Omar Mukbaniani, DSc, Tamara Tatrishvili, PhD & Marc Jean M. Abadie, DSc (Eds.)

35.1 INTRODUCTION

The prospect of widespread use of arsenic compounds began in the 1950s by Georgian researchers and scientists in Georgia. One of the directions was obtaining and producing anthelmintic compounds. Helminthiasis of animals is highly insidious and causes a lot of harm to livestock. This fact forces the veterinary service to seek out and develop effective anthelmintics in the world that will benefit all species of animals and birds.

To fight animal diseases, hundreds of organic – dichlorophen, bithionol, yomesan, aminoachrichine, getol (hetolum), hexachlorophene, and other compounds, etc., as well as inorganic compounds – copper sulfate, carbonic copper, lead arsenate, calcium arsenate, tin arsenate, manganese arsenate, etc., have been used. Arsenates of metals are found to be the most effective against helminths. This is not surprising, since it is known from the very beginning that something that is poisonous in large doses is an effective medication in small doses. Experiments have proven many times that even among arsenic salts only arsenates of different composition and structure are distinguished by their strong anthelmintic nature. Among them are tin and manganese hydroarsenates. For industrial purposes, in some respects, preference was given to tin and zinc hydroarsenates. These drugs were exported from Georgia to Russia and Central Asia, the Transcaucasia, as well as abroad – to Mongolia, Afghanistan, Pakistan, and so on.

The optimal conditions and physiological activities of the compounds have been successfully studied by Georgian scientists. It has been found that these compounds can be used as anthelmintics.

35.2 EXPERIMENTAL

Our goal is to study and make anthelmintic drugs more modern and effective. The object of the study was a mixture of tin hydroarsenate and copper sulfate with some additives.

$$\text{Molecular formula: } SnHAsO_4 \times CuSO_4 \times H_2O$$

The active ingredient is tin hydro-arsenate obtained by the synthesis of 7–10% arsenic acid and tin (II) chloride (protected by copyright). It is a yellowish fine-crystalline substance with a molar mass – 258 g/mol. It is not characterized by taste and smell and does not dissolve in water.

Pharmacological group: anthelmintics in veterinary medicine. It helps sheep and goats to get rid of harmful moniesiosis; improves metabolism in the body. It is of note, that when a certain dose it administered it is not accumulated in parenchymal organs.

The compound obtained by us, obtained in the presence of tin hydrogen arsenate and copper sulfate, is a high-quality and highly effective anthelmintic drug for all types of animals and birds. It has a synergistic action. It belongs to the anti-anoplacephalic agents. It is a light green, fine-crystalline substance (Figure 35.1).

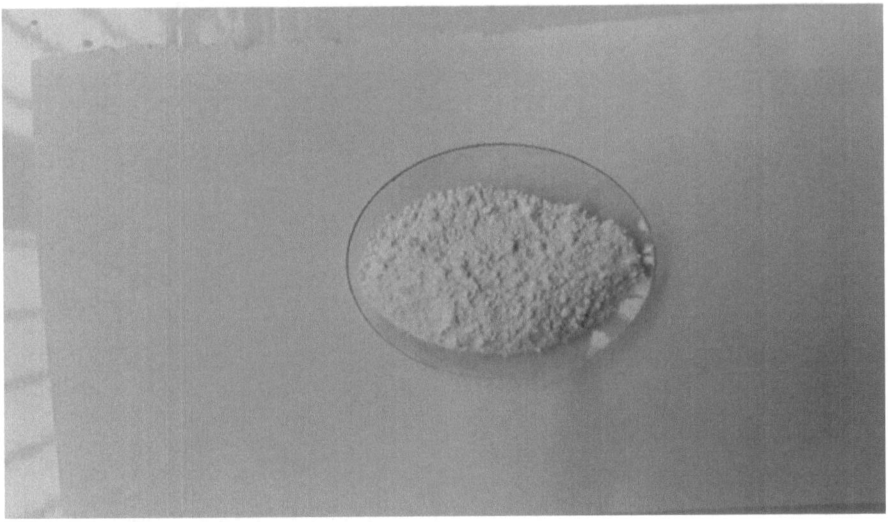

FIGURE 35.1 Stannum (II) hydroarsenate and copper sulfate.

As mentioned above, the area of application of anthelmintics is large. It is being piloted in both cattle and birds. The role of copper sulfate, which is one of the essential microelements for living organisms, is also great. In terms of biological ratio copper is an interesting element in every way. It acts as a catalyst in the oxidation process of the cell. The compound stops bleeding and also oxidizes enzymes, enhances the production of erythrocytes and leukocytes, helps to maintain the immune system, promotes growth and development.

Taking anthelmintics reduces the risk of any residue remaining in the body. It completely frees the body from helminths and kills them. It does not irritate the body.

The object of our study was also the second anthelmintic compound. The study is based on a compound obtained in the presence of zinc hydroarsenate and albendazole.

$$\text{Molecular formula: } ZnHAsO_4 \times C_{12}H_{15}N_3O_2S$$

The technology of obtaining zinc hydroarsenate is based on the action of 11–12% arsenic acid on zinc oxide. The synthesis results in a pharmacopeial zinc hydroarsenate (protected by copyright). The compound obtained by us is a yellowish, fine-crystalline substance with a characteristic odor. The visual appearance of the substance is provided in Figure 35.2.

FIGURE 35.2 Zinc hydroarsenate and albendazole.

35.3 RESULTS AND DISCUSSIONS

The target species are all kinds of animals and birds. It is used for the treatment of gastrointestinal infections caused by roundworms, lung worms, flatworms as well as for the treatment of chrysalis and adult worms. Its composition includes such auxiliary ingredients that ensure the stability of the drug. It increases biological activity in animals and birds and is characterized by a wide range of actions.

Figure 35.3 shows the structural and quantitative behavior of anthelmintic in combination with zinc hydroarsenate albendazole using X-ray fluorescence method.

Modern methods of physicochemical analysis have been used in the study.

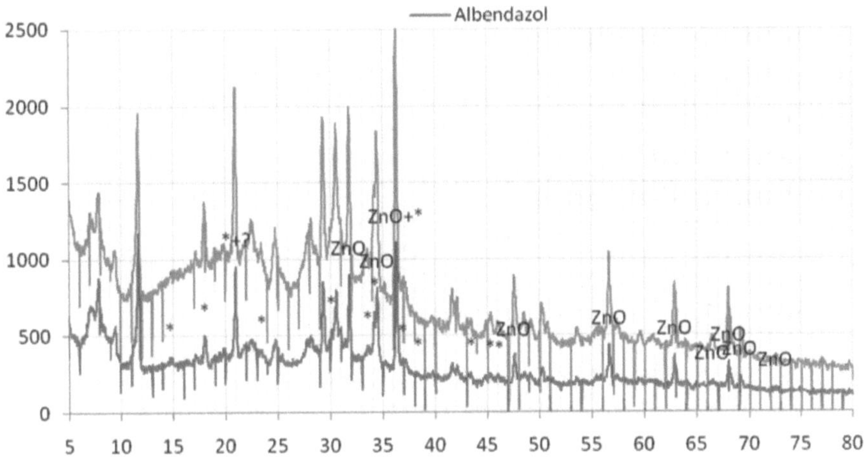

FIGURE 35.3 X-ray fluorescence method of zinc hydroarsenate and anbendazole.

The compound is characterized by fast and safe action; the body gets rid of helminths completely. Above-mentioned nano-compounds are highly effective anti-helminths and they possess the ability of synergic operation. These Nano-compounds are conferred to anti-anoplocephalic product. After obtaining the defined dose, the compounds are not cumulated in the parenchymatous organs. The world's veterinary organizations and services aspire to seek and produce the fastest and highly efficient Anti-helminth compounds, which is eminently crucial for different kinds of animals and birds.

35.4 CONCLUSIONS

Two new anthelmintic compounds have been obtained, which, unlike other anthelmintics, are distinguished by their high quality leading to their fast and safe action. Their production is profitable and the cost is low. It is easy to bring these anthelmintics to the international market as they are modern and contain useful additives promoting the growth of the body. The skeletal system becomes stronger.

KEYWORDS

- **anoplocephalidoses**
- **anthelmintic compounds**
- **anthelmintics**
- **moniosiosis**
- **parenchymatous**
- **pharmacological group**

REFERENCES

1. Lekishvili, N., Gigauri, R., & Rusia, M., (2014). *Arsenic and Stibium Advanced Compounds and Materials with Specific Properties, Monograph*. Tbilisi.
2. Geo. Pat. *Utility Model Method of Obtaining a Tin (II) Hydro Arsenate*. Identification code 14175 / 01.2018.
3. Gigauri, R., Lekishvili, N., & Barbakadze, Kh., (2017). *Advanced Compounds and Materials Based on Natural and Secondary Resources of Georgian Region, 43*(1). Chem. Georgian Nat. Acad. of Sciences.
4. Gegenava, A., & Ugulava, M. (1984). *Methods of Chemical Analysis in Veterinary Medicine*. St.15642052-002-98.
5. Gigauri, R., & Chachava, G., (2004). *Arsenic and the Surrounding Nature*. University Press.

CHAPTER 36

Designing and 3D Printing of Polymer Stents for Chronic Sinusitis

PRZEMYSŁAW GNATOWSKI, MIRIAM PACLER, JUSTYNA KUCIŃSKA-LIPKA, and HELENA JANIK

Department of Polymer Technology, Faculty of Chemistry, Gdańsk University of Technology, Gdańsk, Poland

ABSTRACT

Chronic sinusitis (CRS) occurs when the nose and sinuses are swollen and inflamed due to infection or polyp growths. In some cases, the surgery has to be done, so the weakened sinus has to be reinforced to not allow its shrinkage. Such reinforcement can be done using bioresorbable polymer stent. The aim of the study was to design and 3D print stents and then to evaluate the properties of the obtained printouts in terms of quality and compressive strength. Three different stent models were designed using Autodesk AutoCAD software and as for 3D printing techniques, fused deposition modeling (FDM) and selective laser sintering (SLS) were chosen. Thermoplastic polyurethane (TPU) and Bioflex filaments were used for FDM printing, while standard epoxy resin (Resin BASIC) was used for SLS printing. After obtaining the desired structure of stents, the supports were removed using mechanical methods. Then, the stents were compression tested. The results were differing between project and material and were in the range of 10 to 4,250 kPa. Summarizing our studies, it was concluded that the resin stents showed higher potential for further investigation. The best quality of surface was achieved for the SLS printed stents. However, the material itself was too stiff to apply in the nasal stents, so the resin should be further modified.

Advanced Polymer Structures: Chemistry for Engineering Applications. Omar Mukbaniani, DSc, Tamara Tatrishvili, PhD & Marc Jean M. Abadie, DSc (Eds.)

36.1 INTRODUCTION

Chronic sinusitis (CRS) is one of the most common nasal diseases. It occurs when the nose and sinuses are swollen and inflamed due to infection or polyp growths. It is commonly treated by using drugs based on steroids, but in some cases, surgery has to be done, e.g., to remove polyps. In such cases, the weakened sinus has to be reinforced to prevent its shrinkage. Such reinforcement can be done using bioresorbable polymer stent. The first step to produce a polymer stent for CRS is to design its geometry and select a manufacturing technology. One of the production technologies that were chosen by us was FDM. FDM is an additive manufacturing process that belongs to the material extrusion family and can be used in medicine, for example for wound dressings [1]. Objects are built by selectively depositing melted thermoplastic material in a predetermined path layer-by-layer [2]. The biggest advantage of using 3D printing technology is the ability to print any shape, structure geometry, and the ability to combine it with diagnostic imaging methods such as computer tomography, magnetic resonance, or X-ray radiation to allow patient-specific solutions. There are many drawings of sinus stents described in the patent US 10,471,185 B2 [3]. After creating the design of the stent, the next step is to export the 3D model as an .stl file. The complex part behind STL files is that the surfaces of 3D model are converted into complex network formed from tiny triangles. Such .stl file has to be processed in a program designated for a specific 3D printer, called 'slicer.' At this point, the material from which the designed stent will be printed has to be chosen. Currently, the most common medical filaments for FDM printers are thermoplastic biopolymers (taking into account the application) based on polylactide (PLA) or polycaprolactone (PCL) [4]. They provide the required properties such as biocompatibility, biodegradability, and adequate mechanical strength. Based on used materials, the parameters such as temperature, cooling, and printing speed have to be introduced into slicer software. Software prepares the travel path of the printer nozzle in order to print the model using G-code. G-code is a universal language used for programming manufacturing devices using a series of commands which tell the machine what to do, e.g., move or which temperatures have to be used. In this study, we presented the three different designs of sinus stents, which were then printed using two different technologies: fused deposition modeling (FDM) and SLS. The aim of the study was to design and 3D print stents and then evaluate the properties of the obtained printouts in terms of quality and compressive strength.

36.2 EXPERIMENTAL METHODS AND MATERIALS

For the design of the stents, Autodesk AutoCAD software was used. For FDM printing, Prusa I3 MK3S (single-extruder) printer equipped with 0.4 mm nozzle was used with Devil Design's TPU and F3D Filament's Bioflex biodegradable filament (ester-ether elastomer composition). For SLS printing, the Zortrax Inspire printer was used together with Zortrax's Resin BASIC epoxy resin. As for support material, the same material as the main printing one was used. PrusaSlicer and Z-SUITE software were used as slicer for FDM and SLS printing, respectively. Parameters used for FDM and SLS printing were provided in Tables 36.1 and 36.2, respectively. Compression testing was done using Zwick/Roell Z020 universal tensile machine with a speed of 3 mm/min and a maximum of 15% of stent height measuring range.

TABLE 36.1 Parameters of 3D Printing for the FDM Technique

Printed Material	Layer Height (mm)	Printing Speed (mm/s)	Nozzle Temperature (°C)	Bed Temperature (C)	Support-Model Spacing (mm)	Support Pattern Spacing (mm)	Cooling
PUR	0.15	10	215	40	0.2	1.5	Yes
Bioflex	0.15	10	220	50	0.2	2	No

TABLE 36.2 Parameters of 3D Printing for the SLS Technique

Layer Height (mm)	First 3 Layers UV Exposition Time (s)	UV Exposition Time (s)	Interval Time (s)	Platform Speed (mm/s)	Support Thickness (mm)	Model Tilt Angle (°)
0.05	60	9	5	100	1.5	45

36.3 RESULTS AND DISCUSSION

Three different stent models were prepared. Models were shown in Figure 36.1. The dimensions were 20 × 20 × 20 mm for 1st and 2nd type and 20 × 20 × 12 mm for 3rd type. The single line width was designed as 0.8 mm. The printed stents after support removal were also shown in Figure 36.2 and the stents before support removal were shown in Figure 36.3. Printouts of the 1st and 2nd designs made with TPU and Bioflex filaments, despite the mechanical cleaning, still have a large amount of additional support material. In case of SLS printouts, the support material was much easier to remove, and overall quality of printous was far higher in comparison to FDM technology. The use of twin-extruder FDM could help remove support material, when water soluble material would be used as support. It is also known that printing of elastomer-type material is far more demanding than other common filaments [5]. SLS printouts from Resin BASIC were stiff and brittle which makes them impossible to use as nasal stents. However, there are some promising flexible resins on the market that could be used in exchange of Resin BASIC used in this study. The results of compression testing were presented in Figure 36.1 and Table 36.3. High deviations for FDM printouts were caused by the difficulties in manual support removal process. The highest compressive strength had the printout of 3rd project printed with SLS technique (4,253 ± 870 kPa), which was about eight times higher than that observed value for the TPU filament (542 ± 76 kPa). The printouts of the 1st and 2nd stent models had a similar compression course, and the compressive strength values for the stent model printed with TPU filament and Bioflex were similar. The lowest compression values were found for the printouts of the stent model 1st and 2nd obtained using SLS technique. It was due to the large amount of not removed supports, which affected the results.

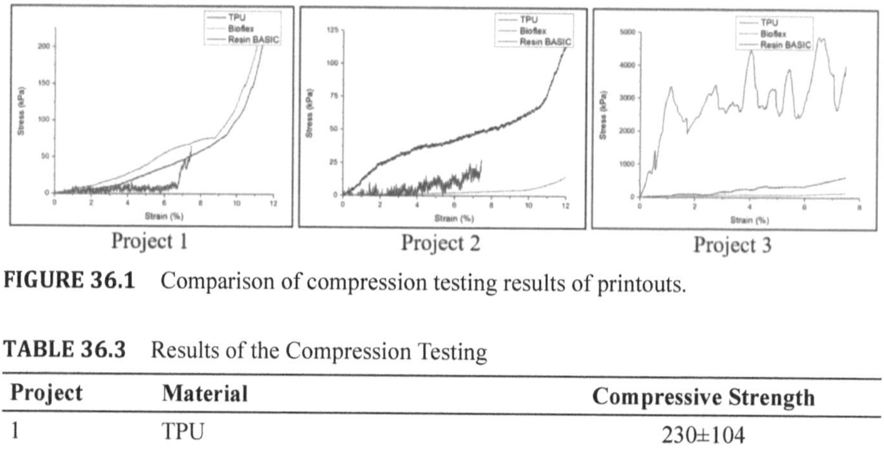

Project 1 Project 2 Project 3

FIGURE 36.1 Comparison of compression testing results of printouts.

TABLE 36.3 Results of the Compression Testing

Project	Material	Compressive Strength
1	TPU	230±104
	Bioflex	252±178
	Resin BASIC	16±12
2	TPU	109±38
	Bioflex	145±18
	Resin BASIC	19±12
3	TPU	542±76
	Bioflex	161±16
	Resin BASIC	4,253±870

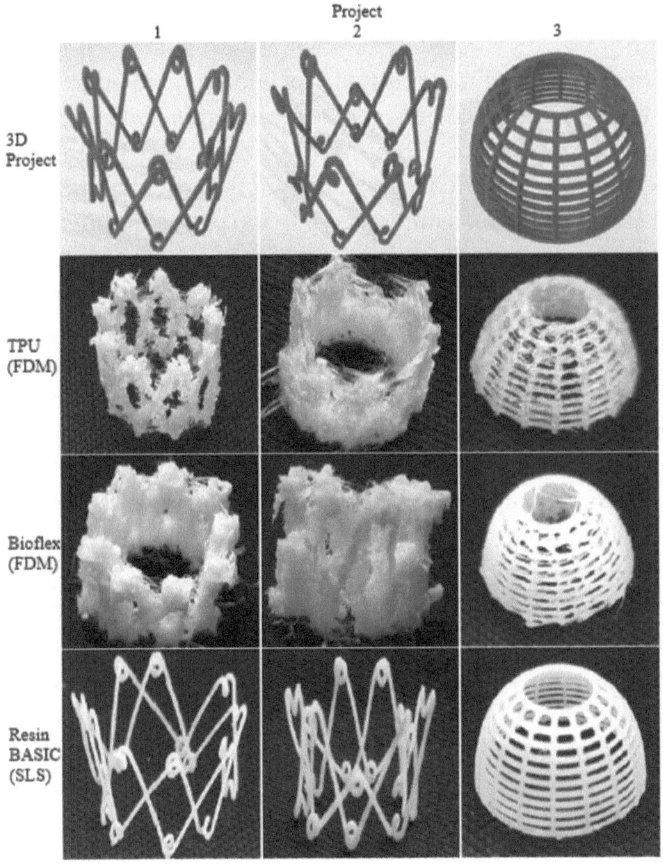

FIGURE 36.2 Comparison of designed and printed stents using different materials and printing technology.

FIGURE 36.3 Comparison of 1st stent project after printing. From left: TPU, bioflex, resin basic.

36.4 CONCLUSION

With the development of materials and technology, medicine is increasingly turning to the production of personalized structures. One of the applications of AM technology may be the printing of tools that help in the patient's recovery to full health, which can be tailored to individual needs. This study focuses on stents that can be implanted into the patient's sinuses after surgery for CRS.

The comparison of the three different stent geometries and three different materials shows that the greatest similarity between the design and the printed model was observed for the printouts obtained using the SLS technique and Resin BASIC. Comparison of the model design with their printouts obtained in a 3D printer using FDM technology, with the use of TPU filament and Bioflex, shows that the printouts were not similar to the projects because they have too much support material left that could not be removed during mechanical cleaning attempts. This proves that the printing precision of FDM technology was too low for such complex geometries. In light of that fact, the compression tests were hard to conclude. For ring-shaped stents, TPU and Bioflex showed similar results, while stent made from resin had much lower compression strength. It was probably caused by the support material that was not fully removed. The dome-shaped stent showed the opposite results: the resin stent offered much higher compression strength in comparison to FDM printed ones. It was found that stents printed with SLS technique using Resin BASIC were too stiff and brittle to be used in medical application. Further studies like modifying the composition of resins or using flexible resins will be taken in the future. The suitability of the FDM technique for stent printing should also be tested using a twin-extruder printer, which would allow for the usage of water-dissolving filaments as supports.

KEYWORDS

- **3D printing**
- **chronic sinusitis**
- **fused deposition modeling (FDM)**
- **polyurethane**
- **selective laser sintering (SLS)**
- **thermoplastic polyurethane**

REFERENCES

1. Szarlej, P., Iga, C., Przemysław, G., Marta, G., Martyna, M., Brillowska-Dąbrowska, A., & Kucińska-Lipka, J., (2021). Composite polyurethane-polylactide (PUR/PLA) flexible filaments for 3D fused filament fabrication (FFF) of antibacterial wound dressings for skin regeneration. *Materials, 14*(20), 6054. https://doi.org/10.3390/ma14206054.
2. Tosto, C., Lorena, S., Eugenio, P., Antonella, P., Giovanni, C., & Gianluca, C., (2021). Fused deposition modeling (FDM): New standards for mechanical characterization. *Macromolecular Symposia, 395*(1), 2000253. https://doi.org/10.1002/masy.202000253.
3. Eaton, D. J., Bin, H., Anthony, J. A., Gail, M. Z., & David, C. G., (2019). *Self-expanding Devices and Methods Therefor*. US10471185B2.
4. Haryńska, A., Iga, C., Paulina, K., Kamil, S., Marcin, Ł., Marta, P., Kucińska-Lipka, J., & Helena, J., (2020). A comprehensive evaluation of flexible FDM/FFF 3D printing filament as a potential material in medical application. *European Polymer Journal, 138*, 109958. https://doi.org/10.1016/j.eurpolymj.2020.109958.
5. Liu, Z., Yanqing, W., Beicheng, W., Chunzhi, C., Yu, G., & Cheng, Y., (2019). A critical review of fused deposition modeling 3D printing technology in manufacturing polylactic acid parts. *The International Journal of Advanced Manufacturing Technology, 102*(9–12), 2877–2889. https://doi.org/10.1007/s00170-019-03332-x.

CHAPTER 37

Polymer Stents Used in Chronic Sinusitis

JUSTYNA KUCIŃSKA-LIPKA, MIRIAM PACLER, PRZEMYSŁAW GNATOWSKI, and HELENA JANIK

Department of Chemical Faculty, Polymer Technological Department, Gdańsk University of Technology, Gdańsk, Poland

ABSTRACT

Chronic sinusitis (CRS) is an upper respiratory disease during which topical drug treatment via the nasal cavity is the most widely used therapeutic strategy. Endoscopic sinus surgery (ESS) is used during CRS; it improves ventilation, drainage, and gives better access to local anti-inflammatory treatment. After surgery, stents can be used for faster recovery. The chapter presents bioabsorbable steroid-eluting sinus stents: Propel and SINUVA, the materials they are made of, and their appearance.

37.1 INTRODUCTION

Chronic sinusitis (CRS) is a chronic inflammatory nasal disease and is diagnosed with or without polyps. It is a common disease with global prevalence rates of 10.9% in Europe. Conservative treatment is sufficient in most cases, but especially in cases of polyps, surgery is necessary. Endoscopic sinus surgery (ESS) is used during CRS, improves ventilation, drainage, and gives better access to the local anti-inflammatory treatment. Despite appropriate and complete sinus surgery, there can be issues after sinus surgery, such as synechiae formation and polyp recurrence. For a faster recovery after the surgery, local medications are used. They are very effective; therefore, treatment with local therapeutic agents is considered as an important type of CRS

Advanced Polymer Structures: Chemistry for Engineering Applications. Omar Mukbaniani, DSc, Tamara Tatrishvili, PhD & Marc Jean M. Abadie, DSc (Eds.)

treatment. The promising way of drug delivery is nasal administration, but it needs a good device that would extend time of drugs in the cavity. To solve these problems, nasal stents have been developed. Sinus stents are designed to be inserted into the sinuses after an ESS to preserve sinus openings during the postoperative healing period and to solve the problems of inflammation and recurrence of polyps. Currently, the Food and Drug Administration (FDA) approved steroid-eluting sinus implants are Propel family products and SINUVA [1, 2].

37.2 MATERIALS USED IN POLYMER STENTS

At first, stents were made of a firm rubber tube, however, the stenosis rate remained high at 30%. Next implants, which used silicone rubber sheets (Silastic), were more promising for long-term results. Stents created from Silastic sheets, resulted in improved mucosal regeneration and decreased inflammation [3]. Currently, FDA approved steroid-eluting sinus implants are built of a bioabsorbable poly(lactide-co-glycolide) (PLGA) polymer coated with the corticosteroid mometasone furoate (MF). Figure 37.1 shows the structural formula of PLGA.

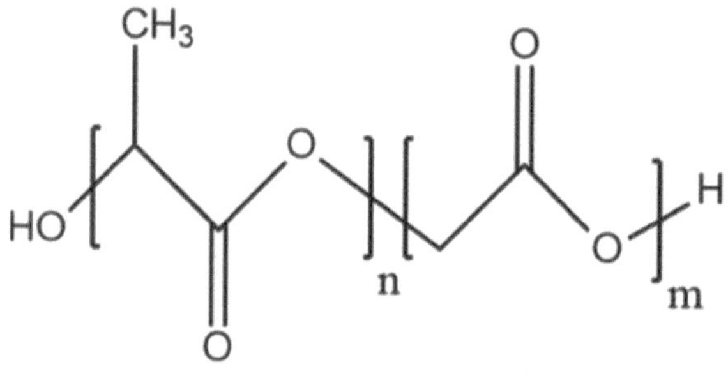

FIGURE 37.1 Structural formula of poly(lactide-co-glycolide).

PLGA has been well recognized for its suitability in drug delivery as a result of its good biocompatibility and ability to achieve complete drug release, as a result of degradation of the polymer matrix. Polyethylene glycol is an anti-protein and anti-inflammatory barrier that affects water retention and promotes tissue biocompatibility [4].

Figure 37.2 shows the structural formula of MF. MF is one of the corticosteroids, which are the most potent anti-inflammatory agents used to control

CRS. It is a well-tolerated, stable molecule that coats the implant and has high potency and low systemic bioavailability. Studies have shown that corticosteroids are safe and improve the symptoms of patients after ESS. MF is the only topical nasal steroid approved by the FDA for the treatment of nasal polyposis. Slow release of MF aims to decrease mucosal inflammation and edema, and to expedite wound healing. Studies revealed that postoperative placement of steroid-eluting stents compared to placebo has a significant effect on endoscopic scores [5–7].

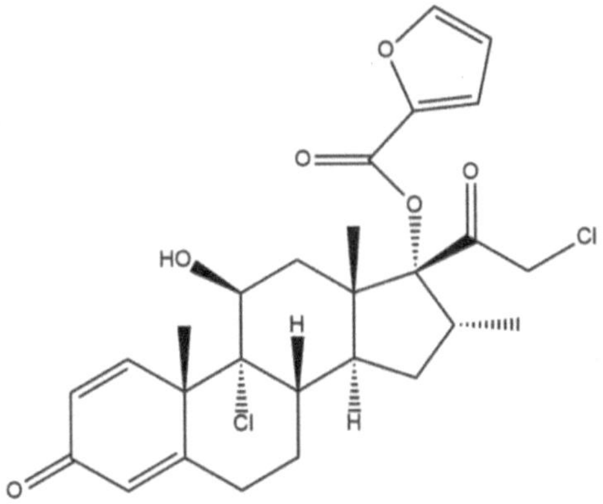

FIGURE 37.2 Structural formula of mometasone furoate.

Propel stents are composed of a bioabsorbable PLGA polymer that is coated with 370 μg of the corticosteroid MF. SINUVA is also made up of PLGA, but it contains 1,350 μg MF [1].

37.3 DESIGN OF STENTS

37.3.1 THE FAMILY OF PROPELS

The Propel family is designed as a short-duration product that can be used immediately after sinus surgery to optimize healing and reduce surgical failure [8]. There are three different types of Propel stents, which differ in height and application [9]:

i. Propel standard: height 23 mm, application for ethmoid sinus;
ii. Propel mini: height 16 mm, application for frontal and ethmoid sinus; and

iii. Propel Contour: height 8 mm, application for frontal and maxillary sinus.

The implant is held in place via the spring-like mechanism of the implant and self-adhesion to the sinus mucosa. Propel is self-expanding, allowing it to fit different sizes and shapes of the sinus ostium. The polymer from which the implants are made helps to control the rate of drug elution from implants. The rate of drug elution is determined by the diffusion mechanism regulated by the polymer morphology, drug concentration, matrix type, chemical composition and coating thickness. These factors allow the Propel stent to administer the dose of drugs in a controlled manner in approximately 30 days [1].

37.3.2 SINUVA

SINUVA is a longer-duration product that delivers more corticosteroids over a longer period of time, to address polyp recurrence after surgery and slow release of drugs in about 3 months. The implant was shown in Figure 37.3, it has a self-expanding, non-obstructive design which allows it to adjust the shape to fit in the operating cavity. The implant has an arched design, which gives it sufficient radial strength to dilate an obstructed ethmoid sinus cavity and to self-remain in the cavity as polypoid tissue recedes. The implant provides immediate onset of action by creating and maintaining an opening in the obstructed middle meatus and ethmoid sinus, thus restoring ventilation and drainage and reducing the sensation of nasal blockage and congestion [10, 11].

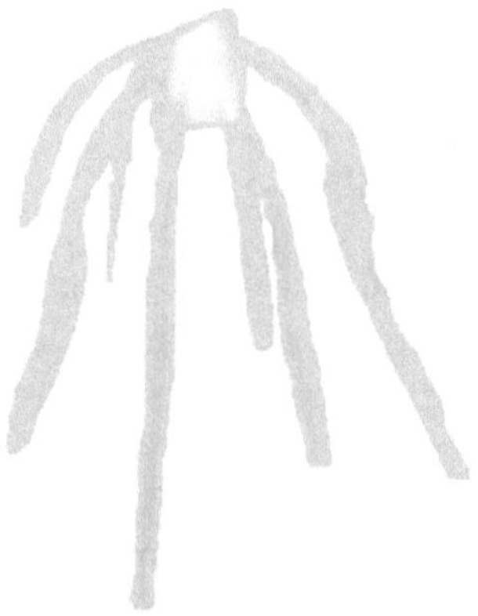

FIGURE 37.3 SINUVA stent.

37.4 CONCLUSION

Stents for CRS, such as Propel or SINUVA are used as a bioabsorbable implants indicated for adults after sinus surgery to maintain sinus patency. They prevent sinus obstruction from adhesions, reduce inflammation and the need for postoperative intervention (e.g., adhesion lysis, oral corticosteroids). The implants have made a significant contribution to reducing postoperative interventions and providing effective management after surgery. The most popular material for stents is PLGA coated in MF corticosteroid.

KEYWORDS

- **chronic sinusitis**
- **endoscopic sinus surgery**
- **Food and Drug Administration**
- **propel**
- **SINUVA**
- **stents**

REFERENCES

1. Tai, J., Lee, K., & Kim, T. H., (2021). Current perspective on nasal delivery systems for chronic rhinosinusitis. *Pharmaceutics, 13*, 246. doi: 10.3390/pharmaceutics13020246.
2. Zhanxiao, H., Bing, Z., Dehui, W., Hongrui, Z., Huankang, Z., Huan, W., Shenquing, W., et al., (2020). Comparison of bioabsorbable steroid-eluting sinus stents versus nasopore after endscopic sinus surgery: A multicenter, randomized, controlled, single-binded clinical trial. *Ear, Nose & Throat Journal*, pp. 2–7. doi: 10.1177/0145561320947632.
3. Lal, D., & Hwang, P. H., (2019). *Frontal Sinus Surgery* (pp. 6, 7). Springer. doi: 10.1007/978-3-319-97/022-6.
4. Zhu, X., & Braatz, R. D., (2014). A mechanistic model for drug release in PLGA biodegradable stent coatings with polymer degradation and erosion, *J. Biomed. Mater. Res. Part A*. doi: 10.1002/jbm.a.35357.
5. Grayson, J. W., & Harvey, R. J., (2019). Topical corticosteroid irrigations in chronic rhinosinusitis. *Int. Forum Allergy Rhinol., 9*, S9–S15. doi: 10.1002/alr.22331.
6. Fokkens, W. J., Lund, V. J., Hopkins, C., et al., (2020). European position paper on rhinosinusitis and nasal polyps 2020. *Rhinology, 58*(suppl S29), 1–464. doi: 10.4193/Rhin20.600.
7. Ourique, A., Contri, R. V., Guterres, S. S., & Beck, R. C., (2011). Set-up of a method using LC-UV to assay mometasone furoate in pharmaceutical dosage forms. *Quimica Nova*. doi: 10.1590/S0100-40422012000400030.

8. Snidvongs, K., & Thanaviratananich, S., (2017). Update on intranasal medications in rhinosinusitis. *Curr. Allergy Asthma Rep.*, 17. doi: 10.1007/s11882-017-0720-3.
9. Propel. (2022). https://propelopens.com (accessed on 02 January 2022).
10. Han, J. K., Forwith, K. D., Smith, T. L., Kern, R. C., Brown, W. J., Miller, S. K., Ow, R. A., Poetker, D. M., Karanfilov, B., Matheny, K. E., et al., (2014). Resolve: A randomized, controlled, blinded study of bioabsorbable steroid-eluting sinus implants for in-office treatment of recurrent sinonasal polyposis. *Int. Forum Allergy Rhinol., 4*, 861–870. doi: 10.1002/alr.21426.
11. Cairns, E., (2017). *Timing of Intersect ENT's Approval is on the Nose*. Evaluate Vantage. https://www.evaluate.com/vantage/articles/news/timing-intersect-ents-approval-nose (accessed on 02 January 2022).

CHAPTER 38

One-Stage Synthesis Method of Triglyceride of Saccharin-6-Carboxylic Acid

E. T. ASLANOVA

Institute of Polymer Materials, Azerbaijan National Academy of Sciences, Sumgait, Azerbaijan

ABSTRACT

With the aim of synthesis of new perpolymers for preparation of thermostable highly branched and lattice polymers, the one-stage synthesis method of triglyceride of saccharin-6-carboxylic acid has been studied. The re-esterification reaction of some aliphatic esters of saccharin-6-carboxylic acid has been carried out. The obtained product is characterized by elemental analysis and methods of IR spectroscopy. It was found that the purposeful product is obtained with a yield 73%.

38.1 INTRODUCTION

It was known that polysulfoimides possess high thermal stability (500–650°C), solubility, stability to radiation, light, acidic, and alkaline hydrolysis. The combination of these properties makes polysulfoimides especially valuable for a number of industries – space and nuclear technology. In this connection, taking into account the advantages of polysulfoimides, it arises a necessity of development of effective methods of the synthesis of new prepolymers on the basis of saccharin monocarboxylic acids [1].

Advanced Polymer Structures: Chemistry for Engineering Applications. Omar Mukbaniani, DSc, Tamara Tatrishvili, PhD & Marc Jean M. Abadie, DSc (Eds.)

There are data in the literature on the synthesis of diesterodisulfoimides by re-esterification of the methyl ester of sulfoimides of saccharin-monocarboxylic acids with aliphatic and aromatic glycols [2, 3].

Based on the above-stated one, we have synthesized a triglyceride of saccharin-6-carboxylic acid as a prepolymer for the preparation of polysulfoimides [4].

The aim of the work is the synthesis of new monomers based on sulfoimide of saccharin-6-carboxylic acid, for their further use in the preparation of highly branched and net heat-resistant polymers.

38.2 EXPERIMENTAL METHODS AND MATERIALS

38.2.1 SYNTHESIS OF TRIGLYCERIDE OF SACCHARIN-6-CARBOXYLIC ACID

A mixture of 40 g (0.02 mol) of ethyl ester of saccharin-6-carboxylic acid, 7.5 g of PbO in 49.6 g (0.5 mol) of 1,2,3-propantriol (glycerine) at stirring was heated to 115–117°C for 4 hours. The reaction mass was dissolved in DMFA and isolated with distilled water.

The obtained coffee-milk color powder product was dried at room temperature and brought to constant weight in vacuum. T_m =167°C.

Similarly, the above-mentioned compounds were synthesized on the basis of ethyl and isopropyl esters of sulfoimide of saccharin-6-carboxylic acid.

The physical-chemical properties of 1,2,3-propantriol and solvents were corresponded to the literature data [5].

The infrared spectra were recorded on Fourier spectrometer LUMOS (firm BRUKER, Germany) in the wavelength range of 600–4,000 cm^{-1}, using an HPVO attachment with ZnSe crystal. The diameter of the crystal is 1 cm. The number of scans of the sample is 24; the measurement duration is 30 seconds.

The elemental analysis has been produced according to the method [6], based on the pyrolytic burning of organic substance in an oxygen flow using the Pregl apparatus.

38.3 RESULTS AND DISCUSSION

The synthesis of triglyceride was carried out by re-esterification of alkyl esters of saccharin-6-carboxylic acid with 1,2,3-propantriol on the following scheme:

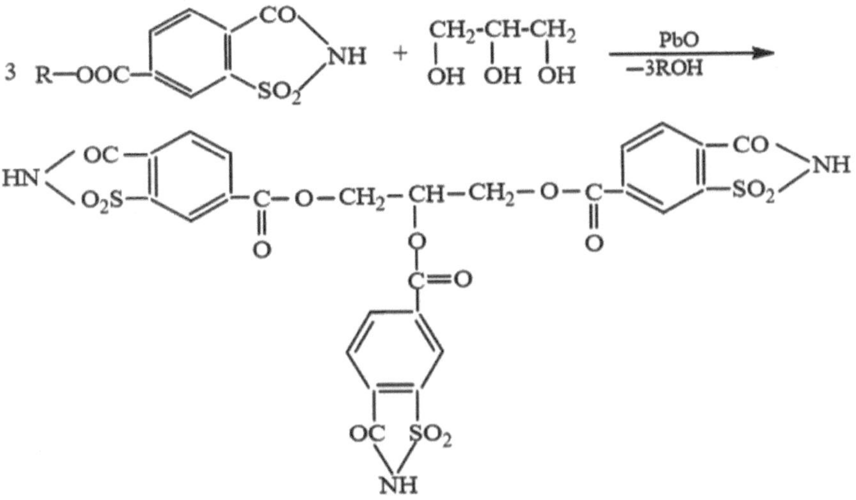

R= $-CH_3$, $-C_2H_5$, $-C_3H_7$

The obtained product is a beige powder, soluble only in aprotic solvents, such as DMFA, DMAA, DMSO, etc.

It was found out that the obtained product corresponded to T_m.=167°C, the yield – 73%. The composition and structure of the obtained compound has been determined by elemental analysis and IR spectroscopy.

The physical constants of products of triglyceride of saccharin-6-carboxylic acid are presented in Table 38.1.

TABLE 38.1 Physical Constants of Triglyceride of Saccharin-6-Carboxylic Acid

Compound	Brutto Formula	Found, % / Calculated, %				M	T_m (°C)	Yield (%)
		C	H	N	S			
Saccharin-6-carboxylic acid triglyceride	$C_{27}H_{17}O_{15}N_3S_3$	44.98 / 45.06	2.46 / 2.36	5.79 / 5.84	13.31 / 13.35	719	167	67

In the IR spectra (Figure 38.1) of the obtained compound the following absorption bands were observed: deformation (720, 395, 1,454 cm^{-1}) and valence vibrations of C–H bond of CH_2 groups; valence (1,644 cm^{-1}) vibrations of C=O bond of the amide; valence vibrations (1,719 cm^{-1}) of C=O bond of the ester; valence (1,154 cm^{-1}) vibrations of C–O bond of the ester;

valence (1,239, 1,283 cm^{-1}) vibrations of SO_2 group; valence (1,019 cm^{-1}) vibrations of S + O-bond; deformation (1,573 cm^{-1}) and valence (3,276 cm^{-1}) vibrations of N–H bond; deformation (617, 678, 750, 859, 1,607 cm^{-1}) vibrations of C–H bond of the substituted benzene ring [7, 8].

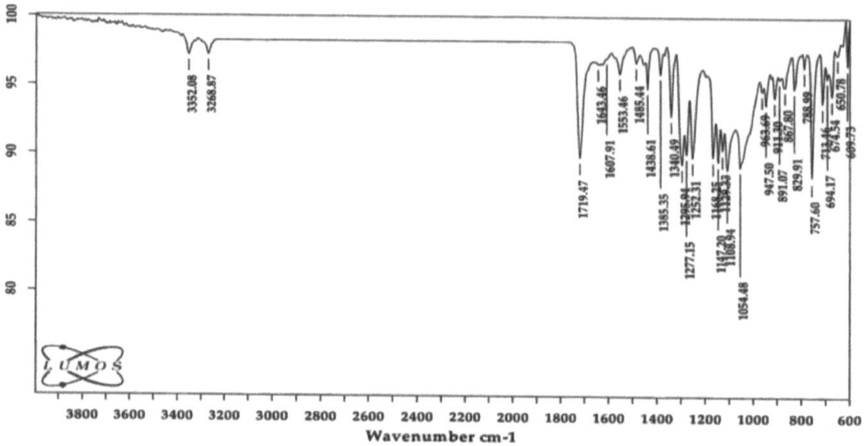

FIGURE 38.1 IR spectra of saccharin-6-carboxylic acid triglyceride synthesized by a two-step method.

The synthesized triglyceride of saccharin-6-carboxylic acid is of interest for its use as a monomer in the preparation of highly branched and net sulfoimide-containing polymers, epoxide resins, and also as a hardener-plasticizer of the industrial epoxide resins.

38.4 CONCLUSION

A one-stage method of synthesis of triglyceride of saccharin-6-carboxylic acid has been studied. The reaction of re-esterification of some aliphatic esters of sulfoimide of saccharin-6-carboxylic acid with 1,2,3-propantriol has been carried out.

It has been found out that the obtained product corresponded to m.p. = 167°C, yield – 73%. The synthesized compound is of interest for its use as a monomer during preparation of highly branched and netlike sulfoimide-containing polymers, epoxide resins, and also the industrial epoxide resins as a hardener-plasticizer.

KEYWORDS

- **1,2,3-propantriol**
- **alkyl esters**
- **aromatic glycols**
- **Fourier spectrometer**
- **IR spectroscopy**
- **saccharin-6-carboxylic acid**

REFERENCES

1. Mironov, V. F., (1974). *Monomers for New Polymer Materials*. NIITEhim., Moscow.
2. Babe, S. G., De Abajo, J., & Fontan, J., (1972). Ordered co-polyamide-imides and co-polyester-imides. *Angew. Makromol. Chem., 21*, 65–67.
3. Aslanov, T. A., Mamedli, U. M., & Guliyev, A. M., (2005). Synthesis of diester disulfimides from saccharin monocarboxylic acids and aromatic diols. *Russian J. Appl. Chem., 78*, 441–442.
4. Aslanov, T. A., Salakhov, M. S., & Efendiev, A. A., (2006). Synthesis of poly(ester sulfoimide) from dichlorocarbonylimide of disaccharindicarboxylic acid and aromatic diols. *Appl. Polym. Sci., 101*, 1687–1691.
5. Gordon, A., & Ford, R., (1976). *Satellite Chemist,* Mir, Moscow.
6. Balandin, V. A., Gurvich, D. B., & Kleshchev, M. S., (1965). *Analytical Polymerization Plastics, Chemistry,* Moscow, Tashkent.
7. Kazitsina, L. A., & Kupletskaya, N. B., (1979). *Application of UF-, IR-, NMR-, and Mass Spectroscopy in Organic Chemistry.* MSU Publishing House, Moscow.
8. Prech, E., Bulmann, F., & Affolter, K., (2006). *Determination of the Structure of Organic Compounds.* Mir, Moscow.

Index

B

C

G

K

L

M

N

O

P

T